Dynamical Collision Theory and Its Applications

Dynamical Collision Theory and Its Applications

Sadhan K. Adhikari

Departamento de Física
Universidade Federal de Pernambuco
Recife, Pernambuco, Brazil

Kenneth L. Kowalski

Department of Physics
Case Western Reserve University
Cleveland, Ohio

ACADEMIC PRESS, INC.
Harcourt Brace Jovanovich, Publishers
Boston San Diego New York
London Sydney Tokyo Toronto

This book is printed on acid-free paper. ∞

ACADEMIC PRESS, INC.
1250 Sixth Avenue, San Diego, CA 92101

United Kingdom Edition published by
ACADEMIC PRESS LIMITED
24-28 Oval Road, London NW1 7DX

Library of Congress Cataloging-in-Publication Data

Adhikari, Sadhan K., date.
Dynamical collision theory and its applications / Sadhan K. Adhikari, Kenneth L. Kowalski.
p. cm.
Includes bibliographical references and index.
ISBN 0-12-044273-6 (acid-free paper)
1. Collisions (Nuclear physics)—Mathematical models. 2. Scattering (Physics)—Mathematical models. I. Kowalski, Kenneth L. II. Title.
QC794.6.C6A24 1991
539.7′5—dc20 90-47161
CIP

Printed in the United States of America
91 92 93 94 9 8 7 6 5 4 3 2 1

To our families,
Ratnabali and Avishek,
and
Audrey, Claudia, and Eric,
and to our parents.

Contents

Preface

Collisions among atomic and subatomic particles that can be described in terms of nonrelativistic quantum mechanics occur throughout nature. Potentials are used in the Schrödinger equation as phenomenological representations of the microscopic electromagnetic or hadronic interactions that mediate the collisions. This model, along with its relativistic generalizations, has been extraordinarily successful in organizing and explaining a host of chemical, atomic, nuclear, and hadronic phenomena, and is likely to remain as useful in the future. Thus, an understanding of the techniques for calculating the predictions of this model for collision processes is relevant to much of the physical sciences. The major objective of this book is to provide an introduction to some of the powerful methods that have evolved in the last two decades for carrying out these calculations to an accuracy sufficient to test the model.

The real problem in all of this, of course, is to deal with collisions that involve bound composites of constituent particles. There are widely differing philosophies, that we call "dynamical strategies," as to how one should go about doing this in a way that is both practical yet comprehensive enough to be scientifically credible. The range of currently popular dynamical strategies is fairly accurately represented by the extant monographs on collision theory.

If we exclude the comprehensive works, e.g., Goldberger (1964), Joachain (1975), and Newton (1982) (We use the last name of the first author and the year of publication as a labeling for our citations throughout the book.), which are concerned mainly with the development of the foundations of the subject rather than its applications, we are left with monographs that represent two distinct dynamical strategies. The more traditional of these, e.g., Tobocman (1961), Mahaux (1969), Austern (1970), Jackson (1970), Hodgson (1971), Wildermuth (1977), Satchler (1980, 1983), Glendenning (1983), and Fano (1986), adapt strategies that are designed to yield dynamical equations that are two-body in nature, and therefore routinely calculable. In many cases, modifications of the basic approximations are introduced in the form of phenomenological effective interactions. This represents the adaptation of still another modeling of the collision process rather than a controlled correction to the original approximation. On the other hand, those books that follow a few-body point of view, e.g., Watson (1967), Schmid (1974), Thomas (1977), and Glöckle (1983a), basically take all the active degrees of freedom into account and aim for an exact solution. The latter approach, of course, rapidly loses its appeal because of sheer complexity as the number of particles goes beyond three. Both approaches have been very fruitful in their special realms.

In our opinion, an exciting development in the past two decades concerns the body of work that has attempted to bridge the calculational and intellectual differences between the "conventional" and the "few-body" strategies. Surely it must be possible to construct controlled corrections to conventional approximations, such as the coupled-reaction channel and resonating-group methods, as well as to forge complete sets of collision integral equations such as those due to Faddeev and Yakubovskii into practical calculational schemes by a well-articulated series of approximations. Although these hopes are not yet fully realized, we have attempted to present a point of view that will provide an introduction to that research by means of a solid grounding in the realizable means of formulating and solving collision integral equations from $N = 2$ on up. It is this point of view that we call "dynamical collision theory" and which is elaborated upon in further detail in Ch. I.

Two-particle potential scattering is reviewed in Ch. II. Then, in Ch. III, we have an extensive review of a variety of methods for calculating off-shell two-body amplitudes as well as approximating them by finite-rank forms. In Ch. IV, the interpretation and applicability of the multichannel, multiparticle Lippmann–Schwinger equations, which form the basis for the conventional approaches, are discussed in considerable detail. A self-contained discussion is given in Ch. V of most of the known N-particle connected-kernel integral equations; their physical predictions are reviewed in Chs. VII and VIII for $N = 3$ and $N = 4$, respectively. We have included descriptions of contemporary field-theoretical and relativistic approaches, such as the Dirac phenomenology for intermediate energy nucleon–nucleus scattering, to the extent that they fit into our general format. The review of the singularity structure of multiparticle amplitudes and the associated dispersion-relation techniques given in Ch. VI is, in part, complementary to that format; on the other hand, the singularity structure associated with the Efimov effect for $N = 3$ plays a crucial role in interpreting the three-nucleon calculational results reported in Ch. VII. Finally, in Ch. IX, we have described what is known at present about the relationship between the conventional (optical potentials, multiple-scattering theories, and the coupled-reaction channel and resonating-group methods) and the few-body approaches.

A number of important topics have been ignored, such as the Coulomb problem and configuration-space methods, that lend themselves to the use of differential equations. At present there is no entirely satisfactory way to incorporate unscreened Coulomb potentials into multiparticle collision theory. We almost exclusively work with collision integral equations, and these are most efficiently dealt with in momentum space. Undoubtedly, we have also inadvertently ignored, or worse, not adequately represented, work that is pertinent to what we do. We sincerely apologize to those authors who have been dealt with in either fashion.

The book is primarily intended for chemists, physicists, and graduate students interested in general scattering theory, intermediate and low-energy hadron and nuclear physics, atomic and molecular physics, statistical mechanics, and physical and quantum chemistry. Moreover, there are a number of topics that may prove interesting to mathematicians as well as particle physicists. Parts, or all, of the book are suitable for use as a primary or secondary text in advanced graduate courses that involve collision theory.

Within the range of subjects that we discuss, there is considerable variation in the level of the presentation. There are a number of reasons why this has occurred that are related to the subject itself, to frequency of use in numerical calculations, and to expectations about reader interest. Whenever possible, we have pointed out what appear to be open questions for further research both in the execution and organization of calculations as well as formal problems. In some chapters, the gap between what is known and what one would like to know is all too obvious.

A number of figures in this book are reproduced from other works. We thank the authors and the American Physical Society, the North-Holland Publishing Company, and Springer-Verlag for granting us permission to use copyrighted material from their publications.

It gives us great pleasure to express our gratitude to our friends and collaborators R. D. Amado, D. Bollé, A. Delfino, L. L. Foldy, A. C. Fonseca,

T. Frederico, E. Gerjuoy, W. Glöckle, R. Goldflam, M. S. Hussein, F. S. Levin, J. A. Lock, H. P. Noyes, T. A. Osborn, A. Picklesimer, S. C. Pieper, W. N. Polyzou, E. F. Redish, D. K. Sharma, E. R. Siciliano, P. C. Tandy, R. M. Thaler, W. Tobocman, and L. Tomio, who have contributed so much in enhancing our knowledge about the topics covered in this book.

We are grateful for the support given throughout the duration of this project by W. Gordon, Chairman of the Case Western Reserve University Physics Department. S. K. A. acknowledges financial support from Conselho Nacional de Desenvolvimento—Científico e Tecnológico of Brazil, which made the collaboration so pleasant. All of the research reported in the book involving K. L. K. was funded, in part, by a series of grants from the U.S. National Science Foundation. Part of the manuscript was written during the sabbatical leave of K. L. K. at the Argonne National Laboratory; the warm hospitality of the High-Energy Division is gratefully acknowledged.

We deeply appreciate the encouragement we received from S. R. A. Canuto, M. A. F. Gomes, and D. L. Karatas. Finally, we thank Mrs. Dorothy M. Straughter for her patience, enthusiasm, and superb professional effort in word processing the manuscript.

One of us (S. K. A.) would like to note his new address and affiliation: Instituto de Física Teôrica, Universidade Estadual Paulista, Rua Pamplona 145, 01405 São Paulo, São Paulo, Brazil.

Sadhan K. Adhikari and Kenneth L. Kowalski

Chapter 1

Scattering Theory

1.1. Introduction

Scattering processes occur at every level of natural phenomena. They often can be described qualitatively in simple, universal terms. It is usually a lot more difficult to characterize them quantitatively, which is the objective of scattering or, as it is sometimes called, collision theory. We are concerned with the development and, especially, the application of that part of this theory that has nonrelativistic quantum mechanics for a fixed number of constituent particles as its dynamical basis. This encompasses those models of atomic and subatomic processes for which the interactions between particles are represented by potentials. We also consider the relativistic generalizations that are possible so long as the explicit effects of particle creation and annihilation are negligible.

The quantum theory of scattering was developed to explain experimental results in atomic, nuclear, and particle physics. As the styles and agendas of these fields diverged, so did their terminologies. In what follows, our terminology has been chosen to cover the broadest range of circumstances with precision.

The mathematical formulation and substantiation of scattering theory has produced breakthroughs for doing practical calculations such as the

Lippmann–Schwinger and Faddeev equations, as well as rigorous work such as proofs of asymptotic completeness. This effort has been accompanied by an unfamiliar terminology and a literature guided very often by its own agenda. Our goal is to render this approach and its results more understandable to other practitioners of scattering theory. Our objectives in this regard are:

(i) To establish that multiparticle scattering theory is soundly based and that definitive solution algorithms exist, in the form of linear integral equations, for calculating any scattering process with a dynamics defined by the Schrödinger equation;
(ii) To make use of an efficient formal machinery for dealing with multiparticle continuum configurations; and
(iii) To indicate how conventional low-order approximations can be placed in a broader framework.

There are a number of interesting and successful approaches in collision theory that at the present time are better described in their own right rather than as special cases of a more comprehensive theory. For example, much of the work on relativistic few-body collisions, Dirac methods in nuclear scattering, and mean-field techniques for heavy-ion collisions is of this kind. Our remaining objectives involve the description of contemporary work on some of these and similar topics.

Scattering refers to the evolution, by means of an interaction of finite duration, of a system from a collection of essentially noninteracting components to another collection of this kind. In the absence of external influences, each of these components is supposed to persist in its free motion for an interval that is long with respect to the interaction time, and is regarded as a *particle* with an internal structure that is of secondary interest asymptotically. The complete interaction among the free particles is called a *collision.*[1] In many simple situations, such as two-particle scattering by a short-range force, the complete interaction characterizing the collision is localized in space and time. When more than two constituents are involved, this is not true even if all of the interconstituent forces are of short range.

A scattering, therefore, involves a redistribution, in some manner, of a group of entities that can be identified with the fundamental constituents of the model of interest, say electrons and nuclei in atomic physics. The freely moving asymptotic components may either be these fundamental constituents or stable composites of them, e.g., atoms or ions in the atomic case. In theories where there is confinement, such as quantum chromodynamics (QCD), only composites appear in the asymptotic configurations.

Evidently, scattering encompasses the entire history of a process from the infinite past to the infinite future. So one can picture the process in a time-independent fashion, if semi-infinite time spans are associated with infinite reaches of space measured from an interaction region. Most of the scattering process, whether gauged by duration or distance, is dynamically trivial. Only collisions engage the dynamics in an interesting way.

We are mainly interested in the means of calculating collision processes by solving the integral equations for the transition probability amplitudes. These equations incorporate the assumed interparticle interactions and also the asympotic boundary conditions. It is this approach to the subject that we call *dynamical collision theory*.

Nonrelativistic quantum mechanics has been used extensively and with considerable phenomenological success for the description of atomic and nuclear collisions. Within this framework, one can formulate well-defined dynamical equations with solutions whose general features are understood. One can also be very specific about how one can go about obtaining these solutions. This is not the case for the relativistic field theories—quantum electrodynamics (QED) and QCD—that provide a microscopic basis for atomic and nuclear collision processes, respectively.

The passage from QED to the description of atomic and molecular interactions in terms of Coulomb and shorter-ranged potentials is reasonably clear in principle, yet quite complicated in practice. None except the simplest features of interatomic forces is presently calculable *ab initio*. So the modeling of such interactions by semi-phenomenological potentials, which are then employed in a Schrödinger or a Dirac equation, represents the dominant style of atomic scattering theory, and is adhered to in this book.

By way of contrast, it is not yet known how to describe hadronic processes starting from QCD. Those models of nuclear processes that have required some statement about the nucleon–nucleon (NN) interaction have usually employed the device of a potential. This has been very successful in codifying the general features of the constituent NN collisions. An important advantage of the NN potential as a repository of the complex interactions among the constituents of the composite nucleons is the simplicity with which one can exploit this information: The potential is just inserted into the Schrödinger equation. We accept this as a means to the dynamical ends pursued in this book.

Relativistic effects and some features of particle annihilation/creation have been successfully modeled in ways that lead to dynamical schemes not too different from what one obtains in nonrelativistic scattering theory. Models for low-energy to medium-energy pion–deuteron scattering are exemplars of these approaches. Generally, pions can be absorbed in nuclei so that one is confronted with manifest particle annihilation. Of course, it is not yet possible to deal with these processes in a fundamental way. Nevertheless, these models provide a vehicle for dealing with annihilation/creation effects while yielding transition amplitudes with the required particle-pole singularities and vertex functions with the correct Lorentz-invariance and internal-symmetry structures.

It has been found that the extra degrees of freedom allowed by describing the nucleon as a Dirac particle facilitate the description of nucleon–nucleus elastic scattering for nucleon energies less than 1 GeV and greater than a few hundred MeV.[2] Some controversy surrounds the success of this so-called Dirac approach, for it seems to depend upon the excitation of virtual antinucleons more readily than is expected to occur at these energies and momentum transfers. Little of this model can be justified as of yet on more fundamental grounds, so one does not have the sort of complete set of dynamical equations that would result if we simply assumed that NN potentials could be used in the Schrödinger equation. The limited set of dynamical equations that do arise in the Dirac approach closely resemble their nonrelativistic counterparts, so very little modification of the conventional solution strategies of nucleon–nucleus phenomenology is required.

In short, all of the different modelings of physical situations that are of interest to us have one feature in common. They all lead to dynamical equations for the transition amplitudes much like the type that arises in nonrelativistic Schrödinger dynamics.

The phenomenologies of atomic, nuclear, and hadronic collisions involve approximations in the modeling of the physics. Usually, further approximations are made to the resulting dynamical equations. Often the two steps are intermingled, so that while successful fits to experimental data can often be achieved, the physical reasons for that success are not always clear.

Approximations to the dynamical equations are inevitable when many particles play an active role. Nonetheless, a well-founded phenomenology should be based upon approximations formulated within a theoretical framework that makes it possible, at least in principle, to obtain successively more accurate solutions of the dynamical equations resulting from a specific model. What should be at question is the dynamical model, not the means for realizing its predictions. Most calculations of atomic and nuclear collision processes fail this criterion by employing truncations that are difficult to incorporate into a more comprehensive framework, so it is often practically impossible to ascertain either the nature or the magnitude of the next order of approximation. Many widely used dynamical schemes are incapable of improvement in their own terms, except by additional parametrization of unknown effects. Typically, this is because an incomplete set of dynamical equations is employed that only partially represents the content of the Schrödinger equation and the physical boundary conditions. In general, such a set is impossible to solve in principle unless it is truncated to the extent that it yields unique solutions.

Facility among research workers with the dynamics of multiparticle collisions, even from the point of view of nonrelativistic Schrödinger quantum mechanics, is still not sufficient to have altered the preceding situation much except for the three-nucleon system. However, considerable progress has been made over the last two decades initiated largely by the work by Faddeev on the three-particle scattering problem.[3] We hope to make the principal features of that progress, and the surprisingly few fundamental ideas that underlie it, accessible to a much wider audience than is presently the case.

In the present chapter, we introduce notation, nomenclature, and those aspects of scattering theory needed for the rest of the book. Although we strive to be self-contained, this is definitely not an introduction to the subject; for that background, we rely on the many extant books.[4] Mathematical approaches to the foundations of collision theory are available, but their methods are seldom relevant to our principal concerns.[5]

1.2. Elastic and Inelastic Scattering

The simplest example of scattering occurs for the system of a single particle that interacts with a finite-range center of force. Prior to the interaction, as well as following it, the particle moves with a constant velocity, although generally in different directions. This is often referred to as *potential scattering* when the interaction can be represented by a one-body potential.

In more complicated situations, the noninteracting systems before and after the collision process need not be the same. For instance, three molecules might scatter into a final configuration consisting of their dissociated atoms. Such processes are often called *reactions*, as opposed to scattering, which is then reserved only for those processes where there is no such change. This terminology is employed less frequently in the theory of collision phenomena, where reactions are regarded as special cases of generic scattering.

We need distinguish only two categories of scattering processes: those that are elastic and those that are not. The initial and final configurations in elastic scattering differ at most in the velocities and spin orientations of each of the particles. So before and after the scattering there are the same (arbitrary) numbers of each species of particle, and each of them is in its original internal state, save for the orientation of its intrinsic angular momentum.

We consistently regard as *inelastic* any scattering process that is not elastic. It is sometimes useful in practice to distinguish *excitation processes* from other inelastic processes with qualitatively different final states. The excited state of a composite system, such as an atom, has the same internal quantum numbers as the ground state, except for mass and its total spin. Nonetheless, excitation and the other inelastic processes are all alike in being incoherent with the elastic scattering, and that is the only relevant property.

1.3. Quantum Theory of Scattering

It is a major task to express the preceding qualitative ideas in the mathematical language of quantum mechanics, and one that we undertake only in outline.[1]

Scattering is distinguished from triviality by the occurrence of a collision that takes place over some finite interval of time. That interval is, of course, not perfectly sharp. Also, when more than two particles are interacting there is the possibility of multiple collisions separated by intervals of arbitrary duration (Fig. 1.1).

Long-range Coulombic forces destroy the picture of collisions neatly localized in space-time. In practice, this presents more of a procedural complication in obtaining solutions rather than a complication of principle, since under realistic circumstances such forces are always screened.[2] The limiting process for dealing with screening in a universal fashion, independent of the screening radius, is delicate and in multiparticle contexts is not fully understood. We suppose that the interaction potentials fall off somewhat faster than a Coulomb potential.

The causality structure of quantum mechanics implies that the state of the system, $|\psi(t)\rangle$, is uniquely determined at the time t by its value at any earlier time, e.g.,

$$|\psi(t)\rangle = e^{-iHt}|\psi(0)\rangle, \tag{1.3.1}$$

where, for simplicity, we use units such that $\hbar = 1$. The time translation is generated by the Hamiltonian H that, we suppose, has no explicit dependence on the time. Scattering theory is concerned with the relationships of the

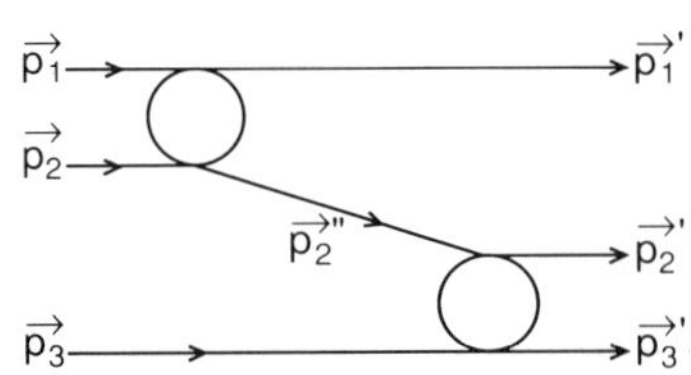

FIG. 1.1. Double scattering in the collision of three free particles. Each collision (circle) can be separated by an arbitrary space–time interval.

limiting configurations of $|\psi(t)\rangle$ as $t \to -\infty$ (infinite past) or $t \to +\infty$ (infinite future).

A nontrivial scattering state is never actually identical to a state that represents a system of noninteracting clusters. The latter state is defined in terms of a time-independent collection q of sets of quantum numbers q_i:

$$q = \{q_1, q_2, \ldots\}, \tag{1.3.2}$$

that determine the kinematical and internal states of the clusters, $i = 1, 2, \ldots$. The kinematical subset of the q_i's refers to either the labels of convenient wave packets or to the energy/momenta in the nonnormalizable limit. Each q is called an *asymptotic configuration*. This is closely related to what we will refer to as a *channel*, in which only the internal quantum numbers of a set of clusters are of interest.

The distinctive feature of scattering states $|\psi(t)\rangle$ that evolve according to Eq. (1.3.1) is that they are *uniquely* determined in terms of either their asymptotic configurations in the infinite past or the infinite future. One obtains, however, very different types of states for each alternative. This requires a precise statement of the nature of the limits $t \to \pm\infty$ of the scattering state called the *asymptotic condition* (Section 1.5).

A system with a Hamiltonian H has bound and scattering states. The bound states are discrete, normalizable, energy eigenstates of H. The scattering states are not so easily categorized except in an idealized limit. Then, their energy "eigenvalues" cover a continuous semi-infinite range and they are not normalizable. The physical scattering states that evolve from wave packets of the noninteracting configurations are normalizable, but they are not energy eigenstates, nor can they be specified typically in the convenient and universal manner that is possible in the limit of sharp energy eigenstates. No matter how we represent the scattering states, the obvious question that arises is whether the bound states along with some subset of the possible scattering states form a complete set. A (qualified) affirmative answer to this question is called the property of *asymptotic completeness* (Section 1.4).

The asymptotic condition and asymptotic completeness are the basis of the quantum theory of scattering, but they are just the beginning of the story if one wishes to calculate scattering amplitudes. We call different ways for achieving this last objective *dynamical strategies* (Section 1.6).

There are two broad classes of such strategies for the dynamical model (the nonrelativistic Schrödinger equation) we are using:

(i) *Time-independent linear integral equations for the (Heisenberg picture) scattering-state vectors or the (off-shell) transition amplitudes*: These equations are usually realized in momentum space, and incorporate all of the boundary conditions implied by the asymptotic condition, as well as the regularity required by normalization constraints.

(ii) *Time-independent differential equations for the (Heisenberg) scattering state in configuration space*: The boundary and regularity conditions are imposed separately on the equations.

Alternative strategies have been actively pursued with varying degrees of success. The so-called group-theoretical approach[3] is another time-independent method with an intriguingly different formalism. Time-dependent Schrödinger techniques have been proposed, but their promise is uncertain.[4] Time-dependent Hartree–Fock and path-integral techniques have been widely used for problems such as heavy-ion collisions in which a large, but finite, number of degrees of freedom are dynamically active simultaneously.[5]

1.4. Asymptotic Completeness: The *S*-Matrix

Before and after they collide, the components of a scattering system consist of two or more freely moving, mutually noninteracting, stable units that we call *clusters.* Each cluster is either an elementary constituent or a stable bound configuration of these constituents. We call the elementary constituents *particles.* They are supposed to have no excited states on the energy scale of interest. The bound, composite clusters may be in any of their discrete energy states, either ground or excited.

Wave-packet states represent the motion of the clusters, and their internal configurations are used to construct the noninteracting multicluster states. The collection of quantum numbers used to label them is denoted as q (cf. Eq. (1.3.2.)).

For each q, there are two very different kinds of scattering states; one has evolved from q in the infinite past, while the other evolves to q in the infinite future. The respective states are designated $|\psi_{\text{in}}(t, q)\rangle$ and $|\psi_{\text{out}}(t, q)\rangle$. Both states satisfy Eq. (1.3.1) for any t. Since $|\psi_{\text{in}}(t, q)\rangle$ ($|\psi_{\text{out}}(t, q)\rangle$) refers to the situation where the system is definitely in the configuration q as $t \to -\infty$ ($t \to +\infty$), the probability amplitude for the scattering $q_{\text{i}} \to q_{\text{f}}$ is

$$\begin{aligned} S(q_{\text{f}}, q_{\text{i}}) &= \langle\psi_{\text{out}}(t, q_{\text{f}})|\psi_{\text{in}}(t, q_{\text{i}})\rangle \\ &= \langle\psi_{\text{out}}(q_{\text{f}})|\psi_{\text{in}}(q_{\text{i}})\rangle, \end{aligned} \tag{1.4.1}$$

where, e.g., $|\psi_{\text{in}}(q_{\text{i}})\rangle = |\psi_{\text{in}}(0, q_{\text{i}})\rangle$. All of the states are presumed to be normalizable, only in the sense of generalized functions when we take the limit of states of sharply defined energies and momenta. The complete array $S(q_{\text{f}}, q_{\text{i}})$ is called the S-matrix.

Now the sets $\{|\psi_{\text{in}}(t, q)\rangle\}$ and $\{|\psi_{\text{out}}(t, q)\rangle\}$, span subspaces $\mathscr{H}_{\text{in}}$ and $\mathscr{H}_{\text{out}}$, respectively, of the complete Hilbert space $\mathscr{H}$, where again we interpret this from a formal, generalized-function viewpoint. Asymptotic completeness means that these subspaces are such that $\mathscr{H}$ decomposes into the direct sums

$$\mathscr{H} = \mathscr{H}_{\text{in}} \oplus \mathscr{H}_{\text{B}}, \tag{1.4.2a}$$

$$\mathscr{H} = \mathscr{H}_{\text{out}} \oplus \mathscr{H}_{B}, \tag{1.4.2b}$$

where $\mathscr{H}_{\text{B}}$ is the subspace spanned by the bound states $|\psi_{\text{B}}\rangle$ of H:

$$H|\psi_{\text{B}}\rangle = E_{\text{B}}|\psi_{\text{B}}\rangle. \tag{1.4.3}$$

The proof of Eqs. (1.4.2a, b) for nonrelativistic scattering with short-ranged potentials is not trivial even for two-particle scattering, but nothing really new enters in until N-particle states with $N \geq 3$ are considered.[1,2]

The immediate consequence of the decompositions (1.4.2a, b) is that there must be a unitary map S called the *scattering operator*, of $\mathscr{H}$ onto itself such that, for each set q and each B, and all t we have[3]

$$|\psi_{\text{in}}(t, q)\rangle = S|\psi_{\text{out}}(t, q)\rangle, \tag{1.4.4}$$

$$|\psi_{\text{B}}\rangle = S|\psi_{\text{B}}\rangle, \tag{1.4.5}$$

where the last equation reflects the stability of the bound states of the system. Because Eq. (1.4.4) holds for any time, we have

$$[S, H] = 0. \tag{1.4.6}$$

Also

$$S(q_{\text{f}}, q_{\text{i}}) = \langle\psi_{\text{out}}(q_{\text{f}})|S|\psi_{\text{out}}(q_{\text{i}})\rangle \tag{1.4.7a}$$

$$= \langle\psi_{\text{in}}(q_{\text{f}})|S|\psi_{\text{in}}(q_{\text{i}})\rangle, \tag{1.4.7b}$$

where we have used the unitarity relations

$$S^\dagger S = SS^\dagger = I, \tag{1.4.7c}$$

and † denotes the Hermitian adjoint.

The overlap of the in- and out-states at any time yields the S-matrix, but there are (generalized) orthonormality relations among the states within each of these sets:

$$\langle\psi_{\substack{\text{in}\\\text{out}}}(t,q')|\psi_{\substack{\text{in}\\\text{out}}}(t,q)\rangle = \delta(q',q), \tag{1.4.8a}$$

$$\langle\psi_{\substack{\text{in}\\\text{out}}}(t,q')|\psi_{\text{B}}\rangle = 0, \tag{1.4.8b}$$

where $\delta(q',q)$ denotes a collective product of Dirac–Kronecker δ-functions. The proof of Eq. (1.4.8a) requires the asymptotic condition discussed in the next section. The validity of Eq. (1.4.8b) is apparent if in the c.m. frame the bound-state energy is below the threshold for scattering; we suppose that the spectrum of H has this property. Given Eqs. (1.4.8a, b), completeness takes the form

$$\sum_{\text{B}}|\psi_{\text{B}}\rangle\langle\psi_{\text{B}}| + \sum_{q}|\psi_{\text{in}}(t,q)\rangle\langle\psi_{\text{in}}(t,q)| = I, \tag{1.4.9}$$

where I is the identity on $\mathscr{H}$, with a similar expression in terms of the out-states.

1.5. The Asymptotic Condition

We now explain what is meant by a system being definitely in a cluster state either as $t \to -\infty$ ("in" state) or $t \to +\infty$ ("out" state) in terms of what is called an *asymptotic condition.*

In order to do this, we have to be more specific about the cluster Hamiltonians and their eigenstates. For N particles, the *cluster Hamiltonian,* H_a, corresponding to a partition a of N particles into a system of non-interacting clusters, is obtained from the complete Hamiltonian by subtracting all of the interactions that connect the particles in different clusters. Therefore, H_a consists of those interactions that are required to form the composite clusters and the kinetic energies of all N particles.

An important example is provided by the systematics of three-particle scattering. If we label the particles by the indices $i, j, k = 1, 2, 3$, the various scattering possibilities are

$$i + (jk) \to i' + (j'k'), \tag{1.5.1a}$$

$$i + (jk) \to 1 + 2 + 3, \tag{1.5.1b}$$

$$1 + 2 + 3 \to 1 + 2 + 3, \tag{1.5.1c}$$

and their time-reverses. Here $i \neq j$, $j \neq k$, $k \neq i$, and i', j', k' is a cyclic permutation of i, j, k, and (jk) refers to a stable, two-particle bound state of j and k. Each of the possible initial and final configurations in the processes

(1.5.1a–c) corresponds to what we have called a *channel*. Channels are associated with the partitions of the three particles into two or three clusters, but there may be more than one channel for each partition.

The scattering from a two- to a three-body channel is typified by the *breakup* process (1.5.1b); the time-reversed collision is sometimes called *recombination*. We refer to the process (1.5.1c) as the three-to-three collision, with similar designations for the other processes for which the numbers of particles in the initial and final states suffice as identifiers.

The complete Hamiltonian H of the three-particle system can be written in terms of the cluster Hamiltonians H_a labeled by each of the distinct partitions a corresponding to the different channels, and the residual interactions V^a between the clusters:

$$H = H_a + V^a, \tag{1.5.2}$$

where $a = \alpha = (i)$ label the three partitions $(i)(jk)$, while $a = 0$ refers to the clustering (1)(2)(3). Thus, we have

$$H_0 = \sum_{i=1}^{3} K_i, \tag{1.5.3}$$

where K_i is the kinetic-energy of particle i,

$$H_\alpha = H_0 + V_\alpha, \tag{1.5.4}$$

and V_α represents the interaction between the two clustered particles in partition α. The complete interaction V also has a partition-degenerate decomposition provided we identify $V^0 \equiv V$ and $V_0 \equiv 0$:

$$V = V_a + V^a. \tag{1.5.5}$$

Two-cluster channels play a distinctive role in scattering theory and its applications because of the space–time localization of two-cluster collisions. The two-cluster partitions that label these channels are often denoted by the lower-case Greek indices $\alpha, \beta, \gamma, \ldots$, as we have done in the three-particle case. When we need to specify the number of clusters m corresponding to a partition a, we use the notation a_m.

We use, with some inevitable exceptions, the lower-case Latin letters beginning with i, namely $i, j, k, \ldots$, to refer to individual particles. The lower-case Latin letters beginning with a, namely, $a, b, c, \ldots$, refer to the distinct partitions of N particles into clusters, with $a = 0$ representing the N-cluster partition and $a = \underline{1}$ the one-cluster partition.

Suppose a collision occurs at time $t = 0$. Then, far into the past or the future the interactions among different clusters are expected to be negligible. Under these circumstances, we can expect that the system is represented by a state vector $|\phi_a(t)\rangle$ whose evolution in time is governed by

$$|\phi_a(t)\rangle = e^{-iH_a t}|\phi_a(0)\rangle, \tag{1.5.6}$$

where H_a is the relevant cluster Hamiltonian. The time evolution of the cluster states that correspond to the initial and final configurations of a scattering process is known because only the positions of the c.m. of each of the clusters are changing. Therefore, each $|\phi_a(t)\rangle$ is a direct product of states representing the c.m. motion and the cluster bound states. In the three-particle case, for example,

$$|\phi_\alpha(t)\rangle = |\text{Total c.m.}\rangle|\alpha_{\text{rel}}(t)\rangle|\hat{\phi}_\alpha\rangle, \tag{1.5.7}$$

where, say for the two-cluster channel $\alpha = 1$, $|\alpha_{\text{rel}}(t)\rangle$ represents the motion of particle 1 relative to the c.m. of the stable bound cluster composed of particles 2 and 3, whose time-independent bound state is $|\hat{\phi}_\alpha\rangle$. Finally, the state $|\text{Total c.m.}\rangle$, whose time dependence we have suppressed, represents the overall c.m. motion; this factors out of any problem with a translationally invariant Hamiltonian.

The states $|\phi_a(t)\rangle$ span a subspace $\mathscr{H}_a$ called the a-channel subspace of $\mathscr{H}$. For $a \neq b$, $\mathscr{H}_a$ and $\mathscr{H}_b$ can overlap nontrivially, but are not identical. Consequently, it is not possible to decompose the full space into a direct product of channel spaces without overcounting. In particular, we note that the N-free-particle channel space $\mathscr{H}_0$, spanned by the eigenstates of H_0, is equivalent to $\mathscr{H}$.

The *full* set of eigenstates of H_a for $a \neq 0$ includes the continuum states that represent scattering among the constituents of each cluster in addition to the channel states $|\phi_a(t)\rangle$. Since H_a is just H with some of the mutual interactions set equal to zero, the discussion of Section 1.4 applies to this situation without modification. The complete orthonormal sets on the full space generated by each of the H_a facilitate the reduction of parts of the N-particle scattering problem in terms of the dynamics of systems with less than N interacting particles.

Asymptotic Condition: We assume that there is a unique solution $|\psi_{\text{in}}(t, a)\rangle$ ($|\psi_{\text{out}}(t, a)\rangle$) of the Schrödinger equation:

$$i\frac{\partial}{\partial t}|\psi_{\substack{\text{in}\\\text{out}}}(t, a)\rangle = H|\psi_{\substack{\text{in}\\\text{out}}}(t, a)\rangle \tag{1.5.8}$$

corresponding to a definite cluster state $|\phi_a(t)\rangle$ in the infinite past (future) such that

$$\lim_{t\to\mp\infty} {}_s \{|\psi_{\substack{\text{in}\\ \text{out}}}(t,a)\rangle - |\phi_a(t)\rangle\} = 0, \tag{1.5.9}$$

where the subscript s denotes *strong* convergence.[1]

Strong convergence is what is usually referred to as (unqualified) *convergence* in defining a Hilbert space.[2] As in Eq. (1.5.9), one often refers to convergent *sequences* of vectors even though t is a *continuous* index. Evidently, t can be regarded as a formal parametrization of an entire class of possible sequences of vectors. There are other types of convergences encountered in scattering theory such as *weak* convergence.[3] Condition (1.5.9) implies weak convergence that, in the present instance, amounts to the equality of the norm of $|\psi_{\text{in}}(t,a)\rangle$, or $|\psi_{\text{out}}(t,a)\rangle$, with the norm of $|\phi_a(t)\rangle$ as $t\to\mp\infty$, which is not very restrictive. Weak convergence is part of the asymptotic condition in field theory, where it refers to the field operators rather than state vectors.[4]

The asymptotic condition combined with asymptotic completeness implies that there are time-independent maps $\Omega_a^{(\pm)}$ from the subspace $\mathscr{H}_a$ such that

$$|\psi_{\substack{\text{in}\\ \text{out}}}(t,a)\rangle = \Omega_a^{(\pm)}|\phi_a(t)\rangle. \tag{1.5.10}$$

The $\Omega_a^{(\pm)}$ are called *Møller wave operators*, and satisfy the isometry conditions

$$\Omega_b^{(\pm)\dagger}\Omega_a^{(\pm)} = \delta_{ab} P_a, \tag{1.5.11}$$

where P_a is the projection onto $\mathscr{H}_a$. These conditions follow from the norm-preserving property (1.5.9) and the orthogonality of the in- or out-states for different partition indices a. The $(\pm)$ superscripting is assigned by a long-standing, physically evocative convention that bears an inverse relationship with the $t\to\mp\infty$ identifying behaviors of the in- and out-states. For instance, the plus sign in $\Omega_a^{(+)}$ refers to the outgoing-wave character of $|\psi_{\text{in}}(t,a)\rangle$ as $t\to+\infty$, and not to the fact that there is an association with the clustering a as $t\to-\infty$. We adopt the standard conventions

$$|\psi_{\text{in}}(0,a)\rangle \equiv |\psi_a^{(+)}\rangle, \tag{1.5.12a}$$

$$|\psi_{\text{out}}(0,a)\rangle \equiv |\psi_a^{(-)}\rangle, \tag{1.5.12b}$$

and we denote $|\phi_a(0)\rangle$ simply as $|\phi_a\rangle$.

We can exploit the existence of the wave operators to define an array of channel-indexed scattering operators S_{ba}, whose matrix elements with respect to the channel states yield the transition amplitudes. That is,

$$\langle\phi_b|S_{ba}|\phi_a\rangle \equiv \langle\psi_b^{(-)}|\psi_a^{(+)}\rangle. \tag{1.5.13}$$

Since

$$\langle \psi_b^{(-)} | \psi_a^{(+)} \rangle = \langle \phi_b | \Omega_b^{(-)\dagger} \Omega_a^{(+)} | \phi_a \rangle, \tag{1.5.14}$$

we see that

$$S_{ba} = P_b \Omega_b^{(-)\dagger} \Omega_a^{(+)} P_a. \tag{1.5.15}$$

1.6. Dynamical Strategies

There is no special way of calculating S-matrix elements given the interactions among the constituents; each alternative is what we call a *dynamical strategy*. We next present a rationale for the general strategy that we have chosen to emphasize.

The Schrödinger equation, along with the asymptotic condition, covers all contingencies, although in a nonspecific manner. A dynamical strategy is a solution algorithm that incorporates the requisite boundary (asymptotic) conditions. For two-body problems, that is achieved using the time-independent Schrödinger equation in configuration space with the interactions expressed in terms of local potentials $V(\mathbf{x})$.[1] This elementary method is not convenient when the potentials are more complicated, which is often the case in realistic situations. It becomes even less advantageous when more degrees of freedom come into play.

An important step towards an alternative strategy was taken by Lippmann and Schwinger (LS)[2], who expressed the scattering problem in terms of formal integral equations. Their original work is strictly correct without substantial qualification only for single-particle scattering, or the equivalent problem of two-particle scattering in the c.m. frame with a translationally invariant interaction. The extension of this work to obtain multichannel LS equations appropriate to more complicated collision problems followed shortly.[3]

For the two-particle case, the LS equation, whether for the wave function or the transition matrix, suffices as a dynamical strategy. The asymptotic condition is fully incorporated and the equation itself is effectively a Fredholm integral equation.

In the multichannel, multiparticle case the LS equations do not represent a fully satisfactory dynamical strategy:

1. The solutions of anything less than some minimal number of LS equations are nonunique.
2. Even if one has a number of LS equations sufficient to avoid manifest nonuniqueness, the kernels of these integral equations contain nonre-

movable Dirac δ-functions. They are said to possess noncompact kernels.[4] There are no established solution algorithms for integral equations of this kind.

The LS equations satisfied by the in-state $|\psi_\alpha^{(+)}(E)\rangle$, in the limit of well-defined energy E, for a collision initiated from a two-cluster channel labeled by the partition α are given by:

$$|\psi_\alpha^{(+)}(E)\rangle = \delta_{\beta\alpha}|\phi_\alpha(E)\rangle + G_\beta^{(+)}(E)V^\beta|\psi_\alpha^{(+)}(E)\rangle, \tag{1.6.1}$$

where

$$G_\beta^{(+)}(E) = (E - H_\beta + i0)^{-1} \tag{1.6.2}$$

is the two-cluster Green function; the notation in Eq. (1.6.2) refers to the limit of $(E - H + i\varepsilon)^{-1}$ as $\varepsilon \to 0+$. The initial asymptotic configuration is represented by the state $|\phi_\alpha(E)\rangle$ that satisfies

$$(E - H_\alpha)|\phi_\alpha(E)\rangle = 0, \tag{1.6.3}$$

where $H_\alpha = H_0 + V_\alpha$ is the channel Hamiltonian and, of course, $|\psi_\alpha^{(+)}(E)\rangle$ satisfies

$$(E - H)|\psi_\alpha^{(+)}(E)\rangle = 0, \tag{1.6.4}$$

with $H = H_\alpha + V^\alpha$.

The Eqs. (1.6.1) hold with α and β interchanged, so any solution of

$$|\psi_\alpha^{(+)}(E)\rangle = |\phi_\alpha(E)\rangle + G_\alpha^{(+)}(E)V^\alpha|\psi_\alpha^{(+)}(E)\rangle \tag{1.6.5}$$

is not unique and may contain an arbitrary admixture of $|\psi_\beta^{(+)}(E)\rangle$ that represents scattering in open rearrangement channels.

Another problem with Eq. (1.6.5) as an integral equation arises from the invariance of its kernel $G_\alpha^{(+)}V^\alpha$ with respect to translations of subsets of fewer than all of the $N > 2$ particles. This disconnected structure yields Dirac δ-function contributions to the momentum-space matrix elements of the kernel that are not overall common factors in the equation. Thus, the kernels of any of the multiparticle LS equations are noncompact.[4] As a practical matter, such an integral equation falls outside the class where the numerical methods appropriate to Fredholm equations are valid.

We conclude that Eq. (1.6.5) does not determine $|\psi_\alpha^{(+)}(E)\rangle$, although it is an exact identity. There are a number of ways to deal with this. The most common one gives up all hope of obtaining an exact $|\psi_\alpha^{(+)}(E)\rangle$ by making approximations that reduce Eq. (1.6.5) to a coupled set of two-*body* equations so that, in effect, no channels with three or more clusters enter its kernel. This equation is then essentially of the Fredholm type, and generally admits unique solutions. What has happened is that one has replaced the real problem by a *model* for which the additional information needed to constrain the solution is no longer needed. This is one realization of what we call the *two-cluster* or *bound-state approximation.* It comprises the essence of many of what we call the conventional dynamical strategies that are widely used in atomic and nuclear physics, sometimes with considerable success. This strategy circumvents the LS problem, but at the price of the means within itself for controlled improvement so the reasons for its successes or failures are not always clear.

The general dynamical inadequacy of Eq. (1.6.5), by itself, coupled with the success of many types of two-cluster approximation schemes that are based upon it carries a sense of contradiction. That sense has pervaded the literature on the applications of collision theory for some time since the work of Epstein, Foldy, Lippmann, and Tobocman indicated that all but the simplest scattering problems present difficulties not addressable by solution strategies then available.[3,5] The cause of these difficulties, as well as a technique for circumventing them that leads to a complete, unique scattering solution, were first identified by Faddeev for three-particle scattering,[6] and then for the N-particle problem by Yakubovskii.[7] This work settled the uniqueness problem and produced so-called connected-kernel equations, without the noncompactness encountered in the LS equations, and forms the basis of our general dynamical strategy.

The work of Faddeev and Yakubovskii seems to have enhanced rather than relieved the preceding sense of contradiction, to the point where it sometimes appears that there are two disjoint ways of going about handling collision problems, each having little relevance to the other. There is no real contradiction if the two-cluster approximation scenarios are viewed for precisely what they are, *viz.*, approximations to a dynamically complete set of scattering integral equations with unique solutions.

The dynamical strategy that is the focus of this book is to formulate well-defined integral equations for the transition amplitudes, such as the Faddeev–Yakubovskii equations, and to propose methods for obtaining their solutions. This aims at solving the problems with the LS equations rather than circumventing them.

1.7. Transition Operators

In this section, we obtain expressions for the transition amplitudes as matrix elements of definite operators with respect to states of sharply defined particle momenta that represent the initial and final asymptotic configurations.

When we have normalized cluster states, the scattering probability amplitudes are the S-matrix elements

$$\langle \phi_b | S_{ba} | \phi_a \rangle = \langle \psi_b^{(-)} | \psi_a^{(+)} \rangle, \tag{1.7.1}$$

where for the moment we use a condensed notation in which all of the variables needed to specify the asymptotic configuration are included within the cluster partition label a. Since the matrix element (1.7.1) contains the trivial occurrence of no collision, it must have the structure

$$\langle \phi_b | S_{ba} | \phi_a \rangle = \delta_{ba} + iR_{ba}, \tag{1.7.2}$$

where R_{ba} vanishes when there is no interaction among the clusters, and so is identified with the probability amplitude for a collision. Unitarity implies that we have

$$i(R_{ba} - R_{ab}^*) = -\sum_c R_{cb}^* R_{ca} = -\sum_c R_{bc} R_{ac}^*, \tag{1.7.3}$$

and therefore the *optical theorem*:

$$2 \operatorname{Im} R_{aa} = \sum_c |R_{ca}|^2. \tag{1.7.4}$$

Equations (1.7.1–1.7.4) provide the basis for the probabilistic interpretation of collisions. However, it is only in the limit of sharply defined particle momenta that experiments can be interpreted independently of the spreads in the momenta of the asymptotic cluster configurations that result from spatial–temporal localization. Therefore, as a practical matter, we are interested in the implications of Eqs. (1.7.1–1.7.4) in this limit, when both δ_{ba} and R_{ba} are replaced by expressions containing Dirac δ-functions that reflect the conservation of the overall and the relevant individual energies and momenta. The collision transition amplitudes are then identified, to within normalization-dependent constants, with R_{ba} after the overall energy–momentum δ-functions have been factored out.

In the limit of well-defined energy E, the scattering states satisfy

$$(E - H) | \psi_a^{(\pm)}(E, \hat{a}) \rangle = 0. \tag{1.7.5}$$

Here $\hat{a}$ refers to any other variables besides the energy and the partition index needed to specify the asymptotic cluster configurations. So, $a, \hat{a}$, and the energy of the asymptotic configuration precisely define what we mean by a *channel*. The cluster, or channel, states $|\phi_a(E, \hat{a})\rangle$ satisfy

$$(E - H_a) | \phi_a(E, \hat{a}) \rangle = 0. \tag{1.7.6}$$

We can decompose $|\psi_a^{(\pm)}(E, \hat{a})\rangle$ into a sum of $|\phi_a(E, \hat{a})\rangle$ and a piece that is solely a consequence of the collision:

$$|\psi_a^{(\pm)}(E, \hat{a})\rangle = |\phi_a(E, \hat{a})\rangle + |\Phi_a^{(\pm)}(E, \hat{a})\rangle, \tag{1.7.7}$$

where we see that

$$(E - H)|\Phi_a^{(\pm)}(E, \hat{a})\rangle = V^a|\phi_a(E, \hat{a})\rangle. \tag{1.7.8}$$

The constraint on $|\Phi_a^{(\pm)}\rangle$ is that it vanish when the intercluster interactions V^a are turned off, so

$$|\Phi_a^{(\pm)}(E, \hat{a})\rangle = G^{(\pm)}(E)V^a|\phi_a(E, \hat{a})\rangle, \tag{1.7.9}$$

where

$$G^{(\pm)}(E) = (E - H \pm i0)^{-1} \tag{1.7.10}$$

is the full Green function corresponding to outgoing-wave (+) or ingoing-wave (−) boundary conditions in the multidimensional configuration space. The notation is similar to that in Eq. (1.6.2).

Only $|\Phi_a^{(\pm)}(E, \hat{a})\rangle$ is relevant for the calculation of transition amplitudes. We use this observation to present a heuristic derivation of the continuum-limit transition amplitude, rather than proceeding from Eqs. (1.7.1–1.7.2) using limits of wave packets of plane wave states.[1]

In association with the change of cluster configurations $a \to b$, we introduce transition operators $T_{ba}(E)$ by means of the identification

$$|\Phi_a^{(+)}(E, \hat{a})\rangle \equiv G_b^{(+)}(E)T_{ba}(E)|\phi_a(E, \hat{a})\rangle, \tag{1.7.11}$$

where

$$G_b^{(+)}(E) = (E - H_b + i0)^{-1} \tag{1.7.12}$$

is the cluster Green function. Note that $T_{ba}(E)$ has no dependence on the $\hat{a}$ variables and that the definition is nonunique to within any operator that annihilates $|\phi_a(E, \hat{a})\rangle$. Because $G_b^{(+)}(E)$ represents outgoing waves in the region of configuration space appropriate to the channel labeled by b, we identify

$$T_{ba}(\hat{b}, \hat{a}; E) \equiv \langle\phi_b(E, \hat{b})|T_{ba}(E)|\phi_a(E, \hat{a})\rangle \tag{1.7.13}$$

as the collision transition amplitude for the process at energy E corresponding to the change $(a, \hat{a}) \to (b, \hat{b})$ in asymptotic configurations. From Eq. (1.7.11), we infer that

$$T_{ba}(\hat{b}, \hat{a}; E) = \langle\phi_b(E, \hat{b})|\vec{G}_b^{(+)}(E)^{-1}|\Phi_a^{(+)}(E, \hat{a})\rangle. \tag{1.7.14}$$

We have introduced the arrow notation to distinguish the left- and right-handed inverses, $\vec{A}^{-1}$ and $\overleftarrow{A}^{-1}$, respectively, of an operator A. When

$\overleftarrow{A}^{-1} \neq \overrightarrow{A}^{-1}$, one must avoid formal manipulations that lead to ambiguities. For example, the product $\overrightarrow{A}^{-1}A\overleftarrow{A}^{-1}$ is ambiguous unless $\overleftarrow{A}^{-1} = \overrightarrow{A}^{-1}$. The domains where $\overrightarrow{A}^{-1}A$ and $A\overleftarrow{A}^{-1}$ are identity maps must be different if $\overleftarrow{A}^{-1} \neq \overrightarrow{A}^{-1}$. In the case of the Green function $G_b^{(+)}(E)$, the reason $\overrightarrow{G}_b^{(+)-1}$ differs from $\overleftarrow{G}_b^{(+)-1}$ arises from the properties of the kinetic-energy part of H_b with respect to the Hermitian adjoint operation in scalar products. Except for certain types of states, $\langle \chi_1 | H_0 \chi_2 \rangle \neq \langle H_0 \chi_1 | \chi_2 \rangle$. Thus, the arrows correspond to $\langle \chi_1 | \overrightarrow{H}_0 | \chi_2 \rangle \equiv \langle \chi_1 | H_0 \chi_2 \rangle$ and $\langle \chi_1 | \overleftarrow{H}_0 | \chi_2 \rangle \equiv \langle H_0 \chi_1 | \chi_2 \rangle$.

We make frequent use of the identities

$$V^a = \overleftarrow{G}_a^{(+)}(E)^{-1} - \overleftarrow{G}^{(+)}(E)^{-1}, \tag{1.7.15a}$$

$$= \overrightarrow{G}_a^{(+)}(E)^{-1} - \overrightarrow{G}^{(+)}(E)^{-1}. \tag{1.7.15b}$$

For example, Eq. (1.7.14) can be rewritten as

$$\begin{aligned} T_{ba}(\hat{b}, \hat{a}; E) = \langle \phi_b(E, \hat{b}) | \overrightarrow{G}_b^{(+)}(E)^{-1} G^{(+)}(E) \overleftarrow{G}_a^{(+)}(E)^{-1} | \phi_a(E, \hat{a}) \rangle \\ - \langle \phi_b(E, \hat{b}) | \overrightarrow{G}_b^{(+)}(E)^{-1} | \phi_a(E, \hat{a}) \rangle. \end{aligned} \tag{1.7.16}$$

However,

$$\langle \phi_b(E, \hat{b}) | \overrightarrow{G}_b^{(+)}(E)^{-1} | \phi_a(E, \hat{a}) \rangle = 0, \tag{1.7.17}$$

because we have cluster rather than propagating states. Even though

$$(E - \overleftarrow{H}_b) | \phi_b(E, \hat{b}) \rangle = 0, \tag{1.7.18}$$

this does not automatically imply the validity of Eq. (1.7.17), unless H_b is effectively self-adjoint in the scalar product (1.7.17). The reason the self-adjointness of H_b is called into question is that the states appearing in Eq. (1.7.17) are not normalizable except in the Dirac δ-function sense. Thus, the surface integral that is picked up in switching the action of $\overrightarrow{G}_b^{(+)}(E)^{-1}$ from $|\phi_a(E, \hat{a})\rangle$ to $\langle \phi_b(E, \hat{b})|$ may not vanish. The validity of Eq. (1.7.17) rests on a property of the channel states called asymptotic orthogonality,[2] where because of the different clustering characteristics of the channel states for $b \neq a$ they vanish in nonoverlapping regions of the asymptotic configuration space. This suffices for the vanishing of the surface integral $\langle \phi_b(E, \hat{b}) | (\overleftarrow{H}_0 - \overrightarrow{H}_0) | \phi_a(E, \hat{a}) \rangle$, which is equal to the left-hand side of Eq. (1.7.17) as a consequence of Eq. (1.7.18).

We remark that paradoxes are encountered when self-adjointness is unwarranted. The most famous one results from the following manipulation:[3]

$$\langle \phi_a(E, \hat{a}') | V^a | \psi_a^{(+)}(E, \hat{a}) \rangle = \langle \phi_a(E, \hat{a}') | G_a^{(+)}(E)^{-1} | \psi_a^{(+)}(E, \hat{a}) \rangle, \tag{1.7.19}$$

where we leave the direction of the action of $G_a^{(+)}(E)^{-1}$ unspecified. The right-hand side of Eq. (1.7.19) vanishes if $G_a^{(+)}(E)^{-1}$ is regarded as self-adjoint (ignoring the $i0$ term). The problem with this is that the left-hand side of Eq. (1.7.19) is the transition amplitude for the process $(a, \hat{a}) \to (a, \hat{a}')$, as we demonstrate shortly, and it definitely does not vanish. We note that $|\psi_a^{(+)}\rangle$ has a *propagating* piece that cancels $G_a^{(+)}(E)^{-1}$ provided we have $\vec{G}_a^{(+)}(E)^{-1}$ in Eq. (1.7.19).

We conclude from Eq. (1.7.16) that the transition operators are any operators on-shell equivalent to the operator product $\vec{G}_b^{(+)}(E)^{-1}G^{(+)}(E)\overleftarrow{G}_a^{(+)}(E)^{-1}$. By *on-shell*, we mean a matrix element, such as in Eq. (1.7.16), where the energies of the two cluster states are equal to the parametric energy occurring in the product of Green functions. *Off-shell* means any other matrix element of the same operator.

The terms on- and off-shell pervade the literature of modern theoretical physics as well as this book. Their diverse meanings depend on the context in which they are used. For example, in relativistic scattering theories the *mass-shell* refers to the situation where the single-particle four-momenta satisfy the Einstein relation $E^2 - \mathbf{p}^2 = m^2$. In virtual states, this relation may not hold, and then the particle is said to be *off its mass-shell.* In nonrelativistic scattering theory of the type we are considering, *particles* are always on their mass-shells. *Matrix elements*, however, may be off-(the-energy)-shell in the manner indicated above when they involve an operator that depends on a parametric energy.

Let us consider

$$\langle \phi_b(E, \hat{b}) | \hat{T}_{ba}(E) | \phi_a(E, \hat{a}) \rangle$$
$$\equiv \langle \phi_b(E, \hat{b}) | \vec{G}_b^{(+)}(E)^{-1} G^{(+)}(E) \overleftarrow{G}_a^{(+)}(E)^{-1} | \phi_a(E, \hat{a}) \rangle. \quad (1.7.20)$$

This is a matrix element of the residue of a double pole of $G^{(+)}(E)$.[4] These poles are made explicit by means of partitioned-generated *resolvent identities*

$$G(z) = G_a(z) + G_a(z)V^aG(z), \quad (1.7.21a)$$

$$G(z) = G_a(z) + G(z)V^aG_a(z), \quad (1.7.21b)$$

which are algebraic relations when z is an arbitrary complex number. Here

$$G(z) \equiv (z - H)^{-1}, \quad (1.7.22a)$$

$$G_a(z) \equiv (z - H_a)^{-1}, \quad (1.7.22b)$$

so that

$$G(z) = G_b(z) + G_b(z)T_{ba}^{(+)}(z)G_a(z), \quad (1.7.23a)$$

$$G(z) = G_a(z) + G_b(z)T_{ba}^{(-)}(z)G_a(z), \quad (1.7.23b)$$

where

$$T_{ba}^{(+)}(z) \equiv V^b + V^b G(z) V^a, \tag{1.7.24a}$$

$$T_{ba}^{(-)}(z) \equiv V^a + V^b G(z) V^a. \tag{1.7.24b}$$

The double poles in $G(z)$ as $z \to E + i0$ arise from the channel Green-function factors in $G_b T_{ba}^{(\pm)} G_a$. The contributions of the single-pole terms on the right-hand sides of Eqs. (1.7.23a, b) from the single Green-function factors vanish in the matrix element (1.7.20). This demonstrates the on-shell equality of the matrix elements $\langle \phi_b(E, \hat{b}) | T_{ba}^{(\pm)}(E) | \phi_a(E, \hat{a}) \rangle$.

Equation (1.7.20) embodies a very general result of scattering theory called the *residue rule*, which relates the transition amplitude to the residues of multiple poles of a full propagator or Green function.[4] It suggests a general definition for a generic transition operator T_{ba} of nonrelativistic scattering theory by means of the decomposition

$$G(z) = f_{ba} G_b(z) + g_{ba} G_a(z) + G_b(z) T_{ba}(z) G_a(z), \tag{1.7.25}$$

where f_{ba} and g_{ba} are numerical quantities that depend only on the channel indices. This definition does not necessarily determine $T_{ba}(z)$ because the formal solution

$$T_{ba}(z) = \vec{G}_b(z)^{-1} [G(z) - f_{ba} G_b(z) - g_{ba} G_a(z)] \overleftarrow{G}_a(z)^{-1}, \tag{1.7.26}$$

is ambiguous in general, unless the factors f_{ba} and g_{ba} are chosen so that the square-bracketed term has an appropriate double-pole structure. We include arrows on the Green-function inverses, although as a practical matter they are needed only if $|\text{Im}\, z| \to 0$. We now review three widely-used possibilities.

Post Form of $T_{ba}(z)$: $f_{ba} = 1$, $g_{ba} = 0$, $T_{ba}(z) = T_{ba}^{(+)}(z)$. The appelation "post" is derived from the asymmetry in Eq. (1.7.24a) favoring the interaction in the final, or post, channel b, i.e.,

$$\langle \phi_b(E, \hat{b}) | T_{ba}^{(+)}(E) | \phi_a(E, \hat{a}) \rangle = \langle \phi_b(E, \hat{b}) | V^b | \psi_a^{(+)}(E, \hat{a}) \rangle. \tag{1.7.27}$$

The identification

$$|\psi_a^{(+)}(E, \hat{a})\rangle = [1 + G^{(+)}(E) V^a] |\phi_a(E, \hat{a})\rangle \tag{1.7.28}$$

follows from Eqs. (1.7.7–1.7.9). Alternatively, we note that

$$|\psi_a^{(+)}(E, \hat{a})\rangle = \lim_{\varepsilon \to 0+} \{[1 + G(E + i\varepsilon) V^a] |\phi_a(E, \hat{a})\rangle\}, \tag{1.7.29}$$

and since

$$i\varepsilon G_a(E + i\varepsilon) |\phi_a(E, \hat{a})\rangle = |\phi_a(E, \hat{a})\rangle, \tag{1.7.30}$$

we obtain

$$|\psi_a^{(+)}(E,\hat{a})\rangle = i\varepsilon G(E + i\varepsilon)|\phi_a(E,\hat{a})\rangle, \tag{1.7.31}$$

where the limit $\varepsilon \to 0+$ is understood.

Prior Form of $T_{ba}(z)$: $f_{ba} = 0$, $g_{ba} = 1$, $T_{ba}(z) = T_{ba}^{(-)}(z)$. Here, we have

$$\langle \phi_b(E,\hat{b})|T_{ba}^{(-)}(E)|\phi_a(E,\hat{a})\rangle = \langle \psi_b^{(-)}(E,\hat{b})|V^a|\phi_a(E,\hat{a})\rangle, \tag{1.7.32}$$

which favors the interaction in the initial, or "prior," channel, and

$$\langle \psi_b^{(-)}(E,\hat{b})| = \langle \phi_b(E,\hat{b})|[V^b G^{(+)}(E) + 1], \tag{1.7.33}$$

or

$$|\psi_b^{(-)}(E,\hat{b})\rangle = [1 + G^{(-)}(E)V^b]|\phi_b(E,\hat{b})\rangle, \tag{1.7.34}$$

is the out-state corresponding to the configuration $\hat{b}$ at $t \to +\infty$. Since this state evolves to a definite asymptotic configuration, the appropriate Green function in this case represents incoming waves rather than outgoing ones, hence the appearance of

$$G^{(-)}(E) = G^{(+)}(E)^\dagger. \tag{1.7.35}$$

We note that the $\varepsilon \to 0+$ form of Eq. (1.7.34) is also consistent with the incoming-wave version of Eq. (1.7.31):

$$|\psi_b^{(-)}(E,\hat{b})\rangle = -i\varepsilon G(E - i\varepsilon)|\phi_b(E,\hat{b})\rangle, \tag{1.7.36}$$

so for an arbitrary partition a we conclude that

$$|\psi_a^{(\pm)}(E,\hat{a})\rangle = G(E \pm i\varepsilon)\vec{G}_a(E \pm i\varepsilon)^{-1}|\phi_a(E,\hat{a})\rangle. \tag{1.7.37}$$

AGS Form of $T_{ba}(z)$: $f_{ba} = g_{ba} = \frac{1}{2}\delta_{ba}$, $T_{ba}(z) = U_{ba}(z)$. Here,

$$U_{ba}(z) = \bar{\delta}_{ba} A_{ba}^b(z) + T_{ba}^{(+)}(z), \tag{1.7.38a}$$

$$U_{ba}(z) = \bar{\delta}_{ba} A_{ba}^a(z) + T_{ba}^{(-)}(z), \tag{1.7.38b}$$

$$\bar{\delta}_{ba} = 1 - \delta_{ba} \tag{1.7.39}$$

is the anti-Kronecker delta function,[5] and

$$A_{ba}^c(z) = \vec{G}_b(z)^{-1}[G_c(z)\bar{\delta}_{ba}]\overleftarrow{G}_a(z)^{-1}. \tag{1.7.40}$$

The transition operators $U_{ba}(z)$, which are not preferential in the roles of the initial and final channels, were first introduced by Alt, Grassberger, and Sandhas (AGS).[6] Originally, no distinction was made between the possible directionalities of the inverses of the channel Green functions $G_a(z)$ and $G_b(z)$. If it is possible to do this, even when $|\mathrm{Im}\, z| \to 0$, then Eqs. (1.7.38a, b) take on the forms that are commonly used in the literature:

$$U_{ba}(z) = \bar{\delta}_{ba}\, G_a(z)^{-1} + T_{ba}^{(+)}(z), \tag{1.7.41a}$$

$$U_{ba}(z) = \bar{\delta}_{ba}\, G_b(z)^{-1} + T_{ba}^{(-)}(z). \tag{1.7.41b}$$

In many of the applications of the AGS operators, there is no difficulty involved in using the simpler forms (1.7.41a, b). The reason why this is so is that the formal constructions $A_{ba}^{c}(z)$ appear only for $b \neq a$, so that manifest pathologies such as $A_{aa}^{a}(z)$ never appear. We note that Eqs. (1.7.41a, b) are equivalent *only* if the directionality of the inverse operators appearing in them are ignored, as may be verified using Eqs. (1.7.15a, b).

Henceforth, we identify the operators given by Eqs. (1.7.41a, b) as *the* AGS transition operators. They possess a number of interesting features that result directly from the symmetrical roles of the initial and final channel indices; we note, however, that

$$U_{ba}(z) \neq U_{ab}(z). \tag{1.7.42}$$

Finally, we relate the on-shell transition operator to the continuum limit of the S-matrix (1.7.1). The in and out states on the right-hand side of Eq. (1.7.1) are wave packets of the continuum states (1.7.28) and (1.7.33), respectively. Using this fact, one identifies the continuum-limit S-matrix elements:

$$\begin{aligned}\langle \psi_b^{(-)}(E', \hat{b}) | \psi_a^{(+)}(E, \hat{a}) \rangle &= \delta_{ba} \langle \phi_b(E', \hat{b}) | \phi_a(E, \hat{a}) \rangle \\ &\quad - 2\pi i\, \delta(E' - E) \langle \phi_b(E', \hat{b}) | T_{ba} | \phi_a(E, \hat{a}) \rangle,\end{aligned} \tag{1.7.43}$$

where

$$\delta_{ba} \langle \phi_b(E', \hat{b}) | \phi_a(E, \hat{a}) \rangle = \delta_{ba}\, \delta(E' - E)\, \delta_{\hat{b}, \hat{a}}, \tag{1.7.44}$$

and $\delta_{\hat{b},\hat{a}}$ denotes a Dirac–Kronecker δ-function. Equation (1.7.43) allows one to make a connection between the scattering observables and the transition matrix (Section A.1b).

1.8. Integral Equations, Subtraction Techniques

Much of collision theory consists in finding suitable integral equations for the transition operators. The most direct way to do this results in the Lippmann–Schwinger (LS) equations for the transition operator $T_{ba}^{(+)}(z)$; similar equations for the other operators can then be obtained by substitution. These equations are based on the resolvent identities (1.7.21a, b). For example, from Eq. (1.7.24a) we see that

$$T_{ba}^{(+)}(z) = V^b + V^b[G_c(z) + G_c(z)V^cG(z)]V^a, \tag{1.8.1}$$

or

$$T_{ba}^{(+)}(z) = V^b + V^bG_c(z)(V^a - V^c) + V^bG_c(z)T_{ca}^{(+)}(z), \tag{1.8.2}$$

for any a, b, c. When $a = b = c$, Eq. (1.8.2) reduces to the uncoupled form

$$T_{aa}^{(+)}(z) = V^a + V^aG_a(z)T_{aa}^{(+)}(z). \tag{1.8.3}$$

The Eqs. (1.8.2) illustrate the two major problems that can be encountered in an arbitrary set of integral equations for the transition operators. First, as is the case with the Eqs. (1.8.2), which possess noncompact kernels, the equations may be unsatisfactory in some respect. Second, the equations may be too complicated for anything except approximate solution. Both problems are very often solved or circumvented by the use of a simple tactic called a *subtraction technique*. This device provides a common base for the many different, seemingly unrelated, special methods used in the literature and throughout this book.

One problem with the Eqs. (1.8.2) is that the kernels $V^bG_c(z)$, and all of their powers, contain Dirac δ-function singularities, in addition to various pole singularities, when they are represented in momentum space. This removes the Eqs. (1.8.2) from the class (Fredholm) of integral equations for which there is a well-defined solution algorithm. Further, when $c = b$, Eq. (1.8.2) becomes

$$T_{ba}^{(+)}(z) = V^bG_b(z)\overleftarrow{G}_a(z)^{-1} + V^bG_b(z)T_{ba}^{(+)}(z), \tag{1.8.4}$$

or, if we interchange a and b,

$$T_{ab}^{(+)}(z) = V^aG_a(z)\overleftarrow{G}_b(z)^{-1} + V^aG_a(z)T_{ab}^{(+)}(z), \tag{1.8.5}$$

If there is a state $|\chi_b(z)\rangle$ such that, for $b \neq a$,

$$V^aG_a(z)\overleftarrow{G}_b(z)^{-1}|\chi_b(z)\rangle = 0, \tag{1.8.6}$$

then

$$T_{ab}^{(+)}(z)|\chi_b(z)\rangle = V^aG_a(z)T_{ab}^{(+)}(z)|\chi_b(z)\rangle, \tag{1.8.7}$$

indicating the possibility of nonuniqueness. There is no reason to expect Eq. (1.8.6) to hold unless $z \to E + i0$ and $|\chi_b(z)\rangle$ is proportional to $|\phi_b(E, \hat{b})\rangle$. This actually happens for any a, $n_a > 1$, when b is a two-cluster partition of $N > 2$ particles and the two-cluster bound states exist.[1] It seems plausible that an attempt to circumvent this problem by solving Eq. (1.8.3) with

Im $z \neq 0$ and then continuing the solution to the real axis from above will be defeated by numerical instabilities as one takes the limit, and so other methods must be sought.

Generally, the integral equations of collision theory are realized in coupled, multidimensional forms. The number and the dimensionality of these equations may frustrate attempts at exact numerical solution.

We consider a method for dealing with all of the preceding problems for a generic collision integral equation of the symbolic form

$$T = B + KT. \tag{1.8.8}$$

A *subtraction technique*[2] is affected using a decomposition

$$K = K_S + K_R. \tag{1.8.9}$$

of the kernel K into a special (S) part K_S and a reduced or residual (R) part K_R. If we introduce auxiliary quantities γ_i by means of the integral equations

$$\gamma_i = I + K_i \gamma_i, \tag{1.8.10}$$

where $i =$ R, S, then

$$T = \gamma_R B + \gamma_R K_S T \tag{1.8.11}$$

and

$$T = \gamma_S B + \gamma_S K_R T \tag{1.8.12}$$

represent alternative subtraction techniques. In what follows we concentrate almost exclusively on Eq. (1.8.11).

Typically, K_S is such that, if γ_R is known, the solution of Eq. (1.8.11) is both well-defined and computationally feasible. For example, in the two-cluster case, $a = \alpha$, Eq. (1.8.3) takes the form

$$T^{(+)}_{\alpha\alpha}(z) = V^{\alpha} + V^{\alpha} G_{\alpha}(z) T^{(+)}_{\alpha\alpha}(z). \tag{1.8.13}$$

We take $K_S = V^{\alpha} G_{\alpha}(z) P_{\alpha}$ and $K_R = V^{\alpha} G_{\alpha}(z) Q_{\alpha}$, where

$$I = P_{\alpha} + Q_{\alpha} \tag{1.8.14}$$

and

$$P_{\alpha} = \sum_{E, \hat{\alpha}} |\phi_{\alpha}(E, \hat{\alpha})\rangle\langle\phi_{\alpha}(E, \hat{\alpha})| \tag{1.8.15}$$

is the projection on the bound-cluster states, and Q_{α} is the projection on the complement of the subspace of these states. Then, Eq. (1.8.11) becomes

$$T^{(+)}_{\alpha\alpha}(z) = U^{\alpha}(z) + U^{\alpha}(z) G_{\alpha}(z) P_{\alpha} T^{(+)}_{aa}(z), \tag{1.8.16}$$

where $U^{\alpha}(z) = \gamma_R V^{\alpha}$, and

$$U^{\alpha}(z) = V^{\alpha} + V^{\alpha} Q_{\alpha} G_{\alpha}(z) U^{\alpha}(z), \tag{1.8.17}$$

is the Feshbach[3] optical potential. The kernel of Eq. (1.8.16) is not invariant with respect to spatial translations of the clusters, so it becomes a well-defined two-body integral equation for $P_{\alpha} T_{\alpha\alpha}(z) P_{\alpha}$ even when $z \to E + i0$.

In the preceding example, we achieved a Fredholm reduction[4] of the non-Fredholm equation (1.8.13). This will be a useful thing to do if K_S represents some dominant features of the dynamics so that γ_R (or U^{α}) can be approximated in a straightforward way.

As a tactical procedure, either K_S or K_R can be chosen to be Fredholm in connection with Eq. (1.8.11). Let us give two examples which recur frequently:

(i) *Generalized Heitler Equations.* Here K_S is a specified and simple enough kernel so that Eq. (1.8.11) is exactly solvable. A frequent criterion for selecting K_S is to subtract off the singularities in K so that K_R is nonsingular. This is used for obtaining auxiliary integral equations (for γ_R) more amenable to calculation, as well as for the construction of unitary formalisms. In the Feshbach optical-potential formalism, e.g., the elastic unitarity cuts are removed from K_R.

(ii) *Connected-Kernel Reductions.* When K contains Dirac δ-function singularities arising from independent, or disconnected, subsystems, it will not be of the Fredholm type unless these singularities are absent in some power K^p, $p > 1$, of K that appears after $p - 1$ iterations of the integral equation (1.8.8). The objective here is to choose K_S so that it is Fredholm in the sense that $(K_S)^p$ is connected for some $p \geq 1$ and γ_R is calculable entirely in terms of subsystem dynamics. Often to carry out this type of Fredholm reduction, T must first be decomposed into auxiliary operators.

Finally, let us discuss the subtraction technique (1.8.12), which encompasses all known forms of the so-called two-potential formula.[5] It is useful in those situations where K_R has a very small effect compared with K_S. For example, if $K = VG_0$, $B = V = V_S + V_R$, with $K_S = V_S G_0$, $K_R = V_R G_0$, and $\gamma_S V_S \equiv T_S$, then Eq. (1.8.12) reduces to

$$T = T_S + (I + T_S G_0)\tilde{T}_R(I + G_0 T_S), \tag{1.8.18a}$$

where

$$\tilde{T}_R = V_R + V_R[G_0 + G_0 T G_0] V_R. \tag{1.8.18b}$$

Equations (1.8.18a, b) are the simplest form of the two-potential formula as applied to one-channel scattering. The applications to multichannel scattering involve the same idea, but take different forms depending upon the choice of off-shell extension.[5]

1.9. Unitarity, Singularities, and Disconnected Structure

The unitarity of the S-matrix leads to constraints on the physical collision amplitudes. In this section, we derive the operator-valued extensions of these constraints.[1,2]

1.9a. Operator Analyticity and Off-Shell Unitarity

Let us first examine the structure of the transition operator

$$T_{ba}^{(+)}(z) = V^b + V^b G(z) V^a \tag{1.9.1}$$

as a function of z. If the matrix elements $\langle \chi_1 | T_{ba}^{(+)}(z) | \chi_2 \rangle$, for arbitrary, fixed states $|\chi_{1,2}\rangle$, are analytic in z, then $T_{ba}^{(+)}(z)$ is said to be *operator* analytic in z. The key tool for determining this structure is the spectral decomposition of $G(z)$:

$$G(z) = \sum \frac{|\psi_d^{(\pm)}(E_d, \hat{d})\rangle\langle\psi_d^{(\pm)}(E_d, \hat{d})|}{z - E_d} + \sum \frac{|\psi_B(\hat{B})\rangle\langle\psi_B(\hat{B})|}{z - E_B}, \tag{1.9.2}$$

as expressed in terms of either the asymptotically complete in $(+)$ *or* the out $(-)$ states (but not both) along with the bound states $|\psi_B(\hat{B})\rangle$ of all of the N-particles. The sums are over all partitions d, the channel energies E_d, the other d-channel variables, $\hat{d}$, the bound states B, and any other variables $\hat{B}$ needed to specify $|\psi_B(\hat{B})\rangle$. The channel energy E_d is a sum of the cluster kinetic energies minus the sum of their binding energies. We note that both E_d and E_B include the c.m. motion, and so run through a continuum of values; they also depend, in general, on $\hat{d}$ and $\hat{B}$, respectively, although we have left this dependence implicit. Only in the N-particle c.m. frame will E_B take on discrete values; henceforth, we suppose all states are referenced to that frame.

From Eq. (1.9.2), we see that $G(z)$, and therefore $T_{ba}^{(+)}(z)$, is operator-analytic in z except along the real z-axis. There it has pole singularities at the discrete bound-state energies $z = E_b$ and branch-cut singularities running from the thresholds $z = E_d^{\text{th}}$ of each of the channels to $z = +\infty$. Similar remarks apply to $T_{ba}^{(-)}(z)$.

We assume that the spectrum of H, as well as that of every H_a, is such that the bound-state energies of any cluster of particles are not positive after all c.m. energies have been subtracted out. Therefore,

$$E_B \leq 0, \tag{1.9.3}$$

$$E_B \leq E_d^{\text{th}} \leq 0, \tag{1.9.4}$$

for all B and for all partitions d in which at least one cluster admits a bound state. The single ($a = 0$) channel that is characterized by an asymptotic configuration of N free particles has a threshold at zero energy in the N-particle c.m. frame.

The unitarity constraints on $T_{ba}^{(\pm)}(z)$ are derived from the representation (1.9.2) by setting $z = E + i0$ and using the distribution-valued Plemelj identities[3]

$$G^{(\pm)}(E) = P\frac{1}{E - H} \mp i\pi\,\delta(E - H), \tag{1.9.5}$$

where P implies that the Cauchy principal-value prescription is to be employed for integrating over the pole singularity. We find that

$$\begin{aligned} G^{(+)}(E) - G^{(+)}(E)^\dagger = &-2\pi i \sum |\psi_d^{(\pm)}(E_d, \hat{d})\rangle\, \delta(E - E_d)\, \langle\psi_d^{(\pm)}(E_d, \hat{d})| \\ &-2\pi i \sum |\psi_B(\hat{B})\rangle\, \delta(E - E_B)\, \langle\psi_B(\hat{B})|, \end{aligned} \tag{1.9.6}$$

which is a spectral decomposition of

$$G^{(+)}(E) - G^{(+)}(E)^\dagger = -2\pi i\,\delta(E - H). \tag{1.9.7}$$

With our spectral assumptions (1.9.3–1.9.4), the bound-state pole contributions to Eqs. (1.9.6–1.9.7) vanish above the threshold for scattering.

Taking $z = E + i0$, we find using Eq. (1.9.7) that

$$T_{ba}^{(+)}(E) - T_{ab}^{(-)}(E)^\dagger = -2\pi i V^b\,\delta(E - H)\,V^a. \tag{1.9.8}$$

This implies much more than the unitarity of the S-matrix, since it is an *off-shell unitarity* relation. This terminology is supported by the observation that, with E above the scattering threshold and $c \neq 1$ in all sums, Eqs. (1.9.6–1.9.7) imply

$$V^b\,\delta(E - H)\,V^a = \sum T_{bc}^{(+)}(E)\,\Delta_c(E)\,T_{ac}^{(+)}(E)^\dagger, \tag{1.9.9a}$$

$$= \sum T_{cb}^{(-)}(E)^\dagger\,\Delta_c(E)\,T_{ca}^{(-)}(E), \tag{1.9.9b}$$

where

$$\Delta_a(E) \equiv \delta(E - E_a)|\phi_a(E_a, \hat{a})\rangle\langle\phi_a(E_a, \hat{a})|. \tag{1.9.10}$$

Therefore, we have

$$T_{ba}^{(+)}(E) - T_{ab}^{(-)}(E)^\dagger = -2\pi i \sum T_{bc}^{(+)}(E)\,\Delta_c(E)\,T_{ac}^{(+)}(E)^\dagger, \tag{1.9.11a}$$

$$= -2\pi i \sum T_{cb}^{(-)}(E)^\dagger\,\Delta_c(E)\,T_{ca}^{(-)}(E). \tag{1.9.11b}$$

On-shell Eq. (1.9.11a), e.g., becomes,

$$T_{ba}(\hat{b}, \hat{a};\, E) - T_{ab}(\hat{a}, \hat{b};\, E)^* = -2\pi i \sum T_{bc}(\hat{b}, \hat{c};\, E)\,\delta(E_c - E)\,T_{ac}(\hat{a}, \hat{c};\, E)^*, \tag{1.9.12}$$

where $T_{ba}(\hat{b}, \hat{a};\, E)$ refers to the on-shell amplitudes (1.7.13) and we have exploited the on-shell equivalence of $T_{ab}^{(-)}(E)$ and $T_{ab}^{(+)}(E)$.

The summations in Eqs. (1.9.9–1.9.12) include an integration over the phase space available in each channel with a total energy E, and sums over any other channel variables needed to specify the internal states of the bound clusters. With a total energy E, a channel $(c, \hat{c})$ is *open* if $E \geq E_c^{\text{th}}$ and *closed* if $E < E_c^{\text{th}}$. Evidently, only open channels contribute to the right-hand sides of Eqs. (1.9.9–1.9.12).

The optical theorem follows from Eq. (1.9.12):

$$\operatorname{Im} T_{aa}(\hat{a}, \hat{a};\, E) = -\pi \sum |T_{ac}(\hat{a}, \hat{c};\, E)|^2\,\delta(E - E_c). \tag{1.9.13}$$

An important feature of the various forms of unitarity is that they can be interpreted as discontinuity relations. Let us use the notation $E\pm \equiv E \pm i0$. We note that

$$T_{ab}^{(-)}(E+) = T_{ba}^{(+)}(E-)^\dagger. \tag{1.9.14}$$

Then, e.g., we have

$$T_{ba}^{(+)}(E+) - T_{ba}^{(+)}(E-) = -2\pi i \sum T_{bc}^{(+)}(E\pm)\,\Delta_c(E)\,T_{ca}^{(-)}(E\mp), \tag{1.9.15}$$

which expresses the discontinuity of $T_{ba}^{(+)}(z)$ across the real z axis above the scattering threshold as a sum of contributions from each of the open channels. This delineates a multicut discontinuity along the real axis in the z-plane when we describe the analytic behavior of $T_{ba}^{(+)}(z)$.[1] Each of the channels contributes a cut. Similarly, one finds

$$T_{ba}^{(-)}(E+) - T_{ba}^{(-)}(E-) = -2\pi i \sum T_{bc}^{(+)}(E\pm)\,\Delta_c(E)\,T_{ca}^{(-)}(E\mp). \tag{1.9.16}$$

The discontinuity relations for the AGS operators $U_{ba}(z)$ differ from the relations for $T_{ba}^{(\pm)}(z)$ by terms that vanish on shell.

For elastic two-body scattering below the inelastic threshold, off-shell unitarity imposes strong and explicit constraints on transition amplitudes in each state of total angular momentum (Section 2.4). Above the inelastic threshold, the unitarity constraints on the scattering are less clear-cut, unless all of the inelastic channels are also of the two-cluster variety.

For collision amplitudes that involve three or more clusters in either the initial or final states, the implications of unitarity are still poorly understood. Also, it is usually impractical to check unitarity in multiparticle calculations, particularly when one has obtained an exact solution for only a subset of the amplitudes connected by unitarity. When the energy is above the multicluster thresholds, unitarity, therefore, is either not an explicit consideration, or else is used only as a constraint for approximations such as the method of the optical potential for two-cluster elastic scattering.

1.9b. Unitarity and Disconnected Structure

Disconnectedness refers to the dynamical independence of two or more collision processes. The simplest example of this occurs in the three-particle scattering depicted in Fig. 1.2. The two-particle disconnected graphs in Fig. 1.2 correspond to the collision, at some arbitrary time, of two particles with the third passing through freely. Evidently, no point in space–time is distinguished as the collision point. Thus, this disconnectedness structure is a

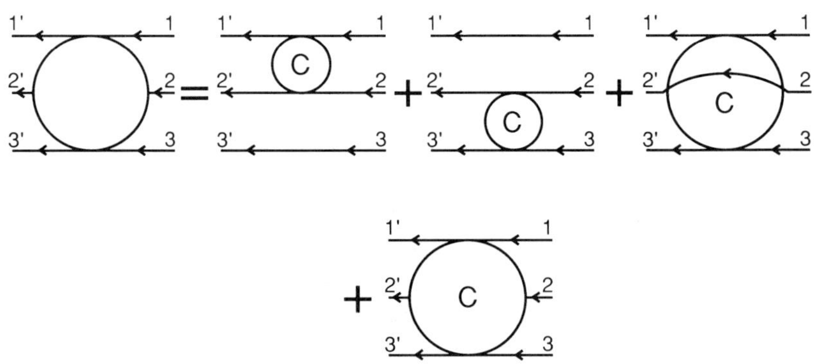

FIG. 1.2. The connectivity structure of the three-to-three particle transition amplitude. The bubbles containing *c* represent either two-particle or three-particle connected amplitudes. The unattached lines represent a noninteracting particle; if the initial and final particles are not all alike then some of the two-body disconnected graphs are not present.

manifestation of the irrelevance of the position in space–time of the third particle with respect to the dynamics of the other two. The momentum-space matrix elements that represent these two-body disconnected parts must therefore contain Dirac δ-functions that represent the separate energy–momentum conservation of the c.m. of each of the independent clusters. This is the only disconnected structure that occurs in three-particle scattering. When there are more than three particles, many more possibilities for disconnected structures arise.[4] Each one of these structures is a unique reflection of the distinct alternatives for relative space–time translational invariances among the subclusters.[5]

We can determine how two-particle disconnected graphs are contained within arbitrary N-particle amplitudes by considering the operator

$$t_i(z) = V_i + V_i G_i(z) V_i, \tag{1.9.17}$$

where i is an $(N-1)$-cluster partition, so V_i is simply a pair interaction between the two clustered particles. We see that t_i satisfies the operator integral equations

$$t_i(z) = V_i + V_i G_0(z) t_i(z), \tag{1.9.18a}$$

$$t_i(z) = V_i + t_i(z) G_0(z) V_i, \tag{1.9.18b}$$

whose substantive content is the set of two-body integral equations for the reduced transition amplitude $\langle \mathbf{k}' | \bar{t}_i(\bar{z}) | \mathbf{k} \rangle$ defined by

$$\langle \mathbf{k}'_1, \ldots, \mathbf{k}'_N | t_i(z) | \mathbf{k}_1, \ldots, \mathbf{k}_N \rangle = \left[\prod_{p \neq r,s} \delta^{(3)}(\mathbf{k}'_p - \mathbf{k}_p) \right] \delta^{(3)}(\mathbf{K}'_{rs} - \mathbf{K}_{rs}) \times \langle \mathbf{k}'_{rs} | \bar{t}_i(\bar{z}) | \mathbf{k}_{rs} \rangle. \tag{1.9.19}$$

Here r, s label the two clustered particles,

$$\mathbf{k}_{rs} = \frac{(m_s \mathbf{k}_r - m_r \mathbf{k}_s)}{m_s + m_r}, \tag{1.9.20a}$$

$$\mathbf{K}_{rs} = \mathbf{k}_r + \mathbf{k}_s, \tag{1.9.20b}$$

the $\mathbf{k}_p$ are momenta, and

$$\bar{z} = z - \sum_{p \neq r,s} \frac{\mathbf{k}_p^2}{2m_p} - \frac{\mathbf{K}_{rs}^2}{2(m_r + m_s)}. \tag{1.9.20c}$$

The operator $\bar{t}_i(\bar{z})$ satisfies integral equations identical in form to Eqs. (1.9.18a, b), except that the Green function is now defined in terms of the relative two-particle kinetic energy rather than H_0. Because of the structure

(1.9.19), $t_i(z)$ is often referred to as a two-particle operator in the N-particle space, as contrasted with the two-particle operator $\bar{t}_i$ that acts only on a two-particle Hilbert space; unless there is danger of ambiguity, we will refer to both as two-particle operators.

On the right-hand side of Eqs. (1.9.11a, b) it is possible for the contributions of two different disconnected pieces to have a particle line in common. The resulting terms represent the discontinuities of the pole singularities in amplitudes of higher connectivity arising from on-shell intermediate propagation. The simplest example of this occurs for $N = 3$ and is depicted in Fig. 1.1, which illustrates a double-scattering part of the three-to-three amplitude; the Dirac δ-function that then appears in the unitarity relation represents the discontinuity across the real line of the pole singularity of $G_0(z)$.

1.9c. Off-Shell Unitarity from Dynamical Integral Equations

The method of Section 1.9a. for obtaining unitarity or discontinuity relations is not useful in practice. Typically, one has a set of exact or approximate dynamical collision integral equations usually for some type of auxiliary amplitudes. Thus, we want to determine the z-plane discontinuity relations, which can always be translated into unitarity-type equations, for generic transition operators denoted by $X(z)$, that satisfy the archetypical operator integral equations:

$$X(z) = B(z) + B(z)g(z)X(z), \tag{1.9.21a}$$

$$X(z) = B(z) + X(z)g(z)B(z). \tag{1.9.21b}$$

Our main assumption is that Eq. (1.9.21a) is a Fredholm equation; then, Eq. (1.9.21b) follows from the formal operator expression $R = (1 - Bg)^{-1}Bg$ for the *resolvent kernel* R, since $X = (I + R)B$. We suppose that z is confined to values slightly above or below the real axis.

If we denote the discontinuity of X across the real z axis as

$$\Delta X \equiv X(+) - X(-), \tag{1.9.22}$$

we find using Eqs. (1.9.21a, b) the following two expressions for it:[6]

$$\Delta X = X(\pm)\Delta g\, X(\mp) + [I + X(\pm)g(\pm)]\,\Delta B[I + g(\mp)X(\mp)]. \tag{1.9.23}$$

The formal manipulations leading to the Eqs. (1.9.23) must be regarded with great care when the original Fredhom assumption fails. This circumstance is usually associated with the appearance of a bound, or a resonant, state of the entire system depending on the type of equation under consideration. A classic example occurs with the two-particle transition operator defined by Eqs. (1.9.18a, b). In this case,

$$\Delta t_i = t_i(\pm)\Delta G_0 t_i(\mp), \tag{1.9.24}$$

since $\Delta V_i = 0$, as contrasted with the result

$$\Delta t_i = t_i(\pm)[-2\pi i\,\delta(E - H_0)]t_i(\mp) + \Delta t_i(B), \tag{1.9.25}$$

obtained from the spectral representation of $G_i(z)$, where

$$\Delta t_i(B) = \sum_B V_i|\phi_i(B)\rangle[-2\pi i\,\delta(E - E_B)]\langle\phi_i(B)|V_i \tag{1.9.26}$$

is the contribution from the two-body bound-state poles at energies E_B. As pointed out by Sloan, the expressions (1.9.25–1.9.26) agree, provided one is careful about interpreting ΔG_0; the zero of ΔG_0 cancels the effect of the bound-state poles in the $t_i(\pm)$ factors in Eq. (1.9.25) to yield the contribution (1.9.26).[7]

The contributions of the two-body bound-state poles in off-shell transition amplitudes are of crucial importance in multiparticle collision problems because the two-body parametric energy $\bar{z}$ has the range $-\infty < \bar{z} \leq z$, which is realizable in intermediate scatterings. One encounters similar bound-state contributions from multiparticle transition operators when they represent subsystem collisions.

1.10. Symmetries

Invariance principles reflect symmetries of the dynamical behavior of a physical system. Symmetry operations are represented by either unitary or antiunitary operators, $U(\xi)$, where ξ refers to whatever parameters are needed to specify the transformations.[1] The decision as to whether $U(\xi)$ is unitary or antiunitary is made by specifying the action of the operator on the observables of the theory. In practice, only symmetry transformations that contain the time-reversal operation are antiunitary.[2]

If the dynamics of a system with a Hamiltonian H are invariant with respect to a set of symmetry transformations $U(\xi)$, then[3]

$$[U(\xi), H] = 0 \tag{1.10.1}$$

for all values of the parameters set ξ. This is preserved throughout the evolution of the system, so it is plausible that[4]

$$[U(\xi), S] = 0, \tag{1.10.2}$$

which is equivalent to the statement that the coefficients $C_{ab}(\xi)$ in the expansions

$$U(\xi)|\psi_a^{(\pm)}\rangle = \sum_b C_{ab}(\xi)|\psi_b^{(\pm)}\rangle \tag{1.10.3}$$

are the same for the in- and out-states.

In many cases, the set of operators $\{U(\xi)\}$ form a group $\mathcal{G}$ of transformations that are continuously connected to the identity $I \equiv U(0)$. Then, the $U(\xi)$ for arbitrary ξ are completely determined in terms of the generators G_j of infinitesimal transformations:

$$U(\xi) = I - i\sum_j \xi_i G_j + O(\xi^2), \tag{1.10.4}$$

where $j = 1, \ldots, p$, corresponding to the p infinitesimal parameters ξ_j. When the ξ_j are all real, the G_j are Hermitian and correspond to observables since then

$$[H, G_j] = [S, G_j] = 0. \tag{1.10.5}$$

Let us list the common continuous geometric transformations and their associated infinitesimal generators:

(i) *Spatial Translations* (the momenta P_i, $i = 1, 2, 3$);
(ii) *Time Translations* (the Hamiltonian H);
(iii) *Spatial Rotations* (the rotation operators J_i, $i = 1, 2, 3$);
(iv) *Space-Time Rotations* (the generators of Lorentz transformations, M_{0i}, $i = 1, 2, 3$).

Because of the conditions (1.10.5), invariance under spatial rotations and translations implies that the total angular momentum and linear momentum are conserved, respectively. Similarly, the identification of the Hamiltonian with the generator of time translations leads to energy conservation when the system is invariant under these translations. The implications of Lorentz invariance are not conveniently expressed in terms of conserved quantities that label the asymptotic configurations; in brief, the Eqs. (1.10.5), with the correspondence $G_i \sim M_{0i}$, imply that the transition probabilities have the same value when evaluated in different inertial frames.[5]

For infinite scattering volumes, space–time translational invariance leads to overall Dirac δ-functions in energy and momentum for the S-matrix elements. When there are at least three particles in both the initial and final states, relative translational invariances among the subgroupings of particles give rise to disconnected structures in the transition amplitudes.

The Dirac δ-functions arise because of the noncompact character of the translational groups. Since the rotation group is compact, the implications of rotational invariance on the structure of transition amplitudes are somewhat different. In this case one encounters a Kronecker-delta behavior in the total angular momentum quantum numbers and very specific dependences upon the angular variables of the asymptotic momenta.

Lorentz invariance implies that the transition amplitudes are invariant functions of their kinematic variables.[5,6]

The two discrete symmetry operations, spatial reflection, or "parity," and time reversal, represent conserved symmetries for both the strong and electromagnetic interactions. Therefore, we suppose that they are exact symmetries for all of the considerations in this book.

Since parity is a binary symmetry, the parity operator U_r is its own inverse, and therefore Hermitian. If $\mathbf{x}_j^{\text{op}}$ and $\mathbf{p}_j^{\text{op}}$ denote the coordinate and momentum operators of the $j = i, \ldots, N$ particles, respectively, then

$$\{U_r, \mathbf{x}_j^{\text{op}}\} = \{U_r, \mathbf{p}_j^{\text{op}}\} = 0, \tag{1.10.6}$$

where $\{A, B\} = AB + BA$ is the anticommutator. The properties (1.10.6) represent a symmetry of the theory if they are preserved throughout the evolution of the system, and they restrict the type of terms that can appear in the Hamiltonian and in the transition amplitude.

Time reversal U_τ is also a binary operation, but an antilinear one so its mathematical structure is somewhat different.[3,4,7] Corresponding to the Eqs. (1.10.6), we have

$$\{U_\tau, \mathbf{p}_j^{\text{op}}\} = 0 \tag{1.10.7a}$$

and

$$[U_\tau, \mathbf{x}_j^{\text{op}}] = 0. \tag{1.10.7b}$$

Again, time reversal is a symmetry of the theory if Eqs. (1.10.7a, b) are preserved throughout the evolution of the system. Note that just as the Eqs. (1.10.6) are consistent with the preservation of the fundamental commutation relations of the theory, $[\mathbf{x}_j^{\text{op}}, \mathbf{p}_j^{\text{op}}] = i\hbar$, under the parity operation, so are Eqs. (1.10.7a, b):

$$U_\tau[\mathbf{x}_j^{\text{op}}, \mathbf{p}_j^{\text{op}}]U_\tau^{-1} = -[\mathbf{x}_j^{\text{op}}, \mathbf{p}_j^{\text{op}}]; \tag{1.10.8a}$$

but

$$U_\tau i\hbar U_\tau^{-1} = -i\hbar, \tag{1.10.8b}$$

because U_τ is antilinear. The antiunitary property of U_τ is expressed in terms of what might be referred to as the antilinear adjoint operation that we denote by $\perp$:

$$U_\tau^{\perp} = U_\tau^{-1}, \tag{1.10.9}$$

where

$$\langle\chi|U_\tau^\perp|\psi\rangle = \langle\psi|U_\tau|\chi\rangle, \tag{1.10.10}$$

for arbitrary states $|\chi\rangle$ and $|\psi\rangle$. Because time reversal is binary, one can always choose the phases such that U_τ is Hermitian with respect to the $\perp$ operation, as well as antiunitary:

$$U_\tau^\perp = U_\tau = U_\tau^{-1}; \tag{1.10.11}$$

and we will suppose that this is always the case.

For any of the various symmetry operations that were represented by unitary operators $U(\xi)$, invariance under the operation implies

$$U^\dagger(\xi)SU(\xi) = S, \tag{1.10.12}$$

where S is the Heisenberg picture S-operator. On the other hand, the antiunitary character of U_τ, which results in changing numbers into their complex conjugates when they are transposed with U_τ, implies that, to within irrelevant phases,

$$U_\tau|\psi_a^{(\pm)}\rangle = |\psi_{a_\tau}^{(\mp)}\rangle, \tag{1.10.13}$$

where a_τ denotes the time-reversed asymptotic configuration. Therefore, we have

$$U_\tau^\perp SU_\tau = S^\dagger, \tag{1.10.14}$$

and so

$$\langle f_\tau|S|i_\tau\rangle = \langle i|S|f\rangle, \tag{1.10.15}$$

which is quite different from what one obtains from Eq. (1.10.12).

The symmetries that we have discussed so far are easy to implement in collision theory and their consequences invariably simplify matters. This simplification also occurs when we exploit the consequences of the permutation symmetries among identical particles in collision theory, but the practical implementation of these symmetries is often difficult. Unlike any of the other symmetries, the implementation of permutation symmetries requires not only a "symmetrical" Hamiltonian, but also a major restriction upon the physical states.

For the sake of simplicity, we consider a single species of particle. Let $U(P)$ be the unitary operator on the N-particle Hilbert space corresponding to a permutation P of the N particles. If $|\psi\rangle$ is any physical state of the system, then $U(P)|\psi\rangle$ must be physically equivalent to it. Furthermore, if A represents any observable consistent with the indistinguishability of the identical particles, then all of the expectation values,

$$\langle\psi|A|\psi\rangle = \langle\psi|U(P)^{\dagger}AU(P)|\psi\rangle, \tag{1.10.16}$$

must be equal for all physical states $|\psi\rangle$ and all P. Equation (1.10.16) implies that A must satisfy

$$[A, U(P)] = 0. \tag{1.10.17}$$

The usual N-particle Hilbert space $\mathscr{H}(N)$ includes many complete sets of commuting observables, whose eigenvectors span the space, but do not satisfy Eq. (1.10.17). Therefore, we conclude $\mathscr{H}(N)$ is "too large" in the identical-particle case. We suppose that on the physical Hilbert space $\mathscr{H}_{\text{phys}}$ all observables satisfy (1.10.17), and consequently any state $|\psi\rangle \in \mathscr{H}_{\text{phys}}$ is an eigenstate of all the permutation operators:

$$U(P)|\psi\rangle = C(P, \psi)|\psi\rangle. \tag{1.10.18}$$

Then, every state in $\mathscr{H}_{\text{phys}}$ is either completely symmetric or completely antisymmetric, i.e., $C(P, \psi) = +1$, for all P, or $C(P, \psi) = \delta(P)$, for all P, respectively. Here $\delta(P) = +1\ (-1)$ for an even (odd) permutation. The choice between the two is provided by the spin-statistics theorem,[8] which associates the symmetric (antisymmetric) alternative with identical particles of integer (half-integer) spin, i.e., bosons (fermions).

The implementation of permutation symmetries is difficult in nonrelativistic collision theory because the full Hilbert space is always taken as a starting point. Second-quantized approaches that formulate scattering theory in Fock space circumvent this, but they have other limitations not encountered in successful applications of Fock space to the many-body (bound-state) problem.[9]

1.11. Permutation Symmetries

In much of atomic and nuclear physics, electrons and nucleons, respectively, suffice as elementary particles. So in any collision process, except of the simplest type, one is confronted with a nontrivial problem of maintaining the symmetry with regard to the exchange of fermions required by the Pauli principle.

Bosonic permutation symmetries are important in atomic and molecular collisions when the electronic degrees of freedom can be ignored. In hadron and nuclear physics, bosons, e.g., pions, occur frequently and they can be treated as such provided that their quark substructure can be ignored.

In view of the considerable and sustained effort that has been devoted to the subject,[1] it is probably safe to assert that "symmetrization," by which we mean either the actual symmetrization of states with respect to the boson variables or antisymmetrization in the case of fermions, is something that should be discussed with great care in collision theory. The contrary opinion,[2] namely, that one can solve the problem of interest and then symmetrize afterward, is probably practical only for two-particle collisions and does not benefit from the simplifications that are obtained by fully incorporating permutation symmetries beforehand. When approximations are required, the "symmetrize-later" strategy must deal with the serious problem of justifying truncations that involve unsymmetrized physics.

Therefore, we regard it as an essential preliminary to any practical calculational scheme to formulate properly symmetrized dynamical equations along with the appropriate boundary conditions. If we symmetrize the collision integral equations both objectives are achieved simultaneously.

Two main techniques have been used for obtaining symmetrized collision integral equations. The most popular approach imposes the permutation symmetry on the unsymmetrized dynamical equations. We call this procedure *explicit symmetrization*. The net effect of this method is a reduction in the number of integral equations and/or a simplification of the inhomogeneous and kernel terms of these equations.[3] The way in which explicit symmetrization can be implemented has been worked out in detail only relatively recently.[4,5] This method is used in most of this book; its main features are described in Section 1.11a, and is based on the work of Bencze and Redish.[4]

The other major technique is the so-called "second-quantized" method, in which the scattering problem is formulated in an appropriate Fock space, thus automatically satisfying the requirements of particle identity. This approach has a number of attractive features. It is widely used in theories of both atomic and nuclear structure, and therefore a multiparticle scattering theory phrased in this way would be expressed in a unifying common language that would facilitate the application of the usual diagrammatic techniques to nonrelativistic scattering theory. Despite this promise and some important early applications,[6] the Fock-space approach has had relatively little impact on either the practical calculation of multiparticle scattering or in the formulation of properly symmetrized collision integral equations. We discuss this technique in Section 1.11b, where we point out what appear to be the major reasons for this lack of effectiveness.

1.11a. Explicit Symmetrization

The usual formulations of nonrelativistic scattering theory ignore any permutation symmetries in the development of sets of collision integral equations for either the wave functions or the transition amplitudes. One is then faced with the task of imposing these symmetries in order to represent the physical situation correctly.

Let us suppose that we have N particles, some of which are identical, and that the Hamiltonian H is symmetric with respect to the permutations of those that are the same. The symmetry group is the permutation group $S(N)$, if all N particles are identical. Generally, we have k distinct species of particles,

$$N = \sum_{i=1}^{k} n_i, \tag{1.11.1}$$

where n_i is the number of particles of type i. Then, the symmetry group S of the system is isomorphic to the direct product of each of the relevant permutation groups $S(n_i)$:

$$S \sim \bigotimes_{i=1}^{k} S(n_i). \tag{1.11.2}$$

Thus, the physical, "symmetrized," Hilbert space, $\mathscr{H}_s(N)$, which is a subspace of the full N-particle space $\mathscr{H}(N)$, has the direct-product structure

$$\mathscr{H}_s(N) = \bigotimes_{i=1}^{k} \mathscr{H}_s(n_i), \tag{1.11.3}$$

where $\mathscr{H}_s(n_i)$ is the Hilbert space of completely symmetrized (bosons) or antisymmetrized (fermions) states of n_i particles.

The incorporation of permutation symmetry with the appropriate statistics is complicated by the numerous partitioning options among the N particles. For each option, there is a channel, or cluster, Hamiltonian; when some particles are identical, not all of these Hamiltonians describe independent subsystems.

Our first task is to sort out the action of the permutation operations on the partition labels. Let L be the subset of all of the partitions, $a, b, \ldots$, of the N particles that are used to label operators that enter into a particular set of collision integral equations. Let $P \in S$ be any element of the symmetry group S. Now S maps L onto itself:

$$P(a) = a', \tag{1.11.4}$$

where $P \in S$ and $a, a' \in L$.

In general, there are a number of permutations of N particles that will leave a particular partition unchanged:

$$S_a \equiv \{P | P \in S, P(a) = a\}. \tag{1.11.5}$$

As we have defined the group S, only particles that are identical are interchanged for any $P \in S$, so there is no possibility of ambiguity concerning the identity of partitions. Evidently, partitions differing by interchanges of particles within a cluster are identified, as are partitions that are related by an exchange of identical clusters of particles. Later, we will see that the cosets of S_a are useful for the analysis of the action of S on the scattering operators.

Any two partitions related by Eq. (1.11.4) refer to groupings of the particles that must be physically equivalent, since only identical particles are exchanged. Thus, the label set $L = \{a\}$ decomposes into disjoint equivalence classes that we denote as $\bar{a}$, $\bar{b}$, etc.:

$$\bar{a} = \{P(a) | \forall P \in S\}. \tag{1.11.6}$$

We anticipate that all physically relevant partition-dependent quantities will be class functions.

Let $\{U(P)\}$ denote a unitary representation of the group S on the unsymmetrized space $\mathscr{H}(N)$. The $U(P)$'s clearly map $\mathscr{H}_s(N)$ onto itself. Any physical N-particle state, $|\psi_{\text{phys}}\rangle \in \mathscr{H}_s(N)$, must satisfy

$$U(P)|\psi_{\text{phys}}\rangle = \delta(P)|\psi_{\text{phys}}\rangle, \tag{1.11.7}$$

for all $P \in S$, where $\delta(P) = +1(-1)$ if an even (odd) number of fermions are exchanged. The projection of $\mathscr{H}(N)$ onto $\mathscr{H}_s(N)$ is

$$\Lambda_S^2 = \frac{1}{|S|} \sum_{P \in S} \bar{U}(P), \tag{1.11.8}$$

where $|S|$ is the order of the group S and $\bar{U}(P) = \delta(P)\,U(P)$. Clearly Λ_S satisfies

$$\Lambda_S^2 = \Lambda_S = \Lambda_S^\dagger, \tag{1.11.9}$$

and

$$U(P)\Lambda_S = \Lambda_S U(P) = \delta(P)\,\Lambda_S. \tag{1.11.10}$$

We see that, for any $|\psi\rangle \in \mathscr{H}_s(N)$,

$$\Lambda_S|\psi\rangle = |\psi\rangle. \tag{1.11.11}$$

In order to proceed further we will need a few formal definitions and some subsidiary mathematical results concerning the group S.[4] A crucial concept is that of a *label-transforming* (L-T) *operator*, $\mathcal{O}^{a,b,\ldots}$ that is defined in terms of the partition indices and that satisfies

$$U(P)\mathcal{O}^{a,b,\ldots}U(P)^{-1} = \mathcal{O}^{P(a),P(b),\ldots}, \tag{1.11.12}$$

for all $P \in S$.

Next we quote [4] the following easily proved result:

Lemma 1. (Lagrange's Theorem). *The $N_{\bar{a}}$ left-cosets $(P_{a'a}) \cdot S_a$, $a' \in \bar{a}$, of S_a are all distinct and exhaust the full group S, so that*

$$N_{\bar{a}} = |S|/|S_a|, \tag{1.11.13}$$

where $|S_a|$ and $|S|$ are the orders of S_a and S, respectively, and where $P_{a'a} \in S$ is such that $P_{a'a}(a) = a'$. (Note that $P_{a'a} \in S_a$ if and only if $a' = a$.)

Lemma 1 implies that, for any $a \in L$,

$$\Lambda_S = \frac{1}{N_{\bar{a}}} \sum_{a' \in \bar{a}} \bar{U}(P_{a'a})R_a = \frac{1}{N_{\bar{a}}} \sum_{a' \in \bar{a}} R_a \bar{U}(P_{aa'}), \tag{1.11.14}$$

where

$$R_a \equiv \frac{1}{|S_a|} \sum_{P \in S_a} \bar{U}(P). \tag{1.11.15}$$

If A^a is any L-T operator, then we see that

$$A^a R_a = R_a A^a. \tag{1.11.16}$$

The counting problems involved in obtaining the factors $|S|$, $|S_a|$, and thus $N_{\bar{a}}$, are taken up in Section A.1c.

We now consider the definition of the antisymmetrized transition operator. We do this in two different ways. The essential complication in collision theory is the enumeration of all of the possibilities allowed by indistinguishability consistent with proper normalization. We recall that a typical asymptotic cluster state $|\phi_a\rangle$ is a direct product of the bound states corresponding to each of the clusters and the kinematic states that describe the c.m. motion of each of the clusters. Suppose then that the in state $|\psi_a^{(+)}\rangle \in \mathscr{H}(N)$ has evolved from a cluster state satisfying

$$|\phi_a\rangle = R_a|\phi_a\rangle, \tag{1.11.17}$$

i.e., each of the bound states entering into $|\phi_a\rangle$ are individually symmetrized.

The physical in-state, $|\psi_{\bar{a}}^{(+)}\rangle \in \mathscr{H}_s(N)$, is constructed from $|\psi_a^{(+)}\rangle$ by assigning equal probabilities and the correct permutation signature to each of the physically equivalent alternatives that correspond to the $N_{\bar{a}}$ distinct configurations that can be obtained from the partition a by permutation:

$$|\psi_{\bar{a}}^{(+)}\rangle = \frac{1}{\sqrt{N_{\bar{a}}}} \sum_{a' \in \bar{a}} \bar{U}(P_{a'a})|\psi_a^{(+)}\rangle. \tag{1.11.18}$$

Any one of the partitions $a \in \bar{a}$ will serve to define the same state $|\psi_{\bar{a}}^{(+)}\rangle$. The one that we single out in any particular calculation is called the *canonical partition* and we denote it, henceforth, as a_0.

Any scattering process can be associated with a transition $\bar{a} \to \bar{b}$ between equivalence classes. There are $N_{\bar{b}}$ final asymptotic configurations, each characterized by cluster states $|\phi_b\rangle$, $b \in \bar{b}$. Thus, the total transition probability will be

$$\sum_{b \in \bar{b}} |\langle (H - E)\phi_b|\psi_{\bar{a}}^{(+)}\rangle|^2 = N_{\bar{b}}|\langle \phi_{b_0}|V^{b_0}|\psi_{\bar{a}}^{(+)}\rangle|^2, \tag{1.11.19}$$

where the equality follows from the fact that $|\phi_b\rangle$ and $|\psi_{\bar{a}}^{(+)}\rangle$ are equal to $U(P_{bb_0})|\phi_{b_0}\rangle$ and $U(P_{bb_0})|\psi_{\bar{a}}^{(+)}\rangle$, respectively, to within a phase. Hence, each of the terms on the left-hand side of Eq. (1.11.19) are equal, corresponding to the fact that each of the partitions $b \in \bar{b}$ represents physically equivalent situations. Now, using Eq. (1.11.18), we see that

$$\sum_{b \in \bar{b}} |\langle (H - E)\phi_b|\psi_{\bar{a}}^{(+)}\rangle|^2 = |\langle \phi_{b_0}|T_{\bar{b}\bar{a}}^{\mathrm{R}}|\phi_{a_0}\rangle|^2, \tag{1.11.20}$$

where

$$T_{\bar{b}\bar{a}}^{\mathrm{R}} = (N_{\bar{b}}/N_{\bar{a}})^{1/2} \sum_{a' \in \bar{a}} T_{b_0 a'} \bar{U}(P_{a'a_0})R_{a_0} \tag{1.11.21}$$

is a correctly symmetrized transition operator. The operators $T_{b_0 a'}$ are transition operators of the usual type introduced in Section 1.7.

Had we begun from the statement of the transition probability in terms of $|\psi_{\bar{b}}^{(-)}\rangle$, we would have arrived at the expression

$$T_{\bar{b}\bar{a}}^{\mathrm{L}} = (N_{\bar{a}}/N_{\bar{b}})^{1/2} \sum_{b' \in \bar{b}} R_{b_0} \bar{U}(P_{b_0 b'})T_{b' a_0}. \tag{1.11.22}$$

Note that $T_{\bar{b}\bar{a}}^{\mathrm{R}}$ and $T_{\bar{b}\bar{a}}^{\mathrm{L}}$ are different operators, although their physical, on-shell matrix elements are identical. The two operators are distinguished by the right (R) or left (L) application of the permutation operators, respectively. The most common ordering convention for collision integral equations corresponds to the use of T^{L}.

We now present an alternative derivation of the symmetrized transition operator (1.11.22). The transition amplitude for the process $\bar{a} \to \bar{b}$ is the on-shell matrix element:

$$T(\bar{b}|\bar{a}) = \langle \bar{\phi}_{\bar{b}}|M|\bar{\phi}_{\bar{a}}\rangle, \tag{1.11.23}$$

where, e.g.,

$$|\bar{\phi}_{\bar{a}}\rangle \equiv B(\bar{a})^{1/2}\Lambda_S|\phi_{a_0}\rangle, \tag{1.11.24}$$

$B(\bar{a})$ is a normalization factor, and

$$M = (H - E)[I + G^{(+)}(E)(H - E)], \tag{1.11.25}$$

where the directionality of the H operator is unimportant when M is used in Eq. (1.11.23). Since $[M, U(P)] = 0$, we find that

$$T(\bar{b}|\bar{a}) = \frac{[B(\bar{b})B(\bar{a})]^{1/2}}{|S|} \sum_{b'} \langle \hat{\phi}_{b'}|M|\phi_{a_0}\rangle, \tag{1.11.26}$$

where

$$\langle \hat{\phi}_b| \equiv \langle \phi_{b_0}|\bar{U}(P_{b_0 b}) \tag{1.11.27}$$

is equal to $\langle \phi_b|$ to within a phase.

If T_{ba} is a transition operator, then on-shell

$$\langle \phi_b|T_{ba}|\phi_a\rangle = \langle \phi_b|M|\phi_a\rangle, \tag{1.11.28}$$

so M can be replaced by $T_{b'a_0}$ on the right-hand side of Eq. (1.11.26). Since, for any b, $b_0 \in \bar{b}$,

$$|S_b| = |S_{b_0}| \equiv |S_{\bar{b}}|, \tag{1.11.29}$$

we find, using Eq. (1.11.27) in conjunction with Eqs. (1.11.17), (1.11.26) and (1.11.28) that

$$T(\bar{b}|\bar{a}) = [B(\bar{b})B(\bar{a})]^{1/2}\frac{|S_{\bar{b}}|}{|S|} \sum_{b' \in \bar{b}} \langle \phi_{b_0}|R_{b_0}\bar{U}(P_{b_0 b'})T_{b'a_0}|\phi_{a_0}\rangle. \tag{1.11.30}$$

It remains to determine the normalization constants $B(\bar{a})$, $B(\bar{b})$. Suppose that we were working with channel states, $|\phi_{a_0}\rangle_{\mathrm{WP}}$ and $|\phi_{b_0}\rangle_{\mathrm{WP}}$, that correspond to normalizable wave packets (WP), i.e.,

$$[\langle \phi_{a_0}|\phi_{a_0}\rangle]_{\mathrm{WP}} = 1. \tag{1.11.31}$$

Then, we expect that for widely separated wave packets for each of the clusters we will have[7]

$$[\langle \phi_{a_0}|U(P)|\phi_{a_0}\rangle]_{\mathrm{WP}} = 0, \tag{1.11.32}$$

for $P \notin S_{a_0}$. So, if we demand that

$$[\langle \bar{\phi}_{\bar{a}}|\bar{\phi}_{\bar{a}}\rangle]_{\mathrm{WP}} = 1, \tag{1.11.33}$$

then

$$B(\bar{a}) = [\{\langle\phi_{a_0}|\Lambda_S|\phi_{a_0}\rangle\}_{\text{WP}}]^{-1}. \qquad (1.11.34)$$

Since for any $|\phi_{a_0}\rangle$, wave packet or not,

$$|\phi_{a_0}\rangle = \bar{U}(P)|\phi_{a_0}\rangle, \qquad (1.11.35)$$

if $P \in S_{a_0}$, we can calculate $B(\bar{a})$ using Eqs. (1.11.31), (1.11.32), and (1.11.35):

$$B(\bar{a}) = |S|/|S_{\bar{a}}| = N_{\bar{a}}. \qquad (1.11.36)$$

Therefore, we see that

$$T(\bar{b}|\bar{a}) = \langle\phi_{b_0}|T^{\text{L}}_{\bar{b}\bar{a}}|\phi_{a_0}\rangle, \qquad (1.11.37)$$

in agreement with our previous identification.

Up to this point we have not made use of any L-T property of the transition operators. In Section 5.4f we establish in detail that all of the transition operators introduced in Section 1.7 are L-T:

$$U(P)T_{ba}U(P)^{-1} = T_{b'a'}, \qquad (1.11.38)$$

where $P(a) = a'$, $P(b) = b'$. The L-T property is needed to prove that the matrix elements (1.11.37) are independent of the choice of the arbitrary, canonical partitions a_0 and b_0 to within an overall phase factor. Let us next illustrate what is involved here.[8]

Define

$$M_{\bar{b}\bar{a}}(b_0, a_0) \equiv \sum_{b \in \bar{b}} R_{b_0}\bar{U}(P_{b_0 b})M_{ba_0}, \qquad (1.11.39)$$

where M_{ba} is a L-T operator. Then, a straightforward, but somewhat lengthy, calculation yields the transformation property[8]

$$\bar{U}(P_{b'_0 b_0})M_{\bar{b}\bar{a}}(b_0, a_0)\bar{U}(P_{a'_0 a_0})^{-1} = M_{\bar{b}\bar{a}}(b'_0, a'_0). \qquad (1.11.40)$$

Therefore, since for $a \in \bar{a}$ we have $|\phi_a\rangle = \bar{U}(P_{aa_0})|\phi_{a_0}\rangle$ to within a phase, we infer that, again to within a phase,

$$\langle\phi_{b_0}|M_{\bar{b}\bar{a}}(b_0, a_0)|\phi_{a_0}\rangle = \langle\phi_{b'_0}|M_{\bar{b}\bar{a}}(b'_0, a'_0)|\phi_{a'_0}\rangle, \qquad (1.11.41)$$

which establishes canonical-partition independence.

The L-T property (1.11.38) is essential for symmetrizing collision integral equations that have the generic form

$$T_{ba} = B_{ba} + \sum_c K_{bc}T_{ca}, \qquad (1.11.42)$$

where the kernel (K_{bc}) and the inhomogeneous (B_{ba}) terms are L-T as well. Let us call

$$X_{\bar{b}c}(b_0) \equiv (N_{\bar{c}}/N_{\bar{b}})^{1/2} \sum_{b \in \bar{b}} R_{b_0} \bar{U}(P_{b_0 b}) X_{bc}, \tag{1.11.43}$$

where c is arbitrary and X_{bc} is label-transforming. Then, Eq. (1.11.42) becomes

$$T_{\bar{b}a}(b_0) = B_{\bar{b}a}(b_0) + (N_{\bar{a}}/N_{\bar{c}})^{1/2} \sum_{c} K_{\bar{b}c}(b_0) T_{ca}. \tag{1.11.44}$$

The operator $T_{\bar{b}a}(b_0)$, of course, is essentially $T^{\mathrm{L}}_{\bar{b}\bar{a}}$. The problem is to deduce a set of integral equations for $T_{\bar{c}a}(b_0)$ from the Eqs. (1.11.44).

We need two preliminary results. First, we obtain the identity

$$X_{\bar{b}c}(b_0) \bar{U}(P_{cc'}) = X_{\bar{b}c'}(b_0) \tag{1.11.45}$$

from definition (1.11.43) using the L-T property. Second, the identity

$$X_{\bar{b}c}(b_0) R_c = X_{\bar{b}c}(b_0) \tag{1.11.46}$$

follows from Eq. (1.11.45) and the definition of R_c.

Now

$$\begin{aligned} \sum_{c} K_{\bar{b}c}(b_0) T_{ca} &= \sum_{\bar{c}} \sum_{c \in \bar{c}} K_{\bar{b}c}(b_0) T_{ca} \\ &= \sum_{\bar{c}} \sum_{c \in \bar{c}} K_{\bar{b}c_0}(b_0) R_{c_0} \bar{U}(P_{c_0 c}) T_{ca}, \end{aligned} \tag{1.11.47}$$

where $c_0 \in \bar{c}$. We see, upon referring back to Eq. (1.11.43), that

$$(N_{\bar{a}}/N_{\bar{c}})^{1/2} \sum_{c} K_{\bar{b}c}(b_0) T_{ca} = \sum_{\bar{c}} \tilde{K}_{\bar{b}\bar{c}} T_{\bar{c}a}(b_0), \tag{1.11.48}$$

where

$$\tilde{K}_{\bar{b}\bar{c}} \equiv K_{\bar{b}c_0}(b_0), \qquad c_0 \in \bar{c}. \tag{1.11.49}$$

Therefore, Eq. (1.11.44) becomes

$$\tilde{T}_{\bar{b}\bar{a}} = \tilde{B}_{\bar{b}\bar{a}} + \sum_{\bar{c}} \tilde{K}_{\bar{b}\bar{c}} \tilde{T}_{\bar{c}\bar{a}}, \tag{1.11.50}$$

where $\tilde{T}_{\bar{b}\bar{a}}$ and $\tilde{B}_{\bar{b}\bar{a}}$ are defined analogously to Eq. (1.11.49) for some choice, $a = a_0$, of canonical partition, and where we have suppressed the dependence on the choice of canonical partitions a_0, b_0, and c_0.

Equation (1.11.50) links together only objects labeled by the equivalence classes $\bar{a}$, $\bar{b}$, and $\bar{c}$; the canonical partition labels are fixed and, in the end, immaterial. This is exactly what we want as a set of dynamical equations involving identical particles. The unknown objects, viz., $\tilde{T}_{\bar{b}\bar{a}}$, have matrix elements that are the physical, symmetrized transition amplitudes. Such integral equations now incorporate only the dynamics, including the boundary conditions, consistent with particle identity and statistics.

We have emphasized the symmetrization of L-T partition-labeled, transition-operator integral equations to the exclusion of some other options. One of these alternatives is the category of equations for quantities labeled by so-called chains of partitions, rather than just one of them. We develop such equations in Ch. 5; their symmetrization requires very little modification of what we have done up to this point.

Also, our restriction to integral equations satisfied by transition operators is easily removed. This should be clear from the fact that only one index of the transition operators plays an active role in the symmetrization procedure. Therefore, all of the preceding machinery can be applied immediately to the integral equations satisfied by the components of state vectors or Green functions, provided everything is L-T.

There is another apparently different, wave-function symmetrization that was originally proposed by Glöckle[9] for the Lippmann–Schwinger equations and later generalized to other sets of wave-function component equations.[10] It is important, therefore, for us to understand the relationship of this approach to other methods.

Here we consider the Glöckle method only for the set of N-particle LS equations,

$$|\psi_\alpha^{(+)}\rangle = \delta_{\beta\alpha}|\phi_\alpha\rangle + G_\beta^{(+)}V^\beta|\psi_\alpha^{(+)}\rangle, \tag{1.11.51}$$

for the state vector $|\psi_\alpha^{(+)}\rangle$ that evolves from a two-cluster state $|\phi_\alpha\rangle$ at an energy, E, that we suppress. Here, β is an arbitrary two-cluster partition; when all such β are considered, the full set (1.11.51) is expected to define $|\psi_\alpha^{(+)}\rangle$ uniquely (*cf*. Section 4.4). A similar remark then applies to Eq. (1.11.52) below for the antisymmetrized state, although the following analysis is independent of these questions of unique determination.

Glöckle argued that the equations that determine the symmetrized state $|\psi_{\bar{\alpha}}^{(+)}\rangle$ defined by Eq. (1.11.18) are

$$|\psi_{\bar{\alpha}}^{(+)}\rangle = \frac{1}{\sqrt{N_{\bar{\alpha}}}} \sum_{\alpha\in\bar{\alpha}} \delta(P_{\alpha\alpha_0})\,\delta_{\beta\alpha} R_\alpha|\phi_\alpha\rangle + G_\beta^{(+)}V^\beta|\psi_{\bar{\alpha}}^{(+)}\rangle, \tag{1.11.52}$$

where α_0 is a canonical partition. It is not at all clear how one could obtain the result (1.11.52), in which the entire content of the symmetrization resides in the inhomogeneous term, using the ideas presented so far in this section.

Instead of the Eqs. (1.11.51), let us begin from the equivalent relations

$$|\psi_\alpha^{(+)}\rangle = G^{(+)}[\overleftarrow{G}_\alpha^{(+)}]^{-1}|\phi_\alpha\rangle, \tag{1.11.53}$$

which certainly hold for the canonical partition α_0, as well as for any α obtained from α_0 by the permutation $P_{\alpha\alpha_0}$. It is necessary to suppose that before symmetrization the cluster states are ordered so that, for any P,

$$|\phi_{P(\alpha_0)}\rangle = U(P)|\phi_{\alpha_0}\rangle. \tag{1.11.54}$$

Then, because of the L-T property

$$U(P)[\overleftarrow{G}_\alpha^{(+)}]^{-1}U(P)^{-1} = [\overleftarrow{G}_{P(\alpha)}^{(+)}]^{-1}, \tag{1.11.55}$$

we see from Eq. (1.11.53) with $\alpha = \alpha_0$ that

$$|\psi_{P(\alpha_0)}^{(+)}\rangle = U(P)|\psi_{\alpha_0}^{(+)}\rangle. \tag{1.11.56}$$

Next, since R_α commutes with both $G^{(+)}$ and $[\overleftarrow{G}_\alpha^{(+)}]^{-1}$, we find from Eq. (1.11.53) that

$$R_\alpha|\psi_\alpha^{(+)}\rangle = G^{(+)}[\overleftarrow{G}_\alpha^{(+)}]^{-1}R_\alpha|\phi_\alpha\rangle. \tag{1.11.57}$$

Then, if we multiply Eq. (1.11.57) by $\delta(P_{\alpha\alpha_0})$, sum over all $\alpha \in \bar{\alpha}$, and use the resolvent identity for $G^{(+)}$, we obtain

$$\sum_{\alpha\in\bar{\alpha}} \delta(P_{\alpha\alpha_0}) R_\alpha|\psi_\alpha^{(+)}\rangle = G_\beta^{(+)} \sum_{\alpha\in\bar{\alpha}} \delta(P_{\alpha\alpha_0})[\overleftarrow{G}_\alpha^{(+)}]^{-1}R_\alpha|\phi_\alpha\rangle$$
$$+ G_\beta^{(+)}V^\beta \sum_{\alpha\in\bar{\alpha}} \delta(P_{\alpha\alpha_0}) R_\alpha|\psi_\alpha^{(+)}\rangle. \tag{1.11.58}$$

This reduces to Eq. (1.11.52) if we use Eq. (1.1.56) in conjunction with the identification (1.11.18). The inhomogeneous terms of Eqs. (1.11.52) and (1.11.58) are brought into congruence by use of identities similar to Eq. (1.8.6).

We see that the crucial step in obtaining the Eqs. (1.11.52), in which the kernel is not intertwined with permutation operators, is the use of Eq. (1.11.57) and the partition-degenerate resolvent identity for $G^{(+)}$. The preceding analysis shows that the Eqs. (1.11.52) involve the same ideas as before, but adapted to the partition-degenerate rather than partition-labeled character of the two-cluster initiated LS equations.

1.11b. Second Quantization

For nonrelativistic systems of identical particles whose dynamics are described by the Schrödinger equation with potentials representing the interactions among the particles, second-quantization introduces no new physics.[11]

What it does achieve is a statement of the dynamics in a particle-number-independent fashion with all symmetrization requirements taken into account automatically. Let us suppose for simplicity in what follows that all of the particles are identical.

Quantum mechanics is then formulated on the Fock space $\mathscr{H}_F$, which is simply a direct sum of all the symmetrized Hilbert spaces, $\mathscr{H}_s(N)$, for arbitrary N, including $N = 0$, which we identify with the vacuum:

$$\mathscr{H}_F = \bigoplus_{N=0}^{\infty} \mathscr{H}_s(N). \tag{1.11.59}$$

We denote our choice for the independent state of $\mathscr{H}_s(0)$ as $|0\rangle$, and we suppose it is normalized to unity, $\langle 0|0\rangle = 1$.

The creation and destruction operators on $\mathscr{H}_F$ map $\mathscr{H}_s(N)$ into $\mathscr{H}_s(N+1)$ and $\mathscr{H}_s(N-1)$, respectively. Classes of these operators are conveniently defined in terms of the eigenvalues of a complete set ζ of single-particle observables, such as the position coordinates $\mathbf{x}$ and the spin λ, or the momenta $\mathbf{k}$ and λ. Bases on $\mathscr{H}_F$ can then be constructed by applying the creation operators to the vacuum.

When $\zeta = (\mathbf{x}, \lambda)$, the symmetrized basis states are given by

$$|\mathbf{x}_1, \lambda_1; \ldots, \mathbf{x}_N, \lambda_N; N\rangle_{(\pm)} = \left(\frac{1}{N!}\right)^{1/2} \psi^\dagger_{\lambda_1}(\mathbf{x}_1) \cdots \psi^\dagger_{\lambda_N}(\mathbf{x}_N)|0\rangle, \tag{1.11.60}$$

where $\psi^\dagger_\lambda(\mathbf{x})$ creates a particle at $\mathbf{x}$ with spin and all other internal quantum numbers represented by λ. Here

$$[\psi_{\lambda_1}(\mathbf{x}_1), \psi^\dagger_{\lambda_2}(\mathbf{x}_2)]_\pm = \delta_{\lambda_1\lambda_2}\delta^{(3)}(\mathbf{x}_1 - \mathbf{x}_2), \tag{1.11.61}$$

where $[A, B]_\pm = AB \pm BA$, and the subscripts $(+)$ and $(-)$ refer to the antisymmetric and the symmetric cases, respectively. We suppress the index λ in what follows.

The Fock-space Hamiltonian is, with a local mutual interaction and no external interactions, given by ($\hbar = 1$)

$$\begin{aligned} H = \frac{-1}{2m} \int d^3x\, \psi^\dagger(\mathbf{x}) \nabla^2 \psi(\mathbf{x}) \\ + \frac{1}{2} \int d^3x \int d^3x'\, \psi^\dagger(\mathbf{x}')\psi^\dagger(\mathbf{x}) V(\mathbf{x}' - \mathbf{x}) \psi(\mathbf{x})\psi(\mathbf{x}'). \end{aligned} \tag{1.11.62}$$

The independence of H on the number of particles is the feature that makes Fock space such a useful device for handling the dynamics of arbitrarily large numbers of identical particles. Ironically, it is this very property that is the source of many of the difficulties that have been encountered in attempting to obtain dynamical collision integral equations using the Fock-space machinery.

For instance, in attempting to facilitate an interaction-picture approach, one encounters the problem that H cannot be split into channel Hamiltonians and residual interactions. Similarly, the operator associated with the energy of the c.m. motion is $\mathbf{P}^2/2mN$, where

$$\mathbf{P} = -i \int d^3x\, \psi(\mathbf{x})\, \nabla\, \psi(\mathbf{x}), \tag{1.11.63}$$

$$N = \int d^3x\, \psi(\mathbf{x})\psi(\mathbf{x}) \tag{1.11.64}$$

are the Fock-space momentum and number operators, respectively; thus, the c.m. Hamiltonian, $H - \mathbf{P}^2/2mN$, is also not an operator one can break down into channel identifications in order to treat recoil motion with facility. This problem is made much more difficult if part of the interaction is in the form of an effective external (mean) field.[11,12]

The upshot of this is that Fock space has been used in practice only for those multiparticle collision processes for which the preceding difficulties either can be handled easily, namely for $N = 2, 3, 4$,[6] or where the questions of channel identifications and recoil are trivial, e.g., the elastic scattering of a projectile from a very heavy composite target, in which case an optical potential is identified with the projectile self-energy in the many-particle medium.[13] Both the fragment-Hamiltonian and recoil problems can be solved in approximate treatments such as mean-field approaches to heavy-ion collisions.[14]

Except for few-particle collision problems, where any results can usually be obtained much more rapidly and transparently by conventional means, the Fock-space approach seems to have limited applicability for the physical problems that are our primary focus. However, when N-particle collisions are confined to one spatial dimension, Fock-space techniques have proven to be exceptionally effective.[15]

For those physical systems where a relativistic field-theoretic description is necessary, one cannot so easily avoid using Fock-space techniques.[16] Basically, all of our previous remarks apply to collision problems in relativistic Fock-space as well. For few-particle collisions, one can formulate Bethe–Salpeter-type equations.[17] For more complicated situations, one again has the optical potential as well as reduction formulas[18] that lead to so-called Low equations reflecting the asymptotic condition and asymptotic completeness.[19] This work has not led to a complete treatment of the collision problem, but rather to frameworks for crafting practical approximations to nucleon– and pion–nucleus scattering.

What appears unassailable about scattering theory as formulated in non-relativistic Fock space concerns the statement and proof of the asymptotic condition and asymptotic completeness in Fock-space language.[11] The treatment of Sandhas[20] is inclusive in this regard and concerns the large-time limits of distributions formed from the basic dynamical component of the Fock-space approach, namely the field operator

$$\psi(\mathbf{x}, t) = e^{iHt}\psi(\mathbf{x})e^{-iHt}. \tag{1.11.65}$$

Unfortunately, it is not clear how one can exploit these results to obtain dynamical collision integral equations in any more efficient way than by ordinary operator techniques. Indeed, it has generally proven to be much harder to obtain the simplest linear integral equations from a Fock-space starting point except if one resorts to diagram summation techniques. It is relevant to conjecture why this is so.

Fock space permits a fully-symmetrized formalism at every step by means of dynamical quantities that do not conserve total particle number N. This results in an N-independent statement of the dynamics, but with a projection on any N-particle subspace that is fully equivalent to the N-particle Schrödinger equation. It appears, however, that the structure of well-defined collision integral equations is very much N-dependent. The combinatorial possibilities for different channels and alternative means of building up an amplitude from the elementary interactions varies nontrivially with N. This complexity is hidden in the Fock-space formalism, and it has proven difficult to extract it. Those methods that have been used seem much more cumbersome than standard operator techniques that work in a projected symmetrized subspace of a fixed-N Hilbert space. On the other hand, for very large N and for limited collision problems such as defining an optical potential, the Fock-space approach has many advantages because of its N-independence.

Appendix (1)

A.1a. Kinematics, Jacobi Coordinates

Reed and Simon[1] have pointed out that N-particle quantum mechanics is complicated for two reasons, one kinematical and one dynamical. Most of this book is concerned with dynamical questions, but in this subsection we confront a few of the kinematical ones that stem from the fact that, except for $N = 2$, there is not a natural choice of coordinate system in the N-body c.m. frame.

There are three common choices of coordinate transformations to the total c.m. coordinates,

$$\mathbf{R}_{\text{c.m.}} = \frac{1}{M_N} \sum_{i=1}^{N} m_i \mathbf{X}_i, \tag{A.1.1}$$

and $N - 1$ additional vector coordinates $\xi_1, \ldots, \xi_{N-1}$. Here, M_N is the sum

$$M_N = \sum_{i=1}^{N} m_i \tag{A.1.2}$$

of the individual particle masses and the $\mathbf{x}_i$ are the particle position vectors in an arbitrary frame. The coordinates ξ_i are constrained by the related requirements that, for all i, j,

$$\mathbf{x}_i - \mathbf{x}_j = \sum_{k=1}^{N-1} C_{ij}^k \xi_k, \tag{A.1.3}$$

where the C_{ij}^k are constants, and that the N-particle kinetic-energy operator H_0 decomposes into a total c.m. part $H_0(\mathbf{R})$ and a remainder $H_0^{\text{rel}}(\xi_1, \ldots, \xi_{N-1})$:

$$H_0 = H_0(\mathbf{R}) + H_0^{\text{rel}}(\xi_1, \ldots, \xi_{N-1}). \tag{A.1.4}$$

There are three commonly used coordinate frames.[1] In *atomic coordinates*, which we shall not use, $\xi_i = \mathbf{x}_i - \mathbf{x}_N$. *Jacobi coordinates* correspond to the choice

$$\xi_i = \mathbf{x}_{i+1} - \frac{1}{M_i} \sum_{i \geq j} m_j \mathbf{x}_j, \tag{A.1.5a}$$

where

$$M_i = \sum_{i \geq j} m_j. \tag{A.1.5b}$$

Then, in coordinate space, we have

$$H_0^{\text{rel}} = -\sum_{i=1}^{N-1} \frac{1}{2\mu_i} \nabla_{\xi_i}^2, \tag{A.1.6a}$$

where

$$\mu_i = m_{i+1} M_i (m_{i+1} + M_i)^{-1}. \tag{A.1.6b}$$

Evidently, there are different choices of Jacobi coordinates depending upon the ordering of the coordinates. Finally, there are *clustered Jacobi coordinates*, which are particularly useful for the explicit description of the clustering properties of N-particle amplitudes.[1,2]

A convenient choice of kinematics is essential in all applications of collision theory. This becomes progressively more difficult to do as the number of particles increases. Fortunately, virtually all practical calculations involve no more than the number of degrees of freedom appropriate to three particles. Thus, we need consider only the cases $N = 2$ and $N = 3$ in detail, and in the latter case the Jacobi coordinates will suffice.

Consider three particles ($\alpha = 1, 2, 3$) of masses m_α, position vectors $\mathbf{x}_\alpha$, momenta ($\hbar = 1$) $\mathbf{k}_\alpha$, and kinetic energies $\mathbf{k}_\alpha^2/2m_\alpha$. Consider first a typical two-particle subsystem, say that comprised of particles 2 and 3. The transformation between $\mathbf{x}_2$ and $\mathbf{x}_3$ and the relative,

$$\boldsymbol{\rho}_{23} = \mathbf{x}_2 - \mathbf{x}_3, \tag{A.1.7a}$$

and c.m.,

$$\mathbf{R}_{23} = \frac{m_2\mathbf{x}_2 + m_3\mathbf{x}_3}{m_2 + m_3}, \tag{A.1.7b}$$

coordinates, respectively, has Jacobian unity. The momenta conjugate to $\boldsymbol{\rho}_{23}$ and $\mathbf{R}_{23}$ are

$$\boldsymbol{\kappa}_{23} = \frac{(m_3\mathbf{k}_2 - m_2\mathbf{k}_3)}{m_2 + m_3}, \tag{A.1.8a}$$

and

$$\mathbf{P}_{23} = \mathbf{k}_2 + \mathbf{k}_3, \tag{A.1.8b}$$

respectively. Also, we have

$$\frac{\mathbf{k}_2^2}{2m_2} + \frac{\mathbf{k}_3^2}{2m_3} = \frac{(\boldsymbol{\kappa}_{23})^2}{2\mu_{23}} + \frac{(\mathbf{P}_{23})^2}{2M_{23}}, \tag{A.1.9}$$

where $\mu_{23} \equiv m_2 m_3/M_{23}$ and $M_{23} \equiv m_2 + m_3$ are the reduced and total masses of the two-body system, respectively.

The position of the three-body c.m. is given by ($M = M_3$)

$$\mathbf{R} = \frac{1}{M}\sum_{\alpha=1}^{3} m_\alpha \mathbf{x}_\alpha. \tag{A.1.10}$$

The two Jacobi coordinates $\boldsymbol{\rho}_{23}$ and

$$\boldsymbol{\zeta}_1 \equiv \mathbf{x}_1 - \mathbf{R}_{23}, \tag{A.1.11}$$

along with $\mathbf{R}$, are related to the $\mathbf{x}_\alpha$ by a linear transformation with unit Jacobian. Note that our choice of Jacobi coordinates corresponds to a different ordering of coordinates and a different notation than is the case for the variables (A.1.5a, b). The momenta conjugate to $\mathbf{R}$, $\boldsymbol{\rho}_{23}$, and ζ_1 are

$$\mathbf{P} = \sum_{\alpha=1}^{3} \mathbf{k}_\alpha, \tag{A.1.12a}$$

$$\mathbf{p}_1 = \frac{1}{M_{23}}(m_3\mathbf{k}_2 - m_2\mathbf{k}_3), \tag{A.1.12b}$$

and

$$\mathbf{q}_1 = \frac{m_1 M_{23}}{M}\left\{\frac{\mathbf{k}_1}{m_1} - \frac{(\mathbf{k}_2 + \mathbf{k}_3)}{M_{23}}\right\}, \tag{A.1.12c}$$

respectively.

It is easy to express the content of the preceding analysis in terms of Jacobi variables in a way that does not single out any two-body subsystem. Let α, β, γ be a cyclic permutation of 1, 2, 3. We define

$$\mu_\alpha \equiv \frac{m_\beta m_\gamma}{m_\beta + m_\gamma}, \qquad \mu^\alpha \equiv \frac{m_\alpha(m_\beta + m_\gamma)}{M}. \tag{A.1.13}$$

Then, corresponding to Eqs. (A.1.12b) and (A.1.12c), we have

$$\mathbf{p}_\alpha = \mu_\alpha\left(\frac{\mathbf{k}_\beta}{m_\beta} - \frac{\mathbf{k}_\gamma}{m_\gamma}\right) \tag{A.1.14a}$$

and

$$\mathbf{q}_\alpha = \mu^\alpha\left\{\frac{\mathbf{k}_\alpha}{m_\alpha} - \frac{\mathbf{k}_\beta + \mathbf{k}_\gamma}{m_\beta + m_\gamma}\right\}, \tag{A.1.14b}$$

respectively. Moreover, we see that

$$E_{\text{kin}} \equiv \sum_{\alpha=1}^{3} \frac{\mathbf{k}_\alpha^2}{2m_\alpha}$$
$$= E_{\text{c.m.}} + \omega_\alpha + v_\alpha, \tag{A.1.15a}$$

for any α, where

$$\omega_\alpha \equiv \frac{1}{2\mu^\alpha}\mathbf{q}_\alpha^2, \qquad v_\alpha = \frac{1}{2\mu_\alpha}\mathbf{p}_\alpha^2, \qquad E_{c.m.} = \frac{1}{2M}\mathbf{P}^2. \tag{A.1.15b}$$

Equations (A.1.15a, b) have operator counterparts, where (A.1.15a), e.g., represents the decomposition of the three-body kinetic-energy operator into the corresponding operators for the total c.m., a relative two-body system, and for a third particle relative to the c.m. of the other two.

Suppressing all internal quantum numbers such as spin, etc., we can construct complete, orthornormal three-body bases, $\{|\mathbf{P}, \mathbf{q}_\alpha, \mathbf{p}_\alpha\rangle_\alpha\}$ for each choice of α. An important quantity in calculations is the overlap matrix ($\alpha \neq \beta$):

$$_\beta\langle \mathbf{q}'_\beta, \mathbf{p}'_\beta | \mathbf{q}_\alpha, \mathbf{p}_\alpha\rangle_\alpha = \delta^{(3)}\left\{\mathbf{p}'_\beta \pm \left[\mathbf{q}_\alpha + \frac{\mu_\beta}{m_\gamma}\mathbf{q}'_\beta\right]\right\}\delta^{(3)}\left\{\mathbf{p}_\alpha \mp \left[\mathbf{q}'_\beta + \frac{\mu_\alpha}{m_\gamma}\mathbf{q}_\alpha\right]\right\}, \tag{A.1.16}$$

where we have factored out $\delta^{(3)}(\mathbf{P}' - \mathbf{P})$, we have assumed that we are in the three-body c.m. frame ($\mathbf{P} = 0$), and we note that the upper (lower) sign applies when α and β are in cyclic (anticyclic) order with respect to some third index $\gamma \neq \alpha, \gamma \neq \beta$.

A.1b. Cross Sections and Observables

It is not trivial to relate transition amplitudes for scattering over infinite space–time volumes to observable quantities. This is the crucial step in connecting the quantum theory of scattering to things that can be measured, and it forms an essential part of the standard treatments of scattering theory.[3] All of these considerations are nondynamical and are now throughly understood. Our agenda is to exploit what we regard as a common body of knowledge rather than developing this material anew once again. Consequently, we confine ourselves to a short listing of the relationships we use more or less explicitly at various places through the book.

Let i be an initial two-cluster state, while f refers to one of a group of final states that may have more than two clusters. We assume that all of the initial and final states have well-defined energy and momentum. A generic transition matrix element then has the form

$$\langle \mathrm{f} | T_{\mathrm{fi}} | \mathrm{i} \rangle = \delta(E_\mathrm{f} - E_\mathrm{i})\,\delta^{(3)}(\mathbf{P}_\mathrm{f} - \mathbf{P}_\mathrm{i})\,M_{\mathrm{fi}}, \tag{A.1.17}$$

where the amplitude M_{fi} is free of all Dirac δ-function singularities. We have used i and f as channel indices in labeling T_{fi}. With the $e^{i\mathbf{k}\cdot\mathbf{x}}/(2\pi)^{3/2}$ type of plane-wave normalization, the transition probability w to the group of states f is then[3]

$$w = 2\pi \sum_{\mathrm{f}} \delta(E_\mathrm{f} - E_\mathrm{i})\,\delta^{(3)}(\mathbf{P}_\mathrm{f} - \mathbf{P}_\mathrm{i})|M_{\mathrm{fi}}|^2. \tag{A.1.18}$$

A generic cross section $\Delta\sigma$ is then

$$\Delta\sigma = w/\Phi_\mathrm{i}, \tag{A.1.19}$$

where Φ_i is the flux of incident projectiles relative to the target.

We consider the collision in the c.m. frame of two spinless particles, where they have a relative wave vector $\mathbf{k}_i$. The differential cross section for elastic scattering into an element of solid angle $d\Omega$ defined by the outgoing relative wave vector $\mathbf{k}_f$ is

$$\frac{d\sigma}{d\Omega}(\mathbf{k}_f, \mathbf{k}_i) = |f(\mathbf{k}_f, \mathbf{k}_i)|^2, \tag{A.1.20}$$

where the scattering amplitude is

$$f(\mathbf{k}_f, \mathbf{k}_i) = -(2\pi)^2 \mu\, t(\mathbf{k}_f, \mathbf{k}_i). \tag{A.1.21}$$

Here, μ is the reduced two-particle mass and $t(\mathbf{k}_f, \mathbf{k}_i)$, which corresponds to M_{fi} in Eq. (A.1.17), is the two-body transition matrix in the c.m. frame. The optical theorem follows from unitarity (Section 1.9), and relates the total cross section σ_T for scattering into all final states, including inelastic ones, that can be initiated from the initial two-body state, to the imaginary part of the forward elastic-scattering amplitude:

$$\sigma_T = \frac{4\pi}{k_i} \operatorname{Im} f(\mathbf{k}_i, \mathbf{k}_i). \tag{A.1.22}$$

The initial and final states in a scattering process are generally labeled by the internal quantum numbers of the particles, such as their intrinsic spins. Given the transition amplitudes, one can calculate the statistical averages of such single-particle observables in the final state given their initial statistical averages. This is facilitated using the density-matrix formalism.[3] With the use of this formalism, various spin observables can be expressed as density-matrix statistical averages of the corresponding spin operators in a straightforward manner, and we refer to the literature for these expressions.

A.1c. Identity Combinatorics

We calculate the various numbers that enter into the expressions for symmetrized operators. We confine ourselves to the only example we utilize in detailed calculations, namely that of two-cluster partitions of two species of identical particles, viz., N "neutrons" and Z "protons." Actually, we explicitly use only the case of a single species ($N = 0$ or $Z = 0$), but the simple generalization to two of them elucidates the combinatorics and obviously covers all of the usual atomic and nuclear situations.

Consider the number $N_{\bar{\alpha}}$ of distinct two-cluster partitions that belong to an equivalence class $\bar{\alpha}$. Any two-cluster partition consists of a cluster of n

neutrons and z protons with $n + z \geq 1$, while the other cluster contains $N - n$ neutrons and $Z - z$ protons. Each pair of integers (n, z) corresponds to an equivalence class of two-cluster partitions. The total number of permutations, $|S|$, is $N!Z!$, but exchanging neutrons or protons within a cluster does not lead to a distinct partition, so this number should be reduced to $\binom{N}{n}\binom{Z}{z}$, where, e.g., $\binom{N}{n} = [N!/n!(N-n)!]$. However, if the two clusters are identical ($n = N/2$, $z = Z/2$), their complete interchange does not yield a new partition and the total number of partitions is reduced by a factor of two. Thus, for $\bar{\alpha} \sim (n, z)$,

$$N_{\bar{\alpha}} = \binom{N}{n}\binom{Z}{z}(1 - \tfrac{1}{2}\delta_{N/2,n}\delta_{Z/2,z}). \tag{A.1.23}$$

We then have that $|S_{\bar{\alpha}}| = N!Z!/N_{\bar{\alpha}}$.

The number of equivalence classes among the $(2^{N+Z-1} - 1)$ two-cluster partitions is identical with the number $P_{=}(N, Z)$ of partitions of the two groups (N, Z) without regard to order into the nonempty clusters (n, z) and $(N - n, Z - z)$. If one distinguishes the order of the two parts, such a partition is called a composition,[4] and the total number of compositions of (N, Z) into two parts is $C(N, Z; 2) = (N + 1)(Z + 1) - 2$. If N and Z are not both even, the two parts can never be identical and there are two compositions for each partition, so

$$P_{=}(N, Z) = \tfrac{1}{2}C(N, Z; 2), \tag{A.1.24}$$

while if both N and Z are even there is only one composition corresponding to the partition into equal clusters, so

$$P_{=}(N, Z) = \tfrac{1}{2}[C(N, Z; 2) - 1] + 1. \tag{A.1.25}$$

For N identical particles ($Z = 0$) there will be $1, 2, 2, 3, \ldots$ equivalence classes corresponding to $N = 3, 4, 5, 6, \ldots$ particles, respectively. The number of equivalence classes is the same as the number of coupled equations that will result from a set of integral equations that couple only the two-cluster partitions.

References and Notes (1)

Section 1.1

1. Our usages of the terms "collision" and "scattering" are consistent with the definitions in Considine (1976). A perfect example of what we would regard as correct and proper nomenclature is provided by the opening sentence of the classic paper by

Ekstein (1956):

> "The general case of scattering, where the reaction products may be different from the colliding systems..."

2. A recent review has been given by Wallace (1987).
3. See Faddeev (1965).
4. Introductions to scattering theory appear in a number of textbooks on quantum mechanics such as the book by Messiah (1965). Specialized monographs vary greatly in their level and scope. See Glöckle (1983a), Goldberger (1964), Rodberg (1967), Watson (1967), Joachain (1975), Newton (1982), and Taylor (1987).
5. A comprehensive bibliography of early mathematically-oriented work on scattering and collision theory is given by Hepp (1972). Subsequently, books have appeared that develop the subject in this way, e.g., Amrein (1977) and Reed (1979). A recent proof of asymptotic completeness for N-particle scattering with short-ranged potentials is given in Sigal (1987) along with extensive references to other work on mathematical studies of N-particle scattering theory.

Section 1.3

1. The exposition of the basic principles in Taylor (1987) is exceptional in its clarity. Comprehensive treatments of the subject are given in Goldberger (1964), Joachain (1975), and Newton (1982).
2. See Goldberger (1964), p. 91, for a discussion of this point. For reviews of the nonrelativistic scattering theory of charged particles see Chandler (1981b, 1987). As an example in three-body scattering see Haftel (1981).
3. The literature on the group-theoretical approach is not large and much of it may be traced from the papers of Alhassid (1986), and Wu (1987).
4. See, e.g., Bjorken (1964), Holz (1988), and Sweeny (1989).
5. A theoretical framework using path-integral techniques for the time-dependent Hartree–Fock method is given by Reinhardt (1982); see also Alhassid (1981), Sandhya Devi (1981). Path-integral reformulations of potential scattering are given in Campbell (1975), Gerry (1980), and Nelson (1983), and references cited therein.

Section 1.4

1. A complete treatment of the three-particle scattering problem including the proof of asymptotic completeness is contained in Faddeev (1965). This work has not, however, provided useful starting points for practical calculations. See Osborn (1971) for an exposition of some of Faddeev's suggestions for practical collision integral equations.
2. The generalization to arbitrary N of the $N = 3$ Faddeev equations was given by Yakubovskii (1967). These equations were used in an attempt to prove asymptotic completeness by Hepp (1969). See also Reed (1979) and Sigal (1987).

3. The S operator is sometimes defined as the adjoint of ours, e.g., Reed (1979) and Taylor (1987). Our definition is the accepted standard in field theory, Lehmann (1955), Itzykson (1980), and we think it important, especially for second-quantized formulations of nonrelativistic scattering, to maintain that association.

Section 1.5

1. The asymptotic condition in nonrelativistic scattering theory is discussed in detail in Amrein (1977), Reed (1979), Newton (1982), and Taylor (1987).
2. The books by Newton (1982) and Taylor (1987) have particularly accessible presentations of the relevant mathematics.
3. See, esp., Newton (1982).
4. The original reference is Lehmann (1955). For more modern treatments see Bjorken (1965), Huang (1975), and Itzykson (1980).

Section 1.6

1. Newton (1982), Taylor (1987).
2. Lippmann (1950).
3. Lippmann (1956).
4. Compactness and related mathematical concepts are discussed in Weinberg (1964), Watson (1967), and Newton (1982). For our purposes it suffices that a compact operator is bounded and it can be approximated arbitrarily closely by an operator of finite rank, namely as a sum of a finite number of operators of the form $|a\rangle\langle b|$.
5. Epstein (1957) and Foldy (1957).
6. Faddeev (1960–1962).
7. Yakubovskii (1967).

Section 1.7

1. Detailed discussions of the progression from wave packets to the idealization of states of well-defined energy and momentum are contained in Ekstein (1956), Goldberger (1964), and Taylor (1987).
2. Goldberger (1964), Joachain (1975).
3. Gell-Mann (1953).
4. Ekstein (1956), Lehmann (1955), Lovelace (1964a, b), Faddeev (1965). Rosenberg (1965), Huang (1975).

5. As a historical curiosity, we note that the ubiquitous anti-Kronecker-delta notation seems to have appeared first in Kowalski (1969).
6. Alt (1967).

Section 1.8

1. The basis for this is a set of relations called the Lippmann identities (Lippmann (1956)) that are discussed in detail in Chs. IV and V. These identities have been proven in the weak topology in Bencze (1982a). For other discussions and applications that are relevant to our presentation, see Benoist-Gueutal (1975, 1978), Kowalski (1977d), L'Huillier (1978), Tobocman (1983a), and Picklesimer (1983a).
2. This terminology and characterization follows that in Kowalski (1977c). For earlier work along these lines and applications to the construction of variational principles for multiparticle scattering, see Rosenberg (1968). Further many-particle applications along with extensive references are contained in Cattapan (1982, 1985a).
3. Feshbach (1958).
4. Given an arbitrary integral equation, Fredholm reduction is the procedure for finding an alternative Fredholm integral equation whose unique solutions are also solutions of the original equation. The technique is described in detail in Muskhelishvili (1953) for Riemann–Hilbert singular integral equations. For applications to collision integral equations and discussions of the nonuniqueness for the reduction, see Faddeev (1960, 1965), Kowalski (1961, 1963b), Sugar (1964), and Broido (1969a).
5. The two-potential formula in potential scattering is generally credited to Gell-Mann and Goldberger (Gell-Mann (1953)). It has been generalized and applied in a number of different ways. An elementary discussion is contained on p. 271 of Taylor (1987). For N-particle and multichannel applications, along with extensive references, see Cattapan (1975), Bencze (1977b), and Kowalski (1978a).

Section 1.9

1. An exposition of the essential ideas of operator analyticity in scattering theory is contained in Lovelace (1964a, b).
2. For investigations of the on-shell analyticity properties of multiparticle transition amplitudes, see Mandelstam (1965), Rubin (1966, 1967a, b), Eden (1966), Federbush (1967), Riahi (1969), Amado (1977), and Dodd (1977).
3. Plemelj (1908), Muskhelishvili (1953).
4. A comprehensive review is contained in Eden (1966).
5. See Wichmann (1963), Hunziker (1965), and Taylor (1966, 1967) for the discussion of the cluster decomposition of the S-matrix from a general point of view.
6. Freedman (1966).
7. Remark by I. H. Sloan quoted in Kowalski (1969).

Section 1.10

1. This is known as Wigner's theorem, Messiah (1965).
2. An exception is the charge-conjugation operation in first-quantized Dirac theory, see Bjorken (1964).
3. Messiah (1965).
4. Goldberger (1964), Joachain (1975), Itzykson (1980).
5. Goldberger (1964).
6. Eden (1966).
7. See, e.g., Streater (1964), Itzykson (1980), p. 157.
8. Streater (1964), Jost (1965).
9. Fetter (1971), Blaizot (1986), Negele (1988), Eisenberg (1989).

Section 1.11

1. Identity effects in collision theory are discussed at length in Goldberger (1964). Modern developments arising from the work of Bencze (1978a, b) are reviewed in Kowalski (1978c, 1984).
2. See, e.g., Newton (1982).
3. Simple examples of this appear in Kerman (1959), Lovelace (1964a, b).
4. Bencze (1975, 1978a, b). Applications and further developments of this formalism appear in Goldflam (1980b, d, e), Kowalski (1981a, 1982c, 1986), Adhikari (1984a).
5. For a rather different treatment of permutation symmetries, see Giraud (1980) and Nagarajan (1983).
6. See, e.g., Skornyakov (1956). The original Amado (1963) model and its descendents, e.g., Fonseca (1976a), use the diagrammatic techniques associated with the second-quantized formalism. The implementation of reduction techniques by Stichel (1964) resulted in $N = 3$, connected-kernel equations similar to those subsequently proposed by Alt (1967). For applications for arbitrary N, see Ballot (1967), Komarov (1969, 1970), where extensive references to previous work can be found. For more recent applications, see Rinat (1977a, b).
7. Goldberger (1964).
8. Kowalski (1981a).
9. Glöckle (1970b).
10. Adhikari (1984a) and references cited therein.
11. For a review of general nonrelativistic Fock-space methods, see Negele (1988). The collision problem is elaborated upon more fully in Redmond (1960), Goldberger (1960, 1964), and Schweber (1962a, b).
12. Redish (1970).
13. Bell (1959, 1962), Blaizot (1986), Negele (1988).
14. Reinhardt (1982), Negele (1988).
15. Thacker (1975, 1976).

16. For an early investigation, see Klein (1957).
17. Salpeter (1951), Schweber (1962a), Freedman (1966), Huang (1975), Kowalski (1979c).
18. Lehmann (1955), Goldberger (1960, 1964), Redmond (1960).
19. See Bhalerao (1980), Celenza (1984) and references cited therein. For reviews and an extensive bibliography see Eisenberg (1980) and Kowalski (1980b).
20. Sandhas (1966).

Appendix (I)

1. Reed (1979), p. 75.
2. Riahi (1969).
3. Comprehensive discussions of how one justifies the relationship of transition amplitudes to cross sections and to other scattering observables are contained in Goldberger (1964) (Chs. 1–4), Joachain (1975) (Ch. 15), Newton (1982) (Ch. 8), and Taylor (1987) (Chs. 3, 7, 17, 22). Goldberger (1964) contains an appendix concerned with the definition of cross sections for three free-particle collisions. A somewhat more rigorous discussion of all of these questions with special attention to the cluster structure of general multiparticle amplitudes is given by Hunziker (1968).
4. See Andrews (1976).

Chapter 2

Potential Scattering

2.1. Introduction

There is no dearth of excellent expositions of the theory of the scattering of a particle by a potential-generated center of force.[1] Nevertheless, much of the material discussed in the extant literature is not directly relevant to our dynamical viewpoint, nor is it a prerequisite for later developments in this book. Thus, we have limited ourselves primarily to those aspects of the subject that are consistent with our overall objectives.

We call single-particle scattering by a potential, which can also represent two particles in their c.m. frame, *potential scattering*. More complicated instances of potential-mediated interactions will be described with a qualifier such as *three-particle* potential scattering, *multiparticle* potential scattering, and so on. Since we deal exclusively with interactions represented by potentials, we very often dispense with the use of the word "potential" altogether when there is no danger of implying anything more general.

Despite the elementary nature of the problem and its long development, potential scattering remains a lively area of research. There are several reasons for this, in addition to its seemingly inexhaustible intrinsic interest. One of them is its role as a theoretical laboratory for ideas about collisions that would be difficult to investigate in more complicated situations. Perhaps the most dramatic instance of this was the discovery by Regge[2] of poles in the complex angular momentum plane, which had great impact in particle physics.[3] Another reason, which is of primary interest to us, arises from the role of two-particle scattering as the simplest component of multiparticle scattering, and so it is needed as input into more complicated problems.

The restrictions we place upon the potentials conform to the discussion of Ch. I. Specifically, the Coulomb and other long-range potentials are excluded unless they are cut off at some distance. Basically, we suppose that the structure of the potential suffices for the validity of the asymptotic condition, asymptotic completeness, and the standard operator formalism of scattering theory.[4]

The potential operator V is the interaction part of the full Hamiltonian

$$H = H_0 + V. \tag{2.1.1}$$

Here, H_0 is the kinetic-energy operator, which for a single particle of mass m has the coordinate-space representation

$$\langle \mathbf{x}'|H_0|\mathbf{x}\rangle = -\frac{1}{2m}\nabla^2_{\mathbf{x}'}\,\delta^{(3)}(\mathbf{x}'-\mathbf{x}), \tag{2.1.2}$$

and $\mathbf{x}$, e.g., is the particle position vector. Alternatively, m may represent a two-particle reduced mass μ with $\mathbf{x}$ the relative position vector. We take $\hbar = 1$ throughout. Two forms of the potential matrix elements are of special interest, namely a local potential,

$$\langle \mathbf{x}'|V|\mathbf{x}\rangle = V(\mathbf{x})\,\delta^{(3)}(\mathbf{x}'-\mathbf{x}), \tag{2.1.3}$$

and a potential of finite rank, which has the generic form

$$\langle \mathbf{x}'|V|\mathbf{x}\rangle = \sum_{i=1}^{N} f_i(\mathbf{x}')g_i(\mathbf{x}). \tag{2.1.4}$$

In the absence of spin, the quantities $V(\mathbf{x})$, $f_i(\mathbf{x})$, and $g_i(\mathbf{x})$ are complex-valued functions of $\mathbf{x}$ that are restricted by symmetries and Hermiticity, $V^\dagger = V$. When the interaction is spin-dependent, $V(\mathbf{x})$ is an operator on the finite-dimensional spin-space. Similar remarks apply to the functions f_i and g_i.

Local potentials possess the virtue of reducing the time-independent, coordinate-space, single-particle, integro-differential Schrödinger equation to a three-dimensional differential equation. A partial-wave decomposition results in an infinite set of ordinary second-order linear differential equations. Any spin–orbit coupling for a fixed total angular momentum involves a finite number of equations. The mathematical problem that remains, that of determining asymptotic solutions of Sturm–Liouville-type differential equations for suitable $V(\mathbf{x})$, exploits a well-developed corpus of mathematics. We remark that the problem of classifying suitable local potentials is relatively straightforward compared to the general nonlocal case.[4]

In addition, local potentials have proven to be efficient phenomenological devices for representing in a simple way what are often very complex microscopic interaction dynamics. Nevertheless, except for the long-ranged parts of the electromagnetic and hadronic interactions, it is difficult to justify their use apart from qualitative arguments such as the similarity of the analyticity properties of the resultant S-matrix elements with those expected on more general grounds.

The preceding indicates why so much of the work has been carried out with local potentials. Yet nonlocal potentials appear naturally when one tries to represent field-theoretic interactions in terms of effective potentials. Moreover, in momentum space, local potentials present few advantages. Finally, many nonlocal potentials of interest can be systematically approximated by finite-rank potentials of the form (2.1.4), for which the solution of the scattering problem reduces to quadrature.

Given a definite interaction potential, the calculation of the on-shell transition amplitudes for two-particle scattering can be carried out using any of a number of efficient numerical techniques. However, one needs far more information than this to represent the off-shell two-body collisions that take place in multiparticle environments. Here one encounters amplitudes for which the initial and final relative energies as well as the parametric energy are all different. Thus, for each state of two-particle relative angular momentum, one requires transition amplitudes that are defined on a three-dimensional, semi-infinite parameter space. The major objective of this chapter is to develop the equations satisfied by such amplitudes, and to investigate their properties.

There are additional incentives for seeking representations of the off-shell two-body transition amplitudes that serve to simplify the calculation of the multiparticle scattering problems in which they occur. Finite-rank representations similar in form to Eq. (2.1.4) have proven to be the most useful in this regard, and considerable literature has appeared on different proposals for obtaining such forms.[5] If the potential itself is of finite rank, so is the off-shell transition amplitude. Moreover, subsequent multiparticle calculations are greatly simplified if the number of terms in the finite-rank expansion is as small as possible.

The off-shell two-nucleon transition amplitudes at energies below the pion-production threshold have been studied extensively.[6] This work has been stimulated in part by the availability of progressively more refined potentials that fit the on-shell data with increased accuracy, for then one can try to differentiate among these various phenomenological models on the basis of different off-shell behaviors.[7] These efforts have been largely inconclusive thus far. In this regard, we remark that within the context of a particular model for a multiparticle process what is meant by the off-shell continuation of an on-shell constituent amplitude is necessarily unambiguous. This meaning is, of course, model dependent.

Early phenomenological models of nucleon–nucleon scattering, in particular, made use of devices such as hard cores, namely infinitely repulsive potentials for distances less than some core radius; outside the core, the interaction is represented by a conventional potential. Refinements, as well as extreme versions, of this idea use boundary-condition stipulations on the wave function at some finite radius as a way of representing unknown dynamics in a simple way. We refer to all models that exploit such devices as *boundary-condition models.*

These models inspired a great deal of effort[8] in regard to the identification of the proper off-shell two-body transition amplitudes, and even were applied to three-particle scattering problems.[9] The general approach has a long history in the treatment of many-particle scattering.[10] Nevertheless, despite this corpus of interesting work, we will not discuss boundary-condition models in any detail. Our major reason for this omission is that these approaches are not truly dynamical, in contradiction to the point of view we have chosen to emphasize.

2.2. Time-Independent Realization of Scattering

The simplicity of potential scattering allows us to elaborate upon the material of Ch. 1 in an elementary context. To begin with, let us explain how the intrinsically time-dependent scattering process can be described in terms of a time-independent formalism.[1] We first obtain an explicit representation for the Møller wave operator, and then we use this to construct the scattering state vectors corresponding to an initial state of well-defined momentum.

Of course, *all* solutions of the Schrödinger equation are dependent upon the time t, even bound-state solutions. In this last case, the time dependence occurs only in the trivial phase $\exp(-iE_B t)$, where E_B is the (discrete) bound-state energy, so all physically interesting probability amplitudes and expectation values are time-independent. This reflects the absolute stability of the bound state corresponding to a single particle permanently confined in a potential well.

Things are different for the scattering of a single particle of mass m by a localized center of force. Far away from the force center, which we can take to be the origin, the appropriate state of the projectile is a wave packet $|\phi(t)\rangle$, which is a normalizable solution of the free Schrödinger equation:

$$i\frac{\partial}{\partial t}|\phi(t)\rangle = H_0|\phi(t)\rangle. \tag{2.2.1}$$

We suppose $|\phi(t)\rangle$ is normalized to unity:

$$\langle\phi(t)|\phi(t)\rangle \equiv \|\phi(t)\|^2 = \int d^3x|\phi(\mathbf{x},t)|^2 = 1, \tag{2.2.2}$$

where $\phi(\mathbf{x},t) = \langle\mathbf{x}|\phi(t)\rangle$ is the coordinate representative of $|\phi(t)\rangle$. The designation of $|\phi(t)\rangle$ as a wave packet follows from the fact that any solution of Eq. (2.2.1) can be expressed as a Fourier transform

$$\phi(\mathbf{x},t) = \int \frac{d^3k}{(2\pi)^{3/2}} a(\mathbf{k})e^{i\mathbf{k}\cdot\mathbf{x}}e^{-iE_k t}, \tag{2.2.3}$$

where $E_k = \mathbf{k}^2/2m$, namely as a packet of waves with wave vectors $\mathbf{k}$ and angular frequencies E_k. The normalization condition (2.2.2) implies that the weight function $a(\mathbf{k})$ must satisfy

$$\int d^3k\,|a(\mathbf{k})|^2 = 1. \tag{2.2.4}$$

Evidently, $|a(\mathbf{k})|^2$ is the probability density for the free particle to have momentum $\mathbf{k}$. When the free-particle state closely represents the idealization of a particle with a well-defined wave vector $\mathbf{k}_0$, then $|a(\mathbf{k})|$ will be sharply peaked about $\mathbf{k}_0$.

Another important attribute of wave packets is that they spread in space with the passing of time. Thus, for a realistic description of the scattering process, it is important to know whether it is possible to prepare packets with sufficiently little spreading under the kinematical conditions of a realistic experiment. In scenarios of this sort, the packets are monitored over the finite flight times from the "injector" to the target and thence to the final detector.[2]

The full dynamical problem is generated by the Schrödinger equation

$$i\frac{\partial}{\partial t}|\psi(t)\rangle = H|\psi(t)\rangle, \tag{2.2.5}$$

where H is given by Eq. (2.1.1). The physical scattering sequence, which begins with the particle (nearly) in an incident (i) free state $|\phi_{\mathrm{i}}(t)\rangle$, is then represented by the ("in") solution of Eq. (2.2.5), $|\psi_i^{(+)}(t)\rangle$, that satisfies the asymptotic condition

$$\lim_{t\to-\infty}{}_{\mathrm{s}} \{|\psi_i^{(+)}(t)\rangle - |\phi_i(t)\rangle\} = 0, \tag{2.2.6}$$

where the limit is taken in the strong sense as discussed in Section 1.5. This accurately reflects the physical situation in that the incident particle is never really free—it merely is influenced by the center of force in a diminishingly small way as one traces out its past history. Equation (2.2.6) is a precise measure of just how the free state is approached in the distant past.

The fact that only asymptotic times enter into the statement of the scattering problem indicate that for interactions with no explicit time dependence, and so conserving energy, it should be possible to formulate time-independent dynamical equations. If V does not explicitly depend on time, we have

$$|\psi_i^{(+)}(t)\rangle = e^{-iHt}|\psi_i^{(+)}(0)\rangle. \tag{2.2.7}$$

The corresponding time-translation behavior of the free state is

$$|\phi_i(t)\rangle = e^{-iH_0t}|\phi_i(0)\rangle. \tag{2.2.8}$$

We see then that Eqs. (2.2.6–2.2.8) are consistent with the mapping

$$|\psi_i^{(+)}(t)\rangle = \Omega^{(+)}|\phi_i(t)\rangle, \tag{2.2.9}$$

where $\Omega^{(+)}$ is the Møller operator (Section 1.5):

$$\Omega^{(+)} = \lim_{t\to-\infty}{}_{\mathrm{s}} \{e^{iHt}e^{-iH_0t}\}. \tag{2.2.10}$$

A sufficient condition for the existence of the limit (2.2.10) is that $\|V\|$ be defined, which for a local potential means that

$$\int d^3x\,|V(\mathbf{x})|^2 < \infty, \tag{2.2.11}$$

and that excludes any potential with a Coulomb tail.

The complete scattering dynamics is determined by the time-independent operator $\Omega^{(+)}$. This is the essential content of the assertion that it is possible to describe the scattering in a time-independent fashion. We note that an important feature of $\Omega^{(+)}$ is that it does not depend on the way the wave packet is prepared.

A major step in determining practical dynamical equations involves moving away from the arbitrariness involved in the specification of wave packets by considering the idealization of states of well-defined energy and perhaps, momentum. For example, corresponding to Eq. (2.2.3) one can write

$$|\phi_{\mathrm{i}}(t)\rangle = \int d^3k\, a_{\mathrm{i}}(\mathbf{k}) e^{-iE_k t}|\mathbf{k}\rangle, \tag{2.2.12}$$

where

$$\langle \mathbf{k}'|\mathbf{k}\rangle = \delta^{(3)}(\mathbf{k}' - \mathbf{k}), \tag{2.2.13}$$

and $|\mathbf{k}\rangle$ corresponds to an idealized state of well-defined (relative) momentum $\mathbf{k}$. The weighting function $a_{\mathrm{i}}(\mathbf{k})$ of the incident (i) state is nondynamical. So, if the limit (2.2.6) is true, it is plausible that

$$|\psi_{\mathrm{i}}^{(+)}(t)\rangle = \int d^3k\, e^{-iE_k t} a_{\mathrm{i}}(\mathbf{k})|\psi_{\mathbf{k}}^{(+)}\rangle, \tag{2.2.14}$$

where

$$(E_k - H)|\psi_{\mathbf{k}}^{(+)}\rangle = 0. \tag{2.2.15}$$

Of course, we have

$$(E_k - H_0)|\mathbf{k}\rangle = 0. \tag{2.2.16}$$

Since the effective one-particle problem is not invariant with respect to translations of the relative coordinates when there is an interaction, the direction of the relative momentum can change. Thus, $|\psi_{\mathbf{k}}^{(+)}\rangle$ represents all of the scattering possibilities that can result from an initial state $|\mathbf{k}\rangle$.

Next we re-express the limit (2.2.10) in a form that corresponds to the plane-wave decompositions (2.2.12) and (2.2.14). First, we make use of the identity

$$\lim_{t\to-\infty} F(t) = \lim_{\varepsilon\to 0+} \left\{\varepsilon \int_{-\infty}^{0} dt\, e^{\varepsilon t} F(t)\right\}, \tag{2.2.17}$$

for a suitably well-behaved function $F(t)$, to rewrite Eq. (2.2.10) as

$$\Omega^{(+)} = \lim_{\varepsilon\to 0+} \varepsilon \int_{-\infty}^{0} dt\, e^{\varepsilon t} e^{iHt} e^{-iH_0 t}, \tag{2.2.18}$$

which has meaning as an integral of the matrix elements of the operators that appear in it. Since these operators are all bounded and $\exp(\varepsilon t)$ is highly convergent at the lower limit of integration, it is certainly plausible that the manifest validity of Eq. (2.2.17) implies the expression (2.2.18).

Using the representation (2.2.18) in Eq. (2.2.9) and employing Eq. (2.2.12), we obtain

$$|\psi_{\mathrm{i}}^{(+)}(0)\rangle = \lim_{\varepsilon\to 0+} \varepsilon \int_{-\infty}^{0} dt\, e^{\varepsilon t} e^{iHt} \int d^3k\, e^{-iE_k t} a_{\mathrm{i}}(\mathbf{k})|\mathbf{k}\rangle, \tag{2.2.19}$$

which becomes, after an interchange in the orders of integration,

$$|\psi_{\mathrm{i}}^{(+)}(0)\rangle = \lim_{\varepsilon\to 0+} \int d^3k\, a_{\mathrm{i}}(\mathbf{k})\{i\varepsilon G(E_k + i\varepsilon)\}|\mathbf{k}\rangle. \tag{2.2.20}$$

The *resolvent operator*, or *Green function*, corresponding to H for a complex parametric energy $E + i\varepsilon$, is given by

$$G(E + i\varepsilon) \equiv (E + i\varepsilon - H)^{-1}, \tag{2.2.21}$$

where the inverse operation exists for $\varepsilon \neq 0$ because of the Hermiticity of H. From Eq. (2.2.14) and Eq. (2.2.20), we infer that

$$|\psi_{\mathbf{k}}^{(+)}\rangle = \lim_{\varepsilon\to 0+} i\varepsilon G(E_k + i\varepsilon)|\mathbf{k}\rangle. \tag{2.2.22}$$

We emphasize that $i\varepsilon G(E + i\varepsilon)$ is *not* to be identified with the E-independent operator $\Omega^{(+)}$. Thus, despite the formal resemblance of Eq. (2.2.9) for $t = 0$ and Eq. (2.2.22), they are very different. Among other things, the states entering into Eq. (2.2.22) are normalizable only in the generalized-function sense. Nevertheless, the expression (2.2.22) should contain everything we require to determine $|\psi_{\mathbf{k}}^{(+)}\rangle$, and therefore any related $|\psi_{\mathrm{i}}^{(+)}(0)\rangle$. In particular, because the limiting procedure $\varepsilon \to 0+$ reflects the asymptotic condition, no further boundary conditions need be imposed. Our objective in the next section is to convert the form (2.2.22) into something for which there exist conventional solution strategies.

The preceding treatment of the in-state carries over to the out-state, $|\psi_{\mathrm{f}}^{(-)}\rangle$, with little difficulty. For a final (f) state $|\phi_{\mathrm{f}}(t)\rangle$, we have

$$|\psi_{\mathrm{f}}^{(-)}(t)\rangle = e^{-iHt}|\psi_{\mathrm{f}}^{(-)}(0)\rangle, \tag{2.2.23}$$

and

$$|\psi_{\mathrm{f}}^{(-)}(t)\rangle = \Omega^{(-)}|\phi_{\mathrm{f}}(t)\rangle. \tag{2.2.24}$$

The wave operator $\Omega^{(-)}$ is given alternatively by

$$\Omega^{(-)} = \lim_{t\to +\infty} (e^{iHt} e^{-iH_0 t}), \tag{2.2.25}$$

$$\Omega^{(-)} = \lim_{\varepsilon\to 0+} \varepsilon \int_{0}^{\infty} dt\, e^{-\varepsilon t} e^{iHt} e^{-iH_0 t}. \tag{2.2.26}$$

Therefore, we conclude that

$$|\psi_{\mathbf{k}}^{(-)}\rangle = \lim_{\varepsilon\to 0+} (-i\varepsilon)G(E_k - i\varepsilon)|\mathbf{k}\rangle. \tag{2.2.27}$$

The two expressions (2.2.10) and (2.2.25) for $\Omega^{(+)}$ and $\Omega^{(-)}$, respectively, lead to a direct evaluation of the S-matrix:

$$S_{\mathrm{fi}} = \langle\psi_{\mathrm{f}}^{(-)}|\psi_{\mathrm{i}}^{(+)}\rangle = \langle\phi_{\mathrm{f}}|\Omega^{(-)\dagger}|\psi_{\mathrm{i}}^{(+)}\rangle. \tag{2.2.28}$$

Using expression (2.2.25), we have

$$S_{\mathrm{fi}} = \lim_{t\to-\infty} \langle\phi_{\mathrm{f}}|\omega(t)|\psi_{\mathrm{i}}^{(+)}\rangle + \left(\lim_{t\to+\infty} - \lim_{t\to-\infty}\right)\langle\phi_{\mathrm{f}}|\omega(t)|\psi_{\mathrm{i}}^{(+)}\rangle, \tag{2.2.29}$$

where

$$\omega(t) = e^{iH_0 t}e^{-iHt}. \tag{2.2.30}$$

Now

$$\lim_{t\to-\infty} \langle\phi_{\mathrm{f}}|\omega(t)|\psi_{\mathrm{i}}^{(+)}\rangle = \langle\psi_{\mathrm{f}}^{(+)}|\psi_{\mathrm{i}}^{(+)}\rangle$$

$$= \delta_{\mathrm{fi}}, \tag{2.2.31}$$

$$\left(\lim_{t\to+\infty} - \lim_{t\to-\infty}\right)\langle\phi_{\mathrm{f}}|\omega(t)|\psi_{\mathrm{i}}^{(+)}\rangle = \int_{-\infty}^{\infty} dt\frac{d}{dt}[\langle\phi_{\mathrm{f}}|\omega(t)|\psi_{\mathrm{i}}^{(+)}\rangle], \tag{2.2.32}$$

and

$$\frac{d}{dt}\langle\phi_{\mathrm{f}}|\omega(t)|\psi_{\mathrm{i}}^{(+)}\rangle = -i\langle\phi_{\mathrm{f}}|V|\psi_{\mathrm{i}}^{(+)}\rangle e^{i(E_{\mathrm{f}}-E_{\mathrm{i}})t}. \tag{2.2.33}$$

Therefore, we have

$$S_{\mathrm{fi}} = \delta_{\mathrm{fi}} - 2\pi i\,\delta(E_{\mathrm{f}} - E_{\mathrm{i}})\langle\phi_{\mathrm{f}}|V|\psi_{\mathrm{i}}^{(+)}\rangle. \tag{2.2.34}$$

Similarly, starting from

$$S_{\mathrm{fi}} = \lim_{t\to-\infty} \langle\psi_{\mathrm{f}}^{(-)}|\omega(t)^{\dagger}|\phi_{\mathrm{i}}\rangle, \tag{2.2.35}$$

we also find that

$$S_{\mathrm{fi}} = \delta_{\mathrm{fi}} - 2\pi i\,\delta(E_{\mathrm{f}} - E_{\mathrm{i}})\langle\psi_{\mathrm{f}}^{(-)}|V|\phi_{\mathrm{i}}\rangle. \tag{2.2.36}$$

Comparing Eqs. (2.2.34) and (2.2.36), we deduce the on-shell ($E_{\mathrm{f}} = E_{\mathrm{i}}$) equivalence

$$\langle\phi_{\mathrm{f}}|V|\psi_{\mathrm{i}}^{(+)}\rangle = \langle\psi_{\mathrm{f}}^{(-)}|V|\phi_{\mathrm{i}}\rangle, \qquad E_{\mathrm{f}} = E_{\mathrm{i}}. \tag{2.2.37}$$

2.3. Lippmann–Schwinger Equation

In this section, we derive the LS equation for $|\psi_{\mathbf{k}}^{(+)}\rangle$ in a manner that will take advantage of the special simplicities (*cf*. Section 1.8) of potential scattering.[1,2]

As a precursor to Eq. (2.2.22), let us define

$$|\psi_{\mathbf{k}}^{(+)}(\varepsilon)\rangle \equiv i\varepsilon G(E_k + i\varepsilon)|\mathbf{k}\rangle. \tag{2.3.1}$$

If we use the identity

$$G(E + i\varepsilon) = G_0(E + i\varepsilon)[I + VG(E + i\varepsilon)], \tag{2.3.2}$$

where

$$G_0(E + i\varepsilon) = (E + i\varepsilon - H_0)^{-1}, \tag{2.3.3}$$

we see from Eq. (2.3.1) that

$$|\psi_{\mathbf{k}}^{(+)}(\varepsilon)\rangle = i\varepsilon G_0(E_k + i\varepsilon)|\mathbf{k}\rangle + G_0(E_k + i\varepsilon)V|\psi_{\mathbf{k}}^{(+)}(\varepsilon)\rangle. \tag{2.3.4}$$

Equation (2.3.4) is a promising starting point for obtaining an integral equation for $|\psi_{\mathbf{k}}^{(+)}\rangle$, but only if we can define the limit $\varepsilon \to 0+$ in an unambiguous manner. The central question is whether or not the relation

$$\lim_{\varepsilon \to 0+} [G_0(E_k + i\varepsilon)V|\psi_{\mathbf{k}}^{(+)}(\varepsilon)\rangle] = G_0^{(+)}(E_k)V|\psi_{\mathbf{k}}^{(+)}\rangle \tag{2.3.5}$$

is true, where $|\psi_{\mathbf{k}}^{(+)}\rangle$ is defined by Eq. (2.2.22) and

$$G_0^{(+)}(E) \equiv \lim_{\varepsilon \to 0+} G_0(E + i\varepsilon). \tag{2.3.6}$$

Equation (2.3.5) is equivalent to the statement that the operations of the mapping of a vector and the limit $\varepsilon \to 0+$ can be interchanged. If they cannot, then Eq. (2.3.4) does not lead to a practical integral equation.

Let us first consider the meaning of the limit implied in Eq. (2.3.6). The Hermiticity of the kinetic-energy operator H_0 (real eigenvalues in a generalized sense) ensures that the inverse operation (2.3.3) is valid so long as $\varepsilon \neq 0$. Thus, we can express $G_0(E + i\varepsilon)$ in spectral form ($\hbar = 2\mu = 1$) using the completeness property of $\{|\mathbf{q}\rangle\}$:

$$G_0(E + i\varepsilon) = \int d^3q \frac{|\mathbf{q}\rangle\langle\mathbf{q}|}{E - q^2 + i\varepsilon}, \tag{2.3.7}$$

where $q = |\mathbf{q}|$. With the normalization $\langle\mathbf{q}|\mathbf{x}\rangle = (2\pi)^{-3/2}\exp(-i\mathbf{q}\cdot\mathbf{x})$, the momentum and coordinate representations of $G_0(E)$ are, respectively,

$$\langle\mathbf{p}'|G_0(E + i\varepsilon)|\mathbf{p}\rangle = \delta^{(3)}(\mathbf{p} - \mathbf{p}')(E - p^2 + i\varepsilon)^{-1} \tag{2.3.8}$$

and

$$\langle \mathbf{x}'|G_0(E + i\varepsilon)|\mathbf{x}\rangle = \int \frac{d^3q}{(2\pi)^3} \frac{e^{i\mathbf{q}\cdot(\mathbf{x}'-\mathbf{x})}}{E - q^2 + i\varepsilon}. \tag{2.3.9}$$

The right-hand side of Eq. (2.3.8) is a well-defined product of generalized functions that involve two sets of independent variables, **p** and **p**′. The meaning of the ε-dependent factor as $\varepsilon \to 0+$ is derived using Eq. (1.9.5), so we can write

$$\langle \mathbf{p}|G_0^{(+)}(E)|\mathbf{p}'\rangle = \delta^{(3)}(\mathbf{p} - \mathbf{p}')(E - p^2 + i0)^{-1}, \tag{2.3.10}$$

with

$$(E - p^2 + i0)^{-1} = P(E - p^2)^{-1} - i\pi\,\delta(E - p^2). \tag{2.3.11}$$

One can use Eq. (2.3.11) to evaluate Eq. (2.3.9), but it is simpler to use contour integration directly. As is appropriate to scattering, we take E to be nonnegative, $E = E_k = k^2$. Then,

$$\langle \mathbf{x}''|G_0(E_k + i\varepsilon)|\mathbf{x}'\rangle = \frac{1}{4\pi^2 R} \int_{-\infty}^{\infty} dq \frac{q \sin qR}{k^2 - q^2 + i\varepsilon}, \tag{2.3.12}$$

where

$$R = |\mathbf{x}'' - \mathbf{x}'|. \tag{2.3.13}$$

The integral in Eq. (2.3.12) is evaluated using Cauchy's theorem yielding, in the limit $\varepsilon \to 0+$,

$$\langle \mathbf{x}''|G_0^{(+)}(E_k)|\mathbf{x}'\rangle = -\frac{1}{4\pi} \frac{e^{ikR}}{R}. \tag{2.3.14}$$

In the usual units, the right-hand side of Eq. (2.3.14) is multiplied by $(2\mu/\hbar^2)$.

We see from Eq. (2.3.14) that the coordinate-space representation of $G_0^{(+)}(E_k)$ results in ordinary rather than generalized functions in the $\varepsilon \to 0+$ limit. The form e^{ikR}/R is, of course, characteristic of an outgoing spherical wave. Some care must be exercised in representing $G_0^{(+)}(E_k)$ as an inverse of the operator $(E_k - H_0 + i\varepsilon)$ as $\varepsilon \to 0+$ when one utilizes the matrix elements of $G_0^{(+)}(E_k)$ in a nonnormalizable basis as in Eq. (2.3.10).

The spectral representation (2.3.7) is consistent with the expectation that, for any ε,

$$[i\varepsilon G_0(E_k + i\varepsilon)]|\mathbf{k}\rangle = |\mathbf{k}\rangle. \tag{2.3.15}$$

Given this and Eq. (2.3.5), we then obtain from the $\varepsilon \to 0+$ limit of Eq. (2.3.4) the LS integral equation for $|\psi_{\mathbf{k}}^{(+)}\rangle$:

$$|\psi_{\mathbf{k}}^{(+)}\rangle = |\mathbf{k}\rangle + G_0^{(+)}(E_k)V|\psi_{\mathbf{k}}^{(+)}\rangle. \tag{2.3.16}$$

Before we turn to the discussion of the implications of Eq. (2.3.16), we return to the question of the validity of Eq. (2.3.5). Now

$$\langle \mathbf{x}|G_0(E_k + i\varepsilon)V|\psi_{\mathbf{k}}^{(+)}(\varepsilon)\rangle = \int d^3x' \langle \mathbf{x}|G_0(E_k + i\varepsilon)|\mathbf{x}'\rangle V(\mathbf{x}')\langle \mathbf{x}'|\psi_{\mathbf{k}}^{(+)}(\varepsilon)\rangle, \tag{2.3.17}$$

where we have assumed a local potential for the sake of simplicity. The singular parts of the ε-dependent factors in Eq. (2.3.17) involve long-ranged ($|\mathbf{x}'| \to \infty$) behavior, but with a short-ranged potential these parts are damped out. This eliminates the only possible cause of trouble, namely a confluence of the singularities of the two factors that would yield results different from the limits taken separately. This implies that the limit of the product of factors is equal to the product of the separate limit of the factors when these factors have V between them.[3]

This heuristic argument can be justified by showing that the solution of Eq. (2.3.16) satisfies the Schrödinger equation and the correct scattering boundary conditions. This indirectly validates Eq. (2.3.5).

In coordinate space Eq. (2.3.16) becomes

$$\psi_{\mathbf{k}}^{(+)}(\mathbf{x}) = \langle \mathbf{x}|\mathbf{k}\rangle + \int d^3x'\, K_k(\mathbf{x}|\mathbf{x}')\psi_{\mathbf{k}}^{(+)}(\mathbf{x}'), \tag{2.3.18}$$

where

$$\psi_{\mathbf{k}}^{(+)}(\mathbf{x}) = \langle \mathbf{x}|\psi_{\mathbf{k}}^{(+)}\rangle, \tag{2.3.19}$$

is the wave function, and

$$K_k(\mathbf{x}|\mathbf{x}') = \int d^3y\, \langle \mathbf{x}|G_0^{(+)}(E_k)|\mathbf{y}\rangle\langle \mathbf{y}|V|\mathbf{x}'\rangle \tag{2.3.20}$$

is the kernel. Since

$$(\nabla_{\mathbf{x}}^2 + k^2)\frac{e^{\pm ik|\mathbf{x}-\mathbf{x}'|}}{|\mathbf{x}-\mathbf{x}'|} = \mp 4\pi\, \delta^{(3)}(\mathbf{x}-\mathbf{x}'), \tag{2.3.21}$$

and

$$(\nabla_{\mathbf{x}}^2 + k^2)\langle \mathbf{x}|\mathbf{k}\rangle = 0, \tag{2.3.22}$$

we see, using the expression (2.3.14) in Eq. (2.3.20), that

$$(\nabla_{\mathbf{x}}^2 + k^2)\psi_{\mathbf{k}}^{(+)}(\mathbf{x}) = \int d^3x' \langle \mathbf{x}|V|\mathbf{x}'\rangle \psi_{\mathbf{k}}^{(+)}(\mathbf{x}'). \tag{2.3.23}$$

This verifies that the solution of the LS equation (2.3.18) satisfies the time-independent Schrödinger equation.

We can determine the asymptotic behavior of $\psi_{\mathbf{k}}^{(+)}(\mathbf{x})$ as $|\mathbf{x}| \equiv r \to \infty$, if we make use of the fact that, for a short-ranged potential, the Green function in the kernel (2.3.20) can be replaced for $r \to \infty$ by the asymptotic form

$$\frac{e^{\pm ik|\mathbf{x}-\mathbf{x}'|}}{|\mathbf{x}-\mathbf{x}'|} \to \frac{e^{\pm ikr}}{r} e^{\mp i\mathbf{k}'\cdot\mathbf{x}'}, \tag{2.3.24}$$

where the limit is taken in the direction of the outgoing wave vector

$$\mathbf{k}' \equiv k\frac{\mathbf{x}}{r}. \tag{2.3.25}$$

We then see from Eq. (2.3.18) that, as $r \to \infty$,

$$\psi_{\mathbf{k}}^{(+)}(\mathbf{x}) \to \frac{1}{(2\pi)^{3/2}}\left\{e^{i\mathbf{k}\cdot\mathbf{x}} + \frac{e^{ikr}}{r} f(\mathbf{k}',\mathbf{k})\right\}. \tag{2.3.26}$$

The exact expression for the *scattering amplitude* $f(\mathbf{k}',\mathbf{k})$ is given below.

The counterpart of Eq. (2.3.16) for the "out" solution $|\psi_{\mathbf{k}}^{(-)}\rangle$ is

$$|\psi_{\mathbf{k}}^{(-)}\rangle = |\mathbf{k}\rangle + G_0^{(-)}(E_k)V|\psi_{\mathbf{k}}^{(-)}\rangle. \tag{2.3.27}$$

In this case, one finds that, as $r \to \infty$,

$$\psi_{\mathbf{k}}^{(-)}(\mathbf{x}) \to \frac{1}{(2\pi)^{3/2}}\left\{e^{i\mathbf{k}\cdot\mathbf{x}} + \frac{e^{-ikr}}{r} f(\mathbf{k},-\mathbf{k}')^*\right\}. \tag{2.3.28}$$

We note that in contrast to the asymptotic form (2.3.26), the non-plane-wave part of $\psi_{\mathbf{k}}^{(-)}(\mathbf{x})$ is an incoming spherical wave. The limit $r \to \infty$ in Eq. (2.3.28) is again taken along $\mathbf{k}'$.

It is not accidental that the same scattering amplitude, but referenced to different directions, occurs in both of the asymptotic forms (2.3.26) and (2.3.28). This can be seen from the identities

$$\langle \mathbf{k}'|V|\psi_{\mathbf{k}}^{(+)}\rangle = \langle \psi_{\mathbf{k}'}^{(-)}|V|\mathbf{k}\rangle = \langle \mathbf{k}'|t(E_k)|\mathbf{k}\rangle, \tag{2.3.29}$$

where

$$t(z) = V + VG(z)V \tag{2.3.30}$$

is the two-particle transition operator for a general, complex energy z and $t(E)$ means $t(E + i\varepsilon)$ as $\varepsilon \to 0+$. Here one uses, for example, Eq. (2.3.1), the reversed form of Eq. (2.3.2),

$$G(E + i\varepsilon) = [I + G(E + i\varepsilon)V]G_0(E + i\varepsilon), \tag{2.3.31}$$

and Eq. (2.3.15). In the calculation leading to the form (2.3.28), one encounters the matrix element

$$\begin{aligned}\langle -\mathbf{k}'|V|\psi_{\mathbf{k}}^{(-)}\rangle &= \langle -\mathbf{k}'|t^{\dagger}(E_k)|\mathbf{k}\rangle \\ &= \langle \mathbf{k}|t(E_k)|-\mathbf{k}'\rangle^{*}.\end{aligned} \tag{2.3.32}$$

Given that

$$f(\mathbf{k}', \mathbf{k}) = -\left(\frac{\mu}{\hbar^2}\right)(2\pi)^2\langle \mathbf{k}'|t(E_k)|\mathbf{k}\rangle, \tag{2.3.33}$$

we then obtain Eq. (2.3.28), where we have restored the reduced mass and $\hbar$ factors.

We conclude that the solutions of the LS equations (2.3.16) and (2.3.27) satisfy the Schrödinger equation and the correct asymptotic boundary conditions. So in the two-particle case the LS equations generate unique scattering wave functions for nonnegative energies.

The relation (2.3.33) allows one to circumvent consideration of the wave function altogether. The elastic differential scattering cross section is given by (*cf*. Eq. (A.1.20))

$$\frac{d\sigma}{d\Omega} = |f|^2. \tag{2.3.34}$$

So the nontrivial aspect of a scattering process, namely the collision, is determined by the matrix elements of the transition operator $t(E_k)$. (For identical particles, the relevant matrix elements are $[\langle \mathbf{k}'|t(E_k)|\mathbf{k}\rangle \pm \langle -\mathbf{k}'|t(E_k)|\mathbf{k}\rangle]$, with $+$ $(-)$ in the boson (fermion) case. This is equivalent to considering the operator $(1 \pm P_{12})t(E_k)$, where P_{12} is the two-particle permutation operator.) These matrix elements are calculated by solving the LS equations,

$$t(z) = V + VG_0(z)t(z), \tag{2.3.35a}$$

$$t(z) = V + t(z)G_0(z)V, \tag{2.3.35b}$$

which follow directly from the definition (2.3.30) with use of the resolvent identities for $G(z)$, and then taking $z = E_k + i0$. A great deal of effort over the past three decades has been expended in devising methods for solving

Eqs. (2.3.35a, b). The major techniques that evolved for doing this are reviewed in Ch. 3.

The validity of Eq. (2.3.15) is established indirectly by our demonstration of the scattering-solution-generating properties of the LS equation for $|\psi_{\mathbf{k}}^{(+)}\rangle$. Let us consider a more direct approach. Now, for small ε,

$$\langle \mathbf{x}|i\varepsilon G_0(E_k + i\varepsilon)|\mathbf{k}\rangle = -\frac{i\varepsilon e^{i\mathbf{k}\cdot\mathbf{x}}}{4\pi(2\pi)^{3/2}}\int d^3R \frac{e^{iKR}}{R} e^{i\mathbf{k}\cdot\mathbf{R}}, \tag{2.3.36}$$

where $\mathbf{R} = \mathbf{x}' - \mathbf{x}$, $R = |\mathbf{R}|$, and $K = [k^2 + i\varepsilon]^{1/2}$. The preceding integral can be evaluated straightforwardly to yield

$$\langle \mathbf{x}|i\varepsilon G_0(E_k + i\varepsilon)|\mathbf{k}\rangle = \left\{-\frac{i\varepsilon}{2k}\left[\frac{1}{K+k} - \frac{1}{K-k}\right]\right\}\langle \mathbf{x}|\mathbf{k}\rangle. \tag{2.3.37}$$

The factor in curly brackets is identically equal to unity for any ε, which establishes Eq. (2.3.15).

To conclude this section, we consider the approach of Gerjuoy[4] for deriving the LS equation (2.3.16) in which $\varepsilon \to 0+$ throughout. Here, the fact that $G_0^{(+)}(E)$ possesses different left- and right-handed inverses, $\vec{G}_0^{(+)}(E)^{-1}$ and $\overleftarrow{G}_0^{(+)}(E)^{-1}$, respectively, is confronted directly. This difference is given by

$$\overleftarrow{G}_0^{(+)}(E)^{-1} - \vec{G}_0^{(+)}(E)^{-1} = \overleftrightarrow{H}_0, \tag{2.3.38}$$

where

$$H_0 \equiv \vec{H}_0 - \overleftarrow{H}_0. \tag{2.3.39}$$

We recall from Section 1.7 that the arrows signify the following operational rules within an inner product for an arbitrary operator A and state vectors $|\psi_{1,2}\rangle$:

$$\langle \psi_1|\vec{A}|\psi_2\rangle \equiv \langle \psi_1|A\psi_2\rangle, \tag{2.3.40a}$$

$$\langle \psi_1|\overleftarrow{A}|\psi_2\rangle \equiv \langle \psi_2|A\psi_1\rangle^*. \tag{2.3.40b}$$

Now, we must have

$$\vec{G}_0^{(+)}(E_k)^{-1}|\psi_{\mathbf{k}}^{(+)}\rangle = V|\psi_{\mathbf{k}}^{(+)}\rangle, \tag{2.3.41}$$

so we find using Eq. (2.3.38) that

$$|\psi_{\mathbf{k}}^{(+)}\rangle = G_0^{(+)}(E_k)\overleftrightarrow{H}_0|\psi_{\mathbf{k}}^{(+)}\rangle + G_0^{(+)}(E_k)V|\psi_{\mathbf{k}}^{(+)}\rangle. \tag{2.3.42}$$

This is equivalent to the LS equation (2.3.16) if and only if

$$G_0^{(+)}(E_k)\overleftrightarrow{H}_0|\psi_{\mathbf{k}}^{(+)}\rangle = |\mathbf{k}\rangle, \tag{2.3.43}$$

or, in terms of the part of $|\psi_{\mathbf{k}}^{(+)}\rangle$ that arises from the collision,

$$|\Phi_{\mathbf{k}}^{(+)}\rangle \equiv |\psi_{\mathbf{k}}^{(+)}\rangle - |\mathbf{k}\rangle, \tag{2.3.44}$$

$$G_0^{(+)}(E_k)\overleftrightarrow{H}_0|\mathbf{k}\rangle = |\mathbf{k}\rangle, \tag{2.3.45a}$$

$$G_0^{(+)}(E_k)\overleftrightarrow{H}_0|\Phi_{\mathbf{k}}^{(+)}\rangle = 0. \tag{2.3.45b}$$

In order to verify Eqs. (2.3.45a, b) explicitly, it is convenient to express the left-hand side of Eq. (2.3.43) in coordinate space:[5]

$$\langle \mathbf{x}|G_0^{(+)}(E_k)\overleftrightarrow{H}_0|\psi_{\mathbf{k}}^{(+)}\rangle = \frac{1}{4\pi}\int d^3x'\left\{\frac{e^{ikR}}{R}\nabla_{\mathbf{x}'}^2\psi_{\mathbf{k}}^{(+)}(\mathbf{x}') - \psi_{\mathbf{k}}^{(+)}(\mathbf{x}')\nabla_{\mathbf{x}'}^2\frac{e^{ikR}}{R}\right\}. \tag{2.3.46}$$

The volume integral in Eq. (2.3.46) can be transformed by using Green's theorem into an integral over the sphere at infinity:[5]

$$\langle \mathbf{x}|G_0^{(+)}(E_k)\overleftrightarrow{H}_0|\psi_{\mathbf{k}}^{(+)}\rangle = \int d\Omega_{\mathbf{x}'}\frac{r'^2}{4\pi}\left\{\frac{e^{ikR}}{R}\frac{\partial\psi_{\mathbf{k}}^{(+)}(\mathbf{x}')}{\partial r'} - \psi_{\mathbf{k}}^{(+)}(\mathbf{x}')\frac{\partial}{\partial r'}\left(\frac{e^{ikR}}{R}\right)\right\}, \tag{2.3.47}$$

where $r' = |\mathbf{x}'|$, $d\Omega_{\mathbf{x}'}$ denotes the measure with respect to the solid angle, and the limit $r' \to \infty$ is left implicit.

The identity (2.3.47) holds separately when $|\psi_{\mathbf{k}}^{(+)}\rangle$ is replaced by either $|\Phi_{\mathbf{k}}^{(+)}\rangle$ or $|\mathbf{k}\rangle$. In the first instance, we see using Eq. (2.3.24) that Eq. (2.3.47) vanishes as $r' \to \infty$ because $\Phi_{\mathbf{k}}(\mathbf{x}') \to e^{ikr'}/r'$ in this limit, and so we recover Eq. (2.3.45b). With the second choice, we establish Eq. (2.3.45a) by also making use of the asymptotic relation,[6] valid as $r' \to \infty$,

$$e^{i\mathbf{k}\cdot\mathbf{x}'} \to \frac{2\pi}{ik}\left\{\delta^{(2)}(\Omega_{\mathbf{k}} - \Omega_{\mathbf{x}'})\frac{e^{+ikr'}}{r'} - \delta^{(2)}(\Omega_{\mathbf{k}} + \Omega_{\mathbf{x}'})\frac{e^{-ikr'}}{r'}\right\}. \tag{2.3.48}$$

This completes Gerjuoy's argument for this case.

2.4. Spectral Representations; Off-Shell Unitarity

We can learn a lot about scattering without ever having to solve any differential or integral equations. The spectral characteristics of the Hamiltonian along with probability conservation imply much of what we wish to know about collisions, as we show in this section in the case of potential scattering.

Equation (2.3.7) is a special case, for parametric energy $z = E + i0$ and when there is no interaction, of the spectral representation for the full Green function $G(z)$. Generally, if we use a complete set of eigenstates of the Hamiltonian H to diagonalize $(z - H)^{-1}$, we obtain

$$G(z) = \sum_B \frac{|\psi_B\rangle\langle\psi_B|}{z - E_B} + \int d^3q \frac{|\psi_{\mathbf{q}}^{(\pm)}\rangle\langle\psi_{\mathbf{q}}^{(\pm)}|}{z - q^2}. \tag{2.4.1}$$

The sum is over the bound states $|\psi_B\rangle$ that are enumerated by the discrete index B:

$$H|\psi_B\rangle = E_B|\psi_B\rangle, \tag{2.4.2}$$

where we recall that we are working in the two-body c.m. frame. For the integral in Eq. (2.4.1), one has the option of using either the in-$[|\psi_{\mathbf{k}}^{(+)}\rangle]$ or the out-$[|\psi_{\mathbf{k}}^{(-)}\rangle]$ states, or a suitable linear combination of both.[1]

We learn from Eq. (2.4.1) that the matrix elements of $G(z)$ are analytic functions of z in the entire complex plane except for simple poles at the real energies E_B and a cut across the positive real z-axis. We assume that E_B is negative. Many physical attributes of the collision process are represented by the analytic properties of $G(z)$ in the complex z-plane that are obtained by using Eq. (2.4.1) in the matrix element of Eq. (2.3.30):

$$\langle\mathbf{p}|t(z)|\mathbf{p}'\rangle = \langle\mathbf{p}|V|\mathbf{p}'\rangle + \sum_B \frac{f_B(\mathbf{p})f_B(\mathbf{p}')^*}{z - E_B} + \int d^3q \frac{\langle\mathbf{p}|V|\psi_{\mathbf{q}}^{(\pm)}\rangle\langle\psi_{\mathbf{q}}^{(\pm)}|V|\mathbf{p}'\rangle}{z - q^2}. \tag{2.4.3}$$

Here the quantity

$$f_B(\mathbf{p}) = \langle\mathbf{p}|V|\psi_B\rangle \tag{2.4.4}$$

is what is very often referred to as a bound-state *form-factor* or *vertex function*. An important feature of Eq. (2.4.3) is that the residues of the bound-state poles of $\langle\mathbf{p}|t(z)|\mathbf{p}'\rangle$ factorize in $\mathbf{p}$ and in $\mathbf{p}'$. Also, we have

$$\langle\mathbf{p}|V|\psi_{\mathbf{k}}^{(+)}\rangle = \langle\mathbf{p}|t(k^2)|\mathbf{k}\rangle, \tag{2.4.5a}$$

$$\langle\psi_{\mathbf{k}}^{(-)}|V|\mathbf{p}\rangle = \langle\mathbf{k}|t(k^2)|\mathbf{p}\rangle, \tag{2.4.5b}$$

where

$$t(k^2) \equiv t(k^2 + i0). \tag{2.4.6}$$

The matrix elements (2.4.5a, b) are what usually are called *half-off-shell amplitudes* in that the (positive) energy corresponding to the momentum of at least one of the plane-wave states is equal to the parametric energy.

When $\mathbf{p}^2 = \mathbf{p}'^2 = k^2$, then $\langle \mathbf{p}|t(k^2)|\mathbf{p}'\rangle$ is said to be *on-shell*; any other matrix element of $t(k^2)$ is said to be *off-shell*. We observe from Eq. (2.4.3) that the off-shell amplitudes for arbitrary z are entirely determined by the positive parametric-energy half-off-shell amplitudes, the energies of the two-body bound states, and the vertex functions $f_B(\mathbf{p})$.[2] There have been a number of attempts to exploit this observation to calculate the off-shell amplitudes, but they have not been actively pursued in recent times.[3]

Off-shell amplitudes occur when one describes a multiparticle process in terms of constituent two-particle collisions. Only overall energy–momentum is required to be conserved and so, in general, the two-particle collisions occur with off-shell kinematics. The corresponding off-shell t-matrix amplitudes are integrated over their parametric energies. Typically, $\operatorname{Re} z$ has the range $-\infty < \operatorname{Re} z < E_{\max}$, where $E_{\max}$ is the maximum energy that can be made available to the two-particle collision consistent with overall energy conservation. The scattering boundary conditions require that $\operatorname{Im} z \to 0+$, so that the pole singularities in Eq. (2.4.3) are integrable.

It is evident from Eq. (2.4.3) that all of the discontinuities of $\langle \mathbf{p}|t(z)|\mathbf{p}'\rangle$ occur across the real z axis, and their representation is derived from Eq. (2.4.3) by letting $z = s + i0$, with s real, but not necessarily positive. Then, using the Hermitian-analyticity property, $t(z^*) = t(z)^\dagger$, we obtain

$$\begin{aligned}\langle \mathbf{p}|[t(s) - t(s)^\dagger]|\mathbf{p}'\rangle = {} & -2\pi i \sum_B f_B(\mathbf{p}) f_B(\mathbf{p}')^* \,\delta(s - E_B) \\ & -2\pi i \int d^3q \,\langle \mathbf{p}|t(q^2)|\mathbf{q}\rangle\, \delta(s - q^2)\, \langle \mathbf{q}|t(q^2)^\dagger|\mathbf{p}'\rangle,\end{aligned} \tag{2.4.7}$$

which is called the two-particle *off-shell unitarity* condition.[4] For $s < 0$, we have discontinuities at only a finite number of points, the bound-state energies E_B.

The unitarity of the S-matrix is related to conservation of probability, and it refers to only on-shell quantities. For two-particle collisions, the off-shell features of Eq. (2.4.7) are not directly physically relevant. However, when such a collision is a constituent of a multiparticle process, then the full off-shell character of Eq. (2.4.7) is needed to ensure *multiparticle* unitarity.

Suppose that $s > 0$, so that Eq. (2.4.7) becomes

$$\langle \mathbf{p}|[t(s) - t(s)^\dagger]|\mathbf{p}'\rangle = -2\pi i \int d^3q \,\langle \mathbf{p}|t(q^2)|\mathbf{q}\rangle\, \delta(s - q^2)\, \langle \mathbf{q}|t(q^2)^\dagger|\mathbf{p}'\rangle. \tag{2.4.8}$$

The nonlinear integral relation (2.4.8) can be reduced to a simpler form if the interaction potential is rotationally invariant, which we will always assume to be the case. In order to do this, as well as for our later work, we outline the consequences of rotational invariance in the present context.

Let $\mathbf{J} = (J_x, J_y, J_z)$ denote the triplet of total angular-momentum operators. Rotational invariance (*cf.* Section 1.10) is simply the statement that

$$[H, \mathbf{J}] = 0, \tag{2.4.9a}$$

which, since

$$[H_0, \mathbf{J}] = 0, \tag{2.4.9b}$$

implies that

$$[V, \mathbf{J}] = 0. \tag{2.4.9c}$$

As a consequence of Eqs. (2.4.9a–c), we see that, at any time t,

$$[e^{iHt}e^{-iH_0t}, \mathbf{J}] = 0, \tag{2.4.10}$$

we conclude that

$$[\Omega^{(\pm)}, \mathbf{J}] = 0, \tag{2.4.11}$$

and therefore

$$[t(z), \mathbf{J}] = 0. \tag{2.4.12}$$

Finally, we note that if $[A, \mathbf{J}] = 0$, then, by Schur's lemma,

$$\langle J', M'|A|J, M\rangle = \delta_{J'J}\,\delta_{M'M}\,A_J, \tag{2.4.13}$$

where the $|J, M\rangle$ are basis vectors in the irreducible representation of the rotation group, labeled by the total angular momentum quantum number J, and its projection M, where $-J \le M \le J$; the quantity A_J is independent of M.

The preceding development is quite general. We now specialize to the two-particle case in the c.m. frame with no spin. A complete set of eigenvectors on the barycentric subspace can be taken as the vectors $\{|k^2, L, M\rangle\}$, where

$$\mathbf{L}^2|k^2, L, M\rangle = L(L+1)|k^2, L, M\rangle \tag{2.4.14a}$$

$$L_z|k^2, L, M\rangle = M|k^2, L, M\rangle, \qquad -L \le M \le +L, \tag{2.4.14b}$$

$$H_0|k^2, L, M\rangle = \frac{\hbar^2 k^2}{2\mu}|k^2, L, M\rangle, \tag{2.4.14c}$$

where we have inserted the appropriate factor of $\hbar^2/2\mu$. The completeness and orthonormality properties are expressed in the relations

$$I = \int dq^2 \sum_{L,M} |q^2, L, M\rangle\langle q^2, L, M|, \tag{2.4.15a}$$

and

$$\langle k'^2, L', M'|k^2, L, M\rangle = \delta(k'^2 - k^2)\,\delta_{L'L}\,\delta_{M'M}, \tag{2.4.15b}$$

respectively, where we keep in mind that the normalization conventions adapted in Eqs. (2.4.15a, b) are arbitrary. In Eq. (2.4.15a) and subsequent equations, the limits on the summations over L and M are $0 \le L < \infty$ and $-L \le M \le L$, respectively.

From the closure relation (2.4.15a), we see that the plane wave

$$\langle \mathbf{x}|\mathbf{k}\rangle = \frac{1}{(2\pi)^{3/2}} e^{i\mathbf{k}\cdot\mathbf{x}} \tag{2.4.16}$$

possesses the formal decomposition

$$\langle \mathbf{x}|\mathbf{k}\rangle = \int dq^2 \sum_{L,M} \langle \mathbf{x}|q^2, L, M\rangle\langle q^2, L, M|\mathbf{k}\rangle. \tag{2.4.17}$$

However, the exponential in Eq. (2.4.16) can be expressed as

$$e^{i\mathbf{k}\cdot\mathbf{x}} = \sum_L (2L+1) i^L j_L(kr) P_L(\tilde{\mathbf{k}}\cdot\tilde{\mathbf{x}}), \tag{2.4.18a}$$

$$= 4\pi \sum_{L,M} i^L j_L(kr) Y_L^M(\tilde{\mathbf{x}}, \tilde{\zeta}) Y_L^M(\tilde{\mathbf{k}}, \tilde{\zeta})^*, \tag{2.4.18b}$$

in terms of the basis functions of the rotation group. Here $Y_L^M(\tilde{\mathbf{x}}, \tilde{\zeta})$, for example, is the usual spherical harmonic; it is a function of the polar and azimuthal angles of $\tilde{\mathbf{x}} \equiv \mathbf{x}/|\mathbf{x}|$ with respect to an axis defined by the arbitrary unit vector $\tilde{\zeta}$. The Legendre polynomial $P_L(\tilde{\mathbf{k}}\cdot\tilde{\mathbf{x}})$, where $\tilde{\mathbf{k}} \equiv \mathbf{k}/|\mathbf{k}|$, is given by

$$\begin{aligned} P_L(\tilde{\mathbf{k}}\cdot\tilde{\mathbf{x}}) &= \left(\frac{4\pi}{2L+1}\right)^{1/2} Y_L^0(\tilde{\mathbf{k}}, \tilde{\mathbf{x}}) \\ &= \left(\frac{4\pi}{2L+1}\right) \sum_M Y_L^M(\tilde{\mathbf{k}}, \tilde{\zeta})^* Y_L^M(\tilde{\mathbf{x}}, \tilde{\zeta}). \end{aligned} \tag{2.4.19}$$

Finally, $j_L(kr)$ is the spherical Bessel function, with $k \equiv |\mathbf{k}|$ and $r \equiv |\mathbf{x}|$.

Comparing Eqs. (2.4.17) and (2.4.18), we see that

$$\langle \mathbf{x}|k^2, L, M\rangle = C_L(k) i^L j_L(kr) Y_L^M(\tilde{\mathbf{x}}, \tilde{\zeta}), \tag{2.4.20}$$

where $C_L(k)$ is a normalization constant that can be determined from Eq. (2.4.15b), the closure relation for the states $|\mathbf{x}\rangle$, and the orthonormality property

$$\int_0^\infty dr\, r^2 j_L(kr) j_L(k'r) = \frac{\pi}{k}\delta(k'^2 - k^2). \tag{2.4.21}$$

After fixing an arbitrary phase, we find that

$$C_L(k) = \left(\frac{k}{\pi}\right)^{1/2}. \tag{2.4.22}$$

Since

$$\langle k'^2, L, M|\mathbf{k}\rangle = \int d^3x\, \langle k'^2, L, M|\mathbf{x}\rangle\langle \mathbf{x}|\mathbf{k}\rangle, \tag{2.4.23}$$

we immediately see that

$$\langle k'^2, L, M|\mathbf{k}\rangle = \left(\frac{2}{k}\right)^{1/2} \delta(k'^2 - k^2) Y_L^M(\tilde{\mathbf{k}}, \tilde{\zeta})^*. \tag{2.4.24}$$

Now

$$\begin{aligned}\langle \mathbf{k}'|\mathbf{k}\rangle &= \delta^{(3)}(\mathbf{k}' - \mathbf{k}) \\ &= \delta^{(2)}(\Omega_{\tilde{\mathbf{k}}'} - \Omega_{\tilde{\mathbf{k}}})\frac{\delta(k' - k)}{k^2},\end{aligned} \tag{2.4.25}$$

so

$$\delta^{(2)}(\Omega_{\tilde{\mathbf{k}}'} - \Omega_{\tilde{\mathbf{k}}}) = \sum_{L,M} Y_L^M(\tilde{\mathbf{k}}', \tilde{\zeta}) Y_L^M(\tilde{\mathbf{k}}, \tilde{\zeta})^*, \tag{2.4.26}$$

where $\Omega_{\tilde{\mathbf{k}}}$ is the element of solid angle associated with the direction $\tilde{\mathbf{k}}$.

A *partial-wave expansion* of a quantity such as $\langle \mathbf{p}|t(z)|\mathbf{p}'\rangle$ simply means

$$\begin{aligned}\langle \mathbf{p}|t(z)|\mathbf{p}'\rangle = \sum_{L',M'} \sum_{L'',M''} \int_0^\infty dk'^2 \int_0^\infty dk''^2\, \langle \mathbf{p}|k'^2, L', M'\rangle \\ \times \langle k'^2, L', M'|t(z)|k''^2, L'', M''\rangle\langle k''^2, L'', M''|\mathbf{p}'\rangle,\end{aligned} \tag{2.4.27}$$

where $t(z)$ is rotationally invariant, so that

$$\langle k'^2, L', M'|t(z)|k''^2, L'', M''\rangle = \delta_{L',L''}\,\delta_{M',M''}\, N_{L'}(k', k'') t_{L'}(k', k''; z), \tag{2.4.28}$$

defines the *partial-wave amplitude* $t_{L'}(k', k''; z)$ to within a z-independent normalization factor $N_{L'}(k', k'')$. We see that the expansion (2.4.27) then simplifies to

$$\langle \mathbf{p}|t(z)|\mathbf{p}'\rangle = \frac{1}{2\pi(pp')^{1/2}}\sum_L (2L+1)N_L(p,p')t_L(p,p';z)P_L(\tilde{\mathbf{p}}\cdot\tilde{\mathbf{p}}'). \quad (2.4.29)$$

For a short-ranged potential, it is straightforward to show that $t_L(p,p';z)$ drops off sufficiently fast as function of L to ensure that the series (2.4.29) converges; we note that any possible bound-state poles occur in a finite number of states of angular momentum.[5]

We choose the normalization convention

$$N_L(p,p') = \frac{(pp')^{1/2}}{\pi}. \quad (2.4.30)$$

Then, we have

$$\langle \mathbf{p}|t(z)|\mathbf{p}'\rangle = \frac{2}{\pi}\sum_{L,M} t_L(p,p';z)Y_L^M(\tilde{\mathbf{p}},\tilde{\zeta})Y_L^M(\tilde{\mathbf{p}}',\tilde{\zeta})^*. \quad (2.4.31)$$

Henceforth, unless explicitly indicated to the contrary, we will be concerned with partial-wave amplitudes at fixed L and for $z = s + i0$, where s is real. We can take advantage of these restrictions by adapting the more compact notation,

$$t_s(p,p') \equiv [t_L(p,p';z)]_{z=s+i0}, \quad (2.4.32)$$

in which the dependence on L is suppressed. In terms of this notation, the partial-wave form of the off-shell unitarity relation (2.4.7) is the algebraic equation

$$t_{k^2}(p',p)^* - t_{k^2}(p,p') = 2ikt_{k^2}(p,k)t_{k^2}(p',k)^*, \quad (2.4.33)$$

when $s = k^2$.

Let us deduce the implications of Eq. (2.4.33). If we set $p = p' = k$ in Eq. (2.4.33), it becomes the on-shell, partial-wave unitarity condition:

$$\operatorname{Im} t_{k^2}(k,k) = -k|t_{k^2}(k,k)|^2. \quad (2.4.34)$$

We infer from this that $t_{k^2}(k,k)$ depends on only a single, real, L-dependent parameter $\delta_L(k)$, called the *phase shift*:

$$t_{k^2}(k,k) = -\frac{1}{k}e^{i\delta_L(k)}\sin\delta_L(k). \quad (2.4.35)$$

To learn any more from off-shell unitarity requires something to be known about the properties of $t_s(p,p')$ under an interchange of p and p'. We assume symmetry under this interchange:

$$t_s(p, p') = t_s(p', p), \tag{2.4.36}$$

which is realized when the interaction is invariant under time reversal (*cf.* Section 1.10)).[6] Then, Eq. (2.4.33) reduces to

$$\operatorname{Im} t_{k^2}(p, p') = -k t_{k^2}(p, k) t_{k^2}(k, p')^*, \tag{2.4.37}$$

which shows that the imaginary part of the fully off-shell amplitude factorizes in the external momenta. Equation (2.4.37) implies that the half- and on-shell amplitudes for a given $s = k^2$ have the same phase, viz.,

$$t_{k^2}(p, k) = f_{k^2}(p, k) t_{k^2}(k, k), \tag{2.4.38}$$

where $f_{k^2}(p, k)$ is purely real. Upon combining Eqs. (2.4.37–2.4.38) we get

$$\operatorname{Im} t_{k^2}(p, p') = f_{k^2}(p, k) \operatorname{Im} t_{k^2}(k, k) f_{k^2}(p', k), \tag{2.4.39}$$

and so we conclude that

$$t_{k^2}(p, p') = f_{k^2}(p, k) t_{k^2}(k, k) f_{k^2}(p', k) + R_{k^2}(p, p'; k), \tag{2.4.40}$$

where $R_{k^2}(p, p'; k)$ is real, symmetric in p and p', and satisfies

$$R_{k^2}(k, p'; k) = R_{k^2}(p, k; k) = 0. \tag{2.4.41}$$

The representation (2.4.40) has a long history.[7] Given the $p \leftrightarrow p'$ symmetry of $t_s(p, p')$, it is a special case of the trivial identity

$$t_s(p, p') = f_s(p, k) t_s(k, k) f_s(p', k) + R_s(p, p'; k), \tag{2.4.42}$$

$$f_s(p, k) \equiv t_s(p, k)/t_s(k, k), \tag{2.4.43}$$

$$R_s(p, p'; k) \equiv t_s(p, p') - \frac{t_s(p, k) t_s(k, p')}{t_s(k, k)}. \tag{2.4.44}$$

The content of off-shell unitarity is to constrain f_s and R_s to be real functions for nonnegative s. The function $f_s(p, k)$ is often referred to as the *half-shell function.*

The representation (2.4.40) "solves" off-shell unitarity in the same sense that the phase-shift parametrization (2.4.35) "solves" the on-shell unitarity relation. There are other "solutions" of off-shell unitarity, which we will consider later.

The representation (2.4.3) for the fully off-shell amplitude, which is sometimes referred to as the Low equation, also implies discontinuity/reality conditions for $t_s(p, p')$ when $s < 0$. We see that

$$\operatorname{Im} t_s(p, p') = -\frac{\pi^2}{2} \sum_B \delta(s - E_B) f_B(p) f_B(p'), \tag{2.4.45}$$

where we have written

$$f_B(\mathbf{p}) = f_B(p) Y_L^M(\tilde{\mathbf{p}}, \tilde{\zeta}), \tag{2.4.46}$$

since each bound state appears in a definite state of angular momentum. Time-reversal invariance implies that the functions $f_B(p)$ are real. We see then that, for nearly all $s < 0$, the off-shell transition amplitude is purely real.

The interesting aspect of what might be called *negative-energy* off-shell unitarity is that its content is partially represented by Eq. (2.4.42), but with k not necessarily related to s. The real functions $f_s(p, k)$ and $R_s(p, p'; k)$ are given by Eqs. (2.4.43) and (2.4.44), respectively; however, $t_s(k, k)$ is not an on-shell amplitude. When $s < 0$, the variable k can take on any nonnegative value; whatever value one chooses, the bound-state poles at $s = E_B$ reside solely in $t_s(k, k)$. So we have

$$f_{E_B}(p, k) = \frac{f_B(p)}{f_B(k)}, \tag{2.4.47}$$

with $R_{E_B}(p, p', k)$ finite. The relation (2.4.42) for $s < 0$, with the discontinuity relation,

$$\operatorname{Im} t_s(k, k) = -\frac{\pi^2}{2} \sum_B \delta(s - E_B) f_B(k)^2, \tag{2.4.48}$$

represents the "solution" to negative-energy off-shell unitarity.

The rank-one approximation to the off-shell amplitude, studied by Kowalski and Noyes (KN),

$$t_s(p, p')_{\mathrm{KN}} \equiv f_s(p, k) t_s(k, k) f_s(p', k), \tag{2.4.49}$$

which is obtained from the canonical form (2.4.42) by setting $R_s(p, p'; E) = 0$, has several interesting properties.[8] We see that $t_s(p, p')_{\mathrm{KN}}$ satisfies off-shell unitarity and time-reversal, contains the entire bound-state-pole content, and is exact for p or p' equal to k. It is important to note that *any* approximation whatsoever to $t_s(p, p')$ with these properties must have the general form (2.4.49). Off-shell unitarity relates the real and imaginary parts of the transition amplitude. This suggests that it may be possible to express $t_s(p, p')$ in terms of purely real auxiliary amplitudes. The earliest such technique was the so-called Heitler formalism, and the auxiliary amplitude was called the

$\hat{K}$-matrix.[9] There are now a host of such techniques.[10] We next describe a very general class of them.

Let s be an arbitrary parametric energy and let $\Gamma_s(p, p'; k)$ be a function of three momenta and s that is defined in terms of $t_s(p, p')$ by the generalized Heitler integral equation (*cf.* Section 1.8)

$$t_s(p, p') = \Gamma_s(p, p'; k) + \Gamma_s(p, k; k) \int_0^\infty dq\, g(q^2, s)\gamma(k, q) t_s(q, p'), \quad (2.4.50)$$

where

$$g(q^2, s) = (2/\pi) q^2 (s + i0 - q^2)^{-1}, \quad (2.4.51)$$

and $\gamma(k, q)$ is a function whose properties will be specified shortly. The integral equation (2.4.50) for t_s has a separable, or rank-one, kernel and so its solution can be reduced to quadrature. We see in fact that

$$t_s(p, p') = \Gamma_s(p, p'; k) + \Gamma_s(p, k; k) I_s(k, p'), \quad (2.4.52)$$

where

$$I_s(k, p') = \frac{\bar{I}_s(k, p')}{1 - \bar{I}_s(k, k)}, \quad (2.4.53)$$

$$\bar{I}_s(k, p') = \int_0^\infty dq\, g(q^2, s)\gamma(k, q)\Gamma_s(q, p'; k). \quad (2.4.54)$$

Therefore,

$$t_s(k, k) = \frac{\Gamma_s(k, k; k)}{1 - \bar{I}_s(k, k)}, \quad (2.4.55)$$

and

$$t_s(p, k) = \frac{\Gamma_s(p, k; k)}{\Gamma_s(k, k; k)} t_s(k, k). \quad (2.4.56)$$

Using the symmetry of $t_s(p, p')$ in p and p', we infer from Eqs. (2.4.52) and (2.4.56) that

$$I_s(k, p') = t_s(k, k) \frac{\Gamma_s(p', k; k)}{[\Gamma_s(k, k; k)]^2} - \frac{\Gamma_s(k, p'; k)}{\Gamma_s(k, k; k)}. \quad (2.4.57)$$

If we insert this expression back into Eq. (2.4.52), we obtain the canonical form (2.4.42), where, in addition to Eqs. (2.4.43–2.4.44), we also have now

$$f_s(p, k) = \frac{\Gamma_s(p, k; k)}{\Gamma_s(k, k; k)}, \quad (2.4.58)$$

and

$$R_s(p, p'; k) = \Gamma_s(p, p'; k) - \frac{\Gamma_s(p, k; k)\Gamma_s(k, p'; k)}{\Gamma_s(k, k; k)}. \tag{2.4.59}$$

The remarkable thing is that we obtain the *same* f_s and R_s for *all* of the different Γ_s. (Note that $\Gamma = t$ for $\gamma \equiv 0$.) We will call this universality *γ-invariance.*

Off-shell unitarity implies that $f_s(p, k)$, and therefore $R_s(p, p'; k)$, are real when $s > 0$, but we do not yet know anything about Γ_s except, from Eq. (2.4.58), that $\Gamma_s(p, k; k)$ and $\Gamma_s(k, k; k)$ have the same phase. Two general classes of the function $\gamma(k, q)$ are interesting.

In order to recover the so-called $\hat{K}$-matrix, we take a distribution-valued subtraction function $\gamma(k, q)_K$, with the properties[10]

$$(s - q^2)^{-1}\gamma(k, q)_K = -i\pi\,\delta(k^2 - q^2), \tag{2.4.60a}$$

when $s = k^2 + i0$, and

$$(s - q^2)^{-1}\gamma(k, q)_K = 0, \tag{2.4.60b}$$

for $s < 0$. Then, the Γ-function generated by γ_K is

$$\Gamma_s(p, p'; k)_K \equiv \hat{K}_s(p, p'), \tag{2.4.60c}$$

which is real and symmetric in p, p', and independent of k. Then, Eq. (2.4.50) is just the off-shell Heitler Equation[11]

$$t_s(p, p') = \hat{K}_s(p, p') - ik\hat{K}_s(p, k)\theta(s)t_s(k, p'). \tag{2.4.60d}$$

where $\theta(s) = 1$ for $s > 0$, and $\theta(s) = 0$ for $s < 0$. In the fully on-shell case, Eq. (2.4.60d) becomes

$$t_{k^2}(k, k) = \hat{K}_{k^2}(k, k) - ik\hat{K}_{k^2}(k, k)t_{k^2}(k, k), \tag{2.4.60e}$$

and so

$$\hat{K}_{k^2}(k, k) = -\frac{1}{k}\tan\delta_L(k). \tag{2.4.60f}$$

We see then that the $\hat{K}$-matrix will be singular at momenta k_r such that $\delta_L(k_r) = (2n + 1)\pi/2$, where n is an integer. These singularities can be shown to be poles,[11] and they correspond to *resonances*[12] in the scattering when $[d\delta_L(k)/dk]_{k_r} > 0$.

The fact that the $\hat{K}$-matrix has no discontinuity across the unitarity cut is one reason $\hat{K}_{k^2}(k, k)$ is analytic in k^2 in the neighborhood of the elastic threshold ($k^2 = 0$).[12] Thus, we can write, alternatively,

$$k^{2L+1}\cot\delta_L(k) = -\frac{1}{a_L} + \frac{1}{2}r_L k^2 + O(k^4), \tag{2.4.61a}$$

or[13]

$$\frac{1}{k}\tan\delta_L(k) = k^{2L}\left\{-a_L + \frac{1}{6}\hat{r}_L^3 k^2 + O(k^4)\right\}. \tag{2.4.61b}$$

The expansion coefficients in the so-called *effective-range expansion* (2.4.61a), *viz.*, a_L and r_L, are called the scattering length and the effective range, respectively, although only for $L = 0$ do these quantities have the dimension of length. It is clear that we must have $\hat{r}_L^3 = -3a_L^2 r_L$. The quantities r_L and $\hat{r}_L$ can be identified with the range of the interaction in the limits of large and small $|a_L|$, respectively. The expansion (2.4.61b) for $-\hat{K}_{k^2}(k,k)$ includes the proper threshold behavior.

We again remark that, by γ-invariance, f_s and R_s are given by Eqs. (2.4.58–2.4.59) with the substitution (2.4.60c). Then, since $\hat{K}_s(p,p')$ is symmetric, we have

$$\hat{K}_s(p,p') = f_s(p,k)\hat{K}_s(k,k)f_s(p',k) + R_s(p,p';k), \tag{2.4.62}$$

with the *same* off-shell functions given in Eqs. (2.4.43–2.4.44) and (2.4.58–2.4.59).[10, 14–16] We have recovered all of the essential features of the off-shell $\hat{K}$-matrix formalism without ever making use of any dynamical integral equations for either t_s or $\hat{K}_s$.

The other class of γ functions are real functions such that

$$\gamma(k,k) = 1. \tag{2.4.63}$$

We can say nothing more concerning the Γ_s-functions in this instance, except to remark that because now

$$\bar{I}_{k^2}(k,p') = \frac{2}{\pi}P\int_0^\infty dq\,\frac{q^2}{k^2-q^2}\gamma(k,q)\Gamma_{k^2}(q,p';k) - ik\Gamma_{k^2}(k,p';k) \tag{2.4.64}$$

for $s = k^2$, it follows from off-shell unitarity that

$$\Gamma_{k^2}(p,p';k)^* = \Gamma_{k^2}(p,p';k). \tag{2.4.65}$$

We show in Ch. 3 that the dynamical integral equation for Γ with a γ-function with the property (2.4.63) has a nonsingular kernel—a property useful for numerical calculation.

Since Eq. (2.4.43) is a special case of Eq. (2.4.58), we see that the half-shell function is singular at points k_0 where

$$t_s(k_0, k_0) = 0. \tag{2.4.66}$$

The identity (2.4.58) implies that $\Gamma_s(p, k_0; k_0)$ is also singular except in the accidental case where $\Gamma_s(k_0, k_0; k_0) = 0$. Now, $t_s(k_0, k_0)$ is nonsingular, so that $R_s(p, p'; k_0)$ must be singular; therefore, the residues of these poles of R must factorize in p and p' in order to cancel the singularities in the other term on the right-hand side of Eq. (2.4.42).[15]

The preceding argument would seem to suggest that if Eq. (2.4.66) is satisfied for k_0^i, $i = 1, \ldots, N$, then the lowest-rank generalization of $t_s(p, p')_{KN}$ that possesses the same characteristics as Eq. (2.4.49) would be of order $N + 1$.[17,18] This is not correct because of the fact that $f_s(p, k)$ clearly must always have the form

$$f_s(p, k) = B_s(p, k) + S_s(p, k), \tag{2.4.67}$$

where $B_s(p, k)$ is nonsingular and satisfies $B_s(k, k) = 1$, and $S_s(p, k)$ contains all of the possible singular structure at the points k_0^i. Then, it follows that the correct generalization of $t_s(p, p')_{\mathrm{KN}}$ is only of rank three for any number of zeros (2.4.66).[19,20] Model calculations[21] have indicated, however, that, except in the vicinity of the points k_0^i, these rank-three generalizations do not generally represent the exact off-shell amplitudes as well as $t_s(p, p')_{\mathrm{KN}}$, and therefore should be used only to interpolate over the points k_0^i.

The rank-one approximation to $t_s(p, p')$ that is obtained from Eq. (2.4.42) by neglecting $\Gamma_s(p, p'; k)$ in the expression (2.4.59) for $R_s(p, p'; k)$ clearly is finite at the zeros (2.4.66).[22] This does not contradict out previous remarks because, unlike a correct generalization of $t_s(p, p')_{\mathrm{KN}}$, this approximation is non-γ-invariant, violates both off-and on-shell unitarity, and is not exact either half-on-shell or on-shell. The non-γ-invariance, however, can be exploited to optimize the approximation, leading to reasonably good agreement with the exact amplitudes.[22]

By analogy with the factorization properties that occur in the neighborhood of the bound-state poles of $t_s(p, p')$, *cf.* Eq. (2.4.3), it has been suggested that the off-shell transition matrix should also factorize near energies $s_r = k_r^2$ corresponding to resonances.[11] Then, one can imagine proposing finite-rank approximations to $t_s(p, p')$ based upon bound-state and resonance-pole dominance. Clearly, $\hat{K}_s(p, p')$ factorizes near $s \simeq s_r$, as can be seen from Eq. (2.4.62), since $R_s(p, p'; k)$ generally is nonsingular there. However, the inference of the pole dominance of $t_s(p, p')$ in the case of a resonance rather than a bound state is more subtle.[15]

The essential fact is that as a function of *complex* s, the resonance poles of $t_s(p, p')$ occur at points $s = s_p$ on the second Riemann sheet that is joined to the physical sheet through the unitarity cut.[23] Assuming that $f_s(p, \sqrt{s})$ and $R_s(p, p'; \sqrt{s})$ are continuous through the cut and analytic in a domain that encloses s_r and s_p, we have then

$$t_s^{\text{II}}(p, p') = f_s(p, \sqrt{s})t_s^{\text{II}}(\sqrt{s}, \sqrt{s})f_s(p', \sqrt{s}) + R_s(p, p'; \sqrt{s}), \qquad (2.4.68)$$

where II refers to the second Riemann sheet. The essential point is that the residue of $t_s^{\text{II}}(\sqrt{s}, \sqrt{s})$ *must* decrease as s_p moves closer to the physical sheet, corresponding to a sharper resonance, in order to keep $t_s(\sqrt{s}, \sqrt{s})$ within the unitarity bound $1/\sqrt{s}$. Thus, as one passes onto the physical sheet, we see that the second-sheet pole term in (2.4.68) is damped by unitarity, while no such constraint applies to R_s. Hence, it is possible for R_s to be comparable to the resonance-pole term off-shell. This has been supported by model calculations.[15]

The net result is that the arguments[11] for resonance-pole dominance based on the off-shell $\hat{K}$-matrix formalism are inconsequential. This is because the infinity resulting from the pole in $\hat{K}_s(p, p')$ at $s = s_r$ is always damped while $R_{s_r}(p, p'; k)$ can be of arbitrary magnitude.

A simple approximation to $t_s(p, p')_{\text{KN}}$ was proposed by Lovelace[11] that presumably incorporates the pole-dominance arguments, namely

$$t_s(p, p')_{\text{KN}} \simeq [N(p)N(p')]^{1/2}D(k)^{-1}, \qquad (2.4.69)$$

where $k = \sqrt{s}$ for both positive and negative s, and where N and D are the functions appearing in the left-hand (N) and right-hand (D) cut-decomposition of $t_s(\sqrt{s}, \sqrt{s}) = N(\sqrt{s})/D(\sqrt{s})$. The zeros (2.4.66) at positive energies cause a different type of trouble with the representation (2.4.69) that corresponds to the choice $f_s(p, k) = [N(p)/N(k)]^{1/2}$, $k = \sqrt{s}$. The difficulty is that the zeros (2.4.66) necessarily require zeros in $N(k)$ and a possible nonremovable change in sign which, because of the square root in Eq. (2.4.69), leads to a violation of off-shell unitarity.

The generalized Heitler equation must be extended even further in order to encompass the same formalisms that have been proposed in the literature.[10,22] One possibility that does this, and is applied in Ch. 3, is

$$t_s(p, p') = \Gamma_s(p, p'; k, k_0) + \Gamma_s(p, k; k, k_0)\int dq \int dq'\, g(q^2, s)$$
$$\times\, \gamma(k, k_0, q, q')t_s(q', p'), \qquad (2.4.70)$$

where k and k_0 are two arbitrary momenta.

2.5. Partial-Wave Dynamical Equations

A great deal can be learned about two-body collisions from the unitarity and analyticity properties of the transition amplitudes. However, in order to calculate these amplitudes, one needs dynamical equations, such as (2.3.16) and (2.3.35) for $|\psi_{\mathbf{k}}^{(+)}\rangle$ and $t(z)$, respectively, that incorporate the interaction and the boundary conditions. In this section, we take the first step towards solving these equations, which is to reduce them to their partial-wave forms. We ignore spin and assume that the interaction potential is rotationally invariant.

In order to do this, we require expressions for V and $G_0^{(+)}(E)$ analogous to Eq. (2.4.31), for instance,

$$\langle \mathbf{p}|V|\mathbf{p}'\rangle = \frac{2}{\pi} \sum_{L,M} V_L(p,p') Y_L^M(\tilde{\mathbf{p}},\tilde{\zeta}) Y_L^M(\tilde{\mathbf{p}}',\tilde{\zeta})^*. \tag{2.5.1}$$

The partial-wave expansion (p.w.e.) of $\langle \mathbf{p}|G_0^{(+)}(E)|\mathbf{p}'\rangle$ is proportional to that for $\delta^{(3)}(\mathbf{p}-\mathbf{p}')$, *cf*. Eq. (2.3.10), namely

$$\delta^{(3)}(\mathbf{p}-\mathbf{p}') = \frac{\delta(p-p')}{p^2} \sum_{L,M} Y_L^M(\tilde{\mathbf{p}},\tilde{\zeta}) Y_L^M(\tilde{\mathbf{p}}',\tilde{\zeta})^*. \tag{2.5.2}$$

In coordinate space for a rotationally invariant local potential ($r = |\mathbf{x}|$),

$$\langle \mathbf{x}|V|\mathbf{x}'\rangle = V(r)\,\delta^{(3)}(\mathbf{x}-\mathbf{x}'), \tag{2.5.3}$$

we also have a p.w.e. equivalent to that for the Dirac δ-function. However, in general,

$$\langle \mathbf{x}|V|\mathbf{x}'\rangle = \sum_{L,M} U_L(r,r') Y_L^M(\tilde{\mathbf{x}},\tilde{\zeta}) Y_L^M(\tilde{\mathbf{x}}',\tilde{\zeta})^*, \tag{2.5.4}$$

where $U_L(r,r')$ is equal to $\delta(r-r')\,[V(r)/r^2]$ in the local limit.

The p.w.e. for the free Green function in coordinate space is

$$\langle \mathbf{x}|G_0^{(+)}(s)|\mathbf{x}'\rangle = \sum_{L,M} G_L^{(+)}(r,r';s) Y_L^M(\tilde{\mathbf{x}},\tilde{\zeta}) Y_L^M(\tilde{\mathbf{x}}',\tilde{\zeta})^*, \tag{2.5.5}$$

where ($\hbar/2\mu = 1$)

$$G_L^{(+)}(r,r';s) = -ikj_L(kr_<)h_L^{(1)}(kr_>) \tag{2.5.6a}$$

for $s = k^2 + i0$ and

$$G_L^{(+)}(r,r';s) = kj_L(ikr_<)h_L^{(1)}(ikr_>), \tag{2.5.6b}$$

for $s = -k^2$. Here $<$ ($>$) denotes the lesser (greater) of the two variables, and $j_L, h_L^{(1)}$ are the usual spherical Bessel functions.

We note that, for $0 \le r, r' < \infty$,

$$|G_L^{(+)}(r, r'; s)|^2 < \frac{\pi}{rr'(2L+1)} \tag{2.5.7}$$

for both positive and negative energies.[1,2] This bound can be used to find large-L bounds on the partial-wave amplitudes $t_L(p, p'; s)$, namely[2]

$$|t_L(p, p'; k^2)| < F_L(p)F_L(p'), \tag{2.5.8a}$$

$$|t_L(p, p'; -k^2)| < \frac{F_L(p)F_L(p')}{(1 - C_L A)} + \sum_B \frac{|f_B^L(p) f_B^L(p')|^2}{|E_B + k^2|}, \tag{2.5.8b}$$

for positive and negative energies, respectively. Here,

$$F_L(p) = (1 - C_L A)^{-1} \int_0^\infty dr\, r^2 [j_L(pr)]^2 |V(r)|, \tag{2.5.8c}$$

for a local potential and L is large enough such that $C_L A < 1$, where $C_L = [\pi/(2L+1)]^{1/2}$ and

$$A = \int_0^\infty dr\, r|V(r)|. \tag{2.5.8d}$$

In Eq. (2.5.8b), $f_B^L(p)$ is a bound-state vertex function. For large enough L, there will be no bound states and $|t_L(p, p'; s)|$ is bounded by a factorizable, s-independent, rapidly-decreasing function of L. This is sufficient to justify the use of a cut-off in L when considering two-particle collisions in a multi-particle setting.

A generalized form of Eq. (2.3.16), which is sometimes referred to as a Bethe–Goldstone[3] or off-shell[2,4] wave function, is often useful:

$$|\psi_{\mathbf{q}}^{(+)}(s)\rangle = |\mathbf{q}\rangle + G_0^{(+)}(s) V |\psi_{\mathbf{q}}^{(+)}(s)\rangle, \tag{2.5.9}$$

where s and q^2 are not necessarily related. If we define the p.w.e. components of the wave function $\langle \mathbf{x}|\psi_{\mathbf{q}}^{(+)}(s)\rangle$ by

$$\langle \mathbf{x}|\psi_{\mathbf{q}}^{(+)}(s)\rangle = \frac{4\pi}{(2\pi)^{3/2}} \sum_{L,M} i^L \psi_L^{(+)}(r, q; s) Y_L^M(\tilde{\mathbf{x}}, \tilde{\zeta}) Y_L^M(\tilde{\mathbf{q}}, \tilde{\zeta})^*, \tag{2.5.10}$$

then we obtain, from Eq. (2.5.9), in the case of a local potential,

$$\psi_L^{(+)}(r, q; s) = j_L(qr) + \int_0^\infty dr'\, r'^2 G_L^{(+)}(r, r'; s) V(r') \psi_L(r', q; s). \tag{2.5.11}$$

For a nonlocal potential, another radial integration enters into Eq. (2.5.11); this complication is not significant in what follows.

Using Eqs. (2.4.31) and (2.5.1), we obtain the partial-wave forms of the LS integral equations (2.3.35a, b) for $z = s + i0$:

$$t_s(p, p') = V(p, p') + \frac{2}{\pi} \int_0^\infty \frac{dq\, q^2}{s - q^2 + i0} V(p, q) t_s(q, p'), \tag{2.5.12a}$$

$$t_s(p, p') = V(p, p') + \frac{2}{\pi} \int_0^\infty \frac{dq\, q^2}{s - q^2 + i0} t_s(p, q) V(q, p'). \tag{2.5.12b}$$

Here

$$t_s(p, p') \equiv t_L(p, p'; s + i0) \tag{2.5.13}$$

and

$$V(p, p') \equiv V_L(p, p'), \tag{2.5.14}$$

where in both instances we have suppressed the L-dependence.

At this point, one could use Eqs. (2.5.12a, b) to derive dynamical integral equations for all of the auxiliary functions, $\Gamma_s(p, p'; k)$, $f_s(p, k)$, $R_s(p, p'; k)$, and $\hat{K}_s(p, p')$, introduced in Section 2.4. We have deferred doing this until Ch. 3 since these equations are so closely coupled to particular solution strategies in two-body collision theory.

The connection between the partial-wave forms of $\langle \mathbf{x} | \psi_{\mathbf{p}'}^{(+)}(s) \rangle$ and $t_s(p, p')$ is often of interest. Since

$$t_s(\mathbf{p}, \mathbf{p}') \equiv \langle \mathbf{p} | V | \psi_{\mathbf{p}'}^{(+)}(s) \rangle = \frac{1}{(2\pi)^{3/2}} \int d^3x\, e^{-i\mathbf{p}\cdot\mathbf{x}} V(r) \langle \mathbf{x} | \psi_{\mathbf{p}'}^{(+)}(s) \rangle, \tag{2.5.15}$$

we obtain

$$t_s(p, p') = \int_0^\infty dr\, r^2 j_L(pr) V(r) \psi_L^{(+)}(r, p'; s). \tag{2.5.16}$$

In the special case $p = p' = k = \sqrt{s}$, this relation reduces to

$$-\frac{1}{k} \sin \delta_L(k)\, e^{i\delta_L(k)} = \int_0^\infty dr\, r^2 j_L(kr) V(r) \psi_L^{(+)}(r, k; k^2). \tag{2.5.17}$$

Using Eqs. (2.5.11) and (2.5.17) and the asymptotic forms of $j_L(kr)$ and $h_L^{(1)}(kr)$ as $kr \to \infty$, we can determine the behavior of the wave function for large r:

$$\psi_L^{(+)}(r, k; k^2) \sim \frac{1}{kr} e^{i\delta_L(k)} \sin\left[kr - \frac{L\pi}{2} + \delta_L(k) \right]. \tag{2.5.18}$$

This expression provides the rationale for the nomenclature *phase shift* for the real parameter $\delta_L(k)$. For $r \to 0$, $\psi_L^{(+)}(r, k; k^2)$ behaves like $j_L(kr)$.

Since $G_L^{(+)}(r, r'; k^2)$ satisfies

$$D_L G_L^{(+)}(r, r'; k^2) = \frac{\delta(r - r')}{r^2}, \tag{2.5.19}$$

where

$$D_L = \frac{1}{r^2}\frac{d}{dr}\left(r^2\frac{d}{dr}\right) + \left[k^2 - \frac{L(L+1)}{r^2}\right], \tag{2.5.20}$$

we obtain, after operating on Eq. (2.5.11) in the case $q = k = \sqrt{s}$, the differential equation

$$D_L\psi_L^{(+)}(r, k; k^2) = V(r)\psi_L^{(+)}(r, k; k^2). \tag{2.5.21}$$

This equation, coupled with the asymptotic condition (2.5.18) and regularity at the origin, can be used as a starting point for the discussion of potential scattering.[5]

2.6. Multichannel Collisions

Collisions between two bodies, which we call A and B, generally produce a multiplicity of final states. Many times only the two-cluster final states, corresponding to the possibilities

$$\mathrm{A} + \mathrm{B} \to \mathrm{A} + \mathrm{B}, \tag{2.6.1a}$$

$$\to \mathrm{A}^* + \mathrm{B}^*, \tag{2.6.1b}$$

$$\to \mathrm{C} + \mathrm{D}, \tag{2.6.1c}$$

of elastic (including spin-flip transitions), inelastic internal excitation, and rearrangement collisions, respectively, are of interest. In this section, we consider collision integral equations that describe the processes (2.6.1a–c) without any explicit treatment of any other inelastic possibilities.[1]

The inelastic processes (2.6.1b) and (2.6.1c) occur only if at least one of A or B and at least one of C or D are composite objects. Therefore, not only do the dynamics underlying these processes involve more than two particles, but there is always the possibility of final states with more than two clusters given enough initial kinetic energy. We want to describe the two-body multichannel dynamical equations that are obtained when the basic multiparticle dynamics is folded into effective two-cluster (but not two-particle) interactions. Examples of how this is done appear frequently in Chs. 5–9.

For the present, we just consider the following denumerable (finite or infinite) set of coupled integral equations for the off-shell transition amplitudes $T_s^{\beta\alpha}(p_\beta, p'_\alpha)$:

$$T_s^{\beta\alpha}(p_\beta, p'_\alpha) = V_s^{\beta\alpha}(p_\beta, p'_\alpha) + \sum_\gamma \int_0^\infty dq_\gamma\, V_s^{\beta\gamma}(p_\beta, q_\gamma) g_\gamma(q_\gamma^2, s) T_s^{\gamma\alpha}(q_\gamma, p'_\alpha), \quad (2.6.2)$$

where

$$g_\gamma(q_\gamma^2, s) = \frac{2}{\pi} q_\gamma^2 [s - q_\gamma^2/2m_\gamma + h_\gamma]^{-1}. \quad (2.6.3)$$

These integral equations refer to two-cluster multichannel collisions in the c.m. frame in a fixed state of *total* angular momentum.

The quantities p'_α, p_β, and q_γ represent the channel relative momenta. The channel Green functions (2.6.3) are defined in terms of the channel reduced masses m_γ and the real number $h_\gamma \geq 0$, which are the total binding energies of the clusters.

A special case of the set (2.6.2) occurs for the elastic scattering of two particles, where at least one particle has spin, such as in nucleon–nucleon or pion–nucleon scattering. In this instance, α, β, and γ represent spin–orbit coupling indices. In general, however, these indices refer to all relevant two-cluster discrete channel indices, including the orbital angular momentum.

The on-shell values of the channel momenta are denoted by k_β, e.g., and are defined by

$$k_\beta \equiv [2m_\beta |s + h_\beta|]^{1/2}, \quad (2.6.4)$$

whether or not the channel is actually open. This last possibility is realized only if

$$s + h_\alpha \geq 0 \qquad \text{(Open Channel)}, \quad (2.6.5)$$

where we recall that s is real to within $i0$. Thus, an open-channel amplitude $T_s^{\beta\alpha}(p_\beta, p'_\alpha)$ is on-shell if $p_\beta = k_\beta$, $p'_\alpha = k_\alpha$, and off-shell otherwise; a closed-channel amplitude is always off-shell.

Multichannel problems are very often expressed as a set of coupled differential equations for wave-function components, rather than the set (2.6.2).[2] The main reason for this is that the truncations of the multiparticle dynamics that are needed to obtain a two-cluster model are frequently formulated in wave-function language. Evidently, the content of such multichannel wave-function equations can always be re-expressed in transition-matrix form, and conversely. The wave-function components are related to the amplitudes $T_s^{\beta\alpha}(p_\beta, p'_\alpha)$ by potential-weighted Fourier–Bessel transforms just as in the single-channel case.

Unitarity questions cannot be addressed unless the singularity structures of the effective-interaction potentials $V_s^{\beta\alpha}(p_\beta, p'_\alpha)$ as a function of s are known. There may be no s-dependence, as for collisions below the inelastic threshold. In such cases, the coupling is due solely to spin–orbit effects, and h_γ is the same for all γ. In contrast, when the energy is above some inelastic threshold, the potentials $V_s^{\beta\alpha}(p_\beta, p'_\alpha)$ are generally complex and discontinuous across the real s-axis, and Eq. (1.9.23) can be used to discuss unitarity.

We can rewrite Eq. (2.6.2) in a matrix notation with respect to the channel indices:

$$T_s(p, p') = V_s(p, p') + \int_0^\infty dq\, V_s(p, q) g(q^2, s) T_s(q, p'), \tag{2.6.6a}$$

where $T_s(p, p')$, for example, is a matrix with elements $T_s^{\beta\alpha}(p_\beta, p'_\alpha)$, and $g(q^2, s)$ has elements $g_\gamma(q_\gamma^2, s)\,\delta_{\gamma\sigma}$. Similarly, we have

$$T_s(p, p') = V_s(p, p') + \int_0^\infty dq\, T_s(p, q) g(q^2, s) V_s(q, p'). \tag{2.6.6b}$$

Using Eq. (1.9.23), we obtain from Eqs. (2.6.6a, b) the discontinuity relation

$$\Delta T_s(p, p') = \int_0^\infty dq\, T_\pm(p, q)\,\Delta g(q^2, s)\, T_\mp(q, p') + [\Delta T]_V, \tag{2.6.7}$$

where $[\Delta T]_V$ represents the contribution from any possible discontinuities across the unitarity cut of the potentials arising from inelasticity,

$$T_\pm(p, p') \equiv [T_s(p, p')]_{s=E\pm i0}, \tag{2.6.8a}$$

and we use the notation

$$\Delta T_s \equiv T_+ - T_-. \tag{2.6.8b}$$

The matrix Δg has its diagonal elements proportional to Dirac δ-functions, so that Eq. (2.6.7) reduces to

$$\Delta T_s(p, p') = -2iT_\pm(p, k)\,\Delta(E, k)\, T_\mp(k, p') + [\Delta T]_V, \tag{2.6.9}$$

where

$$[\Delta(E, k)]_{\gamma\sigma} = 2m_\gamma k_\gamma \theta(E + h_\gamma)\,\delta_{\gamma\sigma}.$$

Operator analyticity of the transition operators with respect to the complex parametric energy implies the Hermitian-analyticity condition

$$[T_+^{\beta\alpha}(p_\beta, p'_\alpha)]^* = T_-^{\alpha\beta}(p'_\alpha, p_\beta), \tag{2.6.10}$$

while time-reversal invariance implies

$$T_+^{\beta\alpha}(p_\beta, p'_\alpha) = T_+^{\alpha\beta}(p'_\alpha, p_\beta). \tag{2.6.11}$$

Using Eqs. (2.6.10) and (2.6.11) we can rewrite the discontinuity relation (2.6.9) as

$$\mathrm{Im}[T_+(p, p')] = -T_+(p, k)\,\Delta(E, k)\,[T_+(k, p')]^* + \frac{1}{2i}[\Delta T]_V, \tag{2.6.12}$$

In the case $[\Delta T]_V = 0$, which means no inelastic channels are open except the two-cluster ones we have accounted for explicitly, Eq. (2.6.12) has implications that are very similar to the single-channel case. In particular, on-shell we have

$$\mathrm{Im}[T_+(k, k)] = -T_+(k, k)\,\Delta(E, k)\,[T_+(k, k)]^*, \tag{2.6.13}$$

where $T_+(k, k)$ is a symmetric matrix whose open-channel submatrix consists of the on-shell multichannel scattering amplitudes. The S-matrix

$$S(k) \equiv I - 2i\,\Delta(E, k)^{1/2}\,T_+(k, k)\,\Delta(E, k)^{1/2} \tag{2.6.14}$$

is unitary and symmetric, and so can be diagonalized by a real orthogonal matrix; therefore, we can introduce eigenphase shifts corresponding to each of the channels.

Also, in the case where $[\Delta T]_V = 0$, we see from Eqs. (2.6.12) and (2.6.13) that

$$\mathrm{Im}[T_+(p, k)] = T_+(p, k)[T_+(k, k)]^{-1}\,\mathrm{Im}[T_+(k, k)]. \tag{2.6.15}$$

This implies that the multichannel generalization of the half-shell function is given by the real matrix

$$f_s(p, k) \equiv T_+(p, k)[T_+(k, k)]^{-1}, \tag{2.6.16a}$$

where it is clear that

$$f_s(p, k) = \mathrm{Im}\ T_+(p, k)\,[\mathrm{Im}\ T_+(k, k)]^{-1} \tag{2.6.16b}$$

$$= \mathrm{Re}\ T_+(p, k)\,[\mathrm{Re}\ T_+(k, k)]^{-1}. \tag{2.6.16c}$$

We again are led to the general structure for $T_+(p, p')$:

$$T_+(p, p') = f_s(p, k)T_+(k, k)\tilde{f}_s(p', k) + R_s(p, p'; k), \tag{2.6.17a}$$

where

$$[\tilde{f}_s(p', k)]_{\alpha\beta} \equiv [f_s(p', k)]_{\beta\alpha}, \tag{2.6.17b}$$

and the real matrix $R_s(p, p'; k)$ satisfies

$$R_s(k, p'; k) = R_s(p, k; k) = 0, \tag{2.6.17c}$$

and the symmetry condition

$$R_s(p, p'; k) = \tilde{R}_s(p', p; k), \tag{2.6.17d}$$

where $\tilde{R}_s$ is the transpose of R_s with respect to the discrete channel indices as in Eq. (2.6.17b).

The preceding matrix notation is deceptively simple because it involves the internal momentum arguments in a nontrivial manner.[3–5] For example, the symmetric matrix

$$(T_+^{-1})^{\alpha\beta}(k_\alpha, k_\beta) \equiv \{[T_+(k, k)]^{-1}\}_{\alpha\beta} \tag{2.6.18a}$$

satisfies

$$\sum_\gamma T_+^{\alpha\gamma}(k_\alpha, k_\gamma)(T_+^{-1})^{\gamma\beta}(k_\gamma, k_\beta) = \delta_{\alpha\beta}, \tag{2.6.18b}$$

$$\sum_\gamma (T_+^{-1})^{\alpha\gamma}(k_\alpha, k_\gamma) T_+^{\gamma\beta}(k_\gamma, k_\beta) = \delta_{\alpha\beta}. \tag{2.6.18c}$$

Thus, we see that

$$[f_s(p, k)]_{\alpha\beta} = \sum_\gamma T_+^{\alpha\gamma}(p_\alpha, k_\gamma)(T_+^{-1})^{\gamma\beta}(k_\gamma, k_\beta) = \sum_\gamma (T_+^{-1})^{\beta\gamma}(k_\beta, k_\gamma) T_+^{\gamma\alpha}(k_\gamma, p_\alpha), \tag{2.6.19a}$$

and

$$[R_s(p, p'; k)]_{\alpha\beta} = T_+^{\alpha\beta}(p_\alpha, p'_\beta) - \sum_{\gamma, \lambda} [f_s(p, k)]_{\alpha\gamma} T_+^{\gamma\lambda}(k_\gamma, k_\lambda)[f_s(p', k)]_{\beta\lambda}. \tag{2.6.19b}$$

Also because of Eqs. (2.6.10) and (2.6.11),

$$f_s(p, k) = T_-(p, k)[T_-(k, k)]^{-1}, \tag{2.6.20}$$

so that

$$f_s(p, k) = T_s(p, k)[T_s(k, k)]^{-1} \tag{2.6.21a}$$

and

$$\tilde{f}_s(p, k) = [T_s(k, k)]^{-1} T_s(k, p), \tag{2.6.21b}$$

for $s = E \pm i0$. Similarly,

$$R_s(p, p'; k) = T_s(p, p') - T_s(p, k)[T_s(k, k)]^{-1} T_s(k, p'). \tag{2.6.22}$$

As indicated in Eq. (2.6.17d), $R_s(p, p'; k)$ has the same symmetry properties as the transition matrix (*cf*. Eq. (2.6.11)).

On the basis of the preceding analysis of off-shell unitarity, we could, as in the single-channel case, at this point introduce the generalized Heitler equation analysis by means of a multichannel counterpart of Eq. (2.4.50) involving multichannel auxiliary amplitudes $\Gamma_s^{\beta\alpha}(p_\beta, p'_\alpha; k_\alpha)$ and functions $\gamma_{\beta\alpha}(k_\beta, q_\alpha)$. There is a greater freedom in choosing the γ-functions in the multichannel case depending on the properties of the Γ_s-functions one desires. For example, for counterparts of the second class of γ-functions (*cf*. Eq. (2.4.63)) one might require $\gamma_{\beta\alpha}(k_\beta, k_\alpha) = \delta_{\beta\alpha}$, but generally $\gamma_{\beta\alpha} \neq 0$ for $\beta \neq \alpha$,[3] or simply[4,5] $\gamma_{\beta\alpha}(k_\beta, q_\alpha) = \delta_{\beta\alpha}\gamma_\alpha(k_\alpha, q_\alpha)$, where $\gamma_\alpha(k_\alpha, k_\alpha) = 1$. A multichannel $\hat{K}$-matrix can be defined in a straightforward way[6] using distribution-valued γ-functions in Eqs. (2.4.60a–f). We can also obtain the generalizations of the γ-invariant representations (2.4.58) and (2.4.59) in a similar manner since $f_s(p, k)$ and $R_s(p, p'; k)$ are universal functions.

2.7. Relativistic Integral Equations

Surprisingly little is actually known about the physical validity of the modeling implicit in the LS description of two-body collisions. The most obvious shortcoming is that this model is nonrelativistic. A closely related problem is the representation of the interaction by a static potential.

The situation is most uncertain in hadronic and nuclear physics, where this modeling is frequently used. Although "reasonable" potentials yield "reasonable" scattering amplitudes via the LS equation, there are elements of circular reasoning in these inferences.

Calculations of atomic processes involving only Coulombic interactions among charged particles are more secure in utilizing the potential concept. On the other hand, the LS equation can only accommodate screened long-range forces. Moreover, when higher accuracies are required, modifications of the simple potential model of the interactions are needed.

In this section, we summarize some of the proposals for the replacement of the LS equations by equations with a similar structure, but a wider range of physical applicability.

The physical basis for going beyond potential scattering is secure for atomic processes where most phenomena are accurately described within QED. In contrast to QCD in the hadronic and nuclear cases, QED is a weakly coupled, nonconfining field theory. Therefore, it is relatively well understood how to calculate physical quantities using QED. For example, two-charged-particle scattering can be described by the Bethe–Salpeter (BS) equation,[1–4] which has the same generic structure as the LS equation, viz.,

$$T = V + VG_0T, \tag{2.7.1}$$

but where the meaning and structure of the potential and Green function are quite different than for simple potential scattering, except for the fact that V is free of any elastic unitarity cuts, while G_0 only has such cuts. All inelastic effects reside in V.

As one would expect, the transition matrix T in Eq. (2.7.1) is a function of the four momenta of each of the initial- and final-state particles. It is also a function of the relativistic (four-component) spin structure of these particles. Virtual processes are now depicted as occurring off the *mass*-shell ($p_0^2 - \mathbf{p}^2 \neq m^2$) rather than off the energy-shell.

It is still unclear on how to progress from QCD to practical collision integral equations for hadronic and nuclear processes. Therefore, most efforts to this end have employed a mix of some type of meson–baryon field theory with phenomenology in order to obtain equations of the form (2.7.1) that have a wider range of validity than the nonrelativistic LS equations.[5]

The greatest activity in this regard has been in two-body hadron–hadron collisions. The derivations of these integral equations vary in their closeness to reality. If Eq. (2.7.1) by itself suffices to define T uniquely, then the "potential" is necessarily an approximation. For a typical nonconfining model field theory, Eq. (2.7.1) is but one of an infinite hierarchy of equations within which V is itself a quantity to be determined. However, even with the drastic truncations in this hierarchy that are needed to obtain a definitive, approximate V, the BS integral equation is still four-dimensional, and after a partial-wave analysis, two-dimensional. Also, both V and G_0 possess complicated singularities in the relevant momentum variables. Nevertheless, there have been a number of detailed calculations based on what can be referred to as the "standard" BS form of Eq. (2.7.1). (By "standard," we mean that only V has been approximated by the lowest-order Feynman graphs, i.e., the so-called ladder approximation.)

In the remainder of this section, we are interested only in the class of three-dimensional relativistic two-body collision integral equations that result from Eq. (2.7.1) by approximating G_0. These integral equations are formally identical to the nonrelativistic LS equation.[6] Then the results of the present chapter, as well as those of the next, can be applied with only minor modifications. For the sake of simplicity, we confine ourselves here only to the case of two equal-mass particles. Some results for the unequal-mass case are quoted in Section 7.6.

The strategy of this reduction is that of a generalized Heitler-type subtraction technique. If, following the terminology of Section 1.8, we call $K_R = VG_0 - K_S$, then K_S is chosen such that the "Heitler" equation

$$T = \gamma_R V + \gamma_R K_S T \tag{2.7.2}$$

involves only a three-dimensional integration in its kernel. Here, $\gamma_R = I + K_R \gamma_R$. Most often[6] K_S has the form Vg, where g is a three-dimensional Green function, so that

$$T = T_R + T_R g T, \tag{2.7.3a}$$

with

$$T_R = V + K_R T_R. \tag{2.7.3b}$$

Usually, the approximation $T_R = V$ is made, although higher-order iterations have been considered.[7]

For the scattering of two spinless particles of mass m the BS equation (2.7.1) for the invariant transition matrix $T(p_1, p_2 | p'_1, p'_2)$ is

$$T(p_1, p_2 | p'_1, p'_2) = V(p_1, p_2 | p'_1, p'_2) + \int d^4q\, V(p_1, p_2 | \tfrac{1}{2}P + q, \tfrac{1}{2}P - q) G_0(P, q).$$
$$\times\, T(\tfrac{1}{2}P + q, \tfrac{1}{2}P - q | p'_1, p'_2), \tag{2.7.4}$$

where

$$G_0(P, q) = -\{2\pi[(\tfrac{1}{2}P + q)^2 - m^2 + i0][(\tfrac{1}{2}P - q)^2 - m^2 + i0]\}^{-1}. \tag{2.7.5}$$

Here p_i, p'_i denote the four momenta of the particles, $P \equiv p_1 + p_2 = p'_1 + p'_2$, and we use the metric convention $p^2 = p_0^2 - \mathbf{p}^2$. We note that, in contrast to the nonrelativistic LS equation, the kernel of Eq. (2.7.4) has double fixed-point pole singularities arising from the Green function (2.7.5).

A class of the approximate Green functions g [*cf.* Eqs. (2.7.3a) and (2.7.5)] is given by[8]

$$g(P, q) = \int_{4m^2}^{\infty} \frac{ds'}{w - s' + i\varepsilon} F(s', w)$$
$$\times\, \delta^{(+)}\left[\left(\frac{1}{2}P' + q\right)^2 - m^2\right] \delta^{(+)}\left[\left(\frac{1}{2}P' - q\right)^2 - m^2\right], \tag{2.7.6}$$

where w is the square of the total energy (P_0^2). Also, $P' \equiv (s'/w)^{1/2} P$, and the notation $\delta^{(+)}$ means that only the positive square root of the δ-function is used. The propagator (2.7.6) possesses the correct elastic unitarity cut provided $F(w, w) = 1$. Then, K_R is continuous across the elastic cut, and the solution of Eq. (2.7.3a) will satisfy elastic unitarity exactly for energies below the inelastic threshold, or for all energies if V is approximated by a potential with no inelasticity cuts. The choices

$$F(s, w)_{\text{BbS}} = 1, \tag{2.7.7a}$$

$$F(s, w)_{\text{T}} = \frac{w^{1/2} + s^{1/2}}{2w^{1/2}} \tag{2.7.7b}$$

yield the equations of Blankenbecler and Sugar (BbS)[9] and Thompson (T),[10] respectively. These equations are three-dimensional, as follows from the fact that in the two-body c.m. frame the integral (2.7.6) reduces to

$$g(P, q)_{\text{c.m.}} = \frac{\delta(q_0)\, F(4\omega_q^2, w)}{\omega_q(w - 4\omega_q^2 + i0)}, \tag{2.7.8}$$

where $\omega_q = (\mathbf{q}^2 + m^2)^{1/2}$. The three-dimensional equations that result from using the expression (2.7.8) in Eq. (2.7.3a) (*cf.* Eq. (2.7.4)) are of exactly the same form as the nonrelativistic LS equations except for relativistic kinematical factors, viz.,

$$\langle \mathbf{p}|T(w)|\mathbf{p}'\rangle = \langle \mathbf{p}|V|\mathbf{p}'\rangle + \int d^3q \frac{\langle \mathbf{p}|V|\mathbf{q}\rangle F(4\omega_q^2, w)}{\omega_q(w - 4\omega_q^2 + i0)} \langle \mathbf{q}|T(w)|\mathbf{p}'\rangle. \tag{2.7.9}$$

Another class of elastic-unitarity-preserving Green functions, $g(P, q)$, is given by Eq. (2.7.6) but with $[\frac{1}{2}(P + q)^2 - m^2]$ and $[(P' - \frac{1}{2}P - k)^2 - m^2]$ as the arguments of the two $\delta^{(+)}$ functions. In this instance,

$$g(P, q)_{\text{c.m.}} = \delta(q_0 + \tfrac{1}{2}\sqrt{w} - \omega_q)\frac{F(4\omega_q^2, w)}{\omega_q(w - 4\omega_q^2 + i0)}, \tag{2.7.10}$$

and the choices (2.7.7a) and (2.7.7b) for $F(s, w)$ yield the three-dimensional equations of Erkelenz and Holinde[11] and Gross,[12] respectively, while Kadychevsky's[13] equation is obtained with the choice $F(s, w)_{\text{K}} = F(w, s)_{\text{T}}$.

The preceding can be extended to the case where one or both of the particles are spin-one-half fermions. A case of considerable interest is that of nucleon–nucleon scattering, which has been investigated using both the three- and four-dimensional forms of the BS equation.[14]

Calculations of scattering of 200–800 MeV nucleons from spin-zero nuclei have been carried out using three-dimensional relativistic BS-type equations.[15] This forms a part of what is referred to as the Dirac phenomenology for nuclear scattering.[16] For this type of application, the derivation and interpretation of the collision integral equations is far less justifiable on fundamental grounds than is the case for hadron–hadron collisions.[5]

References and Notes (2)

Section 2.1

1. The monographs of Joachain (1975) and Newton (1982) contain extensive references as well as comprehensive developments of potential scattering, and should be consulted for the details we omit. Other treatments from various other points of view include Amrein (1977), Goldberger (1964), Rodberg (1967), Reed (1979), and Taylor (1987).
2. Regge (1959).
3. The discovery of Regge poles led to a surge of interest in potential scattering. It also revolutionized the phenomenology of hadronic reactions at high energies, Collins (1977), and initiated a line of developments leading to fundamental string theories, Green (1987).
4. See Amrein (1977), Reed (1979), and Sandhas (1966) for detailed discussions of these issues.
5. A thorough review of work up to 1975 on obtaining finite-rank representations of two-particle off-shell scattering is contained in Srivastava (1975). See Plessas (1986a, b) for recent developments.
6. Off-shell nucleon–nucleon scattering amplitudes were first of interest in calculating corrections to the simple impulse-approximation in nucleon–deuteron elastic scattering, Fulton (1959), Kowalski (1961, 1963a, b). Two recent papers on calculations of nucleon–nucleon off-shell amplitudes that also contain extensive references to previous work are Amos (1988), and Redish (1987a).
7. Several commonly used nucleon–nucleon potentials are contained in Reid (1968), de Tourreil (1975), Lacombe (1980), and Machleidt (1987).
8. An extensive literature on off-shell amplitudes with singular interactions is contained in Srivastava (1975). Representative works are van Leeuwen (1961), Kowalski (1963a), Fuda (1967), Laughlin (1968), Kim (1970), Brayshaw (1971), and Bishop (1973).
9. Brayshaw (1976, 1977).
10. For a recent review, see Barrett (1983).

Section 2.2

1. For explicitly time-dependent treatments, see Bjorken (1964), Holz (1988), and Sweeny (1989).
2. Calculations of this kind are worked out in great detail in Goldberger (1964).

Section 2.3

1. See Ref. 1 of Section 2.1 for references.
2. See also Reed (1979), p. 98.
3. Prugovecki (1971).
4. Gerjuoy (1958).
5. Adhikari (1983b).
6. Messiah (1965), p. 871.

Section 2.4

1. Sometimes the in- and out-states are introduced with equal weights to ensure time-reversal invariance and a truncation of the unitary sum that reflects the fact that the scattering operator commutes with its adjoint. See Goldberger (1960), Kowalski (1965a), and Jackson (1965).
2. This was first pointed out by Noyes (1967, 1969) and later exploited in Baranger (1969) in the case without bound states.
3. See, e.g., Baranger (1969) and Haftel (1970b). The arbitrariness in the choice of bound-state wave functions and the values of the binding energies is established in Kowalski (1971a), which also contains further references.
4. Off-shell unitarity in potential scattering is used to try to justify finite-rank approximations to the transition matrix in Lovelace (1964a). Extensive sets of references concerning the consequences of off-shell unitarity below the inelastic threshold are contained in Wellner (1967) and Bollé (1970) for both potential scattering and relativistic field theory.
5. Bounds on the off-shell partial-wave amplitudes are established in Kowalski (1967).
6. For the consequences of off-shell unitarity and time-reversal symmetry, see Kowalski (1966a, 1973c, 1975), Sobel (1967), Mongan (1969), Bollé (1970), and Sauer (1971). These results should not be confused with Watson's theorem (Watson (1954), Gillespie (1964), Ball (1965), and Kowalski (1965a)) for on-shell multichannel amplitudes.
7. See Srivastava (1975). The first appearance seems to be in Kowalski (1963a, b) for positive energies, and Kowalski (1966b) for negative energies. The independent development by Noyes (1965) emphasized the application of the factorizability properties. The connection to off-shell unitarity is pointed out in Kowalski (1965b).
8. See Noyes (1965) for $s > 0$ and Kowalski (1966b) for $s < 0$.
9. The $\hat{K}$-matrix is alternatively called the reaction or reactance matrix, and is usually considered only on-shell (Taylor (1967)). The off-shell Heitler formalism was introduced by Lovelace (1964a). A review of the properties of the off-shell $\hat{K}$-matrix is given in Srivastava (1975).

10. Adhikari (1989).
11. Lovelace (1964a).
12. Newton (1982).
13. Adhikari (1983c).
14. Kowalski (1961, 1963b).
15. Bollé (1970).
16. Kowalski (1975).
17. Fuda (1969, 1971).
18. Osborn (1969).
19. Kowalski (1972c).
20. Kowalski (1973c).
21. Picker (1975).
22. Stoof (1988).
23. For a discussion of the relationship of the second-sheet poles of $t_s(p, p')$, as a function of s, to the factorization properties of these poles and the analytic properties of $f_s(p, \sqrt{s})$ and $R_s(p, p'; \sqrt{s})$ for certain classes of interactions, see Mennessier (1972), Lovelace (1964a, b), Guennegues (1966), Noyes (1965), Nuttall (1967), and Osborn (1969).

Section 2.5

1. Martin (1962).
2. Kowalski (1967).
3. Bethe (1957, 1963).
4. Srivastava (1975).
5. Messiah (1965).

Section 2.6

1. The treatment in this section is based mainly on that in Kowalski (1968, 1972c, 1975), Adhikari (1979a), and Tomio (1980a). For elastic scattering with spin, see Kowalski (1961, 1963b), Sobel (1967), Mongan (1969), Sauer (1971), and Glöckle (1983a).
2. Satchler (1980, 1983), Austern (1970), Jackson (1970). See also Ch. 9.
3. Kowalski (1972c).
4. Adhikari (1979a).
5. Tomio (1980a, b).
6. See Newton (1982) for the conventional treatment of the multichannel $\hat{K}$-matrix.

Section 2.7

1. Salpeter (1951).
2. See also Nambu (1950), Gell-Mann (1951), Schwinger (1951).
3. We are only interested in the form of the BS equation for the transition matrix in momentum space.
4. There is a considerable literature concerning the BS equation. Detailed reviews are contained in Taylor (1963), Pagnamenta (1968), Broido (1969b), and Nakanishi (1969).
5. The subject of deriving dynamical collision integral equations from a field theory, apart from diagram summation techniques, is relatively unexplored especially for collisions involving composites. For some indication of the complexity of the task and for further references, see Klein (1957), Blankenbecler (1959a), Taylor (1963), Freedman (1966), Huang (1975), Kowalski (1979b, c; 1980b), and Arkhipov (1985).
6. The literature on three-dimensional reductions of the BS equation is in itself quite large, although the essential idea is simple. See Klein (1953, 1954, 1958a, b), Levy (1952), Macke (1953), Logunov (1963), Alessandrini (1965), Blankenbecler (1966), Kadychevsky (1968), Aaron (1968), and Gross (1969). References to other work are contained in Yaes (1971), Klein (1974), Brown (1976), Zmora (1977), and Machleidt (1987, 1988).
7. Woloshyn (1973).
8. See, e.g., Aaron (1968), Yaes (1971), and Brown (1976).
9. Blankenbecler (1966).
10. Thompson (1970).
11. Erkelenz (1972).
12. Gross (1969).
13. Kadychevsky (1968).
14. Brown (1976), Machleidt (1987, 1988).
15. Hynes (1984), Picklesimer (1984).
16. Arnold (1981), Shepard (1983), McNeil (1983a, b), Clark (1983a, b).

Chapter 3

Solution Methods and Techniques: Potential Scattering

3.1. Introduction

The quantum theory of scattering has two partially overlapping goals. One is the establishment of a mathematical framework incorporating the asymptotic condition and probability conservation for the representation of scattering in terms of the S-matrix, Møller operators, and the like. The other is the formulation of general solution algorithms, namely integral equations, for wave functions and off-shell scattering amplitudes. Chapter 2 describes the attainment of both of these goals for the simple physical situation of the scattering of a particle by a potential of suitably short range. In this chapter, we go one step further and investigate specific means for obtaining numerical solutions in this simple situation. It is important for us to remind ourselves why this is not a trivial endeavor.

The general means for finding a complete numerical solution to the problem of scattering by a potential of sufficiently short range should be evident from the development of Ch. 2. With the availability of modern computing power, it might appear that there is little of interest here except choosing among different numerical procedures for affecting a solution. That is certainly the case for relatively uncomplicated problems such as the determination of scattering phase shifts generated by potentials with simple structures.

For those physical circumstances of primary interest to us, matters are not quite so straightforward. Our main concerns involve the scattering of more than two particles that are either free asymptotically or are a part of some composite object, i.e., a cluster. The significant complicating feature is a multiparticle Hamiltonian for the system. Many formulations of such scattering problems are expressed in terms of the individual two-particle collisions of the constituents. These processes occur virtually (off-shell). This allows a much larger kinematical space, with correspondingly more complicated amplitudes for their description, than is the case for on-shell scattering. Thus, different methods are required and we consider some of them in this chapter.[1]

Also, many multiparticle collisions can be viewed as two-cluster, multichannel processes. This leads to a description in terms of sets of coupled two-body equations for two-cluster scattering. Although the objective here is similar to simple potential scattering, namely to obtain on-shell two-body scattering amplitudes, complexities can arise. The effective interactions may be quite complicated and the degree of coupling among different channels may be very high. In these cases, the computational problem can become quite intricate. A large literature exists dealing with approaches to these problems using sets of coupled differential equations.[2] Our development is primarily concerned with the corresponding momentum–space integral equations, although we consider other techniques, such as variational methods, as well.

In Ch. 2, we carried the problem of the scattering of a particle by a short-ranged potential to the point of defining a general procedure for obtaining a solution. Namely, when the two-particle (or center of force) interaction is specified by a potential short ranged enough to permit a truncation in orbital angular momentum, the problem can always be reduced to the question of determining the solution of a finite number of coupled one-dimensional inhomogeneous Fredholm integral equations.[3] These equations possess the generic form

$$F_{ij}^{s}(x, y) = B_{ij}^{s}(x, y) + \sum_{k} \int_{0}^{\infty} dz\, K_{ik}^{s}(x, z) F_{kj}^{s}(z, y), \tag{3.1.1}$$

where x, y, and z are nonnegative real variables. The discrete indices i, j, k include the orbital angular momentum L, which may or may not be conserved. The continuous variable s is typically a parametric energy with possible values running over the entire real line; it plays a passive role in the solution process. Nonetheless, the analytic structures of the kernels K_{ij}^{s}, the inhomogeneous (or "Born") terms B_{ij}^{s}, and, consequently, the solutions, F_{ij}^{s}, may change dramatically as a function of s.

For integral equations of the form (3.1.1), the real variable y is passive. Very often in scattering theory, $F^s_{ij}(x, y)$ also satisfies a set of integral equations that have the conjugate form

$$F^s_{ij}(x, y) = \tilde{B}^s_{ij}(x, y) + \sum_k \int_0^\infty dz\, F^s_{ik}(x, z)\tilde{K}^s_{kj}(z, y), \tag{3.1.2}$$

in which y now assumes an active role.

Equations of the forms (3.1.1) and (3.1.2) appear in the representation of more general collision processes besides simple potential scattering. These representations are invariably approximate and involve the reduction of multiparticle problems to a description by a set of coupled effective two-body equations. The reductions of complexity are justified differently depending on the physical situation, but result in the discarding of intractable aspects of the physics that are assumed to be dynamically insignificant. The resulting integral equations often present difficult numerical problems arising from the complexity of their inhomogeneous and kernel terms along with large discrete index sets. These difficulties, however, are relatively trivial compared to the enormous increase in computing power needed to handle the collision integral equations in more than one continuous variable that appear in conjunction with the collision of three or more particles if no approximations are made.

Even ordinary two-particle potential scattering is far from routine when one requires fully off-shell two-particle transition operators. This is equivalent to the specification of $F^s_{ij}(x, y)$ over the full semi-infinite domains of x and y and, typically, a semi-infinite domain, $-\infty < s < s_0$ of s, where s_0 corresponds to some maximal energy.

Most of the development of this chapter is directed towards the solution of the LS integral equation

$$t = V + VG_0^{(+)}t, \tag{3.1.3}$$

which in momentum space for a given L becomes

$$t_s(q, q') = V(q, q') + \frac{2}{\pi}\int_0^\infty \frac{dp\, p^2}{s - p^2} V(q, p)t_s(p, q'), \tag{3.1.4}$$

for the off-shell partial-wave amplitudes $t_s(q, q')$. We have suppressed the index L and, henceforth, when no limits of integration are given they are assumed to run from 0 to ∞. We assume that $V(q, q') = V(q', q)$, so that $t_s(q, q')$ possesses the same symmetry. It is not obvious that the kernel of Eq. (3.1.4) is of the Fredholm type, namely compact, because its norm does not exist when s approaches the positive real axis. However, it is possible to show that, for potentials of shorter than Coulombic range, the kernel of Eq. (3.1.4) is in fact compact on a suitable space of functions even when $s \to |s| + i0$.[4] In this regard, we note that one of the various equivalent definitions of compactness is that such operators, or "kernels" in the general sense of the term, can be uniformly approximated by operators (kernels) of finite rank. This attribute will be exploited at times in the sequel.

3.2. General Calculational Questions

A direct approach to solving Eq. (3.1.4) is to reduce it to an approximate set of linear algebraic equations by means of a quadradure rule. Let us describe what this implies in terms of the generic counterpart to Eq. (3.1.4), viz.,

$$F(x) = B(x) + \int_a^b dz\, K(x,z) F(z), \tag{3.2.1}$$

which is consistent with the notation introduced along with Eq. (3.1.1). If the domain of integration is subdivided into N "mesh" points $\{y_i\}$, then Eq. (3.2.1) can be reduced to the algebraic set

$$F_i = B_i + \sum_{j=1}^{N} K_{ij} \Delta_j F_j, \tag{3.2.2}$$

where, e.g., $F_i = F(x_i)$ and Δ_j is a weight function.

With a properly constructed meshing algorithm, the solution of Eq. (3.2.2) approaches $F(x)$ uniformly when $N \to \infty$, if Eq. (3.2.1) is Fredholm. Evidently, the magnitude of N needed in practice depends on the method of selecting quadrature points as well as the degree of precision desired.

The kernel of Eq. (3.1.4) is singular at $p = k$ when $s = k^2 + i0$, so special care is needed for the discretization of the integral. Given a definite discretization algorithm for principal-value (p.v.) integrals, the singularity problem can be handled by using the decomposition of $(k^2 - p^2 + i0)^{-1}$ into its p.v. and δ-function parts. Then, one can deal with Eq. (3.1.4) directly or in terms of an auxiliary $\hat{K}$-matrix equation that has the same form, but with a p.v. rather than a $+i0$ prescription at the pole singularity when $s > 0$.

Another way of handling the $s > 0$ fixed-point singularity in an equation like (3.1.4) is to subtract it off by expressing $t_s(q, q')$ in terms of the solution of an auxiliary nonsingular integral equation. The technique is a standard one for reducing certain types of singular integral equations arising in Cauchy–Hilbert problems to Fredholm form;[1] its application to off-shell collision integral equations was carried out in a different manner.[2] There are an infinite number of ways of affecting the subtraction, with each of them leading to a distinct auxiliary nonsingular integral equation. Other criteria must be imposed to favor a particular subtraction prescription.

The singularity problem can be circumvented by contour rotation. Here instead of the positive real line in Eq. (3.1.4), one integrates along some contour C in the complex p-plane that begins at $p = 0$ and ends at $|p| = \infty$, along a definite route. So, by Cauchy's theorem, we have

$$\int_0^\infty \frac{dp\, p^2}{k^2 - p^2 + i0} V(q,p)t_s(p,q') = \int_C \frac{dz\, z^2}{k^2 - z^2} V(q,z)t_s(z,q'), \qquad (3.2.3)$$

where z denotes the complex extension of p on C, provided no singularities of $V(q,z)t_s(z,q')$ are crossed in rotating from the line $(0,\infty)$ to C.[3] This method is not particularly advantageous for simple two-particle scattering. Its importance is that it can be applied to scattering integral equations with moving-point singularities in the kernel that depend upon both *q and p*. Such equations are encountered in multiparticle scattering problems.

Once a method for handling the fixed-point singularity in Eq. (3.1.4) has been decided upon, one still is confronted with solving a Fredholm integral equation such as Eq. (3.2.1). Fredholm theory provides a general solution algorithm in the form of an infinite series,[4] but this is seldom practical. The simplest practical approach is to use the discretization (3.2.2), which is a definitive systematic prescription for calculating $(I - K)^{-1}$.

An entirely different strategy is to use what is known, alternatively, as the iteration, Neumann, perturbation, or Born series of Eq. (3.2.1), viz., in operator form,

$$F = B + KB + K^2B + \cdots. \qquad (3.2.4)$$

The quantity KB is called the *second Born term*, and so on for higher-order terms. This method is practical only if the powers of K become progressively weaker. If this is not the case, the information contained in the first few Born terms can sometimes be exploited more efficiently by means of the Padé[5] or other techniques that accelerate the convergence of the Born series.

Variational techniques provide a flexible nonperturbative approach. This device can also be used to generate the method of separable expansions, which is especially useful for representing the solutions of the off-shell two-particle amplitudes that appear in few-particle scattering integral equations. This is because the dimensionality of some types of N-particle scattering integral equations is reduced if separable expansions are used for the off-shell two-body (and m-body, $m < N$, in general) transition amplitudes.

In practice, a solution strategy is selected by means of criteria tied into the ultimate use of the t-matrix elements. For example, if these matrix elements are used as input into a three-particle scattering problem, it may be highly advantageous to use the method of separable expansions even with a possible sacrifice in accuracy compared to other methods.

The formal Fredholm solution to Eq. (3.2.1) does not always exist. This happens if and only if there is a solution of the associated homogeneous equation

$$F_0(x) = \int_a^b dy\, K(x, y) F_0(y), \tag{3.2.5}$$

which means that $(I - K)^{-1}$ does not exist.[4] The eigenfunction problem implied by Eq. (3.2.5), in connection with the t-matrix equation (3.1.4), is the mathematical manifestation of the appearance of bound states in the two-particle system. These typically appear for $k^2 \leq 0$, i.e., below or at the threshold for scattering; bound states in the continuum can occur for certain potentials, but this is atypical.[6]

The classic textbook rendition of the two-particle bound-state problem features the differential form of the Schrödinger equation. In scattering theory, the bound-state problem naturally appears in association with the homogeneous form of an integral equation such as Eqs. (3.2.1–3.2.2). However, the corresponding eigenfunction calculation can be unwieldly, especially for large N. A more efficient method based on integral equations, obtained using subtraction techniques, can be used.

As a final detail we note that, since the range of the magnitude of the momentum covers the entire nonnegative portion of the real axis, we encounter an infinite integral in Eq. (3.1.4). Compactness ensures us that the kernel decays rapidly to zero as $p \to \infty$. Thus, we can always introduce a cutoff, or upper limit, on the ranges of the p and q:

$$0 \leq \{p, q\} \leq \Lambda. \tag{3.2.6}$$

The compactness of the kernel also implies that the exact solution is approached uniformly as $\Lambda \to \infty$. In practice, it is often more convenient to map the line $(0, \infty)$ into a finite interval by means of a Möbius, or bilinear, transformation such as

$$y = \frac{p - c}{p + c}, \tag{3.2.7}$$

where the images of the points $p = 0, c, +\infty$ are $y = -1, 0, +1$, respectively, as part of the mapping of $0 \leq p < \infty$ into $-1 \leq y < 1$.

After using Eq. (3.2.7), we see that Eq. (3.1.4) takes the form (3.2.1) with $a = -1$, $b = +1$. The integral in (3.2.1) can then be discretized over a set of points that are selected, for example, by means of a Gauss–Legendre quadrature procedure.[7] Typically, one chooses an equal number of points in each of the half intervals $(-1, 0)$ and $(0, 1)$. Evidently, the largest quadrature point in the y interval determines Λ in p-space, but the choice of parameter c in Eq. (3.2.7) is entirely arbitrary. Generally, particular values for c are dictated by the special features of the kernel resulting from the potential and the accuracy that is sought.

3.3. Singularity Removal

3.3a. Principal-Value Prescription

For $s = k^2 + i0$, the kernel of the integral equation (3.1.4) has a pole singularity at $p = k$. Although the integration appearing there is well defined, care is required in practical calculations because the $i0$-limit is to be carried out *after* performing the integration.

This problem can be circumvented by using the decomposition

$$\int_0^\infty \frac{dp\, p^2}{k^2 - p^2 + i0} f(p) = P \int_0^\infty \frac{dp\, p^2}{k^2 - p^2} f(p) - i\frac{\pi}{2} k f(k), \tag{3.3.1}$$

where $f(p)$ is a suitably well-behaved function of p. Another limiting process is implicit in the p.v. prescription, but one that can be more easily adapted to discretization. For example, Sloan[1] pointed out that for the mapping (3.2.7), with $c = k$ and $k \neq 0$, if the quadrature points are chosen symmetrically about $y = 0$, consistency with the p.v.-limit is maintained.

If we use Eq. (3.3.1) along with Sloan's prescription, the fully off-shell version of Eq. (3.1.4) can be transformed into a finite-dimensional matrix equation:

$$t_{ij} = V_{ij} + \sum_{n=1}^{M} \frac{2}{\pi} \frac{V_{in} p_n^2 \Delta_n}{p_0^2 - p_n^2} t_{nj} - i p_0 V_{i0} t_{0j}, \tag{3.3.2}$$

where, e.g.,

$$t_{ij} = t(p_i, p_j), \tag{3.3.3}$$

p_i, p_j, $i, j = 1, \ldots, M$, refer to integration mesh points, $p_0 = k$, and Δ_n is the weight function associated with the quadrature rule. It is understood that the p_i are chosen so that the mesh points in y-space are symmetrically placed around $y = 0$, which itself is not a mesh point. The linear equations (3.3.2) can be solved, e.g., by matrix inversion. This method does not yield particularly stable or precise numerical results.

More satisfactory results are obtained if Sloan's method is applied directly to the $\hat{K}$-matrix equation which, as we shall see, satisfies Eq. (3.1.4), but with a p.v. prescription for the integral. This yields the set of real matrix equations

$$\hat{K}_{ij} = V_{ij} + \sum_{n=1}^{M} \frac{2}{\pi} V_{in} \frac{p_n^2 \Delta_n}{p_0^2 - p_n^2} \hat{K}_{nj}, \tag{3.3.4}$$

where the notation is similar to that employed in Eqs. (3.3.2) and (3.3.3). The caret is used to distinguish the $\hat{K}$-matrix from the generic symbol for the kernel of an arbitrary integral equation that we frequently employ. A significant difference between Eq. (3.3.2) and Eq. (3.3.4) is the absence of the asymmetrical on-shell term on the right-hand side of the latter. After solving the Eqs. (3.3.4), the t-matrix is calculated by means of the Heitler equation.

The Sloan method, which yields good results when the choice $c = k$ in Eq. (3.2.7) provides an efficient mapping, not only removes the singularity in Eq. (3.1.4), it also gives a definite prescription for choosing the integration mesh points. This procedure, however, is not expected to be good in some situations. For example, at low energies, $k \to 0$, half of the integration points accumulate at $k = 0$, which results in an uneconomic distribution of mesh points. Also, for realistic nucleon–nucleon interactions with a soft core, the kernel of Eq. (3.1.4) does not fall off rapidly for large momentum and the cutoff Λ has to be large; then, the choice $c = k$ in Eq. (3.2.7) corresponds to taking a negligibly small number of mesh points for large momenta. In both of these situations, the accuracy of the Sloan method is not high, unless the number of mesh points is increased substantially or a more complex distribution of integration mesh points is employed. This motivates the Fredholm-reduction technique of Section 3.3b. Then, c can be chosen independently of k and this gives an extra flexibility in distributing the integration mesh points.

Another way of dealing with the p.v. singularity uses the identity

$$P \int_0^\infty \frac{dp}{k^2 - p^2} = 0 \tag{3.3.5}$$

to rewrite the $\hat{K}$-matrix equation for $s = k^2$ and fixed L as[2]

$$\hat{K}_{k^2}(q, q') = V(q, q') + \frac{2}{\pi} \int \frac{dp}{k^2 - p^2} \times [p^2 V(q, p) \hat{K}_{k^2}(p, q') - k^2 V(q, k) \hat{K}_{k^2}(k, q')]. \tag{3.3.6}$$

There is no need for the p.v. prescription since the *integral*, but not the kernel, in Eq. (3.3.6) is nonsingular at $p = k$. The discretization of Eq. (3.3.6) can now proceed normally, if the point $p = k$ is avoided when choosing the mesh points. Nevertheless, the manifest asymmetry of Eq. (3.3.6) with respect to the on-shell point ($p = k$) limits the precision and stability of the solutions that can be calculated in this way.[3] It is because of that asymmetry, which brings in $\hat{K}_{k^2}(k, q')$ in order to cancel the singularity, that Eq. (3.3.6) does not have the canonical form of the nonsingular integral equations we consider next, but rather that of a class of equations discussed in Section 3.3c.

3.3b. Fredholm Reduction

A definitive way of dealing with the positive-s, fixed-point singularities in Eq. (3.1.4), or its $\hat{K}$-matrix counterpart, involves the complete removal of the $i0$-prescription from the part of the solution procedure that involves solving an integral equation. The basic idea is to express $t_{k^2}(q, q')$ in terms of the solution of an auxiliary Fredholm integral equation with a nonsingular kernel that is independent of the $i0$-prescription.[4] That prescription of course must reappear somewhere in order to obtain $t_{k^2}(q, q')$, but here it does so only in the relatively innocuous form of a quadrature involving $(k^2 - p^2 + i0)^{-1}$ and the auxiliary amplitude, where it can be handled reliably.

Actually, the preceding method applies to Eq. (3.1.4) for arbitrary[5] s and to the homogeneous equation for the bound-state wave function,[6] where the singularity-removal feature is irrelevant.

The general scheme[7] for constructing a class of auxiliary nonsingular equations begins by introducing a so-called *subtraction function* $\gamma(k, q)$, which is a suitably well-behaved function of q that depends parametically on a subtraction point $k \geq 0$, and which satisfies the constraint

$$\gamma(k, k) = 1, \tag{3.3.7}$$

as in Eq. (2.4.63). There is a close relationship between the generalized Heitler formalism of Ch. 2 and the present development.

Next, we can rewrite the integral equation (3.1.4) for the fully off-shell amplitude $t_s(q, q')$ as

$$\begin{aligned} t_s(q, q') = V(q, q') + V(q, k)\frac{2}{\pi}\int \frac{dp\, p^2}{s - p^2}\gamma(k, p)t_s(p, q') \\ + \int dp\, A_s(q, p; k)t_s(p, q'), \end{aligned} \tag{3.3.8}$$

where

$$A_s(q, p; k) = \frac{2}{\pi}p^2[V(q, p) - V(q, k)\gamma(k, p)]\frac{1}{s - p^2}. \tag{3.3.9}$$

Because of Eq. (3.3.7), the kernel $A_s(q, p; k)$ is nonsingular at the point $p = k$, if $s = k^2$. In fact,

$$A_{k^2}(q, k; k) = -\left(\frac{1}{\pi}k\right)\left\{\frac{\partial}{\partial p}[V(q, p) - V(q, k)\gamma(k, p)]\right\}_{p=k}. \tag{3.3.10}$$

The manipulation (3.3.8) has the consequence that $t_s(q, q')$ can be expressed in terms of the solution $\Gamma_s(q, q'; k)$ of the manifestly nonsingular integral equation

$$\Gamma_s(q, q'; k) = V(q, q') + \int dp\, A_s(q, p; k)\Gamma_s(p, q'; k), \tag{3.3.11}$$

as we next demonstrate.

The proof of the preceding assertion is simplified by expressing Eqs. (3.3.11) and (3.3.8) in the condensed forms

$$\Gamma = V + A\Gamma, \tag{3.3.12}$$

and

$$t = V + V \cdot \gamma g t + At, \tag{3.3.13}$$

respectively. The dot in Eq. (3.3.13) denotes the on-shell restriction and simple algebraic multiplication, *cf.* Eq. (3.3.8). Other products refer to integration over p with measure dp. The factor g represents $(2/\pi)p^2(s - p^2)^{-1}$. Since A is at least formally Fredholm, we suppose that $(I - A)^{-1}$ exists. Then, we have

$$\Gamma = (I - A)^{-1}V, \tag{3.3.14}$$

and so

$$t = (I - A)^{-1}[V + V \cdot \gamma g t], \tag{3.3.15}$$

or

$$t = \Gamma + \Gamma \cdot \gamma g t. \tag{3.3.16}$$

We recognize this as the Heitler equation (2.4.50); therefore, $t_s(q, q')$ takes on the form defined by Eqs. (2.4.42, 2.4.58–2.4.59).

Of course, one can take Eqs. (3.3.15) and (3.3.16) as starting points, as in the Heitler formalism, and then one immediately infers Eq. (3.3.14), and so Eq. (3.3.12). The connection between the resolvent kernel $r_A \equiv A(I - A)^{-1}$ and Γ_s allows one to characterize the fully off-shell structure of t_s in different ways. We note that

$$\Gamma = (I + r_A)V, \tag{3.3.17}$$

and

$$r_A = (\Gamma - \Gamma \cdot \gamma)g. \tag{3.3.18}$$

Using Eqs. (3.3.17–3.3.18), we can establish the identity

$$\Gamma - \frac{\Gamma \cdot \Gamma}{(\cdot\, \Gamma\, \cdot)} = r_A g^{-1} - \left[\frac{\Gamma}{(\cdot\, \Gamma\, \cdot)}\right] \cdot r_A g^{-1}, \tag{3.3.19}$$

which we will write out explicitly below in order to unify different versions of the Γ-matrix formalism.

We now combine the preceding results with the generalized Heitler analysis of Section 2.4. The result is that $t_s(q, q')$ can be expressed entirely in terms of the auxiliary amplitudes $\Gamma_s(q, q'; k)$:

$$\begin{aligned} t_s(q, q') &= f_s(q, k) t_s(k, k) f_s(q', k) \\ &\quad + \left\{ \Gamma_s(q, q'; k) - \frac{\Gamma_s(q, k; k)\Gamma_s(k, q'; k)}{\Gamma_s(k, k; k)} \right\}, \end{aligned} \tag{3.3.20}$$

where the *half-shell function*

$$f_s(q, k) = \frac{\Gamma_s(q, k; k)}{\Gamma_s(k, k; k)} \tag{3.3.21}$$

is a universal function of Γ_s. The only place the $i0$-prescription can enter in is as a quadrature in the representation of the on-shell matrix element

$$t_s(k, k) = \Gamma_s(k, k; k)\left[1 - \frac{2}{\pi}\int \frac{dp\, p^2}{s - p^2}\gamma(k, p)\Gamma_s(p, k; k)\right]^{-1}, \tag{3.3.22}$$

where $s = k^2 + i0$. The half-on-shell t-matrix is given by

$$t_s(q, k) = f_s(q, k) t_s(k, k). \tag{3.3.23}$$

When s is real and negative, Eqs. (3.3.20–3.3.22) still hold although there is no motivation, as there was in the real-positive-energy case, for any particular connection between s and the subtraction point $k \geq 0$.[5] A fixed value of k independent of s and the functional relationship $k = |s|^{1/2}$ have both been used in calculations.[6,8] Independently of the choice of this connection, the representation (3.3.22) can sometimes provide us with an integral constraint for the occurrence of a bound-state pole in $t_s(q, q')$ when $s = E_B$, a (negative) bound-state energy, viz.,

$$\frac{2}{\pi}\int \frac{dp\, p^2}{E_B - p^2}\gamma(k, p)\Gamma_{E_B}(p, k; k) = 1, \tag{3.3.24}$$

provided that it is possible to satisfy this condition.

The purely off-shell part of $t_s(q, q')$, namely $R_s(q, q'; k)$, can be written in a variety of equivalent forms. First, we recall Eq. (2.4.59),

$$R_s(q, q'; k) = \Gamma_s(q, q'; k) - \frac{\Gamma_s(q, k; k)\Gamma_s(k, q'; k)}{\Gamma_s(k, k; k)}, \tag{3.3.25}$$

which reflects the γ-invariance of R_s. Note, however, that R_s depends upon k. Alternatively, from the identity (3.3.19), we have the γ-invariant form

$$R_s(q, q'; k) = \frac{\pi}{2}[r_A(q, q'; k) - f_s(q, k)r_A(k, q'; k)]\frac{s - q'^2}{q'^2}. \tag{3.3.26}$$

The forms (3.3.25–3.3.26) are equivalent because r_A and Γ_s satisfy integral equations with the same kernel. Note, however, that the form (3.3.26) is obtained using the dynamical equations while Eq. (3.3.25) follows from the Heitler analysis alone.

We observe that the expression (3.3.26) simplifies if $r_A(k, q'; k) = 0$, which implies that $A_s(k, q'; k) = 0$. This can take place for arbitrary k only if the subtraction function is the one introduced by Kowalski, Feldman, and Noyes (KFN):[4,9,10]

$$\gamma_{\mathrm{KFN}}(k, p) = \frac{V(k, p)}{V(k, k)}, \tag{3.3.27}$$

which is distinguished by a number of unique consequences. First, it defines the integral equation for the half-shell function f_s:

$$f_s(q, k) = \frac{V(q, k)}{V(k, k)} + \int dp\, \Lambda_s(q, p; k) f_s(p, k), \tag{3.3.28}$$

where

$$\Lambda_s(q, p; k) = [A_s(q, p; k)]_{\mathrm{KFN}}. \tag{3.3.29}$$

Second, we see that

$$t_s(k, k) = V(k, k)\left[1 - \frac{2}{\pi}\int \frac{dp\, p^2}{s - p^2} V(k, p) f_s(p, k)\right]^{-1}, \tag{3.3.30}$$

so that the poles of $t_s(k, k)$ are unambiguously associated with the vanishing of the denominator of Eq. (3.3.30), a conclusion that is not so obvious with arbitrary subtraction functions. Third, with Eq. (3.3.27), the residual function assumes its simplest form:

$$R_s(q, q'; k) = \frac{\pi}{2} r_{\mathrm{KFN}}(q, q'; k)\frac{(s - q'^2)}{q'^2}, \tag{3.3.31}$$

where r_{KFN} is the resolvent kernel corresponding to Λ_s. Thus, the integral equation for R_s is given by

$$R_s(q, q'; k) = V(q, q') - \frac{V(q, k)V(k, q')}{V(k, k)} + \int dp\, \Lambda_s(q, p; k) R_s(p, q'; k), \quad (3.3.32)$$

which differs only in its inhomogeneous term from (3.3.28); the inhomogeneous term in Eq. (3.3.32) is exactly what one would obtain by taking $\Gamma_s = V$ in Eq. (3.3.25). Each order of the iteration series of Eq. (3.3.32) vanishes when either q or q' is equal to k.

It is interesting to point out that the integral equation (3.3.28) can be solved exactly for a square-well potential.[4] Evidently, an exact closed-form solution (in terms of spherical Bessel functions) exists for Eq. (3.3.32) in this case as well. These exact solutions are useful for studying the qualitative behavior of the off-shell functions f_s and R_s.

The subtraction function $\gamma(k, p) = (p/k)^L$ used by Brown *et al.*[11] and others[6,8,12] is of special interest because of the relationship of Γ_s in this instance for local potentials to the Jost function.

In general, Γ_s-matrix formalism can be readily applied to the relativistic two-particle equations considered in Section 2.7 to deal with fixed-point singularities. In fact, however, most of the applications of this method to both the one- and two-dimensional forms of these equations have employed the subtraction function (3.3.27) to facilitate numerical work.[13] In addition, a number of interesting formal results have been generated, such as the connection between the iteration series of Eq. (3.3.28) for $f_s(q, k)$ and the $[1, N]$ Pade approximants for $t_s(k, k)$, and the counterpart of Eq. (3.3.20) for the Bethe–Salpeter amplitude, but using a fully off-shell generalization of $f_s(q, k)$.[14,15] (The fully off-shell Γ_s-function for γ_{KFN} had been introduced previously.[4])

3.3c. *Alternative Subtraction Schemes*

A number of alternative subtraction techniques have been proposed that differ, at least in form if not in content, from the class considered in the preceding section. Each has distinctive properties that are of special interest.

The treatment of Section 2.4 can be carried out even if the subtraction function $\gamma(k, q)$ does not satisfy Eq. (3.3.7). An important example is the $\hat{K}$-matrix formalism which corresponds to the choice (2.4.60a–b). Then, we have

$$\Gamma_s(q, q'; k) = \hat{K}_s(q, q'), \quad (3.3.33)$$

with $k = |s|^{1/2}$ for all real s, namely the off-shell $\hat{K}$-matrix that satisfies

$$\hat{K}_s(q, q') = V(q, q') + \frac{2}{\pi} P \int \frac{dp\, p^2}{s - p^2} V(q, p) \hat{K}_s(p, q'). \tag{3.3.34}$$

Here, P denotes the p.v. prescription for s positive. At real-negative energies, the p.v. prescription is not needed. As we remarked in Section 2.4, the t-matrix formalism is recovered with the trivial choice $\gamma \equiv 0$, which also does not satisfy Eq. (3.3.7).

What appears to be a distinctive subtraction technique was introduced by Bolsterli[16] specially to handle the p.v. singularity in Eq. (3.3.34). If $\phi(q, k)$ is a function such that $\phi(k, k) = 1$, then we consider a subtraction procedure that begins from the definition[17]

$$\zeta_s(q, q'; k) \equiv \hat{K}_s(q, q') - \phi(q, k) \hat{K}_s(k, q'). \tag{3.3.35}$$

One finds that the auxiliary quantity (3.3.35) satisfies

$$\begin{aligned} \zeta_s(q, q'; k) = {} & V(q, q') - \gamma_{\mathrm{B}}(k, q) V(k, q') \\ & + \frac{2}{\pi} P \int \frac{dp\, p^2}{s - p^2} [V(q, p) - \gamma_{\mathrm{B}}(k, q) V(k, p)] \zeta_s(p, q'; k). \end{aligned} \tag{3.3.36}$$

Here, the subtraction function is given by

$$\gamma_{\mathrm{B}}(k, q) = \frac{\phi(q, k) - I_s(q, k)}{1 - I_s(k, k)}, \tag{3.3.37}$$

with

$$I_s(q, k) = \frac{2}{\pi} P \int \frac{dp\, p^2}{s - p^2} V(q, p) \phi(p, k). \tag{3.3.38}$$

Note that although the kernel of Eq. (3.3.36) is singular at $p = k$ if $s = k^2$, because $\zeta_s(k, q'; k) = 0$ the entire integral is well-defined. The form of Eq. (3.3.36) suggests a close connection with the Γ_s-matrix formalism as is evident from the identity

$$\zeta_s(q, q'; k) = \frac{\pi}{2} r_A^{(B)}(q', q; k) \frac{(s - q^2)}{q^2}, \tag{3.3.39}$$

where $r_A^{(B)}(q', q; k)$ denotes the resolvent kernel generated by the kernel $A_s^{(B)}(q', q; k)$ given by Eq. (3.3.9) with the subtraction function (3.3.37).

The Γ_s-matrix corresponding to $A_s^{(B)}$ satisfies the remarkable identity[17]

$$I_s(k, k) = \frac{2}{\pi} P \int \frac{dp\, p^2}{s - p^2} \gamma_B(k, p) \Gamma_s(p, k; k), \tag{3.3.40}$$

which implies that

$$\hat{K}_s(k,k) = [1 - I_s(k,k)]^{-1}\Gamma_s(k,k;k). \tag{3.3.41}$$

This provides an obvious exception to the criterion (3.3.24) for the appearance of a bound state for negative-real s, where the p.v. prescription is irrelevant and $\hat{K}_s(q,q')$ and $t_s(q,q')$ coincide. In these instances, $\Gamma_s(p,k;k)$ has poles at the bound-state energies.

The subtraction technique based upon the starting point (3.3.35) can be applied to $t_s(q,q')$, which then yields still another new feature: a complex γ-function given by the Eqs. (3.3.37) with the integral (3.3.38) without the p.v. prescription defined in terms of complex s.

This last observation allows us to provide a concise momentum-space expression for the several mixed coordinate- and momentum-space representations of the off-shell amplitudes that have been proposed since the earliest investigations of off-shell scattering.[18] For example, we have, from Eqs. (3.3.35) and (3.3.39),

$$\hat{K}_s(q,q') = \phi(q,k)\hat{K}_s(k,q') + \frac{(s-q^2)}{q^2}\left\{\frac{\pi}{2}r_A^{(B)}(q',q;k)\right\}. \tag{3.3.42}$$

For $s < 0$ we note that $r_A^{(B)}(q',k;k) = 0$.

In the case $\phi(q,k) = 1$ and $s = k^2 + i0$, Srivastava and Sprung[19] have expressed the $r_A^{(B)}$ term in Eq. (3.3.42) as an integral of the difference of interacting and free coordinate-space standing-wave functions. This has proven useful in the analysis of the sensitivity of the off-shell behavior to the wave function within the range of the potential. The expression provides a generalization of these results for all real s, arbitrary $\phi(q,s)$, and a momentum-space integral equation for the residual term.

Picker *et al.*[20] favor the choice $\phi(q,k) = (q/k)^L$ for each L-state, and also consider the transition matrix, rather than the $\hat{K}$-matrix. We immediately see that

$$t_s(q,q') = \phi(q,k)t_s(k,q') + \frac{(s-q^2)}{q^2}\left\{\frac{\pi}{2}r_A^{(B)}(q',q;k)\right\}, \tag{3.3.43}$$

but in this instance $r_A^{(B)}$ will be complex for $s = k^2 + i0$.

The expressions (3.3.42) and (3.3.43) complement the coordinate-space representations for the residual terms.[18–20] Either form can be used to find alternative expressions for the f_s and R_s functions, including the representation (3.3.26). A coordinate-space version of the integral equation (3.3.28) was used in the first calculations,[4,21] of the $f_s(q,k)$ function; this equation is not used in the considerations of Refs. 18–20.

Prior to Bolsterli,[16] Taylor[22] had introduced a technique that is equivalent to the starting point (3.3.35) with the same choice as Bolsterli for ϕ, namely $\phi(q,k) = 1$ and with an auxiliary function equal to $\zeta_s(q,k;k)(q-k)^{-1}$. Taylor also showed how this approach can be extended to two-dimensional integral equations with the same fixed-point singularity in each of the two variables as in the case of the Bethe–Salpeter equation. That is, instead of a singular integral equation with $(s-p^2)^{-1}$ in its kernel, one deals with a kernel containing the product $[(s-p^2)(s-q^2)]^{-1}$, where p and q are the two integration variables. The removal of fixed-point singularities in the two-dimensional Bethe–Salpeter equation after partial-wave analysis was considered independently by Graves-Morris.[23]

We next introduce quite a different type of subtraction scheme that shares with the Bolsterli–Taylor method the property of yielding auxiliary functions that satisfy nonsingular integral equations, but with singular kernels. Let $\bar{\gamma}(k,k_0,p)$ be some arbitrary real function of the three independent wave vectors k, k_0, and p, and define

$$\hat{I}_s(k,k_0) \equiv \frac{2}{\pi}\int \frac{dp\,p^2}{s-p^2}\bar{\gamma}(k,k_0,p). \tag{3.3.44}$$

Then we can rewrite Eq. (3.1.4) as

$$\begin{aligned} t_s(q,q') &= V(q,q') + V(q,k)\hat{I}_s(k,k_0)t_s(k,q') \\ &\quad + \frac{2}{\pi}\int \frac{dp\,p^2}{s-p^2}[V(q,p)t_s(p,q') - V(q,k)\bar{\gamma}(k,k_0,p)t_s(k,q')], \end{aligned} \tag{3.3.45}$$

The standard version of this idea involves choosing $\bar{\gamma}(k,k_0,q)$ so that $\hat{I}_s(k,k_0)$ vanishes. Then, Eq. (3.3.45) in nonsingular, albeit with a singular kernel. For example, when the manipulation leading to (3.3.45) is carried out on the $\hat{K}_s$-matrix equation (3.3.34) for s positive, one obtains Eq. (3.3.6) with the choice $\bar{\gamma}(k,k_0,p) = (k/p)^2$, by virtue of the identity (3.3.5).

When $\hat{I}_s(k,k_0) \neq 0$, the solution of Eq. (3.3.45) can be expressed in the form of the Heitler equation (2.4.70) in terms of the solution of the auxiliary integral equation

$$\begin{aligned} \Gamma_s(q,q';k,k_0) &= V(q,q') + \frac{2}{\pi}\int \frac{dp\,p^2}{s-p^2}[V(q,p)\Gamma_s(p,q';k,k_0) \\ &\quad - V(q,k)\bar{\gamma}(k,k_0,p)\Gamma_s(k,q';k,k_0)], \end{aligned} \tag{3.3.46}$$

provided we make the identification

$$\gamma(k,k_0;p,p') = \bar{\gamma}(k,k_0,p)\,\delta(p'-k). \tag{3.3.47}$$

Because of the relationship to the Heitler equation, the universal representation (3.3.19–3.3.21) holds with Γ_s given by the solution of Eq. (3.3.46). Although the integral in Eq. (3.3.46) is nonsingular, its kernel is a highly singular, distribution-valued quantity.

The k-independent choice

$$\bar{\gamma}(k, k_0, q) = \theta(k_0 - q) \tag{3.3.48}$$

for the subtraction function yields a Γ_s-matrix equivalent to that proposed by Stoof *et al.*[24] for s positive.

For the k_0-independent choice

$$\bar{\gamma}(k, k_0, q) = (k/q)^2 \tag{3.3.49}$$

we obtain

$$\hat{K}_s^c(q, q') = V(q, q') + \frac{2}{\pi}\int \frac{dp\, p^2}{s - p^2}\left[V(q, p)\hat{K}_s^c(p, q') - \frac{k^2}{p^2}V(q, k)\hat{K}_s^c(k, q')\right], \tag{3.3.50}$$

where we have used a superscript c to indicate that $\hat{K}_s^c$ is a different analytic continuation to complex s than the quantity defined by Eq. (3.3.34).[25] For $s > 0$, Eqs. (3.3.6), (3.3.34), and (3.3.50) are all identical and define the same quantity, but this is not the case for arbitrary complex s.

3.3d. *Calculation of Off-Shell Amplitudes*

The singularity-removal techniques of this section confront the problem of calculating exact off-shell amplitudes[26] given a potential, and then representing that information in terms of the universal off-shell functions $f_s(q, k)$ and $R_s(q, q'; k)$. This contrasts sharply with the agenda for the remainder of the chapter, which is concerned with approximation methods. Thus, it is appropriate here for us to review the results of extant exact computations of off-shell amplitudes. The most extensive body of work of this kind is for the nucleon–nucleon (NN) system, and so we confine ourselves to this case.

Although total spin (S) is conserved in NN interactions, there are substantial spin–orbit couplings in the triplet ($S = 1$) states. The extension of the subtraction formulation to this case is straightforward. The reason for this, as can be seen by referring to Section 2.6, is that the fixed-point singularities, which determine the subtraction points, are the same in each (L, S) pair. Thus, for fixed total angular momentum J, all of the preceding considerations hold for $S = 1$ with the adaptation of a 2×2 matrix notation.[4]

Equations (2.6.16) and (2.6.17) provide a universal representation for the off-shell transition amplitudes both for positive and negative energies. The off-shell $\hat{K}$-matrix has an identical representation in terms of the *same* functions f_s and R_s. This fact is often obscured in calculations that employ $\hat{K}$-matrices in an effort to work with real amplitudes.[27–30] For example, instead of the matrices $f_s(q,k)$ for the spin–triplet case, the real so-called "f-ratios" $\{[\hat{K}_{k^2}(q,k)]_{\beta\alpha}/[\hat{K}_{k^2}(k,k)]_{\beta\alpha}\}$ are introduced,[27–30] where the channel indices α, β represent the quantum numbers (J, S, L, I), with $I = 0, 1$ denoting the isospin states.

Since (*cf*. Eq. (2.6.16)) the matrix elements

$$[f_s(q,k)]_{\beta\alpha} = \{\hat{K}_s(q,k)[\hat{K}_s(k,k)]^{-1}\}_{\beta\alpha}, \tag{3.3.51a}$$

$$= \{t_s(q,k)[t_s(k,k)]^{-1}\}_{\beta\alpha}, \tag{3.3.51b}$$

where

$$\sum_{\beta}[\hat{K}_s(k,k)]_{\alpha\beta}\{[\hat{K}_s(k,k)]^{-1}\}_{\beta\gamma} = \delta_{\alpha\gamma}, \tag{3.3.51c}$$

are purely real, the motivation for introducing the nonuniversal "f-ratios" in the first place is immaterial. The "f-ratios" and the matrix elements $[f_s(q,k)]_{\beta\alpha}$ are related by simple determinantal relations; however, the f_s-functions provide a direct measure of the off-shell behavior of both the $\hat{K}_s$- and t_s-matrices in the same manner, a feature not possessed by the "f-ratios." The 2×2, real matrix, half-shell function f_s was used in early calculations of off-shell, $S = 1$, NN scattering.[4]

The f_s-functions are only part of the off-shell story. This is, of course, recognized in most calculations, but the possibility of presenting the essential behavior of the full off-shell amplitudes in terms of the real, universal residual function $R_s(q, q'; k)$ seems to have been over-looked.[19,27–30] It is clear from Sections 2.4, (2.6), and 3.3b that, e.g.,

$$R_s(q, q'; k) = \hat{K}_s(q, q') - \hat{K}_s(q,k)[\hat{K}_s(k,k)]^{-1}\hat{K}_s(k, q'), \tag{3.3.52}$$

as in Eq. (2.6.22). The numerical results of some calculations[27–30] could possibly be more easily interpreted if the results for the $\hat{K}_s$-matrices were organized using the expression (3.3.52). This is strongly recommended for future calculations irrespective of the incidential features of the computational method, such as the particular singularity-removal technique that is employed.

The functions f_s and R_s represent different aspects of off-shell behavior. Roughly speaking, one requires $f_{k^2}(q,k)$ for *all* $q, k \geq 0$ to calculate $R_s(q, q'; k)$ at a single point s, k, q, q'. This illustrates very strikingly the difference in the information content of the two functions, but it does not imply that one must *actually* calculate all of the half-shell functions in order to compute R_s as implied in some investigations.[28] This is obvious from the fact that $f_s(q, k)$ and $R_s(q, q'; k)$ satisfy integral equations with identical kernels and with inhomogeneous terms that differ in a minor way from a computational standpoint. Since the various factors in the inhomogeneous terms (*cf.* Eqs. (3.3.28) and (3.3.32)) all must be calculated to determine the (common) kernel, the same program will determine both quantities. Alternatively, one can use any one of the Γ_s-matrices; the important thing is to organize the computational output in terms of the f_s and R_s functions.

It is equally important to do this when testing approximation methods, such as finite-rank representations of the transition matrix.[27] Usually, approximations of this sort are made because one wants simple input for multiparticle calculations, not because the exact transition matrices are difficult to calculate. Therefore, in assuming the accuracy of approximate representations, one should compare the predicted f_s and R_s functions with the exact results. If the approximation violates unitarity and/or time-reversal invariance, other means of comparison may be needed.

Since $R_s(q, q'; k)$ vanishes for $q, q' = k$, the function

$$\kappa_s(q, q'; k) \equiv \frac{q^2}{(s - q^2)} R_s(q, q'; k) \frac{q'^2}{(s - q'^2)} \tag{3.3.53}$$

might be a sharper indicator of the fully off-shell behavior of the transition matrix, and thus provide a better means of differentiating among different interactions. It is interesting to note that κ_s satisfies the nonsingular integral equation

$$\kappa_s(q, q'; k) = \lambda_s(q, q'; k) + \frac{2}{\pi} \int dp \frac{(s - p^2)}{p^2} \lambda_s(q, p; k) \kappa_s(p, q'; k), \tag{3.3.54a}$$

with the symmetrical kernel

$$\lambda_s(q, q'; k) = \frac{q^2}{(s - q^2)} \{V(q, q') - V(q, k)[V(k, k)]^{-1} V(k, q')\} \frac{q'^2}{(s - q'^2)}. \tag{3.3.54b}$$

We obtain Eqs. (3.3.54a, b) from the multichannel counterpart of Eq. (3.3.32) with the subtraction function[4,31]

$$\gamma_{\mathrm{KFN}}(k, p) = V(k, p)[V(k, k)]^{-1}. \tag{3.3.55}$$

The order of the potential matrices in the $S = 1$ version of Λ_s is obvious from Eq. (3.3.54b), and the inhomogeneous term in Eq. (3.3.28) is now $V(q, k)[V(k, k)]^{-1}$.

A major conclusion of the extensive calculations[32] carried out in Refs. 27–30 is that there are relatively few differences in the off-shell behaviors generated by different realistic NN interactions. Various interpretations of this have been proposed.[28,33,34] It is possible that this general picture will still hold after the additional analyses suggested above (comparisons of the f_s and the R_s, or κ_s, functions) have been completed.

The preceding conclusion is consistent with the calculation of proton–proton bremsstrahlung, $p + p \to p + p + \gamma$, by Workman and Fearing[35] using the Bonn[36] and Paris[37] potentials. This reaction has been long conjectured to be the cleanest process for probing the half-off-shell behavior of the NN amplitude.[38] Until recently, this promise was unfulfilled because of the lack of experiments of sufficient accuracy. The experiment of Kitching *et al.*[39] achieved an order of magnitude increase in accuracy over previous work and also succeeded in measuring the analyzing power for the first time. These results were decisive in demonstrating a strong disagreement with the purely on-shell ($f_s = 1$) NN description of this process, but were unable to distinguish between the Bonn and Paris potentials, both of which yielded fits in good agreement with experiment. These results represent the first direct measurement of the off-shell behavior of the NN interaction, along with evidence that modern potential models of that interaction are capable of describing that behavior.

We conclude this section concerning the use of propagator-singularity-free auxiliary functions to remark about the zero-energy kinematic singularities that appear with their use for $L \neq 0$.[8] The problem arises because the partial-wave amplitude of the potential in the state L has the structure

$$V_L(q, q') = q^L \bar{V}_L(q, q') q'^L, \tag{3.3.56}$$

where the reduced matrix element $\bar{V}_L(q, q')$ is nonzero as $q \to 0$ or $q' \to 0$. Thus, ratios such as $V_L(q, k)/V_L(k, k)$ are not defined as $k \to 0$, nor are the half-shell functions $f_s^L(p, k) = \Gamma_s^L(p, k; k)/\Gamma_s^L(k, k; k)$ for $L \neq 0$. Of course, these singularities cancel out in the various t-matrix expressions into which the auxiliary functions enter, so that they are harmless.[4] Nevertheless, in order to avoid possible numerical instabilities, one can factor the singularities out to begin with by introducing new auxiliary functions in which the kinematic structure (3.3.56) is explicated:

$$\Gamma_s^L(q, q'; k) = q^L \bar{\Gamma}^L(q, q'; k) q'^L. \tag{3.3.57}$$

Then, $\bar{\Gamma}_s^L(q, q'; k)$ satisfies Eq. (3.3.11) with the kernel (3.3.9) provided $V(q, q') \to \bar{V}_L(q, q')$, $\gamma(k, p) \to (k/p)^L \gamma_L(k, p)$, and $(2/\pi)p^2 \to (2/\pi)p^{2L+2}$. It is assumed that $\gamma_L(k, p)p^{-L}$ is defined as $p \to 0$. Then,

$$f_s^L(q, k) = \left(\frac{q}{k}\right)^L \bar{f}_s^L(q, k), \tag{3.3.58}$$

where $\bar{f}_s^L(q, k) = \bar{\Gamma}_s^L(q, k; k)/\bar{\Gamma}_s^L(k, k; k)$. Also, since

$$R_s^L(q, q'; k) = q^L \left\{ \bar{\Gamma}_s^L(q, q'; k) - \frac{\bar{\Gamma}_s^L(q, k; k)\bar{\Gamma}_s^L(k, q'; k)}{\bar{\Gamma}_s^L(k, k; k)} \right\} q'^L, \tag{3.3.59}$$

$\bar{f}_s^L(q, k)$, and $\bar{R}_s^L(q, q'; k) \equiv (qq')^{-L} R_s^L(q, q'; k)$ satisfy integral equations similar to the one for $\bar{\Gamma}_s^L(q, q'; k)$ corresponding to the choice $\gamma_L(k, p) = V_L(k, p)/V_L(k, k)$. This procedure eliminates kinematic singularities.

3.4. Solution by Iteration

There are several alternatives to the direct matrix inversion of either the original singular or a nonsingular integral equation. Some of these can produce high-precision numerical results with comparatively little computational effort (time). Perhaps the simplest is solution by iteration, represented schematically in Eq. (3.2.4), which we next describe in more detail.

For the sake of definiteness, we consider only the t-matrix equation (3.1.3), which we write in operator form as

$$t = V + Kt, \tag{3.4.1}$$

where

$$K = VG_0^{(+)}. \tag{3.4.2}$$

Successive iterations of Eq. (3.4.1) yield the Born or Neumann series

$$t = V + KV + K^2V + \cdots. \tag{3.4.3}$$

The calculation of $t_s(q, q')$ using Eq. (3.4.3) requires only successive matrix vector multiplications as compared to matrix inversion or other algebraic means of solving sets of linear equations. However, in many physical situations, the series (3.4.3) either diverges or converges too slowly to be of practical value. Evidently, it is crucial to have analytic criteria for the convergence characteristics of Eq. (3.4.3).

To this end, we consider the eigenvalue condition

$$\lambda K(E)|\chi(E, \lambda)\rangle = \eta(E, \lambda)|\chi(E, \lambda)\rangle, \tag{3.4.4}$$

where we have explicated the dependence upon the parametric energy E that enters into the propagator $G_0^{(+)}(E)$. The parameter λ provides a means of adjusting the overall strength of the interaction, with V regarded as a fixed operator. The relevant t-matrix then satisfies

$$t(\lambda) = \lambda V + \lambda K t(\lambda), \tag{3.4.5}$$

rather than Eq. (3.4.1). For a given λ, if Eq. (3.4.4) possesses eigenvalues

$$\eta(E_B, \lambda) = 1 \tag{3.4.6}$$

for some energies E_B, then the homogeneous solution of Eq. (3.4.5) has nontrivial solutions signaling the appearance of bound states (typically with $E_B < 0$), the nonuniqueness of the solution of Eq. (3.4.5), and the vanishing of the Fredholm determinant; thus, the Fredholm solution is not defined. Generally, the necessary and sufficient condition for the convergence of the iteration series of Eq. (3.4.5) at an energy E is that all the eigenvalues satisfy[1]

$$|\eta(E, \lambda)| < 1. \tag{3.4.7}$$

For a sufficiently weak interaction, Eq. (3.4.7) is satisfied for all E and in the limit $\lambda \to 0$, $\eta(E, \lambda) \to 0$. For $|E|$ larger than a value depending on λ, which we denote as $E_c(\lambda)$, Eq. (3.4.7) is satisfied, which corresponds to the well-known result that the Born series converges at high energies.

Let us describe the behavior, as $|\lambda|$ is increased from zero, of the eigenvalues $\eta(E, \lambda)$ at a fixed $E < 0$ for a pure attractive-type potential that is capable of sustaining bound states when it is of sufficient strength. When $|\lambda|$ is sufficiently large, say $|\lambda_1|$, the largest eigenvalue, $\eta_1(E, \lambda)$, will have increased in magnitude so that

$$\eta_1(E, \lambda_1) = 1, \tag{3.4.8a}$$

but with

$$\eta_i(E, \lambda_1) < 1, \tag{3.4.8b}$$

for all of the other eigenvalues $i \neq 1$. Equations (3.4.8a, b) mean that there is a single bound state with binding energy $|E|$.

As $|\lambda|$ is increased further, there is a value, $\lambda = \lambda_2$, such that the next largest eigenvalue, η_2, becomes unity:

$$\eta_2(E, \lambda_2) = 1, \tag{3.4.9a}$$

with

$$\eta_1(E, \lambda_2) > 1, \tag{3.4.9b}$$

$$\eta_i(E, \lambda_2) < 1, \tag{3.4.9c}$$

for $i \neq 1, 2$. Equations (3.4.9a–c) mean that there are two bound states with binding energy greater than or equal to $|E|$. We have enumerated the eigenvalues according to the order of their attainment of magnitude unity as $|\lambda|$ is increased.

Typically, we are interested in the behavior of $\eta(E, \lambda)$ with respect to E for a fixed λ corresponding to a definite choice of potential. If a system has n bound states, then

$$|\eta_i(0, \lambda)| > 1 \tag{3.4.10}$$

for $i = 1, \ldots, n$, so at the threshold for scattering ($E = 0$) the Born series diverges and it will diverge for $E < E_c(\lambda)$. For $E > E_c(\lambda)$ the interaction is effectively weakened to the point where Eq. (3.4.7) is satisfied for all eigenvalues, and the Born series converges.

Many realistic potentials are strongly repulsive at short distances. Then, there can be eigenvalues with magnitudes greater than unity that do not correspond to bound states, but, nevertheless, imply that the Born series diverges. This arises with potential representations of the NN interaction.

One way to circumvent the divergence difficulties arising from either bound states or strongly repulsive potentials is to segregate those features into a subsidiary problem that can be solved exactly. Then, hopefully, the iteration technique can be applied to the remainder of the problem. The *Schmidt method* for doing this is a special case of the generalized Heitler subtraction technique of Section 1.8.[1] It is based on the decomposition of K into a separable (finite-rank) part K_S, and nonseparable remainder K_R:

$$K = K_S + K_R. \tag{3.4.11}$$

Then, the analysis of Section 1.8 applied to Eq. (3.4.1) yields

$$t = \Gamma_R V + \Gamma_R K_S t, \tag{3.4.12}$$

where

$$\Gamma_R = 1 + K_R \Gamma_R. \tag{3.4.13}$$

Since K_S is separable, if Γ_R is known Eq. (3.4.12) can be solved by quadrature. The objective in practice is to choose K_S as such a good approximation that the iteration solution for Γ_R gives results of the desired accuracy. For compact K, it is always possible, in principle, to find a K_S such that the eigenvalues of K_R are all less than unity. In practice, however, the criteria for choosing an appropriate K_S are seldom obvious.

As an example, suppose K has only a single eigenvalue greater than unity and it corresponds to a bound state $|B\rangle$ that satisfies

$$(H_0 + V)|B\rangle = E_B|B\rangle. \tag{3.4.14}$$

A rank-one potential that generates the same bound state is clearly

$$V_{\mathrm{S}} = \frac{V|B\rangle\langle B|V}{\langle B|V|B\rangle}. \tag{3.4.15}$$

This suggests a decomposition (3.4.11) of the form[1]

$$K_{\mathrm{S}} = V_{\mathrm{S}} G_0^{(+)}, \tag{3.4.16}$$

$$K_{\mathbf{R}} = (V - V_{\mathrm{S}}) G_0^{(+)}, \tag{3.4.17}$$

in Eq. (3.4.1). For a potential V, which supports a single bound-state, it is not difficult to show that $K_{\mathbf{R}}$ has no eigenvalues corresponding to a bound state.

The preceding simple example is difficult to generalize. A problem arises when K has more than one eigenvalue of magnitude greater than unity, especially when some of these eigenvalues correspond to strong repulsive interactions rather than bound states. The structure of K_{S} will then be more complicated than (3.4.16) because an energy-independent, rank-one potential such as V_{S} can sustain only a single-bound state. Thus, in general, the effective interactions entering into K_{S} and $K_{\mathbf{R}}$ need to be energy-dependent, or of higher rank.

The Schmidt–Weinberg[1] technique certainly provides a systematic procedure for constructing a finite-rank K_{S} in terms of all the eigenvectors corresponding to eigenvalues that can attain magnitudes equal to or greater than unity. This is one way of extending the simple example summarized in Eqs. (3.4.14–3.4.17). The extension we consider next retains the rank-one form for V_{S}, but allows the "form-factors" to be energy dependent.

One method for determining such a K_{S} utilizes the formalism developed in Section 3.3b. We see that a decomposition of the form (3.4.11) is implicit in (3.3.8) with the partial-wave amplitudes corresponding to K_{S} given by

$$K_{\mathrm{S}}(p, q) = \frac{2}{\pi} \frac{V(p, k)\gamma(k, q)q^2}{(k^2 - q^2 + i0)}, \tag{3.4.18}$$

so that

$$K_{\mathbf{R}}(p, q) = A_{k^2}(p, q; k). \tag{3.4.19}$$

Evidently, both $K_{\mathbf{R}}(p, q)$ and $K_{\mathrm{S}}(p, q)$ depend on the parametric energy k^2 and the angular momentum L in an important way, even though we have not indicated this dependence. The primary question is whether the decomposition (3.4.18–3.4.19) is effective in attaining the original goal of the entire approach. Namely, does the iteration solution corresponding to $K_{\mathbf{R}}$ as given by Eq. (3.4.19) converge?

The answer is always affirmative for a local potential with the Brown *et al.* (BFLS)[2] choice

$$\gamma_{\text{BFLS}}(k, q) = \left(\frac{q}{k}\right)^{L}, \tag{3.4.20}$$

a result, proven by Coester[3] and Blasczak and Fuda,[4] independent of the number of eigenvalues greater than unity in magnitude.

When the potential is not local, there is no obvious candidate for a suitable subtraction function. Moreover, even for a local potential, there is no indication of how fast the relevant iteration series converges. It is likely that different choices of subtraction functions will be more effective for different problems.

Another choice of subtraction function can be motivated as follows.[5] Let us note that

$$\int dq \left\{ \int dp \, [A_{k^2}(p, q; k)]^2 \right\} < 1, \tag{3.4.21}$$

which suffices for the convergence of the iteration series, can be regarded as a constraint on A as a functional of the subtraction function. While we cannot implement Eq. (3.4.21) exactly, it does suggest we look for $\gamma(k, q)$ such that, for first-order changes in γ,

$$\delta \int dp \, \omega(p)[V(p, q) - V(p, k)\gamma(k, q)]^2 = 0, \tag{3.4.22}$$

for arbitrary functions $\omega(p)$. Equation (3.4.22) is satisfied if γ has the form

$$\gamma_{\text{VAR}}(k, q) \equiv \left\{ \int dp \, \omega(p) V(p, k) V(p, q) \right\} \bigg/ \left\{ \int dp \, \omega(p) [V(p, k)]^2 \right\}. \tag{3.4.23}$$

The function $\omega(p)$ is chosen so as to enhance or suppress behaviors of the potential in certain regions, and such that both integrals in Eq. (3.4.23) are convergent. In practice $\omega(p)$ is taken to be a simple analytic form such as p^n, where $n = 1, 2, \ldots$.

3.5. Convergence

In this section, we present three different methods of accelerating the convergence of the iteration algorithm discussed in Section 3.4.

3.5a. *Iteration–Subtraction Method*

The original subtraction procedure of Section 3.3 leads to the expression of the solution of the LS equation in terms of the solution of the auxiliary equation (3.3.11) for $\Gamma_s(q, q'; k)$. Let us suppose that this amounts to a "weakening" of the LS kernel that ensures the iteration series for $\Gamma_s(q, q'; k)$ either converges or at least diverges slower than for the original equation (3.1.4). It seems plausible to expect that if additional subtractions are made to weaken successive auxiliary kernels, the corresponding convergence rates progressively increase. We next show how this can be done.

We introduce another subtraction function $\gamma_1(k_1, q)$ with respect to a new subtraction point such that

$$\gamma_1(k_1, k_1) = 1. \tag{3.5.1}$$

Then, carrying out the standard subtraction algorithm on Eq. (3.3.11) using γ_1, we obtain[1]

$$\Gamma_s(q, q'; k) = V(q, q') + A_s(q, k_1; k) \int dp\, \gamma_1(k_1, p)\Gamma_s(p, q'; k) + \int dp\, A_s^1(q, p; k, k_1)\Gamma_s(p, q'; k), \tag{3.5.2}$$

where $s = k^2$ and

$$A_s^1(q, p; k, k_1) = A_s(q, p; k) - A_s(q, k_1; k)\gamma_1(k_1, p). \tag{3.5.3}$$

Since $A_s(q, p; k)$ is nonsingular, we see that, in contrast to Eq. (3.3.10),

$$A_s^1(q, k_1; k; k_1) = 0, \tag{3.5.4}$$

which is consistent with a weakening of the kernel.

The form of Eq. (3.5.2) suggests that we introduce the auxiliary integral equation

$$\Gamma_s^1(q, q'; k, k_1) = V(q, q') + \int dp\, A_s^1(q, p; k, k_1)\Gamma_s^1(p, q'; k, k_1), \tag{3.5.5}$$

so that

$$\Gamma_s(q, q'; k) = \Gamma_s^1(q, q'; k, k_1) + Y_s^1(q; k, k_1) \int dp\, \gamma_1(k_1, p)\Gamma_s(p, q'; k), \tag{3.5.6}$$

where

$$Y_s^1(q; k, k_1) = \frac{2}{\pi} k_1^2 [\Gamma_s^1(q, k_1; k, k_1)$$

$$- \Gamma_s^1(q, k; k, k_1)\gamma(k, k_1)] \frac{1}{k^2 - k_1^2}. \tag{3.5.7}$$

Given $\Gamma_s^1(q, q'; k, k_1)$, the solution of Eq. (3.5.6) for $\Gamma_s(q, q'; k)$ is trivial because its kernel is separable.

This procedure can be continued indefinitely. The present algorithm begins by solving Eq. (3.5.5) by iteration; in this regard, we note that the function γ_1 can be chosen analogously to Eq. (3.4.23), but with $V(p, k), \ldots,$ replaced by $A_s(p, k), \ldots,$ and $\omega(p)$ by a suitable weighting function $\omega_1(p)$. The t-matrix is obtained by using the solution of Eq. (3.5.6) in connection with Eqs. (3.3.20–3.3.21).

3.5b. *Continued-Fraction Method*

Horacek and Sasakawa[2] have developed a continued-fraction method that utilizes only the terms generated in the Born series of the LS equation for the t-matrix.

Let us suppose we wish to calculate $t_s(p, q)$ for two specific values of the momenta, say p and q; the parametric energy is presumed fixed at some positive value k^2. Then, we define, using a bra-ket notation for a given L, an auxiliary potential

$$V_1 = V - \frac{V|q\rangle\langle p|V}{V(p, q)}, \tag{3.5.8}$$

which generates a t-matrix t_1 from the LS equation

$$t_1 = V_1 + V_1 G_0^{(+)} t_1, \tag{3.5.9}$$

where we note that $\langle p|V_1$ and $V_1|q\rangle$ vanish. It follows that

$$t_1 = t - \frac{t|q\rangle\langle p|V}{V(p, q)}(1 + G_0^{(+)} t_1). \tag{3.5.10}$$

If we introduce

$$\tau_1 \equiv \langle p|(VG_0^{(+)}V + VG_0^{(+)} t_1 G_0^{(+)} V)|q\rangle, \tag{3.5.11a}$$

then we see from Eq. (3.5.10) that

$$\tau_1 = t(p,q) - V(p,q) - \frac{t(p,q) - V(p,q)}{V(p,q)}\tau_1, \tag{3.5.11b}$$

so that

$$t(p,q) = \frac{[V(p,q)]^2}{V(p,q) - \tau_1}. \tag{3.5.12}$$

Here we have suppressed the dependence of t, t_1, τ_1, etc., on the energy parameter s.

Since t_1 satisfies a LS equation, the preceding procedure can be repeated indefinitely by introducing operators t_i that satisfy[2,3]

$$t_i = V_i + V_i G_0^{(+)} t_i, \tag{3.5.13}$$

where

$$V_{i+1} = V_i - \frac{V_i|q_i\rangle\langle p_i|V_i}{V_i(p_i, q_i)}, \tag{3.5.14}$$

with $i = 0, 1, \ldots, N-1$, $V_0 = V$, and $t_0 = t$, and

$$|q_{i+1}\rangle = G_0^{(+)} V_i |q_i\rangle, \tag{3.5.15a}$$

$$\langle p_{i+1}| = \langle p_i| V_i G_0^{(+)}, \tag{3.5.15b}$$

with $|q_0\rangle = |q\rangle$ and $\langle p_0| = \langle p|$. The repeated use of the (3.5.10–3.5.12) procedure yields

$$\tau_i = V_{i-1}(p_{i-1}, q_i) + t_i(p_i, q_i), \tag{3.5.16a}$$

where

$$t_i(p_i, q_i) = [V_i(p_i, q_i)]^2 [V_i(p_i, q_i) - \tau_{i+1}]^{-1}. \tag{3.5.16b}$$

Thus, τ_1, as it appears in Eq. (3.5.12), can be calculated recursively by means of Eqs. (3.5.16a, b).

Note that the level of approximation is fixed by setting $\tau_N = 0$. The iteration scheme based on Eqs. (3.5.16a, b) depends parametrically on $\langle p|$ and $|q\rangle$, and converges for small values of N in many instances.[2]

We note, in general, that the $t_i(p_i, q_i)$ are complex, unlike $\Gamma_s(p, q; k)$, which is purely real. Evidently, a purely real iteration scheme can be formulated for the $\hat{K}$-matrix.

Horacek and Sasakawa[2] have pointed out that the above method of defining a continued fraction can be used in other contexts, for example, for the wave function or the full Green function, and this leads to slightly different ways of solving the scattering equations.

In this context it is appropriate to mention the closely related method of Oryu[4] for calculating the off-shell t-matrix elements in finite-rank forms. The method is based on the Γ-matrix method of Section 3.3 with the subtraction function γ_{KFN} of Eq. (3.3.27). In this case, the residual off-shell term $R_s(q, q'; k)$ satisfies the LS equation (3.3.32), so this procedure can be applied indefinitely. For example, R_s can be written in the form (3.3.20) as the sum of a separable term and a nonseparable remainder, provided that the new subtraction point k_1 is not taken as the on-shell momentum and a KFN subtraction function of the form (3.3.27) is used. The calculational scheme is defined by setting the nonseparable remainder to zero. Oryu and collaborators have successfully applied this method for solving the two-nucleon and three-nucleon scattering problems.

3.5c. *Padé Method*

The Padé approximant technique is especially suited to utilize the Neumann series of an integral equation to approximate the exact solution irrespective of whether the series is slowly convergent or even divergent. It can be applied either to the integral equations for $t_s(q, q')$ or $\Gamma_s(q, q'; k)$.

Let us review the basic ideas that underlie the Padé method.[5,6] Typically one has a series representation for a quantity f of the form

$$f = \sum_{j=0}^{\infty} f_j. \tag{3.5.17a}$$

When this series diverges, it has a formal meaning as the $\lambda = 1$ case of a series

$$f(\lambda) = \sum_{j=0}^{\infty} \lambda^j f_j \tag{3.5.17b}$$

that can be generated according to some systematic algorithm, and that converges for some domain of λ that may or may not include the point $\lambda = 1$. The $[M, N]$ Padé approximant to $f(\lambda)$ is the rational fraction

$$f^{[M,N]}(\lambda) = A_N(\lambda)/B_M(\lambda). \tag{3.5.18}$$

It is constructed by writing

$$A_N(\lambda) = \sum_{j=0}^{N} a_j \lambda^j, \tag{3.5.19a}$$

$$B_M(\lambda) = 1 + \sum_{j=1}^{M} b_j \lambda^j, \tag{3.5.19b}$$

and then demanding that the expansion of $f^{[M,N]}(\lambda)$ in powers of λ coincide with the first $N - M - 1$ terms of the series (3.5.17b). This yields $(N - M - 1)$ equations in an equal number of unknowns, viz., the a_j's and the b_j's. One finds that

$$A_N(\lambda) = \begin{vmatrix} f_{N-M+1} & f_{N-M+2} & \cdots & f_{N+1} \\ f_{N-M+2} & f_{N-M+3} & \cdots & f_{N+2} \\ \vdots & \vdots & \vdots & \vdots \\ f_N & f_{N+1} & \cdots & f_{N+M} \\ \sum_{i=M}^{N} f_{i-M}\lambda^i & \sum_{i=M-1}^{N} f_{i-M+1}\lambda^i & \cdots & \sum_{i=0}^{N} f_i\lambda^i \end{vmatrix} \tag{3.5.20a}$$

and

$$B_M(\lambda) = \begin{vmatrix} f_{N-M+1} & f_{N-M+2} & \cdots & f_{N+1} \\ f_{N-M+2} & f_{N-M+3} & \cdots & f_{N+2} \\ \vdots & \vdots & & \vdots \\ f_N & f_{N+1} & \cdots & f_{N+M} \\ \lambda^M & \lambda^{M-1} & \cdots & 1 \end{vmatrix}. \tag{3.5.20b}$$

For certain classes of $f(\lambda)$, it can be proven that

$$f^{[M,N]}(\lambda) \to f(\lambda) \tag{3.5.21}$$

as $N, M \to \infty$.

The preceding method requires the calculation of the determinants (3.5.20), and this entails round-off errors that increase with increasing rank. It is sometimes better to employ a special version of the Padé method that takes the form of a continued fraction

$$f(\lambda) = 1 + \cfrac{\alpha_1^{(0)}\lambda}{1 + \cfrac{\alpha_2^{(0)}\lambda}{1 + \cfrac{\alpha_3^{(0)}\lambda}{1 + \cdots}}}, \tag{3.5.22}$$

where $f(\lambda)$ has been normalized so that $f(0) = 1$. The termination of the continuation by setting $\alpha_j^{(0)} = 0$ for j larger than some integer N must yield an expression identical with some Padé approximant, since then the right-hand side of Eq. (3.5.22) can be expressed as a ratio of two polynomials.

The continued-fraction algorithm can be characterized by the prescription

$$f_j(\lambda) = 1 + [\lambda\alpha_j^{(0)}/f_{j+1}(\lambda)], \tag{3.5.23a}$$

where $j = 1, 2, \ldots,$ and

$$f(\lambda) = f_1(\lambda). \tag{3.5.23b}$$

This recurrence relation can be put in a more tractable form if we write

$$f_j(\lambda) = g_j(\lambda)/g_{j+1}(\lambda), \tag{3.5.24}$$

so that it is equivalent to

$$g_j(\lambda) - g_{j+1}(\lambda) = \alpha_j^{(0)} \lambda g_{j+2}(\lambda), \tag{3.5.25}$$

where we can set

$$g_j(0) = g_2(\lambda) = 1, \tag{3.5.26}$$

without any loss of generality.

The new recurrence relations (3.5.25) can be solved in a straightforward manner. For $j = 1$, we have

$$g_1(\lambda) - g_2(\lambda) = f_1\lambda + f_2\lambda^2 + \cdots = \alpha_1^{(0)} \lambda g_3(\lambda). \tag{3.5.27}$$

Using Eq. (3.5.26), we find

$$\alpha_1^{(0)} = f_1, \tag{3.5.28}$$

$$g_3(\lambda) = \frac{1}{f_1} \sum_{j=1}^{\infty} f_j \lambda^{j-1}. \tag{3.5.29}$$

For $j = 2$, we have

$$g_2(\lambda) - g_3(\lambda) = -\frac{1}{f_1} \sum_{j=1}^{\infty} f_{j+1} \lambda^j = \alpha_2^{(0)} \lambda g_4(\lambda), \tag{3.5.30}$$

so that

$$\alpha_2^{(0)} = -f_2/f_1, \tag{3.5.31}$$

$$g_4(\lambda) = \frac{1}{f_2} \sum_{j=2}^{\infty} f_j \lambda^{j-2}. \tag{3.5.32}$$

Similarly, taking $j = 3$, we obtain

$$g_3(\lambda) - g_4(\lambda) = \sum_{j=1}^{\infty} \left(\frac{f_{j+1}}{f_1} - \frac{f_{j+2}}{f_2} \right) \lambda^j = \alpha_3^{(0)} \lambda g_5(\lambda), \tag{3.5.33}$$

so that

$$\alpha_3^{(0)} = \frac{f_2}{f_1} - \frac{f_3}{f_2},$$

$$g_5(\lambda) = \frac{1}{f_2/f_1 - f_3/f_2} \sum_{j=2}^{\infty} \left(\frac{f_j}{f_1} - \frac{f_{j+1}}{f_2} \right) \lambda^{j-2}.$$

Finally, we need to consider the case $j = 4$, for which

$$g_4(\lambda) - g_5(\lambda) = \sum_{j=1}^{\infty} \left(\frac{f_{j+2}}{f_2} - \gamma_{j+2} \right) \lambda^j = \alpha_4^{(0)} \lambda g_6(\lambda), \tag{3.5.34}$$

where

$$\gamma_{j+2} = \left(\frac{f_{j+2}}{f_1} - \frac{f_{j+3}}{f_2} \right) \Big/ \left(\frac{f_2}{f_1} - \frac{f_3}{f_2} \right). \tag{3.5.35}$$

We infer that

$$\alpha_4^{(0)} = \frac{f_3}{f_2} - \gamma_3. \tag{3.5.36}$$

The preceding determination of the $\alpha_j^{(0)}$'s can be formalized in a simple way. We introduce additional auxiliary quantities $\alpha_j^{(i)}$, where $i = 0, 1, \ldots$. Then, for $j \geq 3$,

$$\alpha_j^{(i)} = \frac{\alpha_{j-2}^{(i+1)}}{\alpha_{j-2}^{(0)}} - \frac{\alpha_{j-1}^{(i+1)}}{\alpha_{j-1}^{(0)}}, \tag{3.5.37a}$$

where

$$\alpha_1^{(i)} = f_{i+1}, \tag{3.5.37b}$$

$$\alpha_2^{(i)} = -f_{i+2}/f_1. \tag{3.5.37c}$$

The Padé approximants $f^{[M,N]}(\lambda)$ are related to the continued fractions $f_L(\lambda)$ defined by Eq. (3.5.22) with $\alpha_{L+1}^{(0)} = 0$ by

$$f^{[M,M]}(\lambda) = f_{2M}(\lambda), \tag{3.5.38a}$$

$$f^{[M,M-1]}(\lambda) = f_{2M-1}(\lambda), \tag{3.5.38b}$$

for integers M. As $M \to \infty$, both forms of approximation are expected to converge to the same function. We emphasize that the continued-fraction form is distinctive in *not* dealing with determinants, and thus is expected to have less round-off error.

3.5d. Numerical Test of the Rate of Convergence

The methods presented in this section are efficient for obtaining high-precision numerical results for the solution of LS-type scattering equations. For the computation of on-shell phase shifts, all of these methods usually use the iterative Neumann series solution of a manifestly nonsingular integral equation, and not the matrix-inversion technique. If a convergent solution could be obtained with the use of a small number of terms in the Neumann series, the accumulated error is expected to be small. On the other hand, if a large number of terms of a slowly convergent or divergent Neumann series (as in the Padé technique) are used, one expects low accuracy. In practical cases, the dimension of the discretized matrix equation is usually large, and so direct matrix inversion has a large round-off error. It is relevant to ask which of the preceding three iterates of the Neumann series has the greatest accuracy. The appearance of each $G_0^{(+)}$ corresponds to an integration, and hence an iteration of the LS equation. The different methods are to be compared by the order of the multiple integrations one has to perform in order to achieve a given accuracy. For instance, the Nth iteration of the continued-fraction method makes use of $2N$ iterations of the original LS equation.

All three methods have been used in solving different problems. In order to make a comparison of the rates of convergence obtained with the three methods, we apply them to the solution of the LS equation with the Reid NN $^1S_0$ interaction. Because of a strong soft core, this potential provides a stringent numerical test. The kernel of the LS equation in this case has large repulsive eigenvalues, so the original Neumann series diverges badly at zero energy; the ratio of successive terms is -15.

Table 3.1 lists the results for the $^1S_0$ phase shift obtained with three different methods using the Reid potential. The iteration–subtraction method (ISM) was employed using two subtractions with functions of the form (3.4.23); since these subtractions have been performed optimally, the convergence is rapid.

The continued-fraction method employs successive subtractions in the kernel, and terminates with the Born approximation in the final equation with N subtractions in the kernel. Since the subtractions have not been done optimally in this last case, the initial convergence is slow; eventually, it is expected to converge faster than the ISM. This expectation is based on the presumption that one subtraction in the kernel reduces its eigenvalues faster than two iterations of a convergent kernel. Also, it is realized that one subtraction in the kernel involves two integrals, and two iterations also involve two integrals. Hence a single subtraction in the method of continued fractions should be compared numerically with two iterations of the ISM.

The success of the methods rest on the promise of finding a converged solution to a desired accuracy after a small number of iterations; otherwise, the accumulated numerical error may jeopardize the final accuracy. From Table 3.1, it is seen that the ISM has excellent initial convergence because the two subtractions in this method have been optimally chosen for this purpose. We present results for two versions of the continued-fraction method,[2,7] both

TABLE 3.1. The Nth order terms for the on-shell t-matrix of the conventional Neumann series and the phase shifts calculated with three other methods at $E_{\text{c.m.}} = 12$ MeV for the Reid $^1S_0$-potential. Continued fraction:* Horacek (1985),** Ishikawa (1987b). Padé: Glöckle (1983a). Iteration subtraction: Adhikari (1986).

N	Neumann series (on-shell t-matrix)	Continued fraction (phase shift)	Padé method (phase shift)	Iteration subtraction (phase shift)
2	$-0.937 \times 10^1 - i0.427 \times 10^0$	−0.0098* 0.41509**	0.7983	−0.06174
4	$-0.225 \times 10^4 - i0.526 \times 10^3$	0.9849* 0.60430**	2.151	0.78171
6	$-0.539 \times 10^6 - 0.236 \times 10^6$	0.8293* 0.85141**	2.227	0.85958
8	$-0.124 \times 10^9 - 0.841 \times 10^8$	0.8272* 0.85053**	0.9079	0.86058
10	$-0.272 \times 10^{11} - i0.268 \times 10^{11}$	0.8613* 0.86063**	0.8610	0.86063
12	$-0.556 \times 10^{13} - i0.798 \times 10^{13}$	0.8618* 0.86066**	0.8607	0.86063
14	$-0.120 \times 10^{16} - i0.226 \times 10^{16}$	0.86065**	0.8607	0.86063

of which have poorer convergence properties than the ISM. From the results in Table 3.1, we see that numerical instabilities appear in the continued-fraction method, and the final result oscillates around the correct value, whereas the ISM converges faster, and does not present these oscillations. The rate of convergence obtained with the Padé technique is only fair compared to the other two methods.

The use of Padé techniques has a long tradition in many areas of applied mathematics, chemistry, and physics,[6] including scattering problems. For example, it has been successfully used by Cornelius, Glöckle, and Witala[8] for solving the three-nucleon scattering problem above the breakup threshold with the Bonn and the Paris NN potentials. The other two methods of this section have not had widespread application in solving realistic collision problems, and perhaps deserve more attention in the future.

The ISM has several advantages over the commonly used Padé technique.[1] Firstly, there are serious questions on the uniformity of convergence in the Padé method because of the appearance of spurious poles in the denominator. Secondly, it is numerically more tedious to construct the Padé approximants using the iterative solution than to construct the solution using the ISM. Finally, the rate of convergence of the method of Padé approximants depends in a complicated way on the structure of the equations. Tjon and Nieland[9] have noted that, for higher partial waves of the LS equation, the solutions obtained using Padé approximants converge rather poorly, possibly because of the threshold factors in these equations. The higher-partial-wave kernels are weaker compared to the S-wave kernel and the ISM has been demonstrated to converge much faster for higher partial waves.

3.6. Collision Integral Equation Approach to the Bound-State Problem

We next present a numerical method for solving the two-body bound-state problem that is intimately related to the Γ-matrix equations of Section 3.3b for collisions. The usual textbook prescriptions for calculating the two-body bound-state wave function in configuration space is practical only for local potentials. Configuration space is therefore cumbersome for the realistic potentials that are used to represent hadron–hadron interactions that are often most conveniently expressed in momentum space. This is also usually the case for the effective interactions between two composite objects that are widely employed in atomic and nuclear physics.

After a partial-wave projection, the momentum-space integral equation for the bound-state wave-function $\psi(p, E)$ is

$$\psi(p, E) = \frac{2}{\pi}(E - p^2)^{-1} \int dq\, q^2 V(p, q)\psi(q, E), \tag{3.6.1}$$

where $E < 0$, $\hbar = 2\mu = 1$, μ is the reduced mass, and we suppress all angular momentum labels.

First, we describe a method due to Malfliet and Tjon[1] for solving Eq. (3.6.1), which can be written schematically as

$$\psi = K(E)\psi, \tag{3.6.2}$$

where $K(E)$ is the integral-operator kernel. When the largest (in modulus) eigenvalue of K is attractive in nature and corresponds to a bound state, it is useful to consider the Neumann series

$$\psi = \sum_n \psi_n \equiv \sum_n K^n(E)\phi_0, \tag{3.6.3}$$

where ϕ_0 is any wave function. If the kernel K has at least one eigenvalue greater than or equal to unity, then

$$\lim_{n\to\infty} \frac{\psi_{n+1}}{\psi_n} = \eta_{\max}, \tag{3.6.4}$$

where $\eta_{\max}$ is the largest eigenvalue. It is easy to verify the limit (3.6.4) in a representation where K is diagonal. At the bound-state energy, Eq. (3.6.2) suggests that $\eta_{\max} = 1$. Hence, one should vary the energy and determine at what energy the limit (3.6.4) is satisfied with $\eta_{\max} = 1$ to a desired accuracy. In the present case, this energy corresponds to the bound-state energy, and $\lim_{n\to\infty} \psi_n$ is the bound-state wave function. This method has also been successfully applied to the three-nucleon problem.[2]

Usually, one needs a large value of n in order to achieve convergence; in practice, this means there will be an associated accumulation of numerical error. Also, the kernel K of Eq. (3.6.2) may have repulsive eigenvalues at the bound-state energy larger than unity in magnitude, which are to be subtracted out before applying the method. One such subtraction method has been suggested by Glöckle.[3] The original Malfliet–Tjon method has similar difficulties in finding the excited states. An alternative approach that is free of all of these difficulties is to apply the idea of Section 3.3b to the bound-state problem.

Our strategy for dealing with Eq. (3.6.1) will be to convert it into the form of an inhomogeneous integral equation characteristic of a scattering problem. In doing this, we keep in mind that Eq. (3.6.1) has solutions only for particular values of E and that $\psi(p, E)$ remains undefined to within a multiplicative constant.

Suppose we decompose the potential as

$$V(p, q) = f_1(p) f_2(q) + V^{\mathrm{R}}(p, q), \tag{3.6.5}$$

which for some form-factors $f_1(p)$ and $f_2(q)$ defines the generally nonseparable remainder $V^{\mathrm{R}}(p, q)$. Let us choose the wave function normalization

$$\frac{2}{\pi} \int dq\, q^2 f_2(q) \psi(q, E) = 1. \tag{3.6.6}$$

Inserting Eq. (3.6.5) into Eq. (3.6.1) and using Eq. (3.6.6), we find that

$$\psi(p, E) = [(E - p^2)^{-1} f_1(p)] + (E - p^2)^{-1} \frac{2}{\pi} \int dq\, q^2 V^{\mathrm{R}}(p, q) \psi(q, E). \tag{3.6.7}$$

We note that the original equation (3.6.1) has solutions only for certain energies that are identified with the finite number of bound states of the system. On the other hand, the inhomogeneous equation (3.6.7) has a unique solution for all E except those for which its homogeneous form admits a nontrivial solution. Generally, these exceptional energies will not coincide with the actual bound-state energies.

By construction, any solution of Eq. (3.6.7) that satisfies the constraint (3.6.6) is a solution of Eq. (3.6.1). The strategy is to guess E, solve Eq. (3.6.7), and then check if Eq. (3.6.6) is satisifed. If it is, the trial energy-eigenvalue is the exact energy and the solution of Eq. (3.6.7) is the desired wave function; otherwise, one makes a new guess for E and repeats the process. If one knows the bound-state energies beforehand, the wave functions can be calculated from Eq. (3.6.7) straightaway.

We note that, since the energies of interest are negative, $(E - p^2)^{-1}$ is nonsingular. The objective of any subtraction procedure such as (3.6.5) is then only to weaken the kernel so that the resultant inhomogeneous equation (3.6.7) has a rapidly convergent Neumann series solution. From the discussion of Section 3.3b, we know that possibly a good choice for V^{R} to achieve this is[4]

$$V^{\mathrm{R}}(p,q) = V(p,q) - V(p,k)\gamma(k,q), \tag{3.6.8}$$

where k is some arbitrary momentum. Then, Eqs. (3.6.6) and (3.6.7) become

$$\frac{2}{\pi}\int dq\, q^2\gamma(k,q)\psi(q,E) = 1, \tag{3.6.9a}$$

and

$$\psi(p,E) = (E - p^2)^{-1}\left\{V(p,k) + \frac{2}{\pi}\int dq\, q^2[V(p,q) - V(p,k)\gamma(k,q)]\psi(q,E)\right\}, \tag{3.6.9b}$$

respectively. From Eqs. (3.3.22) and (3.6.9a), we infer that, at the bound-state, we have $E = E_B$ and[5]

$$\psi(p,E) = \frac{1}{E_B - p^2}\Gamma_{E_B}(p,k;k), \tag{3.6.10}$$

provided the Γ-matrix does not have bound-state poles. The bound-state energy is determined by the condition (3.3.24).

The preceding integral-equation approach for bound-state calculations has been applied by Dzhibuti and Tsiklauri.[6] It has also been used by Sasakawa and Ishikawa,[7] but with a different form of $V^{\mathrm{R}}(p,q)$ than (3.6.8); these authors solved their counterpart of Eq. (3.6.7) using the continued-fraction method of Section 3.5b.

In the first application of this method, the subtraction function (3.4.23) was utilized, although, in this case, the parameter k was not identified with an on-shell momentum.[4] Instead, k was chosen to be a constant independent of E; it is possible that taking $k = |E|^{1/2}$, for example, might be more efficient.[5,8]

3.7. Variational Principles for the Scattering Amplitudes

Variational formulations are very useful for calculating approximate scattering amplitudes. We do not attempt to review all of the many variational principles (v.p.'s) that have been proposed, but rather confine ourselves to those that seem to be promising for future developments.

The extant v.p.'s relating to the transition operator t are either Hulthén–Kohn or Schwinger types.[1–4] The former type refers to the combination $(t - V)$ while the latter only to t.

The original Hulthén–Kohn v.p. is the simplest. It does not involve $G_0^{(+)}$, but possesses the drawback of requiring that the trial wave functions have the same nonnormalizable asymptotic behavior as the exact scattering wave functions. All Schwinger-type v.p.'s, as well as the more complicated versions of the Hulthén–Kohn variety, involve $G_0^{(+)}$ explicitly, and can employ normalizable L^2 trial wave-functions.

These v.p.'s can be written in two different, equivalent forms: (a) as a linear combination of different exact expressions for the unknown quantity, and (b) as a ratio involving different exact expressions for the unknown quantity. In our discussion, we are concerned only with the first form; most of the same ideas can be applied to the second form with equivalent results.

3.7a. *Schwinger Variational Principles*

The original Schwinger v.p. is based on the identity

$$[\langle \mathbf{p}'|t|\mathbf{p}\rangle] = \langle \mathbf{p}'|V|\psi_{\mathbf{p}}^{(+)}\rangle + \langle \psi_{\mathbf{p}'}^{(-)}|V|\mathbf{p}\rangle - \langle \psi_{\mathbf{p}'}^{(-)}|(V - VG_0^{(+)}V)|\psi_{\mathbf{p}}^{(+)}\rangle, \tag{3.7.1}$$

where

$$|\psi_{\mathbf{p}}^{(+)}\rangle = |\mathbf{p}\rangle + G_0^{(+)}V|\psi_{\mathbf{p}}^{(+)}\rangle, \tag{3.7.2a}$$

$$\langle \psi_{\mathbf{p}'}^{(-)}| = \langle \mathbf{p}'| + \langle \psi_{\mathbf{p}'}^{(-)}|VG_0^{(+)}. \tag{3.7.2b}$$

The square-bracket notation in Eq. (3.7.1) is used to signify an expression that will be of interest variationally and, typically, is realized only approximately.

The variational aspect of Eq. (3.7.1) is derived from the following observation. Suppose we were to calculate $[\langle \mathbf{p}'|t|\mathbf{p}\rangle]$ using the trial wave functions

$$|\psi_{\mathbf{p}}^{(+)}\rangle_{\mathrm{T}} = |\psi_{\mathbf{p}}^{(+)}\rangle + |\delta\psi_{\mathbf{p}}^{(+)}\rangle, \tag{3.7.3a}$$

$$_{\mathrm{T}}\langle\psi_{\mathbf{p}'}^{(-)}| = \langle\psi_{\mathbf{p}'}^{(-)}| + \langle\delta\psi_{\mathbf{p}'}^{(-)}|, \tag{3.7.3b}$$

which differ from the exact wave functions by the variations $|\delta\psi_{\mathbf{p}}^{(+)}\rangle$ and $\langle\delta\psi_{\mathbf{p}'}^{(-)}|$. Then,

$$[\langle \mathbf{p}'|t|\mathbf{p}\rangle]_{\mathrm{T}} = \langle \mathbf{p}'|t|\mathbf{p}\rangle - \langle\delta\psi_{\mathbf{p}'}^{(-)}|(V - VG_0^{(+)}V)|\delta\psi_{\mathbf{p}}^{(+)}\rangle. \tag{3.7.4}$$

Here, $[\langle \mathbf{p}'|t|\mathbf{p}\rangle]$ is calculated from Eq. (3.7.1) using the wave functions (3.7.3a, b); the fact that it differs from the exact transition matrix by a quantity that is of second order in the variations is the statement of the v.p. That is, $[\langle \mathbf{p}'|t|\mathbf{p}\rangle]_{\mathrm{T}}$ and $\langle \mathbf{p}'|t|\mathbf{p}\rangle$ are identical to within first-order variations of the wave functions; this would not be the case if we had used instead

$$[\langle \mathbf{p}'|t|\mathbf{p}\rangle] = \langle \mathbf{p}'|V|\psi_{\mathbf{p}}^{(+)}\rangle, \tag{3.7.5}$$

for example. We note that the preceding discussion, as well as the one to follow, applies equally well to the scattering in each state of orbital angular momentum.

Other v.p.'s can be obtained from Eq. (3.7.1). For instance, we can rewrite this expression as

$$[\langle \mathbf{p}'|t|\mathbf{p}\rangle] = \langle \mathbf{p}'|\chi_{\mathbf{p}}^{(+)}\rangle + \langle\chi_{\mathbf{p}'}^{(-)}|\mathbf{p}\rangle - \langle\chi_{\mathbf{p}'}^{(-)}|(V^{-1} - G_0^{(+)})|\chi_{\mathbf{p}}^{(+)}\rangle, \tag{3.7.6}$$

where

$$|\chi_{\mathbf{p}}^{(+)}\rangle = V|\psi_{\mathbf{p}}^{(+)}\rangle, \tag{3.7.7a}$$

$$\langle\chi_{\mathbf{p}'}^{(-)}| = \langle\psi_{\mathbf{p}'}^{(-)}|V, \tag{3.7.7b}$$

and, clearly,

$$|\chi_{\mathbf{p}}^{(+)}\rangle = V|\mathbf{p}\rangle + VG_0^{(+)}|\chi_{\mathbf{p}}^{(+)}\rangle, \tag{3.7.8a}$$

$$\langle\chi_{\mathbf{p}'}^{(-)}| = \langle \mathbf{p}'|V + \langle\chi_{\mathbf{p}'}^{(-)}|G_0^{(+)}V. \tag{3.7.8b}$$

Equation (3.7.6) is stationary with respect to first-order variations in $|\chi_{\mathbf{p}}^{(+)}\rangle$ and $\langle\chi_{\mathbf{p}'}^{(-)}|$. There are an infinite number of Schwinger-type v.p.'s that are related to each other in a similar manner. For example, consider the identity

$$[\langle \mathbf{p}'|t|\mathbf{p}\rangle] = \langle \mathbf{p}'|V|\psi_{\mathbf{p}}^{(+)}\rangle + \langle\chi_{\mathbf{p}'}^{(-)}|\mathbf{p}\rangle - \langle\chi_{\mathbf{p}'}^{(-)}|(I - G_0^{(+)}V)|\psi_{\mathbf{p}}^{(+)}\rangle. \tag{3.7.9}$$

The expression (3.7.9) is stationary with regard to deviations of $|\psi_{\mathbf{p}}^{(+)}\rangle$ and $\langle\chi_{\mathbf{p}'}^{(-)}|$ from their exact values.

3.7b. Hulthén–Kohn Variational Principles

The canonical Hulthén–Kohn v.p. for the t-matrix is based on the identity

$$[\langle \mathbf{p}'|t|\mathbf{p}\rangle] = \langle \mathbf{p}'|V|\mathbf{p}\rangle + \langle \mathbf{p}'|V|\hat{\chi}_{\mathbf{p}}^{(+)}\rangle + \langle \hat{\chi}_{\mathbf{p}'}^{(-)}|V|\mathbf{p}\rangle - \langle \hat{\chi}_{\mathbf{p}'}^{(-)}|\{(G_0^{(+)})^{-1} - V\}|\hat{\chi}_{\mathbf{p}}^{(+)}\rangle, \quad (3.7.10)$$

where

$$|\hat{\chi}_{\mathbf{p}}^{(+)}\rangle = G_0^{(+)}|\chi_{\mathbf{p}}^{(+)}\rangle, \quad (3.7.11a)$$

$$\langle \hat{\chi}_{\mathbf{p}'}^{(-)}| = \langle \chi_{\mathbf{p}'}^{(-)}|G_0^{(+)}. \quad (3.7.11b)$$

The expression (3.7.10) for the t-matrix is stationary with respect to variations of the states (3.7.11a, b).

If Eq. (3.7.10) is regarded as a variational expression for $\langle \mathbf{p}'|(t - V)|\mathbf{p}\rangle$ rather than the t-matrix alone, we notice a similarity between its structure and that of Eq. (3.7.1). Each right-hand expression has three terms, each of which is equal to the quantity on the left or its negative. Also, each expression has one term that is quadratic in the wave functions, so that it is possible to have the first-order corrections vanish without having an exact wave function.

Other Hulthén–Kohn variational expressions for the t-matrix can be easily derived by rewriting Eq. (3.7.10). For example, we see that the expressions

$$[\langle \mathbf{p}'|(t - V)|\mathbf{p}\rangle] = \langle \mathbf{p}'|V|\hat{\chi}_{\mathbf{p}}^{(+)}\rangle + \langle \chi_{\mathbf{p}'}^{(-)}|G_0^{(+)}V|\mathbf{p}\rangle - \langle \chi_{\mathbf{p}'}^{(-)}|(I - G_0^{(+)}V)|\hat{\chi}_{\mathbf{p}}^{(+)}\rangle \quad (3.7.12)$$

and

$$[\langle \mathbf{p}'|(t - V)|\mathbf{p}\rangle] = \langle \mathbf{p}'|VG_0^{(+)}|\chi_{\mathbf{p}}^{(+)}\rangle + \langle \chi_{\mathbf{p}'}^{(-)}|G_0^{(+)}V|\mathbf{p}\rangle - \langle \chi_{\mathbf{p}'}^{(-)}|[G_0^{(+)} - G_0^{(+)}VG_0^{(+)}]|\chi_{\mathbf{p}}^{(+)}\rangle \quad (3.7.13)$$

are stationary with respect to variations of $\langle \chi_{\mathbf{p}'}^{(-)}|$, $|\hat{\chi}_{\mathbf{p}}^{(+)}\rangle$, and $|\chi_{\mathbf{p}}^{(+)}\rangle$ about their exact values. The expressions (3.7.10), (3.7.12), and (3.7.13) are trivally related to each other by a change in the structure of the trial wave functions. One can continue in this way indefinitely.

3.7c. Variational Principles of Higher Order

The Schwinger and Hulthén–Kohn v.p.'s refer to the matrix elements of t and $t - V$, respectively. Evidently, one can continue this progression and consider v.p.'s for t minus the first N terms, $N = 0, 1, 2, \ldots$, of the Born series for

t.[5] For example, the expression

$$[\langle \mathbf{p}'|t|\mathbf{p}\rangle] = \langle \mathbf{p}'|(V + VG_0^{(+)}V)|\mathbf{p}\rangle + \langle \mathbf{p}'|VG_0^{(+)}|\hat{\psi}_{\mathbf{p}}^{(+)}\rangle$$
$$+ \langle \hat{\psi}_{\mathbf{p}'}^{(-)}|G_0^{(+)}V|\mathbf{p}\rangle - \langle \hat{\psi}_{\mathbf{p}'}^{(-)}|(V^{-1} - G_0^{(+)})|\hat{\psi}_{\mathbf{p}}^{(+)}\rangle \quad (3.7.14)$$

is stationary with respect to variations of the quantities $|\hat{\psi}_{\mathbf{p}}^{(+)}\rangle$ and $\langle \hat{\psi}_{\mathbf{p}'}^{(-)}|$, where

$$|\hat{\psi}_{\mathbf{p}}^{(+)}\rangle = VG_0^{(+)}V|\psi_{\mathbf{p}}^{(+)}\rangle, \quad (3.7.15a)$$

$$\langle \hat{\psi}_{\mathbf{p}'}^{(-)}| = \langle \psi_{\mathbf{p}'}^{(-)}|VG_0^{(+)}V. \quad (3.7.15b)$$

Thus, Eq. (3.7.14) really amounts to a v.p. for $[t - (V + VG_0^{(+)}V)]$, as does the expression

$$[\langle \mathbf{p}'|t|\mathbf{p}\rangle] = \langle \mathbf{p}'|(V + VG_0^{(+)}V)|\mathbf{p}\rangle + \langle \mathbf{p}'|VG_0^{(+)}V|\eta_{\mathbf{p}}^{(+)}\rangle$$
$$+ \langle \eta_{\mathbf{p}'}^{(-)}|VG_0^{(+)}V|\mathbf{p}\rangle - \langle \eta_{\mathbf{p}'}^{(-)}|[V - VG_0^{(+)}V]|\eta_{\mathbf{p}}^{(+)}\rangle, \quad (3.7.16)$$

where

$$|\eta_{\mathbf{p}}^{(+)}\rangle = G_0^{(+)}V|\psi_{\mathbf{p}}^{(+)}\rangle, \quad (3.7.17a)$$

$$\langle \eta_{\mathbf{p}'}^{(-)}| = \langle \psi_{\mathbf{p}'}^{(-)}|VG_0^{(+)}. \quad (3.7.17b)$$

The expression (3.7.16) is stationary with respect to small variations of $|\eta_{\mathbf{p}}^{(+)}\rangle$ and $\langle \eta_{\mathbf{p}'}^{(-)}|$.

This classification of v.p.'s on the basis of the Born series is relatively new. We remark again that within each of the Schwinger or the Hulthén–Kohn classes, there are an infinite number of v.p.'s that are all related to one another by changes in the structure of the trial functions. The usefulness of these v.p.'s for solving the LS equation is examined in the next section.

3.8. Separable Expansions and Degenerate-Kernel Schemes

The earmark of a compact operator is that it can be approximated uniformily by sums of separable operators. Other synonyms for such an operator are "degenerate" or "finite rank." We have noted previously that the kernel of the two-particle LS equation (3.1.3) is compact for a class of potentials that is physically relevant; a rank-N approximation to the kernel reduces Eq. (3.1.3) to a matrix equation of rank N. Such an approximation is attractive only if N is much less than the number of quadrature points that would be needed to affect a numerical solution in the standard fashion to the same degree of accuracy.

If the kernel of Eq. (3.1.3) is approximated by a finite-rank form, the resulting t-matrix would also be of finite-rank except for the inhomogeneous term, the potential V. If the latter is also approximated by a finite-rank form, the resulting t-matrix is of finite rank and of great utility in rendering three-particle and other few-particle problems numerically tractable.

3.8a. Generalized Separable-Expansion Scheme

3.8a.1. General Considerations

Let us suppose, for the sake of illustration, that V has the simplest rank-one form:[1]

$$V = |f\rangle\langle f|, \tag{3.8.1}$$

where $|f\rangle$ is some normalizable "form-factor" or "vertex function" vector. Thus,

$$\langle \mathbf{p}'|V|\mathbf{p}\rangle = \langle \mathbf{p}'|f\rangle\langle f|\mathbf{p}\rangle. \tag{3.8.2}$$

If V is rotationally invariant, $\langle \mathbf{p}'|f\rangle$, for example, is a function only of $|\mathbf{p}'|$, and Eq. (3.8.1) generates scattering only in the $L = 0$ angular-momentum state, and so is realistic only for very low energies.

The solution of Eq. (3.1.3) with the potential (3.8.1) is

$$t = |f\rangle J\langle f|, \tag{3.8.3}$$

where

$$J^{-1} = 1 - \langle f|G_0^{(+)}|f\rangle. \tag{3.8.4}$$

A zero of J^{-1} at a particular energy corresponds to a bound state.

At the bound-state energy $E_B \equiv -B$, the t-matrix has a pole and its residue is separable. This pole term is given by

$$t_B \equiv \frac{V|B\rangle\langle B|V}{E + B + i0}, \tag{3.8.5}$$

which follows from Eq. (2.4.3). The separable approximation $t \simeq t_B$, which is accurate in the vicinity of the pole, does not even satisfy on-shell unitarity for $E > 0$.

The following separable-expansion scheme generalizes the rank-one potential (3.8.2) to a rank-N approximate potential V_N defined by[2]

$$V_N = \sum_{n,m=1}^{N} V|u_n\rangle D_{nm}\langle v_m|. \tag{3.8.6a}$$

The matrix D is defined through its inverse elements

$$(D^{-1})_{mn} \equiv \langle v_m | u_n \rangle, \qquad n, m = 1, \dots, N, \tag{3.8.6b}$$

where $|u_n\rangle$ and $\langle v_n|$ with $n = 1, \dots, N$, are two sets of linearly independent functions whose choice is only constrained at this point by the requirement that the $N \times N$ matrix in Eq. (3.8.6b) is nonsingular. The matrix D depends on N; for example, D_{11} for $N = 1$ is not identical with D_{11} for $N = 2$. Now V_N has the property of being exact when it operates on any linear combination of the N selected functions $|u_1\rangle, \dots, |u_n\rangle$, since we see that

$$V_N | u_n \rangle = V | u_n \rangle, \qquad n = 1, \dots, N. \tag{3.8.7}$$

If we use the approximation $V \simeq V_N$ in the kernel and the inhomogeneous term of Eq. (3.1.3), we find that $t \simeq t_N$, where

$$t_N \equiv \sum_{n,m=1}^{N} V | u_n \rangle J_{nm} \langle v_m |, \tag{3.8.8}$$

and where J depends on N, and is defined by

$$(J^{-1})_{mn} \equiv \langle v_m | (1 - G_0^{(+)} V) | u_n \rangle, \qquad n, m = 1, \dots, N. \tag{3.8.9}$$

At a bound state generated by the potential V_N, the matrix J^{-1} is singular and t_N has a pole.

As $N \to \infty$, we have $V_N \to V$ and, consequently, $t_N \to t$. The computation of t is affected by using the truncated sum (3.8.8).

For numerical calculations one needs explicit choices for the functions $|u_n\rangle$ and $\langle v_m|$. Some possibilities are:[2]

$$|u_n\rangle = |f_n\rangle, \qquad \langle v_m| = \langle f_m | V; \tag{3.8.10a}$$

$$|u_n\rangle = G_0^{(+)} |f_n\rangle, \qquad \langle v_m| = \langle f_m | G_0^{(+)} V; \tag{3.8.10b}$$

$$|u_n\rangle = G_0^{(+)} |f_n\rangle, \qquad \langle v_m| = \langle f_m |; \tag{3.8.10c}$$

$$|u_n\rangle = |f_n\rangle, \qquad \langle v_m| = \langle f_m |; \tag{3.8.10d}$$

$n, m = 1, \dots, N$, where the f_n's are, in turn, some suitably chosen real normalizable functions.

The choice (3.8.10a) leads to the following symmetric t-matrix:

$$t_N = \sum_{n,m=1}^{N} V | f_n \rangle J_{nm} \langle f_m | V, \tag{3.8.11a}$$

with

$$(J^{-1})_{mn} = \langle f_m | (V - V G_0^{(+)} V) | f_n \rangle. \tag{3.8.11b}$$

We see that t_N satisfies the LS equation

$$t_N = V_N + V_N G_0^{(+)} t_N, \tag{3.8.12a}$$

with a Hermitian potential

$$V_N = \sum_{n,m=1}^{N} V|f_n\rangle D_{nm}\langle f_m|V, \tag{3.8.12b}$$

where

$$(D^{-1})_{mn} = \langle f_m|V|f_n\rangle, \tag{3.8.12c}$$

and $|f_n\rangle$ and $\langle f_n|$ are duals of each other. Because V_N is Hermitian, t_N satisfies off-shell unitarity.

Most of the commonly known separable expansions, especially for the two-particle t-matrix, use the form (3.8.11a, b). Specific choices of f_m lead to different expansions. The interesting aspect of the particular form (3.8.12b) for V_N is duality:

$$V_N|f_m\rangle = V|f_m\rangle, \tag{3.8.13a}$$

$$\langle f_m|V_N = \langle f_m|V. \tag{3.8.13b}$$

This means that V_N is *exact* in the space spanned by $|f_m\rangle$'s. This is relevant because usually we do not need to calculate the t-matrix in the whole Hilbert space. For example, if we use t_N in the kernels of multiparticle collision integral equations, it operates on functions with certain analytic properties. If the $|f_m\rangle$'s are chosen to span this special class of functions, then t_N will be a good approximation to t in this context. A good pointwise agreement between the momentum-space elements of t and t_N is not usually required in practice. Next, we consider some specific choices of the $|f_m\rangle$.

3.8a.2. Unitary Pole Expansion (UPE)

The UPE corresponds to the choice:[2–4]

$$V|f_m\rangle = |\psi_m(-B)\rangle, \tag{3.8.14a}$$

$$\langle f_n|V = \langle\bar{\psi}_n(-B)|, \tag{3.8.14b}$$

where $|\psi_m(-B)\rangle$ and $\langle\bar{\psi}_n(-B)|$ are defined by

$$|\psi_m(-B)\rangle = \lambda_m(-B)VG_0^{(+)}(-B)|\psi_m(-B)\rangle, \tag{3.8.15a}$$

$$\langle\bar{\psi}_n(-B)| = \lambda_n(-B)\langle\bar{\psi}_n(-B)|G_0^{(+)}(-B)V, \tag{3.8.15b}$$

and are assumed to be normalized such that

$$\langle\bar{\psi}_n(-B)|G_0^{(+)}(-B)|\psi_m(-B)\rangle = \delta_{mn}, \tag{3.8.15c}$$

with $m, n = 1, \ldots, N$. Typically, B is the binding energy of a two-particle ground state, but if there is no bound state, B is usually taken to be zero. Then, we have

$$t_N(E) = \sum_{n,m=1}^{N} |\psi_n(-B)\rangle J_{nm}(E) \langle \bar{\psi}_m(-B)|, \tag{3.8.16a}$$

with

$$[J^{-1}(E)]_{mn} = \lambda_m(-B)\,\delta_{mn} - \langle \bar{\psi}_m(-B)| G_0^{(+)}(E) |\psi_n(-B)\rangle. \tag{3.8.16b}$$

The lowest-order term t_1 was advocated by Fuda and is called the unitary-pole approximation (UPA).[5] The full expansion, that was suggested by Harms,[3] satisfies off-shell unitarity.

The UPE has been found to converge well for nucleon–nucleon potentials with a soft core. It has been widely used as input for solving three-nucleon problems. One difficulty is that the rate of convergence is very sensitive to the order in which one includes the eigenfunctions (3.8.15a, b) in the expansion (3.8.16a). For example, ordering according to decreasing eigenvalues may not be optimal. Generally, if one is not careful about ordering, the UPE may be slowly convergent.[6]

3.8a.3. *Hilbert–Schmidt Expansion (HSE)*

The HSE corresponds to the nondual choice:[7]

$$V|f_m(E)\rangle = |\psi_m(E)\rangle, \tag{3.8.17a}$$

$$\langle f_n(E)|V = \langle \bar{\psi}_n(E)|, \tag{3.8.17b}$$

where now the ψ's are the energy-dependent eigenfunctions

$$\lambda_m(E) V G_0^{(+)}(E) |\psi_m(E)\rangle = |\psi_m(E)\rangle, \tag{3.8.18a}$$

$$\lambda_n(E) \langle \bar{\psi}_n(E)| G_0^{(+)}(E) V = \langle \bar{\psi}_n(E)|, \tag{3.8.18b}$$

normalized such that

$$\langle \bar{\psi}_n(E)| G_0^{(+)}(E) |\psi_m(E)\rangle = \delta_{mn}. \tag{3.8.18c}$$

We then have

$$t_N(E) = \sum_{n=1}^{N} \frac{|\psi_n(E)\rangle \langle \bar{\psi}_n(E)|}{[\lambda_n(E) - 1]}. \tag{3.8.19}$$

This approximation was proposed by Weinberg.[7] It is expected to produce a faster convergence rate than the UPE and has been employed in various contexts; however, it is certainly more complicated than the UPE. One needs to calculate the eigenfunctions and eigenvalues of the kernel at many energies in the HSE, in contrast to the UPE. Moreover, it fails to satisfy off-shell unitarity, a serious drawback in multiparticle contexts.

3.8a.4. Ernst–Shakin–Thaler Expansion (ESTE)

Here one takes the $|f_m\rangle$'s to be eigenstates (either bound or continuum) of the full Hamiltonian $(H_0 + V)$.[8] Specifically, the $|f_m\rangle$'s are identified with the bound-state wave functions $|B_m\rangle$, or the scattering solutions $|\psi_{E_m}\rangle$, of the equation

$$|\psi_{E_m}\rangle = |\phi_{E_m}\rangle + G_0^{(+)}(E_m)V|\psi_{E_m}\rangle \tag{3.8.20}$$

at a finite number of energies $E_m \geq 0$. The resulting $t_N(E)$ has the correct bound-state pole positions and residues, as in the UPA, if all of the bound-state wave functions are included in the $|f_m\rangle$'s. Also, using Eqs. (3.1.3) and (3.8.11a, b), one can verify that

$$t(E_m)|\phi_{E_m}\rangle = t_N(E_m)|\phi_{E_m}\rangle, \tag{3.8.21a}$$

$$\langle\phi_{E_m}|t(E_m) = \langle\phi_{E_m}|t_N(E_m). \tag{3.8.21b}$$

Namely, at the preselected positive energies, the t-matrix is exact half-on-shell; this ensures that the approximation satisfies off-shell unitarity. The method, although it requires the construction of the expansion functions numerically, converges rapidly, if one is careful about choosing the energies E_m. This is because the ESTE method forces the wave function to be exact at selected energies, and so provides a convenient interpolation to other energies. However, interpolation methods have a well-known defect in numerical analysis; namely, oscillations tend to build up in the neighborhood of the interpolation points as the number of these points is increased. Oscillations of this kind were first observed in Pieper's application[9] of the ESTE method, and were studied in detail by Plessas and collaborators,[10] who found that the oscillations can be very large, unless the interpolation energies are carefully chosen.

3.8a.5. Bateman Expansion (BE)

For the Bateman[11] expansion, the $|f_m\rangle$'s are eigenstates of the free Hamiltonian H_0, viz.,

$$|f_m\rangle = |\phi_{E_m}\rangle, \qquad H_0|\phi_{E_m}\rangle = E_m|\phi_{E_m}\rangle, \tag{3.8.22}$$

where the $|\phi_{E_m}\rangle$ are the incident momentum eigenstates. This choice is simple to use and has acceptable numerical convergence properties. It is interesting to point out that the BE method is identical to the ESTE method with $V = 0$ in Eq. (3.8.20), for example.

3.8a.6. Analytic Separable Expansion (ASE)

This method uses a set of analytic or quasi-analytic functions in Eq. (3.8.8) or in Eq. (3.8.11a, b), and has certain advantages.[2,12,13] If the analytic expansion functions are appropriately chosen, and this is relatively easy to do, this method produces rapid convergence. One does not have the problem of calculating the expansion functions as in the UPE, HSE, or ESTE methods. This method has one further advantage. One of the purposes of making a finite-rank approximation to the t-matrix is to simplify the kernels of the Faddeev equations. In addition, at positive energies, one way of avoiding the singularities of the kernel of the Faddeev equation is to rotate the momentum-integration contour into the complex-momentum plane; this continuation is more easily executed if the finite-rank t-matrix has simple analytic form-factors. In their application of the ESTE method, Plessas *et al.*[14] approximate the expansion functions by simple analytic form-factors, and thus combine the virtues of both the ESTE and the ASE.

3.8a.7. Other Separable Expansions and the Relationship with the Schwinger Variational Principle

Each version of the preceding separable expansions for the t-matrix can be derived from one of the v.p.'s of Section 3.7. For example, the separable expansion (3.8.11) is a consequence of the Schwinger v.p. (3.7.1) with the choice of trial wave functions[15]

$$|\psi_{\mathbf{p}}^{(+)}\rangle = \sum_{n=1}^{N} a_n(\mathbf{p})|f_n\rangle, \tag{3.8.23a}$$

$$\langle\psi_{\mathbf{p}'}^{(-)}| = \sum_{m=1}^{N} \langle f_m|b_m(\mathbf{p}'), \tag{3.8.23b}$$

where a_n and b_m are constants to be determined by the variational requirement, and $\{|f_m\rangle\}$ is a set of expansion states. We have used the same set of states $|f_m\rangle$ in both of Eqs. (3.8.23a, b); this is not essential, but it has the virtue of making the final result symmetric.

In order to determine the constants $b_m(\mathbf{p}')$ and $a_n(\mathbf{p})$, we substitute Eqs. (3.8.23a, b) into Eq. (3.7.1), and require that the resulting expression be stationary with respect to variations of $a_n(\mathbf{p})$ and $b_m(\mathbf{p}')$. It then follows that

$$b_m(\mathbf{p}') = \sum_{n=1}^{N} \langle \mathbf{p}'|V|f_n\rangle J_{nm}, \tag{3.8.24a}$$

$$a_n(\mathbf{p}) = \sum_{m=1}^{N} J_{nm}\langle f_m|V|\mathbf{p}\rangle, \tag{3.8.24b}$$

where J_{mn} is defined by Eq. (3.8.11b), and so

$$|\psi_{\mathbf{p}}^{(+)}\rangle = \sum_{n,m=1}^{N} |f_n\rangle J_{nm}\langle f_m|V|\mathbf{p}\rangle, \tag{3.8.25a}$$

$$\langle\psi_{\mathbf{p}'}^{(-)}| = \sum_{n,m=1}^{N} \langle \mathbf{p}'|V|f_n\rangle J_{nm}\langle f_m|. \tag{3.8.25b}$$

When these trial functions are substituted in Eq. (3.7.1), we obtain the separable expansion (3.8.11a).

Let us consider next the choice given by Eq. (3.8.10b) of expansion functions that leads to

$$t_N = \sum_{n,m=1}^{N} VG_0^{(+)}|f_n\rangle J_{nm}\langle f_m|G_0^{(+)}V, \tag{3.8.26a}$$

where

$$(J^{-1})_{mn} = \langle f_m|(G_0^{(+)}VG_0^{(+)} - G_0^{(+)}VG_0^{(+)}VG_0^{(+)})|f_n\rangle, \tag{3.8.26b}$$

In this case, V_N is not Hermitian at positive energies, and so t_N does not satisfy unitarity. A Hermitian V_N is obtained if we take the principal-value parts of the $G_0^{(+)}$ operators in Eq. (3.8.10b). The expansion (3.8.26a, b) is complicated because more integrations are needed to evaluate Eq. (3.8.26b) than for Eq. (3.8.11b). Nevertheless, this expansion has been useful in situations where V and $G_0^{(+)}$ have complicated singularity structures. For example, the three-body scattering equation can be written in the LS-form (3.1.3), but with a V that is both energy dependent and complex at positive energies. Then, since the expansion functions are dressed by the operators $VG_0^{(+)}$ and $G_0^{(+)}V$ in Eq. (3.8.26a), this form is better able to reproduce the singularity structure of the t-matrix in a low-rank expansion. This attribute has been exploited in constructing separable expansions of the three-body t-matrix.[16] The expansion (3.8.26a, b) is derivable from the Schwinger v.p. for an appropriate choice of trial functions.[2]

With the expansion functions (3.8.10c), we obtain[2]

$$t_N = \sum_{n,m=1}^{N} VG_0^{(+)}|f_n\rangle J_{nm}\langle f_m|, \tag{3.8.27a}$$

where

$$(J^{-1})_{mn} = \langle f_m|(G_0^{(+)} - G_0^{(+)}VG_0^{(+)})|f_n\rangle. \tag{3.8.27b}$$

In this case, V_N is neither symmetric nor Hermitian, so the t-matrix is not even unitary on-shell. It can be argued that, if the $|f_n\rangle$'s are appropriately chosen, this asymmetric form for t_N is expected to be a good approximation if it operates on $G_0^{(+)}$ times a function having the singularity structure of V.[2] This means that t_N as given by Eq. (3.8.27a, b) should be a good approximation to the exact t-matrix when used in the kernels of few-particle collision integral equations. This conjecture gains some support when t_N, as given by Eq. (3.8.27a, b), is substituted into the kernel of the LS equation, $t = V + tG_0^{(+)}V$, to generate the approximate t-matrix:

$$t'_N = V + \sum_{n,m=1}^{N} VG_0^{(+)}|f_n\rangle J_{nm}\langle f_m|G_0^{(+)}V, \tag{3.8.28}$$

which is a good approximation to t.[2,17] The expansion (3.8.27a, b) has never been tested in the kernel of the Faddeev equation, although it has been suggested that it should be effective in that context.[2]

Calculations have been carried out with some of the choices (3.8.10a–d) that do not lead to a Hermitian V_N, resulting in a nonunitary t_N. Nevertheless, rapid convergence is obtained for the nucleon–nucleon off-shell t-matrix with small unitarity violations.[2]

Expansion (3.8.8) with the choice (3.8.10d) has the great virtue of simplicity, since it requires at least one less integration in order to evaluate J, via Eq. (3.8.9), than any of the other choices. The accurate results obtained with this scheme shows that absence of exact symmetry is not necessarily a barrier to numerical calculations.[18]

Expansion (3.8.8) follows from the Schwinger-type v.p. (3.7.9) with the trial functions[19]

$$|\psi_{\mathbf{p}}^{(+)}\rangle = \sum_{n=1}^{N} a_n(\mathbf{p})|u_n\rangle, \tag{3.8.29a}$$

$$\langle\chi_{\mathbf{p}'}^{(-)}| = \sum_{m=1}^{N} \langle v_m|b_m(\mathbf{p}'). \tag{3.8.29b}$$

The constants $a_n(\mathbf{p})$ and $b_m(\mathbf{p}')$ are determined by the stationary requirement. In this way, we obtain

$$a_n(\mathbf{p}) = \sum_{m=1}^{N} J_{nm}\langle v_m|\mathbf{p}\rangle, \tag{3.8.30a}$$

$$b_m(\mathbf{p}') = \sum_{n=1}^{N} \langle \mathbf{p}'|V|u_n\rangle J_{nm}, \tag{3.8.30b}$$

with J^{-1} of Eqs. (3.8.30a, b) defined by Eq. (3.8.9), which leads to the expansion (3.8.8).

3.8b. Degenerate-Kernel Schemes

If one is only interested in the numerical representation of the t-matrix, the expansion (3.8.6a) for V_N can be used in the kernel, but not for the inhomogeneous term of Eq. (3.1.3). The resultant approximate t-matrix satisfies

$$t_N = V + V_N G_0^{(+)} t_N, \tag{3.8.31}$$

with

$$t_N = V + V \sum_{n,m=1}^{N} |u_n\rangle J_{nm} \langle v_m| G_0^{(+)} V, \tag{3.8.32}$$

where J_{nm} is again given by Eq. (3.8.9), and is very useful in practice. This method is known as the degenerate-kernel scheme.[20]

There are two simple choices for the expansion functions:

$$|u_n\rangle = |f_n\rangle, \qquad \langle v_m| G_0^{(+)} = \langle f_m|; \tag{3.8.33a}$$

$$|u_n\rangle = G_0^{(+)} |f_n\rangle, \qquad \langle v_m| = \langle f_m|; \tag{3.8.33b}$$

$n, m = 1, \ldots, N$, among the infinitely many possibilities.

The choice (3.8.33a) leads to the symmetric t-matrix

$$t_N = V + \sum_{n,m=1}^{N} V |f_n\rangle J_{nm} \langle f_m| V, \tag{3.8.34a}$$

where

$$(J^{-1})_{mn} = \langle f_m| (E - H) |f_n\rangle. \tag{3.8.34b}$$

This expansion is actually usually applied to the $\hat{K}$-matrix, in which case it is understood that the expansion functions are real and satisfy the principal-value prescription. When the t-matrix is written in the general form (3.8.32), we expect $\det(J^{-1})$ to vanish at the bound-state poles of t_N. But Eqs. (3.8.34a, b) do not involve a Green function and have a problem associated with the vanishing of $\det(J^{-1})$ at (spurious) energies other than for bound states. For positive energies, $(E - H)$ has a continuous spectrum that includes zero, so that it can easily happen that one eigenvalue of the finite matrix is small, especially if the number of basis functions is large. If this happens, the matrix J^{-1} becomes nearly singular, which causes difficulties with the expansion

(3.8.34a, b). In all other versions of the expansions considered in this chapter, the matrix J^{-1} becomes singular only when $VG_0^{(+)}$ has an eigenvalue near unity, which is the condition of a true or continuum bound state.[21] It has been pointed out by Abdel-Rauof[22] that special care is needed in choosing the functions $|f_m\rangle$ of Eqs. (3.8.34a, b) in order to use this expansion in practical calculations. Also, anomalies have been observed in the Schwinger variational phase-shifts due to the appearance of continuum bound states at positive energies.[21]

The expansion (3.8.34a, b) can be shown to be a consequence of the Hulthén–Kohn v.p. (3.7.10) with the variational functions[19]

$$|\hat{\chi}_{\mathbf{p}}^{(+)}\rangle = \sum_{n=1}^{N} a_n(\mathbf{p})|f_n\rangle, \tag{3.8.35}$$

$$\langle\hat{\chi}_{\mathbf{p}'}^{(-)}| = \sum_{m=1}^{N} b_m(\mathbf{p}')\langle f_m|. \tag{3.8.36}$$

By choosing complex expansion functions in Eqs. (3.8.34) satisfying the outgoing-wave asymptotic boundary condition one can readily calculate the t-matrix elements.[23] In the space of complex functions the inversion of the matrix J^{-1} of Eq. (3.8.34b) is not expected to present any problem associated with a vanishing determinant. Consequently, expansion (3.8.34a) is not expected to present the anomalies of the Kohn method for the $\hat{K}$-matrix. This so called complex Kohn v.p. for the t-matrix produces very good convergence for exponential and Yukawa-type potentials and for certain atomic scattering processes.

It was later shown[24] that even the complex Kohn v.p. exhibits anomalous behavior like the Schwinger v.p.[21] However, these anomalies in the Schwinger or the complex Kohn v.p.'s are accidental in nature[21] and occur for a special class of potentials or expansion functions. They should be rare in practice and hence not of concern in numerical calculations. The usual Kohn v.p., on the other hand, exhibits anomalies for any potential and expansion function and are so frequent that they have significantly limited the usefulness and applicability of this method.[24]

The expansion functions (3.8.33b) yield the symmetric t-matrix

$$t_N = V + \sum_{n,m=1}^{N} VG_0^{(+)}|f_n\rangle J_{nm}\langle f_m|G_0^{(+)}V, \tag{3.8.37}$$

with J defined by Eq. (3.8.27b), which is identical to the t'_N of Eq. (3.8.28) that we derived in a completely different way. (It can also be derived from v.p. (3.7.13).) The approximation (3.8.37) was suggested and tested numerically by

Sloan and Brady for both the two- and the three-particle systems.[20] Though this t_N is symmetric, it does not even satisfy on-shell unitarity.

Expansion functions other than (3.8.33a, b) have been numerically successful. We do not consider them here, and refer the interested reader to original articles.[4,18–21]

One can develop degenerate-kernel schemes based on the other v.p.'s treated in last subsections. For example, the v.p. (3.7.16) leads to the approximation[19]

$$t_N = V + VG_0^{(+)}V + \sum_{n,m=1}^{N} VG_0^{(+)}V|f_n\rangle J_{nm}\langle f_m|VG_0^{(+)}V, \qquad (3.8.38)$$

with J given by Eq. (3.8.11b). This expansion has been suggested and applied by Takatsuka and McKoy.[25]

3.8c. Separable Expansions for Energy-Dependent Potentials

The discussion of the preceding subsections was limited to energy-independent potentials. In many physically relevant situations, the potential operator depends on the parametric energy E. This dependence naturally appears when some dynamical degree of freedom is suppressed in the process of deriving an effective interaction between two composite objects. Some examples of this are meson-theoretic nucleon–nucleon potentials, where the mesonic degrees of freedom are suppressed; the optical-model potential based on the Feshbach approach, where the irrelevant nucleonic degrees of freedom are suppressed; and the nucleon–deuteron potential that is used in four-nucleon calculations.

In such cases, the direct application of the preceding techniques leads to separable expansions for the t-matrix with energy-dependent form-factors. One can easily apply Eq. (3.8.26a) in this case as has been done by Sofianos, McGurk, and Fiedeldey for nucleon–deuteron scattering amplitudes that enter into a four-nucleon calculation.[16] Such a rank-N t-matrix is known in the literature as the energy-dependent pole expansion (EDPE). It would be extremely useful if one could construct a similar expansion, but with energy-independent form-factors. We next show how to achieve this goal.

The energy-dependent potential $V(E)$ is assumed to appear in the following LS-type equation:

$$t(E) = V(E) + V(E)G_0^{(+)}(E)t(E). \qquad (3.8.39)$$

We approximate[19] $V(E)$ by the following rank-N potential:

$$V_N(E) = \sum_{i,j=1}^{N} |g_i\rangle K_{ij}(E)\langle \bar{g}_j|, \tag{3.8.40a}$$

$$[K^{-1}(E)]_{ji} = \sum_{n,m=1}^{N} \langle \bar{g}_j|f_n\rangle C_{nm}(E)\langle \bar{f}_m|g_i\rangle, \tag{3.8.40b}$$

$$[C^{-1}(E)]_{mn} = \langle \bar{f}_m|V(E)|f_n\rangle, \tag{3.8.40c}$$

where g_i, $\bar{g}_i f_i$, $\bar{f}_i$, $i = 1, 2, \ldots, N$, represent suitable sets of functions. In the limit $N \to \infty$ in the space of functions g_i, $\bar{g}_j$, we have

$$\langle \bar{g}_j|V_N(E)|g_i\rangle = \langle \bar{g}_j|V(E)|g_i\rangle. \tag{3.8.41}$$

If one solves Eq. (3.8.39) with the rank-N potential (3.8.40a), one obtains the following rank-N t-matrix:[19]

$$t_N(E) = \sum_{i,j=1}^{N} |g_i\rangle D_{ij}(E)\langle \bar{g}_j|, \tag{3.8.42a}$$

$$[D^{-1}(E)]_{ji} = [K^{-1}(E)]_{ji} - \langle \bar{g}_j|G_0^{(+)}(E)|g_i\rangle, \tag{3.8.42b}$$

with $K^{-1}(E)$ defined by (3.8.40b). In the expansions (3.8.40a) and (3.8.42a), the energy dependence remains only in the matrices K_{ij} and D_{ij} if the functions g_i and $\bar{g}_j$ are energy independent. Two interesting choices can be made for the functions g_i and $\bar{g}_j$. One is a generalization of the ESTE and the other of the UPE, with energy-independent form factors in both cases.

3.8c.1. Generalized ESTE (GESTE)

The rank-N ESTE for the case of an energy-independent potential has the property of being exact half-on-shell at N selected energies. The same can be achieved in the case of an energy-dependent potential if one uses in Eq. (3.8.42) the expansion states

$$\begin{aligned} |g_i\rangle &= V(E_i)|\psi_i^{(+)}\rangle, \\ \langle \bar{g}_j| &= \langle \psi_j^{(-)}|V(E_j), \\ |f_i\rangle &= |\psi_i^{(+)}\rangle, \\ \langle f_j| &= \langle \psi_j^{(-)}|. \end{aligned} \tag{3.8.43}$$

Here the scattering wave functions at energies $E_i = k_i^2$ are defined by

$$\begin{aligned} |\psi_i^{(+)}\rangle &= |\mathbf{k}_i\rangle + G_0^{(+)}(E_i)V(E_i)|\psi_i^{(+)}\rangle, \\ \langle \psi_i^{(-)}| &= \langle \mathbf{k}_i| + \langle \psi_i^{(-)}|V(E_i)G_0^{(+)}(E_i), \end{aligned} \tag{3.8.44}$$

for $i = 1, 2, \ldots, N$. Then the rank-N t-matrix (3.8.42a) becomes:

$$t_N(E) = \sum_{i,j=1}^{N} V(E_i)|\psi_i^{(+)}\rangle D_{ij}(E)\langle\psi_j^{(-)}|V(E_j), \tag{3.8.45a}$$

$$[D^{-1}(E)]_{ji} = \sum_{n,m=1}^{N} \langle\psi_j^{(-)}|V(E_j)|\psi_n^{(+)}\rangle C_{nm}(E)\langle\psi_m^{(-)}|V(E_i)|\psi_i^{(+)}\rangle$$

$$- \langle\psi_j^{(-)}|V(E_j)G_0^{(+)}(E)V(E_i)|\psi_i^{(+)}\rangle, \tag{3.8.45b}$$

$$[C^{-1}(E)]_{mn} = \langle\psi_m^{(-)}|V(E)|\psi_n^{(+)}\rangle. \tag{3.8.45c}$$

We next show that

$$t_N(E_n)|\mathbf{k}_n\rangle = t(E_n)|\mathbf{k}_n\rangle, \tag{3.8.46a}$$

$$\langle\mathbf{k}_n|t_N(E_n) = \langle\mathbf{k}_n|t(E_n), \tag{3.8.46b}$$

where n refers to any one of the N selected energies. Using Eq. (3.8.45a), we rewrite the left-hand side of Eq. (3.8.46a) as

$$t_N(E_n)|\mathbf{k}_n\rangle = \sum_{i,j=1}^{N} V(E_i)|\psi_i^{(+)}\rangle D_{ij}(E_n)\langle\psi_j^{(-)}|V(E_j)|\mathbf{k}_n\rangle. \tag{3.8.47}$$

Now, Eq. (3.8.42b) at $E = E_n$ becomes

$$\sum_{j=1}^{N} D_{ij}(E_n)\{[K^{-1}(E_n)]_{jn} - \langle\psi_j^{(-)}|V(E_j)G_0^{(+)}(E_n)V(E_n)|\psi_n^{(+)}\rangle\} = \delta_{in}, \tag{3.8.48}$$

where with the functions (3.8.43) we have

$$[K^{-1}(E_n)]_{jn} = \sum_{i,m=1}^{N} \langle\psi_j^{(-)}|V(E_j)|\psi_i^{(+)}\rangle C_{im}(E_n)\langle\psi_m^{(-)}|V(E_n)|\psi_n^{(+)}\rangle, \tag{3.8.49}$$

or, because of Eq. (3.8.45c),

$$[K^{-1}(E_n)]_{jn} = \langle\psi_j^{(-)}|V(E_j)|\psi_n^{(+)}\rangle. \tag{3.8.50}$$

Equation (3.8.50), when substituted into Eq. (3.8.48), yields

$$\sum_{j=1}^{N} D_{ij}(E_n)\langle\psi_j^{(-)}|V(E_j)\{1 - G_0^{(+)}(E_n)V(E_n)\}|\psi_n^{(+)}\rangle = \delta_{in}, \tag{3.8.51}$$

which, with the LS equation (3.8.44), becomes

$$\sum_{j=1}^{N} D_{ij}(E_n)\langle\psi_j^{(-)}|V(E_j)|\mathbf{k}_n\rangle = \delta_{in}. \tag{3.8.52}$$

When this is used in the right-hand side of Eq. (3.8.47), one obtains Eq. (3.8.46a); Eq. (3.8.46b) follows in a similar fashion. This means that the rank-N t-matrix t_N of Eq. (3.8.45a) is a generalization of the ESTE[8] to the case of energy-dependent potentials, with energy-independent form-factors, and which is exact half-on-shell at N selected energies.[26]

3.8c.2. Generalized UPE (GUPE)

A similar generalization of the UPE follows from Eqs. (3.8.42a, b) with a special choice of expansion functions.[19] This can be achieved as with an energy-independent potential by employing eigenfunctions of the kernel of the LS equation (3.8.39) at the bound-state energy $E = -B$ that satisfy

$$\begin{aligned} \lambda_i V(-B) G_0^{(+)}(-B)|\psi_i(-B)\rangle &= |\psi_i(-B)\rangle, \\ \lambda_j \langle\bar{\psi}_j(-B)|G_0^{(+)}(-B)V(-B) &= \langle\bar{\psi}_j(-B)|, \end{aligned} \tag{3.8.53}$$

and are normalized according to

$$\langle\bar{\psi}_j(-B)|G_0^{(+)}(-B)|\psi_i(-B)\rangle = \delta_{ij}. \tag{3.8.54}$$

Then, one uses, in Eqs. (3.8.42a, b),

$$\begin{aligned} |g_i\rangle &= |\psi_i(-B)\rangle, \\ \langle\bar{g}_j| &= \langle\bar{\psi}_j(-B)|, \\ |f_i\rangle &= G_0^{(+)}(-B)|\psi_i(-B)\rangle, \\ \langle f_j| &= \langle\bar{\psi}_j(-B)|G_0^{(+)}(-B), \end{aligned} \tag{3.8.55}$$

to obtain

$$t_N(E) = \sum_{i,j=1}^{N} |\psi_i(-B)\rangle D_{ij}(E)\langle\bar{\psi}_j(-B)|, \tag{3.8.56a}$$

$$[D^{-1}(E)]_{ji} = B_{ji}(E) - \langle\bar{\psi}_j(-B)|G_0^{(+)}(E)|\psi_i(-B)\rangle, \tag{3.8.56b}$$

$$[B^{-1}(E)]_{ji} = \langle\bar{\psi}_j(-B)|G_0^{(+)}(-B)V(E)G_0^{(+)}(-B)|\psi_i(-B)\rangle, \tag{3.8.56c}$$

which is the GUPE.[27]

The preceding discussion illustrates the flexibility, generality, and usefulness of the separable expansion scheme (3.8.42a, b), which is equivalent to repeated applications of the Schwinger-type v.p.[19]

The GUPE of Casel *et al.* has been applied to a three-nucleon t-matrix, with reasonable convergence properties, but the EDPE of Sofianos *et al.* has better convergence.[28] The GESTE appears very attractive, but it has not yet been used in practical calculations. As in the energy-independent case, the GESTE may converge faster than either the GUPE or the EDPE. This conjecture is based on a comparison of the relative rates of convergence of the UPE and the ESTE for energy-independent potentials.

3.9. Numerical Applications of Expansion Methods

We review now the numerical applications of expansion methods in nuclear and atomic/molecular physics for both single-channel and multichannel collisions.

The numerical solution of two-body collision integral equations can, of course, be obtained by the more precise techniques developed in Sections 3.2–3.6. However, the purpose behind the use of the expansion techniques is not to compete with these integral equation methods in numerical accuracy, but rather to obtain a reasonably accurate result with relatively little effort and/or to obtain convenient finite-rank approximations.

On the other hand, although expansion methods yield approximate solutions easily, the convergence to the exact result can be rather slow. The reason for this could be the use of a (set of) finite-norm L^2 function(s) to expand non-normalizable scattering wave functions. The set used in practical calculations can be linearly dependent, and so overcomplete on a subspace. Osborn[1] has commented about the theoretical difficulties of the convergence of these methods, but Sloan and Gray[2] show that these concerns are not of practical consequence.

In nuclear physics, expansion methods, especially the separable type, have been widely applied to NN scattering.[3] This is because a separable expansion to the NN t-matrix greatly facilitates the task of solving the dynamical three-nucleon problem by reducing the dimensionality of the partial-wave-projected momentum-space collision integral equations from two to one. The purpose behind the use of the degenerate-kernel scheme in NN scattering, or in using any of these methods in Nd scattering calculations, is to find an approximate solution of the model with the least effort.[4] This is also the motivation behind the use of these methods with the dynamical equations one encounters in atomic/molecular physics, or in chemistry. In the latter situations, an accurate numerical solution is neither called for, because of drastic approximations made in the formulation of the dynamical model, nor easily obtainable because of the large number of coupled channels involved.[5]

3.9a. Applications in Nuclear Physics

The NN system, although it involves at most only two coupled channels, is not numerically trivial because of a (soft) repulsive-core interaction. In momentum-space treatments, this means that the very-large momentum components of the interaction contribute substantially to low-energy results.

Consequently, a large number of momentum mesh points extending over a large range of momenta are needed for an accurate discretization of the infinite integral appearing in the LS equation.

3.9a.1. Separable Expansions

Interest in constructing a separable expansion for the NN t-matrix started with the theoretical study by Weinberg on the Hilbert–Schmidt expansion scheme, which is also often known as the Weinberg method.[6] This particular method requires that the eigenfunctions of the kernel of the LS equation be precalculated at each energy even for an energy-independent NN interaction. Finding these eigenfunctions is actually numerically more complicated than solving the LS equation, so the Weinberg method has not been widely applied to NN scattering. The systematic construction of separable expansions with diverse, realistic NN interactions started with the development of the UPE by Harms. Levinger and collaborators constructed the UPE for different NN interactions, and then solved the trinucleon bound-state problem with the resulting t-matrices.[7] It should be noted that the UPE is a special case of the Weinberg expansion when the eigenfunctions of the kernel of the LS equation at a particular energy (the bound-state energy) are utilized for the expansion of the t-matrix at all energies. Hence, the form-factors of the expansion are energy-independent in the case of the UPE.

The numerical activity in the construction and use of separable t-matrix expansions increased after the theoretical studies by Ernst, Shakin, Thaler (EST) and by Adhikari and Sloan (AS).[8] The ESTE method proposed by EST has better convergence properties than the UPE. A rank-N ESTE for the t-matrix is exact half-on-shell at $m \leq N$ preselected positive energies, and possesses the correct bound-state residues and poles. If the preselected energies are suitably chosen, this scheme provides a good interpolation for the t-matrix to other energies and momenta.

A general ASE expansion scheme proposed by AS has as special cases all the other known expansions and includes criteria for obtaining rapid convergence. It emphasizes the use of explicit analytic representations of the separable-expansion functions. This is useful if one is interested in solving the three-body problem by the method of contour rotation above the threshold for scattering. Calculations exploiting these ideas have yielded significantly better convergence properties than the UPE.[9] Afnan and Birrell[10] solved the trinucleon bound-state problem with the Reid soft-core potential using one version of the ASE for the two-nucleon t-matrices; they also found superior convergence characteristics compared with the UPE method.

These successes motivated Plessas and coworkers[11] to construct separable expansions for the NN t-matrix that combine the virtues of the ESTE and ASE methods. Specifically, they constructed low-rank EST expansions for realistic NN potentials, where the form-factors were expressed as linear combinations of simple analytic functions. Then Koike, Plessas *et al.* solved the Nd scattering problem above the breakup threshold using these types of separable expansions of the NN potentials proposed by the Bonn and the Paris groups.[12]

In spite of these successes, there are numerical problems with the ESTE method. This method produces reasonably good results with a small number of expansion terms, but final convergence can be slow and problematical. In fact, large oscillations in the NN phase-shift are observed with increasing rank of the separable expansion.[13] These oscillations have been identified with the occurrence of a spurious resonance(s).[11]

There has been renewed interest in constructing separable expansions of realistic local potentials. Baldo, Ferreira, and Streit[14] have constructed such an expansion for the Reid and the Argonne V_{14} NN-potentials using Gamow functions in the ASE method. They obtained rapid convergence and claimed that this approach does not lead to the spurious resonances that can appear in the ESTE method.

Separable expansions have also been studied by Bund, Canton, Hartt, and their collaborators using different approaches and potentials. We refer the interested reader to the original articles.[15]

We conclude this overview with a some comment on the expansion method of Oryu,[16] which has been successfully used both for generating separable expansions for the NN t-matrix and in solving Nd scattering in the Amado model. Though Oryu's technique sometimes yields faster convergence than some of the other methods, it should be realized that it uses energy-dependent expansion functions, in contrast to some of the other schemes. In this approach, the LS equation must be presolved at each energy under consideration. The reason for this is that Oryu's method is basically a systematic extension of the Kowalski–Noyes approximation of Eq. (2.4.49) using the nonsingular integral equations of Section 3.3. The $f_s(p, k)$ functions, e.g., serve as form-factors for the term containing the bound-state poles, and it certainly is energy dependent.

We also remark that the first attempt at the representation, along with primitive, but encouraging, numerical tests, of the t-matrix to arbitrary accuracy in terms of an expansion of separable terms was carried out by Kowalski and Feldman[17] by means of a separable expansion of the kernel Λ_s (Eq. (3.3.29)). This yields an expansion method, like Oryu's, that is exact half-on-shell at *all* positive energies, along with the other features of the Kowalski–Noyes rank-one approximation.

3.9a.2. Degenerate-Kernel Schemes

Sections 3.7 and 3.8 contain methods that do not lead to a separable expansions for the t-matrix. As we have pointed out in general, these methods are not very precise, although they are reasonably accurate and involve little effort. These approaches are usually called degenerate-kernel schemes and are related to a Hulthén–Kohn v.p. Applications of these methods to scattering problems have been made by Adhikari, Brady, Sloan, and Rawitscher.[18]

3.9b Applications in Atomic/Molecular Physics

Compared to the few-nucleon problems in nuclear physics, the few-body problems in atomic/molecular physics are much more complicated to deal with both theoretically and numerically. The principal theoretical difficulty concerns the correct treatment of the long-range Coulombic forces. This also leads to a numerical problems. One is confronted with large numbers of channels in typical physical situations. An accurate modeling of the physics must therefore incorporate a large number of channels and, consequently, a large set of coupled scattering integral equations. A simple approximate solution is very useful as an alternative to exact solution in these cases. Hence the v.p.'s and the related expansions of Sections 3.7 and 3.8 are widely used for atomic/molecular collision problems in physics and chemistry.[19]

The early attempts to solve atomic/molecular scattering processes made use of the usual Schwinger or Kohn v. p. for the wave functions rather than the t- or $\hat{K}$-matrices. Here, the scattering wave function is expanded in a set of simple functions, and their coefficients are determined variationally. Such applications of the variational method to scattering problems have been more difficult than might have been anticipated.[20] Since continuum wave functions are not square integrable, the integrals that appear are not easy to handle. Also, it is not simple to determine whether the result has converged to its correct value or not. A closely related problem is the existence of (spurious) singularities in phase shifts when they are calculated variationally.

Some of these problems can be circumvented by using the relevant separable-expansion or degenerate-kernel schemes of Section 3.8. The separable-expansion method makes use of normalizable L^2 functions and is entirely equivalent to the Schwinger v.p. The use of normalizable functions eliminates some of the preceding numerical problems and (spurious) singularities.

The separable-expansion scheme has been used successfully by McKoy and collaborators in a diverse class of atomic/molecular collision problems, and they have found that this way of applying the Schwinger v.p. leads to good convergence.[21] Applications were made to single-channel electron–atom scattering, and to complicated multichannel problems ranging from electron–molecule collisions to molecular photoionization. Both the Schwinger and Hulthén–Kohn v.p.'s, as well as v.p.'s of higher order, were used.[22] One version of the Hulthén–Kohn v.p. that was used involved the inverse of the full Green's function $(E - H)^{-1}$; this may lead to spurious resonances (*cf.* Section 3.8b).

There have been many applications of these techniques to the study of electron–atom and atom–diatom scattering in atomic/molecular physics by Kouri, Truhlar *et al.*[23,24] They have established that, even in cases where the modeling involves several hundred coupled equations, the methods of Section 3.8 usually lead to converged results of the desired precision. In their applications of the separable-expansion[23] and degenerate-kernel schemes,[24] they found better convergence properties with the latter method.

At this point, two remarks made by Adhikari and Tomio[25] in their study of v.p.'s are worth emphasizing since they do not seem to be fully appreciated by the community. First, for each separable-expansion or degenerate-kernel scheme for the solution of the LS equation, there is a v.p. from which it can be derived. Hence, all these schemes are actually variational, although usually, many of them are termed nonvariational, see, e.g., Staszewska and Truhlar.[23] Second, often claims have been made concerning the superiority of one variational method over another by making numerical calculations with different methods using the same set of expansion functions.[26] Such a comparison is not to the point, because in these methods different functions are expanded using a set of trial functions; in order to find a rapidly convergent expansion, the set of trial functions must be modified from one method to another. Usually the same set of expansion functions are used in such a comparison, and the conclusions so reached will be dependent on the model studied and the set of expansion functions used. Such studies may well prove to be very interesting numerically, but they will not provide any physical insight about why a particular method failed or how to improve the convergence. It has been emphasized by Adhikari and Tomio that success of these methods depends on an appropriate choice of the expansion functions; this may require the expansion functions to have essential mathematical properties and asymptotic behavior similar to the unknown function. This point was earlier made by Adhikari and Sloan in relation to their solution of three-body scattering equations,[27] namely, the inclusion of a subenergy singularity in the expansion functions vastly improves the convergence properties.

However, most of the calculations with v.p.'s used the simplest version of the Schwinger or the Kohn v.p. Though there is no fundamental reason for preferring one method over the other (apart from the difficulties associated with the anomalies of the Kohn v.p. and those associated with the more complicated integrals in the Schwinger v.p. involving the Green function), the Kohn v.p. has produced faster convergence, whenever a comparison between the convergence rates of the two methods were made.[26] From these studies it seems that though the choice of an optimal set of expansion functions in the Kohn or the complex Kohn v.p. is under control, this choice in the Schwinger method is a more difficult task.

References and Notes (3)

Section 3.1

1. Two-particle off-shell scattering is reviewed in Watson (1967) and Srivastava (1975).
2. Austern (1970), Glendenning (1983), Satchler (1983), Fano (1986).
3. The mathematics that is relevant here concerns compactness, Schmidt classes, and Fredholm-type integral equations. Some useful books on integral equations are Smithies (1958), Mikhlin (1964), Pagorezelski (1966), and Hoheisel (1968). Very readable reviews of the relevant background mathematics are given by Weinberg (1963), Lovelace (1964b), and Newton (1982).
4. See, Lovelace (1964a, b) for a proof of compactness in suitably defined Banach spaces.

Section 3.2

1. Muskhelishvili (1953).
2. It seems as if the first application of Fredholm reduction to two-particle off-shell collision integral equations appears in Kowalski (1961). See also Brown (1963), Kowalski (1963a, b), Sasakawa (1963).
3. This method for off-shell two-particle equations along with the various analyticity properties is discussed by Brayshaw (1968a, b).
4. Jost (1951), Smithies (1958), Newton (1982).
5. Padé (1892), Baker, Jr. (1970).
6. Bolsterli (1969), Bagchi (1977, 1979), Mulligan (1981).
7. See, e.g., Press (1986).

Section 3.3

1. Sloan (1968b). See Redish (1987a) for an application of this method to off-shell nucleon–nucleon scattering.
2. Haftel (1970a).
3. For limitations of the Haftel–Tabakin approach see Brown (1976), Glöckle (1983a), and Redish (1987a). See, however, Amos (1988) for a successful numerical study on off-shell nucleon–nucleon scattering using this method.
4. Kowalski (1961, 1963a, b).
5. Kowalski (1966b).
6. Adhikari (1981b, 1987a, 1988a).
7. Coester (1971), Kowalski (1972c), Adhikari (1979a), Tomio (1980a, b).
8. Bartnik (1986, 1987), Frank (1988), Gibson (1989).
9. Noyes (1965).
10. Kowalski (1965b).
11. Brown (1963).
12. Blasczak (1973).
13. See, e.g., Ahmadzadeh (1967), Levine (1966, 1967), and Thompson (1970, 1971).
14. Cohen (1969).
15. Haymaker (1968).
16. Bolsterli (1974, 1977).
17. Kowalski (1974a).
18. Fulton (1959).
19. Srivastava (1970, 1975).
20. Picker (1971, 1975), Redish (1972).
21. The f_s-function is denoted as ϕ in Ref. 4. The f_s notation became standard after the work in Refs. 9, 10.
22. Taylor (1963).
23. Graves-Morris (1966).
24. Stoof (1988).
25. Roriz (1988).
26. There are alternative ways to calculate these amplitudes, e.g., see Reiner (1976).
27. Haidenbauer (1986a), Schwarz (1986).
28. Redish (1987a).
29. Loiseau (1985).
30. Amos (1988).
31. Kowalski (1968).
32. Refs. 27–30 contain numerous references to other calculations of the off-shell characteristics of NN transition matrices. See also Ref. 19 for a complete review of the situation prior to 1975.
33. Redish (1987b).
34. See Fearing (1984) for a general discussion of the problem of interpreting off-shell properties.

35. Workman (1986).
36. Machleidt (1987).
37. Lacombe (1980).
38. See Fearing (1987) for a review and other references.
39. Kitching (1986).

Section 3.4

1. Weinberg (1963, 1964).
2. Brown (1963).
3. Coester (1971). This is relevant to earlier work by Sasakawa (1963).
4. Blasczak (1973).
5. Tomio (1981).

Section 3.5

1. Tomio (1981).
2. Horácek (1983, 1984, 1985).
3. Adhikari (1986).
4. Oryu (1974, 1983, 1986).
5. Padé (1892).
6. Baker, Jr. (1970), Glöckle (1983a).
7. Ishikawa (1987a, b).
8. Witala (1988).
9. Tjon (1970).

Section 3.6

1. Malfliet (1968).
2. Malfliet (1970), Afnan (1977a), Birrell (1978).
3. Glöckle (1983a).
4. This method is first proposed in Adhikari (1981b).
5. This specific connection was implicit in Adhikari (1981b), but was explicitly noted independently in Adhikari (1987a) and Bartnik (1986). An equivalent relation involving the bound-state vertex functions is contained in Kowalski (1966b).
6. Dzhibuti (1984).
7. Sasakawa (1986).
8. This choice is proposed in Kowalski (1966b), where, in addition, the complete freedom of the connection between E and k for negative E is pointed out.

Section 3.7

1. Hulthén (1944, 1948).
2. Kohn (1948).
3. Schwinger (1947a, b).
4. Reviews of v.p.'s are contained in Callaway (1978), Abdel–Raouf (1982, 1984a), Delves (1972), Abdallah, Jr. (1974), Nesbet (1975, 1979), Gerjuoy (1983), Lucchese (1986), and Staszewska (1987).
5. Higher-order v.p.'s are considered by Takatsuka (1981a, b), Adhikari (1987b), and Staszewska (1987).

Section 3.8

1. See, for example, Amado (1970b).
2. Adhikari (1975a).
3. Harms (1970).
4. For reviews on separable expansions, see Levinger (1974), Srivastava (1975), Plessas (1986a, b), and Ferreira (1986).
5. Fuda (1968).
6. Afnan (1973a, b).
7. Weinberg (1963).
8. Ernst (1973a, b).
9. Pieper (1974).
10. Haidenbauer (1983, 1984).
11. Bateman (1922).
12. Adhikari (1975b).
13. Adhikari (1975c).
14. Plessas (1986a, b), Haidenbauer (1983, 1984).
15. The $N = 1$ treatment was given by Gerjuoy (1954), and the general treatment by Adhikari (1975c). Essentially the same proof appears in Zubarev (1976); see also Sugar (1964).
16. Sofianos (1979), See, also, Fonseca (1983).
17. Sloan (1974).
18. Sloan (1972a), Adhikari (1975a).
19. Adhikari (1987b).
20. Sloan (1972a, 1974), Newton (1982).
21. Adhikari (1990b).
22. Sloan (1972a), Abdel–Rauof (1984b).
23. Miller (1987), McCurdy (1987), and Zhang (1988b).
24. Lucchese (1989).
25. Takatsuka (1981a).
26. The GESTE is arrived at by Pearce (1987) by a different technique.

27. The GUPE has been suggested in Casel (1982) from a different viewpoint.
28. See, e.g., Fonseca (1986a, b) and Fiedelday (1987).

Section 3.9

1. Osborn (1973).
2. Sloan (1973).
3. For a summary of these applications, see Levinger (1974), Plessas (1986a, b), and Ferreira (1986); see, also, Adhikari (1987b).
4. Applications of degenerate-kernel scheme in nuclear physics are to be found in Sloan (1972a, 1974), Rawitscher (1982, 1984, 1989), and Adhikari (1975d, 1987b).
5. For an account of applications in atomic and molecular physics, see Staszewska (1987). The number of coupled channels in these cases can be of the order of 1000.
6. Weinberg (1963, 1964).
7. See, e.g., Levinger (1974).
8. Ernst (1973a, b), Adhikari (1974a, 1975a–d).
9. Adhikari (1975a–d), Pieper (1974).
10. Afnan (1977a), Birrell (1978).
11. Haidenbauer (1983, 1984, 1986a).
12. Koike (1987a), Plessas (1986a, b).
13. Pieper (1974).
14. Baldo (1985, 1986, 1987a, b), Ferreira (1986).
15. Canton (1987, 1988), Hartt (1987), Bund (1980, 1985).
16. Oryu (1974, 1983).
17. Kowalski (1963b).
18. Adhikari (1975d, 1987b), Sloan (1972a, 1974), Brady (1974), Rawitscher (1982, 1984, 1989).
19. See, for example, Ref. 4 to Section 3.7.
20. Callaway (1978).
21. For an account of these applications see, Lucchese (1986).
22. Takatsuka (1981b, c), Watson (1979), Lucchese (1980).
23. Zhang (1988a), Staszewska (1987).
24. Schwenke (1988), Staszewska (1987).
25. Adhikari (1987b).
26. Staszewska (1987), Callaway (1980), Lucchese (1986), Takatsuka (1981a–c), Miller (1987), Zhang (1988b).
27. Adhikari (1975d).

Chapter 4

Many-Particle Lippmann–Schwinger Equations

4.1. Introduction

In Ch. 2, we dealt with an integral equation (Eq. (2.3.18)) whose unique solution yields the entire content of the single-channel, two-particle scattering problem. While there are many ways of obtaining these solutions, as we saw in Ch. 3, there are no other questions of principle.

For multiparticle collisions, it is a lot more complicated to formulate scattering integral equations that have unique solutions, let alone solve them. The problems one encounters in each of these endeavors are major concerns of the rest of this book. We begin in this chapter by considering the multiparticle Lippmann–Schwinger (LS) equations,[1,2] and seeing whether they constitute a consistent set of dynamical equations for N (≥ 3) distinguishable particles.[3]

We suppose that any subgroupings of m of these N particles, $2 \leq m \leq N$, can form stable bound clusters. It may be that some, or all, subgroupings of two or more of the N particles cannot form stable bound states. Then, as explained in Ch. 1, there are no asymptotic channels associated with these subgroupings. We will see that there is an LS equation for each state vector corresponding to a nontrivial asymptotic channel. The two-cluster channels, however, play a distinguished role in the subsequent analysis; this fact highlights the principal useful features of the LS equations, as well as their limitations.

4.2. Homogeneous and Inhomogeneous Equations

We confine ourselves initially solely to scattering processes with two clusters in the initial state. The LS equation for the (in-) state vector $|\psi_\alpha^{(+)}(E)\rangle$ that has evolved from an incident state $|\phi_\alpha(E)\rangle$ corresponding to the two-cluster partition α is, in the notation of Ch. 1,

$$|\psi_\alpha^{(+)}(E)\rangle = |\phi_\alpha(E)\rangle + G_\alpha^{(+)}(E)V^\alpha|\psi_\alpha^{(+)}(E)\rangle, \tag{4.2.1}$$

where

$$H_\alpha|\phi_\alpha(E)\rangle = E|\phi_\alpha(E)\rangle, \tag{4.2.2a}$$

$$H|\psi_\alpha^{(+)}(E)\rangle = E|\psi_\alpha^{(+)}(E)\rangle. \tag{4.2.2b}$$

The state $|\phi_\alpha(E)\rangle$ is the direct product of the two-cluster bound states and free states representing their translational motion. The α-channel Green function is

$$G_\alpha^{(+)}(E) = (E + i0 - H_\alpha)^{-1}. \tag{4.2.3}$$

An important stimulus for the development of complete and consistent dynamical integral equations for multiparticle collisions was the observation by Epstein and, independently, by Foldy and Tobocman (EFT) that, for any energy E above the threshold for scattering, the solution of Eq. (4.2.1) need not be unique.[1] This is related to the existence of nontrivial solutions $|\chi(E)\rangle$ of the homogeneous form of Eq. (4.2.1),

$$|\chi(E)\rangle = G_\alpha^{(+)}(E)V^\alpha|\chi(E)\rangle, \tag{4.2.4}$$

for energies above the scattering threshold. This is in striking contrast to the two-particle case, where the only nonuniqueness is associated with the appearance of a two-particle bound state, which is usually not realized in the scattering region.

The implications of this result, although even now not widely appreciated, are of major significance. It implies that any attempt to calculate the exact scattering state from Eq. (4.2.1) *alone* will fail, because the equation itself does not sufficiently constrain the physics. Therefore, the only way in which Eq. (4.2.1) can be used in isolation as a calculational tool is if the dynamics is approximated to such a degree that the resultant equation yields a unique prediction. The simplest example of this is the Born approximation

$$|\psi_\alpha^{(+)}(E)\rangle \simeq |\phi_\alpha(E)\rangle + G_\alpha^{(+)}(E)V^\alpha|\phi_\alpha(E)\rangle. \tag{4.2.5}$$

Other techniques involve approximations that, in effect, reduce Eq. (4.2.1) to an integral equation with the degrees of freedom and the uniqueness property of the two-body case. Many widely used calculational techniques fall into this category.

The EFT demonstration of LS nonuniqueness was based on formal manipulations of singular operators, limited only to two-cluster initial states, and provided little intuitive insight as to why the nonuniqueness occurred. Later, Gerjuoy[2] used more conventional mathematical methods, in the representation associated with the configuration space of the independent-particle coordinate vectors, to establish the same nonuniqueness property. His analysis indicates that solutions of Eq. (4.2.1) are not unique because when particle rearrangement can take place the scattered part of the state vector,

$$|\Phi_\alpha^{(+)}(E)\rangle \equiv |\psi_\alpha^{(+)}(E)\rangle - |\phi_\alpha(E)\rangle, \tag{4.2.6}$$

need not be outgoing in the multidimensional configuration space in the way one would expect from $G_\alpha^{(+)}(E)$. This suggests that unique solutions cannot be obtained unless additional boundary conditions are imposed on $|\psi_\alpha^{(+)}(E)\rangle$. However, when no rearrangement channels are open, the solution to Eq. (4.2.1) is expected to be unique, $|\Phi_\alpha^{(+)}(E)\rangle$ and $G_\alpha^{(+)}(E)$ are outgoing in the same manner, and the only solutions of the homogeneous equation occur at the discrete c.m. energies of the bound states of the full N-particle system.

Gerjuoy's proof of nonuniqueness, however, depends on certain assumptions about the asymptotic behavior of the solutions to the multiparticle Schrödinger equation, as well as of the Green functions. These assumptions have not been established rigorously, as, for example, in the analysis of Faddeev.[3] But they are consistent with everything that is known about these asymptotic behaviors. In the case of three-particle scattering, Glöckle[4] has obtained these asymptotic forms, and his results agree with Gerjuoy's assumptions. Solvable three-particle model studies have produced results consistent with Gerjuoy's analysis.[5,6]

We remark that the nonuniqueness problem is pertinent only when we consider a *single* LS equation. Glöckle[7] has argued that a full set of LS equations satisfied by $|\psi_\alpha^{(+)}(E)\rangle$ defines it uniquely.

The nonuniqueness problem of LS equations is very subtle, indeed. Possible loopholes in Gerjuoy's and Lippmann's analyses have been pointed out by Mukherjee,[8] by Benoist-Gueutal,[9] and by Sawada and Sasakawa.[10] There have been a number of refutations[11] of these criticisms, in support of the

EFT–Gerjuoy–Lippmann analyses. In view of the sustained interest in these questions and their importance to collision theory, we discuss the relevant aspects of these analyses, pointing out their strengths along with their weak points.

Specifically, EFT infer nonuniqueness using the Lippmann identities,[12] whose validity depends upon the vanishing of certain volume integrals involving the outgoing-wave Green functions in the $\varepsilon \to 0+$ limit. On the other hand, Gerjuoy infers nonuniqueness on the basis of the vanishing of certain surface integrals over the sphere at infinity. We investigate in detail the relationship between these two types of integrals and whether they really vanish in the appropriate limits.

The nonuniqueness controversy is associated with ambiguities in the meaning of the limit $\varepsilon \to 0+$ implicit in Eq. (4.2.1). Now, Eq. (4.2.1) is presumed to be an appropriate limit of

$$|\psi_\alpha(E + i\varepsilon)\rangle = |\phi_\alpha(E)\rangle + G_\alpha(E + i\varepsilon)V^\alpha|\psi_\alpha(E + i\varepsilon)\rangle, \tag{4.2.7}$$

where $G_\alpha(E + i\varepsilon) = (E - H_\alpha + i\varepsilon)^{-1}$. We examine various ways of taking this limit. This analysis, coupled with the use of conventional configuration-space techniques, allows us to relate Gerjuoy's surface integrals and Lippmann's identities, thereby validating both devices as well as the result that Eq. (4.2.1) does not determine $|\psi_\alpha^{(+)}(E)\rangle$ uniquely when rearrangement is possible.

Now we know that (*cf*. Eq. (1.7.31))

$$|\psi_\alpha^{(+)}(E)\rangle = \lim_{\varepsilon \to 0+} \{i\varepsilon G(E + i\varepsilon)|\phi_\alpha(E)\rangle\}, \tag{4.2.8}$$

where $G(E + i\varepsilon) = (E - H + i\varepsilon)^{-1}$, is related to $G_\beta(E + i\varepsilon)$ by the resolvent identity

$$G(E + i\varepsilon) = G_\beta(E + i\varepsilon) + G_\beta(E + i\varepsilon)V^\beta G(E + i\varepsilon), \tag{4.2.9}$$

where β is a two-cluster rearrangement channel. We then obtain

$$|\psi_\alpha^{(+)}(E)\rangle = \lim_{\varepsilon \to 0+} \{i\varepsilon G_\beta(E + i\varepsilon)|\phi_\alpha(E)\rangle + G_\beta(E + i\varepsilon)V^\beta[i\varepsilon G(E + i\varepsilon)|\phi_\alpha(E)\rangle]\}. \tag{4.2.10}$$

If we assume that in the last term on the right-hand side of Eq. (4.2.10) the limit of the product is equivalent to the product of the limits, we obtain

$$|\psi_\alpha^{(+)}(E)\rangle = \lim_{\varepsilon \to 0+} \{i\varepsilon G_\beta(E + i\varepsilon)|\phi_\alpha(E)\rangle\} + G_\beta^{(+)}(E)V^\beta|\psi_\alpha^{(+)}(E)\rangle. \tag{4.2.11}$$

The homogeneous LS equations for $|\psi_\alpha^{(+)}(E)\rangle$,

$$|\psi_\alpha^{(+)}(E)\rangle = G_\beta^{(+)}(E)V^\beta|\psi_\alpha^{(+)}(E)\rangle, \qquad \beta \neq \alpha, \tag{4.2.12}$$

and the inhomogeneous LS equation (4.2.1) follow from Eq. (4.2.11) with the application of the so-called Lippmann identities[1,2,12]

$$\lim_{\varepsilon\to 0+} i\varepsilon G_\alpha(E+i\varepsilon)|\phi_\beta(E)\rangle = \delta_{\alpha\beta}|\phi_\beta(E)\rangle. \tag{4.2.13}$$

The nonuniqueness of Eq. (4.2.1) is manifested in the homogeneous equations

$$|\psi_\beta^{(+)}(E)\rangle = G_\alpha^{(+)}(E)V^\alpha|\psi_\beta^{(+)}(E)\rangle, \qquad \beta \neq \alpha, \tag{4.2.14}$$

that are obtained from Eq. (4.2.12) by interchanging α and β. Thus, for any scalar λ, the linear combination

$$|\psi_{\alpha\beta}(\lambda)\rangle = |\psi_\alpha^{(+)}(E)\rangle + \lambda|\psi_\beta^{(+)}(E)\rangle \tag{4.2.15}$$

is a solution of Eq. (4.2.1).

It is important to emphasize that it is the rearrangement channel wave functions $|\psi_\beta^{(+)}(E)\rangle$, $(\beta \neq \alpha)$, and *not* the channel wave functions for multi-cluster partitions a, $|\psi_a^{(+)}(E)\rangle$, which satisfy the homogeneous equations (4.2.14) that are manifestly responsible for the nonuniqueness of solutions to Eq. (4.2.1). Instead, $|\psi_a^{(+)}(E)\rangle$ satisfies an *inhomogeneous* equation with the Green function $G_\alpha^{(+)}(E)$. This can be seen by using Eq. (4.2.9) in conjunction with

$$|\psi_a^{(+)}(E)\rangle = \lim_{\varepsilon\to 0+} \{i\varepsilon G(E+i\varepsilon)|\phi_a(E)\rangle\} \tag{4.2.16}$$

to obtain

$$|\psi_a^{(+)}(E)\rangle = \lim_{\varepsilon\to 0+} \{i\varepsilon G_\alpha(E+i\varepsilon)|\phi_a(E)\rangle\} + G_\alpha^{(+)}(E)V^\alpha|\psi_a^{(+)}(E)\rangle, \tag{4.2.17}$$

where we have again assumed that in the last term the limit of the product is equivalent to the product of the limits. The first term on the right-hand side of Eq. (4.2.17) generates a scattering state of the multiparticle system when some of the interactions are set equal to zero, and does not typically vanish.

The homogeneous version of Eq. (2.3.18), in contrast to Eq. (4.2.1), does not have a scattering solution. If it did, it would violate the boundary condition (2.3.26), which requires a plane wave in addition to the outgoing wave—an impossibility for a solution of the single-channel homogeneous equation. On the other hand, Eq. (4.2.14) implies that $|\psi_\beta^{(+)}(E)\rangle$ is purely outgoing in channel $\alpha \neq \beta$ and also has the outgoing-wave asymptotic behavior, plus the two-cluster plane wave, in channel β. The plane wave is annihilated by virtue of Eq. (4.2.13).

The derivation of Eqs. (4.2.1) and (4.2.14) depends on both the Lippmann identities and the ε-limiting process entailing the replacement of the limit of a product by the product of the limits; we have established neither one except by plausible analogy with the single-channel case. With these points in mind, we move onto a more detailed discussion of these equations.

4.3. Alternative Interpretations of the LS Equations

A formal operator algebra derivation of Eq. (4.2.1), as in the preceding section, suggests that this equation can be interpreted either as

$$|\psi_\alpha(E)\rangle_1 = |\phi_\alpha(E)\rangle + \lim_{\varepsilon\to 0+} [G_\alpha(E + i\varepsilon)V^\alpha|\psi_\alpha(E)\rangle_1], \tag{4.3.1}$$

or as

$$|\psi_\alpha(E)\rangle_2 = |\phi_\alpha(E)\rangle + \left[\lim_{\varepsilon\to 0+} G_\alpha(E + i\varepsilon)\right] V^\alpha|\psi_\alpha(E)\rangle_2. \tag{4.3.2}$$

The kernel terms in Eqs. (4.3.1–4.3.2) can be expressed as integrals in configuration space. In Eq. (4.3.1), we maintain $\varepsilon \neq 0$ in $G_\alpha(E + i\varepsilon)$ until this integral is evaluated whereas, in Eq. (4.3.2), the limit is taken first in $G_\alpha(E + i\varepsilon)$, and then the integration is performed.

On the other hand, Eq. (4.2.1) can also be interpreted as the limit of Eq. (4.2.7) with

$$|\psi_\alpha(E)\rangle_3 \equiv \lim_{\varepsilon\to 0+} |\psi_\alpha(E + i\varepsilon)\rangle, \tag{4.3.3}$$

with $|\psi_\alpha(E + i\varepsilon)\rangle$ defined by Eq. (4.2.7) for $\varepsilon \neq 0$.

The preceding three interpretations yield state vectors $|\psi_\alpha(E)\rangle_i$, $i = 1, 2, 3$, that are not necessarily identical. However, of this triplet, Eq. (4.3.2) appears to be equivalent to Eq. (4.2.1) because (*cf.* Eq. (4.2.3))

$$G_\alpha^{(+)}(E) \equiv \lim_{\varepsilon\to 0+} G_\alpha(E + i\varepsilon). \tag{4.3.4}$$

In the remainder of this section we closely follow the analysis of Gerjuoy and Adhikari,[1,2] which uses Gerjuoy's[3] configuration-space methods. All of the equations to follow are to be understood as being in the configuration-space representation, so the bra-ket notation is dropped. The real-energy argument E of the wave functions ψ, ϕ, Φ, etc., is suppressed unless necessary.

4.3a. Equivalence between the Schrödinger and the LS Equation and the Nonuniqueness Problem

Let us examine the implications and consequences of Eq. (4.3.2). The wave function ψ_α corresponding to the state (4.3.2) satisfies the Schrödinger equation

$$(E - H)\psi_\alpha = 0, \tag{4.3.5}$$

and has the form $\psi_\alpha = \phi_\alpha + \Phi_\alpha$, where ϕ_α is the incoming wave and the scattered part Φ_α asymptotically contains spherically outgoing waves in all channels. In what follows, a generic ψ always denotes a solution of Eq. (4.3.5); similarly a generic Φ refers to $\psi - \phi$, where ϕ is the relevant incident wave (if any).

Next, we note that the Green function $G_\alpha^{(+)}(E)$ satisfies the correct outgoing-wave boundary condition for channel α and the relations

$$(E - \vec{H}_0 - V_\alpha)G_\alpha^{(+)}(E) = I, \tag{4.3.6}$$

$$G_\alpha^{(+)}(E)(E - \overleftarrow{H}_0 - V_\alpha) = I. \tag{4.3.7}$$

In operator equations such as (4.3.6–4.3.7), it is important to denote by an arrow on H_0 the direction in which the differential operators in H_0 act (*cf.* Section 1.7).

Let Eq. (4.3.7) operate from the left on an arbitrary Schrödinger wave function ψ. When the resultant equation is combined with

$$(E - H_\alpha)\psi = V^\alpha\psi, \tag{4.3.8}$$

we obtain

$$\psi = S_\infty[G_\alpha^{(+)}(E), \psi] + G_\alpha^{(+)}(E)V^\alpha\psi. \tag{4.3.9a}$$

Here (*cf.* Eqs. (2.3.38–2.3.43)),

$$\begin{aligned} S_\infty[G_\alpha^{(+)}(E), \psi] &\equiv G_\alpha^{(+)}(E)\overleftrightarrow{H}_0\psi \\ &= G_\alpha^{(+)}(E)(\vec{H}_0 - \overleftarrow{H}_0)\psi \end{aligned} \tag{4.3.9b}$$

is the surface integral over the generally multidimensional sphere at infinity, S_∞, in configuration space. This follows from the general definition,

$$S_\infty[X, Y] \equiv \int [d^3x][X(\mathbf{x})\vec{H}_0 Y(\mathbf{x}) - Y(\mathbf{x})\vec{H}_0 X(\mathbf{x})], \tag{4.3.10}$$

after applying Green's theorem in **x**-space. Here, **x** represents a complete set of coordinate vectors that specify the relative positions between the particles; the multidimensional volume element is denoted as $[d^3x]$. From Eq. (4.3.9a), we realize that ψ is also a solution of the real-energy LS equation (4.2.1) if and only if

$$S_\infty[G_\alpha^{(+)}(E), \psi] = \phi_\alpha. \tag{4.3.11}$$

Moreover, when this condition is satisfied by different independent solutions, ψ's, they are also (nonunique) solutions of the LS equation.

Proceeding as before, but now using Eqs. (4.2.2a) and (4.3.7), we *deduce*[1] that

$$S_\infty[G_\alpha^{(+)}(E), \phi_\alpha] = \phi_\alpha. \tag{4.3.12}$$

Thus, any ψ also satisfies the LS equation (4.3.2) if its scattered part satisfies

$$S_\infty[G_\alpha^{(+)}(E), \Phi] = 0. \tag{4.3.13a}$$

We call Eq. (4.3.13a) the everywhere-outgoing (E.0.) condition.

In the case of potential scattering, $G_0^{(+)}(E)$ and Φ behave similarly at infinity and this ensures that Eq. (4.3.13a) is satisfied (*cf*. Eq. (2.3.45b)). The situation is different for multiparticle scattering or even for two-particle collisions in the laboratory coordinate system. In the latter case, the E.O. scattered part Φ of a wave function ψ having the incident wave ϕ does not behave like the outgoing Green function $G_0^{(+)}(E)$ at infinity along arbitrary directions because it includes a plane wave that represents the c.m. motion. For systems of three or more particles, the asymptotic behaviors of $G_\alpha^{(+)}(E)$ and the scattered part Φ associated with a physically correct ψ can be different even in the c.m. system. Nevertheless, it has been *shown*[1] that Eq. (4.3.13a) is satisfied by such a Φ. In particular, if Φ_α and Φ_β denote the E.O. scattered parts of the wave functions ψ_α and ψ_β with incident waves ϕ_α and ϕ_β, respectively, in the α and $\beta(\neq\alpha)$ channels, then

$$S_\infty[G_\alpha^{(+)}(E), \Phi_\alpha] = 0, \tag{4.3.13b}$$

$$S_\infty[G_\alpha^{(+)}(E), \Phi_\beta] = 0. \tag{4.3.13c}$$

Also, since channel $\beta(\neq\alpha)$ is assumed to be a nontrivial rearrangement of channel α, $G_\alpha^{(+)}(E)$ has a vanishing projection on the bound-state eigenfunction characterizing the β-channel in the asymptotic region. Therefore, it is also expected (*assumed*) that

$$S_\infty[G_\alpha^{(+)}(E), \phi_\beta] = 0. \tag{4.3.13d}$$

Thus, we have, from Eqs. (4.3.13c, d),

$$S_\infty[G_\alpha^{(+)}(E), \psi_\beta] = 0, \tag{4.3.14a}$$

whereas Eqs. (4.3.12) and (4.3.13b) imply

$$S_\infty[G_\alpha^{(+)}(E), \psi_\alpha] = \phi_\alpha. \tag{4.3.14b}$$

Equations (4.3.14a) and (4.3.14b) imply

$$S_\infty[G_\alpha^{(+)}(E), \psi_{\alpha\beta}(\lambda)] = \phi_\alpha, \tag{4.3.14c}$$

where the linear combination $\psi_{\alpha\beta}(\lambda)$ is defined as in Eq. (4.2.15).

Equation (4.3.14c) therefore implies (recall Eqs. (4.3.11) and (4.3.13a)) that $\psi_{\alpha\beta}(\lambda)$ satisfies Eq. (4.3.2). Hence, if the *assumed* relation (4.3.13d) really holds, the real-energy LS equation (4.3.2) has nonunique solutions.

4.3b. *Equivalence between Lippmann's Limits and Gerjuoy's Surface Integrals*

We mentioned before that Lippmann's discussion of the LS equation is based on certain real-energy limits known as Lippmann identities. Gerjuoy's configuration-space treatment, on the other hand, is based on certain surface integrals in multidimensional configuration space at infinity. We now establish a connection between these two approaches.

In general, Eq. (4.3.1) is not identical with Eq. (4.3.2). If they were the same, we would have, for a generic ψ,

$$\lim_{\varepsilon\to 0+} [G_\alpha(E+i\varepsilon)V^\alpha\psi] = \left[\lim_{\varepsilon\to 0+} G_\alpha(E+i\varepsilon)\right]V^\alpha\psi, \tag{4.3.15a}$$

or, in explicit notation,

$$\lim_{\varepsilon\to 0+}\int [d^3x']G_\alpha(\mathbf{x},\mathbf{x}';E+i\varepsilon)V^\alpha(\mathbf{x}')\psi(\mathbf{x}',E)$$
$$=\int [d^3x']\left[\lim_{\varepsilon\to 0+} G_\alpha(\mathbf{x},\mathbf{x}';E+i\varepsilon)\right]V^\alpha(\mathbf{x}')\psi(\mathbf{x}',E). \tag{4.3.15b}$$

Equations (4.3.15a, b) have been established for potential scattering;[4] for three-particle systems, arguments have been given for their correctness, but for a general many-particle system they are not easy to prove.[1] We assume, henceforth, that the product of integrals on the left-hand side of Eq. (4.3.15b) is uniformly convergent as $\varepsilon \to 0+$, so that the interchange of this limit and the configuration-space integrations is justified. We next explore some consequences of Eqs. (4.3.15a, b) assuming they are correct.

Equation (4.3.15b) is equivalent to

$$\lim_{\varepsilon\to 0+}\left[\lim_{R\to\infty}\int^R [d^3x']G_\alpha(\mathbf{x},\mathbf{x}';E+i\varepsilon)V^\alpha(\mathbf{x}')\psi(\mathbf{x}',E)\right]$$
$$=\lim_{R\to\infty}\left[\int^R [d^3x']\left[\lim_{\varepsilon\to 0+} G_\alpha(\mathbf{x},\mathbf{x}';E+i\varepsilon)\right]V^\alpha(\mathbf{x}')\psi(\mathbf{x}',E)\right], \tag{4.3.16a}$$

where the integrations are confined to spheres with radii R that eventually tend to infinity. On the right-hand side of Eq. (4.3.16a), the fact that the $\varepsilon \to 0+$ limit exists for any finite R makes it possible to take this limit outside the integral, and this leads to

$$\lim_{\varepsilon\to 0+}\lim_{R\to\infty} F(\mathbf{x},R;\varepsilon) = \lim_{R\to\infty}\lim_{\varepsilon\to 0+} F(\mathbf{x},R;\varepsilon), \tag{4.3.16b}$$

where

$$F(\mathbf{x}, R; \varepsilon) \equiv \int^{R} [d^3x'] G_\alpha(\mathbf{x}, \mathbf{x}'; E + i\varepsilon) V^\alpha(\mathbf{x}') \psi(\mathbf{x}', E) \tag{4.3.16c}$$

or, schematically,

$$F(R; \varepsilon) \equiv \{G_\alpha(E + i\varepsilon) V^\alpha \psi\}_R. \tag{4.3.16d}$$

The notation $\{\cdots\}_R$ refers to a volume integral confined to a sphere of radius R.

Now, in general, ψ and ϕ_α satisfy

$$(E + i\varepsilon - H_\alpha)\psi = i\varepsilon\psi + V^\alpha\psi, \tag{4.3.17}$$

$$(E + i\varepsilon - H_\alpha)\phi_\alpha = i\varepsilon\phi_\alpha. \tag{4.3.18}$$

Equations (4.3.16–4.3.18) have an interesting consequence: The surface integrals (4.3.13a–d) and (4.3.14a–c) are identical to the real-energy limits of certain volume integrals first studied by Lippmann. This establishes a connection between the operator-algebra[5] and configuration-space[6] approaches to multiparticle collision theory.

In order to demonstrate this equivalence, we operate on Eq. (4.3.17) from the left by $G_\alpha(E + i\varepsilon)$, and restrict the intermediate integration to a sphere of radius R. This leads to

$$F(R; \varepsilon) = -S_R[G_\alpha(E + i\varepsilon), \psi] + \psi - i\varepsilon\{G_\alpha(E + i\alpha)\psi\}_R, \tag{4.3.19}$$

where $S_R[G_\alpha, \psi]$ now refers to a surface integral evaluated over the sphere of radius R, and not at infinity. As with Eq. (4.3.16a), there are two distinct ways of evaluating the terms in Eq. (4.3.19). If we first take the $\varepsilon \to 0+$ limit and then the $R \to \infty$ limit in Eq. (4.3.19), the last term of this equation vanishes, and we find that

$$\psi = S_\infty[G_\alpha^{(+)}(E), \psi] + G_\alpha^{(+)}(E) V^\alpha \psi. \tag{4.3.20a}$$

If, on the other hand, we first take the $R \to \infty$ limit before taking the $\varepsilon \to 0+$ limit in Eq. (4.3.19), the surface integral vanishes in the $R \to \infty$ limit because $G_\alpha(E + i\varepsilon)$, $\varepsilon \neq 0$, decays exponentially as $R \to \infty$. Then, Eq. (4.3.19) yields

$$\psi = \lim_{\varepsilon \to 0+} i\varepsilon G_\alpha(E + i\varepsilon)\psi + G_\alpha^{(+)}(E) V^\alpha \psi. \tag{4.3.20b}$$

The last terms on the right-hand sides of Eqs. (4.3.20a, b) are identical by virtue of the identity (4.3.16b). So, whenever Eq. (4.3.16b) holds, we have the following connection between Gerjuoy's surface integral and Lippmann's limit:

$$S_\infty[G_\alpha^{(+)}(E), \psi] = \lim_{\varepsilon \to 0+} i\varepsilon G_\alpha(E + i\varepsilon)\psi, \tag{4.3.21}$$

where the left-hand side is evaluated on the sphere at infinity and the right-hand side is integrated over all space. Equations (4.3.14a, b) and (4.3.21) lead to the Lippmann identites:

$$\lim_{\varepsilon\to 0+} i\varepsilon G_\alpha(E+i\varepsilon)\psi_\alpha = \phi_\alpha, \tag{4.3.22a}$$

$$\lim_{\varepsilon\to 0+} i\varepsilon G_\alpha(E+i\varepsilon)\psi_\beta = 0, \qquad \alpha \neq \beta. \tag{4.3.22b}$$

There are other, similar identities which can be derived. For example, if we perform the same manipulations needed to derive Eq. (4.3.19) on Eq. (4.3.18), rather than on Eq. (4.3.17), we obtain

$$S_R[G_\alpha(E+i\varepsilon), \phi_\alpha] = \phi_\alpha - i\varepsilon\{G_\alpha(E+i\varepsilon)\phi_\alpha\}_R. \tag{4.3.23}$$

If we first take the $\varepsilon\to 0+$ limit and then the $R\to\infty$ limit, we obtain Eq. (4.3.12). However, the left-hand side of Eq. (4.3.23) vanishes for $R\to\infty$ with $\varepsilon \neq 0$, so (*cf*. Eq. (2.3.15))

$$i\varepsilon G_\alpha(E+i\varepsilon)\phi_\alpha = \phi_\alpha, \tag{4.3.24}$$

valid for *any* ε. Equations (4.3.12) and (4.3.24) imply that

$$S_\infty[G_\alpha^{(+)}(E), \phi_\alpha] = i\varepsilon G_\alpha(E+i\varepsilon)\phi_\alpha, \tag{4.3.25}$$

which establishes again a one-to-one equivalence between Gerjuoy's surface integral and Lippmann's limit.

In order to derive other identities, we next note that $(E - H_\beta)\phi_\beta = 0$ can be rewritten as

$$(E - H_\alpha + i\varepsilon)\phi_\beta = -(V^\beta - V^\alpha)\phi_\beta + i\varepsilon\phi_\beta. \tag{4.3.26}$$

In the same way that we deduced Eq. (4.3.19), we find from Eq. (4.3.26) that

$$D(R;\varepsilon) = -S_R[G_\alpha(E+i\varepsilon), \phi_\beta] + \phi_\beta - i\varepsilon\{G_\alpha(E+i\varepsilon)\phi_\beta\}_R, \tag{4.3.27a}$$

where

$$D(R;\varepsilon) \equiv \{G_\alpha(E+i\varepsilon)(V^\alpha - V^\beta)\phi_\beta\}_R, \tag{4.3.27b}$$

and it remains to prescribe the order of the $R\to\infty$ and $\varepsilon\to 0+$ limits. If we take the $\varepsilon\to 0+$ limit first, the last term on the right-hand side of Eq. (4.3.27a) vanishes; then, taking the $R\to\infty$ limit we have

$$\lim_{R\to\infty}\lim_{\varepsilon\to 0+} D(R;\varepsilon) = -S_\infty[G_\alpha^{(+)}(E), \phi_\beta] + \phi_\beta. \tag{4.3.28a}$$

On the other hand, if we interchange the order of the $R\to\infty$ and $\varepsilon\to 0+$ limits, the surface integral in (4.3.27a) vanishes and we get instead

$$\lim_{\varepsilon\to 0+}\lim_{R\to\infty} D(R;\varepsilon) = -\lim_{\varepsilon\to 0+} i\varepsilon G_\alpha(E+i\varepsilon)\phi_\beta + \phi_\beta. \tag{4.3.28b}$$

If we assume that

$$\lim_{\varepsilon \to 0+} \lim_{R \to \infty} D(R; \varepsilon) = \lim_{R \to \infty} \lim_{\varepsilon \to 0+} D(R; \varepsilon), \tag{4.3.28c}$$

we find another relationship similar to Eqs. (4.3.21) and (4.3.25):

$$S_\infty[G_\alpha^{(+)}(E), \phi_\beta] = \lim_{\varepsilon \to 0+} i\varepsilon G_\alpha(E + i\varepsilon)\phi_\beta, \tag{4.3.29a}$$

which, in view of Eq. (4.3.13d), yields the Lippmann identity

$$\lim_{\varepsilon \to 0+} i\varepsilon G_\alpha(E + i\varepsilon)\phi_\beta = 0. \tag{4.3.29b}$$

Equations (4.3.24) and (4.3.29b) together with Eqs. (4.3.22a, b) imply that

$$\lim_{\varepsilon \to 0+} i\varepsilon G_\alpha(E + i\varepsilon)\Phi_\gamma = 0, \tag{4.3.30}$$

for all γ. Finally, Eqs. (4.3.21), (4.3.25), and (4.3.29a) lead to

$$S_\infty[G_\alpha^{(+)}(E), \Phi_\gamma] = \lim_{\varepsilon \to 0+} i\varepsilon G_\alpha(E + i\varepsilon)\Phi_\gamma. \tag{4.3.31}$$

In conclusion, the presumed validity of the limit interchange identities (4.3.16b) and (4.3.28c) allows us to establish a connection between Gerjuoy's surface integrals at infinity and the $\varepsilon \to 0+$ limits involved in the Lippmann's identities.

4.3c. Complex-Energy LS Equation and the Outgoing-Wave Boundary Condition

Next, we would like to explore the implications of the LS equation (4.2.7). Operating with $(E - \vec{H}_0 - V_\alpha + i\varepsilon)$ on both sides of this equation, we obtain

$$(E - \vec{H}_0 - V + i\varepsilon)\psi_\alpha(E + i\varepsilon) = i\varepsilon\phi_\alpha, \tag{4.3.32}$$

which has a unique solution for ψ_α, because $E + i\varepsilon$ does not lie in the spectrum of the Hermitian operators H or H_α. In fact, Eq. (4.3.32) is solved by

$$\psi_\alpha(E + i\varepsilon) = i\varepsilon G(E + i\varepsilon)\phi_\alpha, \tag{4.3.33}$$

where $G(E + i\varepsilon) \equiv (E + i\varepsilon - H)^{-1}$ is the unique nonsingular total Green function. Hence, we conclude that the complex-energy LS equation (4.2.7) has a unique solution, in contrast to the real-energy versions (4.3.1) or (4.3.2).[1,2] Also, the function $|\psi_\alpha(E)\rangle_3$ defined by Eq. (4.3.3) is the unique limit of the sequence of functions $|\psi_\alpha(E + i\varepsilon)\rangle$.[1]

In order to express $\psi_\alpha(E + i\varepsilon)$ in more useful form, we rewrite Eq. (4.3.32) as

$$(E - \vec{H}_0 - V + i\varepsilon)\Phi_\alpha(E + i\varepsilon) = V^\alpha \phi_\alpha, \tag{4.3.34}$$

so that

$$\Phi_\alpha(E + i\varepsilon) = G(E + i\varepsilon)V^\alpha \phi_\alpha, \tag{4.3.35}$$

or

$$\psi_\alpha(E + i\varepsilon) = \phi_\alpha + G(E + i\varepsilon)V^\alpha \phi_\alpha. \tag{4.3.36}$$

The limit implied by Eq. (4.3.3) is now given by

$$\psi_\alpha(E)_3 = \phi_\alpha + G^{(+)}(E)V^\alpha \phi_\alpha, \tag{4.3.37}$$

where we have assumed that

$$\lim_{\varepsilon \to 0+} [G(E + i\varepsilon)V^\alpha \phi_\alpha] = \left[\lim_{\varepsilon \to 0+} G(E + i\varepsilon)\right] V^\alpha \phi_\alpha, \tag{4.3.38}$$

with

$$G^{(+)}(E) \equiv \lim_{\varepsilon \to 0+} G(E + i\varepsilon). \tag{4.3.39}$$

We can show that the expression (4.3.37) is the unique solution of the Schrödinger equation (4.2.2b) whose incoming part is ϕ_α and whose scattered part $\Phi_\alpha(E)$ obeys the E.O. condition

$$S_\infty[G^{(+)}(E), \Phi_\alpha(E)] = 0. \tag{4.3.40}$$

In order to prove this, we need the counterpart of Eqs. (4.3.6–4.3.7) for $G^{(+)}(E)$,

$$(E - \vec{H}_0 - V)G^{(+)}(E) = G^{(+)}(E)(E - \overleftarrow{H}_0 - V) = I, \tag{4.3.41a}$$

and the Schrödinger equation (4.2.2b) rewritten in the form

$$(E - \vec{H}_0 - V)\Phi_\alpha(E) = V^\alpha \phi_\alpha(E). \tag{4.3.41b}$$

Then, we obtain, in the usual way,

$$\Phi_\alpha(E) = S_\infty[G^{(+)}(E), \Phi_\alpha(E)] + G^{(+)}(E)V^\alpha \phi_\alpha(E), \tag{4.3.42}$$

or

$$\psi_\alpha(E) = S_\infty[G^{(+)}(E), \Phi_\alpha(E)] + \phi_\alpha(E) + G^{(+)}(E)V^\alpha \phi_\alpha(E). \tag{4.3.43}$$

Comparing Eq. (4.3.43) and Eq. (4.3.37), we see that $\psi_\alpha(E)_3$ is the unique solution $\psi_\alpha(E)$ of the Schrödinger equation (4.2.2b) with incoming part $\phi_\alpha(E)$ and whose scattered part $\Phi_\alpha(E)$ satisfies the E.O. condition (4.3.40).

It is interesting to comment now that the existence of the limit in Eq. (4.3.3) need not imply that

$$\psi_\alpha(E)_3 = \phi_\alpha + \lim_{\varepsilon\to 0+} [G_\alpha(E + i\varepsilon)V^\alpha\psi_\alpha(E + i\varepsilon)], \tag{4.3.44}$$

because there is no guarantee that the limit on the right-hand side involving the yet unknown function $\psi_\alpha(E + i\varepsilon)$ exists. But, after our analysis, we now know that the limit $\varepsilon \to 0+$ in Eq. (4.3.3) yields $\psi_\alpha(E)_3$ of Eq. (4.3.37), which satisfies the Schrödinger equation. The function $\psi_\alpha(E + i\varepsilon)$ on the right-hand side of Eq. (4.3.44) does have a limit, viz., $\psi_\alpha(E)_3 = \psi_\alpha(E)$, where we now drop the subscript 3. Thus, making the usual assumption that the sequence of infinite integrals on the right-hand side of Eq. (4.3.44) converges uniformly, we conclude that the limit in this equation is indeed meaningful, so that

$$\begin{aligned}\lim_{\varepsilon\to 0+} [G_\alpha(E + i\varepsilon)V^\alpha\psi_\alpha(E + i\varepsilon)] &= G_\alpha^{(+)}(E)V^\alpha\psi_\alpha(E)\\ &\equiv \left[\lim_{\varepsilon\to 0+} G_\alpha(E + i\varepsilon)\right]V^\alpha\left[\lim_{\varepsilon\to 0+} \psi_\alpha(E + i\varepsilon)\right].\end{aligned} \tag{4.3.45}$$

Hence, Eq. (4.3.3) yields the desired solution of the Schrödinger equation, provided that Eq. (4.3.45) holds and $\psi_\alpha(E)_3$ has the well-behaved limit $\psi_\alpha(E)$.

Condition (4.3.45) is also relevant for potential scattering[4] (Ch. 2) and connected-kernel N-particle equations[7,8] (Ch. 5). In both cases, the scattering integral equations have the same structure as Eqs. (4.2.7), (4.3.1–4.3.2), but now the kernel is compact and the solution is unique provided that the equality (4.3.45) holds.

Equation (4.3.45) implies that Eqs. (4.2.7) and (4.3.3) should reduce in the $\varepsilon \to 0+$ limit to Eqs. (4.3.1) or (4.3.2). The function $\Phi_\alpha(E)$ corresponding to the solution of Eq. (4.3.43) must be given by

$$\Phi_\alpha(E) = G_\alpha^{(+)}(E)V^\alpha\psi_\alpha(E). \tag{4.3.46}$$

However, we have seen that whenever a solution to the Schrödinger equation also satisfies the LS equation (4.3.1) or (4.3.2), i.e., whenever its scattered part is given by Eq. (4.3.46), then Eq. (4.3.13a) must hold.[1] Hence Eq. (4.3.13b) immediately follows for $\Phi_\alpha(E)$ given by Eq. (4.3.46) and we have now *deduced*[1] that the same $\Phi_\alpha(E)$ that satisfies the E.O. condition (4.3.40) also satisfies the E.O. condition in the form (4.3.13b).

It is now easy to realize that, just as we found Eq. (4.3.12), we can now *deduce* from Eqs. (4.3.41a) and (4.2.2b) that, for the α-channel wave function, we have

$$S_\infty[G^{(+)}(E), \psi_\alpha(E)] = \psi_\alpha(E) \tag{4.3.47a}$$

without any further assumptions, while Eqs. (4.3.40) and (4.3.47a) imply that

$$S_\infty[G^{(+)}(E), \phi_\alpha(E)] = \psi_\alpha(E). \tag{4.3.47b}$$

Similarly, for the β-channel, we find

$$S_\infty[G^{(+)}(E), \psi_\beta(E)] = \psi_\beta(E), \tag{4.3.47c}$$

$$S_\infty[G^{(+)}(E), \phi_\beta(E)] = \psi_\beta(E). \tag{4.3.47d}$$

If we believe Eq. (4.3.14a), a function ψ_β, which apparently satisfies the E.O. condition involving channel Green functions $G_\alpha^{(+)}(E)$, can have an incoming part and, as Eq. (4.3.47c) shows, need not satisfy the E.O. condition with respect to the full Green function $G^{(+)}(E)$. Indeed, one such $\psi_\beta(E)$ is the wave function with an incoming part $\phi_\beta(E)$ and an E.O. part $\Phi_\beta(E)$ that obeys

$$S_\infty[G^{(+)}(E), \Phi_\beta(E)] = 0, \tag{4.3.48}$$

i.e., the β channel analogue of Eq. (4.3.40). Finally, it can be demonstrated that Eq. (4.3.48) implies Eq. (4.3.13c) which, together with Eq. (4.3.13d), yields Eq. (4.3.14a).[1]

4.3d. Conclusion

Under plausible assumptions about the convergence of the infinite integrals in Eqs. (4.3.1–4.3.2) and (4.2.7), these three equations are identical in the limit $\varepsilon \to 0+$. Under the same assumptions, one solution of the real energy LS equation (4.3.2) will be the solution ψ_α of the Schrödinger equation (4.3.5) whose scattered part Φ_α satisfies the E.O. condition (4.3.40) and also (4.3.13b). However, essentially the same outgoing conditions (4.3.48) and (4.3.13c) are satisfied by the E.O. wave Φ_β in the β-channel, which is a rearrangement of the α-channel. In addition, if we accept the plausible results (4.3.13d) and (4.3.14a), then Eq. (4.3.14c) follows immediately, and so does the conclusion that the real-energy LS equation (4.2.1) has nonunique solutions.

We have discussed two different approaches to the multiparticle LS equations by Lippmann and Gerjuoy. In the course of proving their equivalence, we have established a one-to-one correspondence between the real-energy limits that occur in Lippmann identities and the surface integrals appearing in Gerjuoy's approach.

Unlike the real-energy LS equations, Eqs. (4.3.1–4.3.2), the complex-energy LS equation (4.2.7) has a unique solution ($\varepsilon \neq 0$), and the limit (4.3.3) yields the solution satisfying the E.O. condition expressed by Eq. (4.3.40). This is because the Green function is quadratically integrable in the complex-energy plane. Any attempt, however, to solve this equation in the complex-energy plane and then take the limit (4.3.3) is bound to fail because of numerical instabilities.

The plausible assumptions we have made about asymptotic behaviors of the Green functions and solutions of the Schrödinger equation have been established in the context of solvable three-particle models. One such model consists of three particles in one spatial dimension interacting via pairwise attractive δ-function potentials. This model, which was first studied by McGuire,[9] has analytical solutions for the wave functions ψ_α and ψ_β. One can then verify directly that (a) both the homogeneous and the inhomogeneous LS equations are true, and that (b) the Lippmann identities and Gerjuoy's surface integrals have the values conjectured originally.[6] Also, for a three-particle system in three dimensions with the use of established asymptotic forms for the wave functions and various Green functions, Eqs. (4.3.14a–c) can be verified directly, and hence both the homogeneous and inhomogeneous LS equations then follow.[10]

The discussion presented so far applies to the scattering of the N-particle system initiated by a two-cluster state. The consideration of a collision process initiated by an n-cluster initial state ($n > 2$) is more complicated, in general, because of convergence difficulties with the integrals in this case (Section 4.5).

4.4. The Basic Set of LS Equations

We now know that for multiparticle collisions a single inhomogeneous LS scattering integral equation does not fully constrain a solution to satisfy the physical boundary conditions. Asymptotically, $\psi_\alpha^{(+)}$ should have outgoing spherical waves in all channels in addition to the incident plane wave in channel α. The reason the solutions of the inhomogeneous LS equation (4.2.1) are not unique is that there are nontrivial solutions of its homogeneous form. All solutions of Eq. (4.2.4) are solutions of the Schrödinger equation (now denoted by the generic ψ), but obey boundary conditions for two-cluster rearrangement channels $\beta \neq \alpha$. These rearrangement channels are eliminated in situations where the two-body bound state(s) of pair β are not allowed; then, there are no nontrivial scattering solutions of Eq. (4.2.4) and the solution of Eq. (4.2.1) is unique.

The situation is however different for $\varepsilon \neq 0$, in which case the homogeneous equation

$$|\psi(E + i\varepsilon)\rangle = G_\alpha(E + i\varepsilon)V^\alpha|\psi(E + i\varepsilon)\rangle, \tag{4.4.1}$$

has no nontrivial solution. We can verify this by operating with $(E + i\varepsilon - \vec{H}_0 - V_\alpha)$ on Eq. (4.4.1), reducing it to

$$(E - \vec{H}_0 - V + i\varepsilon)|\psi(E + i\varepsilon)\rangle = 0. \tag{4.4.2}$$

As H has a real spectrum, Eq. (4.4.2), and thus Eq. (4.4.1), is satisfied only trivially.

The correct boundary conditions for α-channel scattering can also be imposed automatically if we consider the basic set of LS equations satisfied by $|\psi_\alpha^{(+)}(E)\rangle$, where we include all $\beta \neq \alpha$:[1]

$$|\psi_\alpha^{(+)}(E)\rangle = |\phi_\alpha(E)\rangle + G_\alpha^{(+)}(E)V^\alpha|\psi_\alpha^{(+)}(E)\rangle, \tag{4.4.3}$$

$$|\psi_\alpha^{(+)}(E)\rangle = G_\beta^{(+)}(E)V^\beta|\psi_\alpha^{(+)}(E)\rangle. \tag{4.4.4}$$

Now we know that a Schrödinger wave function ψ also satisfies Eqs. (4.4.3–4.4.4) if and only if Eq. (4.3.11) and

$$S_\infty[G_\beta^{(+)}(E), \psi] = 0 \tag{4.4.5}$$

hold for the incident channel α and for all rearrangement channels $\beta \neq \alpha$, respectively. Recalling Eqs. (4.3.12–4.3.13), we realize that the conditions (4.4.5) exclude all incoming waves ϕ_β, $\beta \neq \alpha$, from ψ. Hence, if a wave function ψ also satisfies the set (4.4.3–4.4.4), it must have the form $\psi \equiv \psi_\alpha = \phi_\alpha + \Phi_\alpha$.

The set (4.4.3–4.4.4), which is usually called the basic set of LS equations, selects the unique α-channel solution $|\psi_\alpha^{(+)}(E)\rangle$ with an incident wave in channel α and outgoing waves in the asymptotic regions of all channels, β; there is no need to impose further boundary conditions. It is important to stress that this is true because the homogeneous form of the set (4.4.3–4.4.4), viz.,

$$|\psi_\alpha^{(+)}(E)\rangle = G_\alpha^{(+)}(E)V^\alpha|\psi_\alpha^{(+)}(E)\rangle, \tag{4.4.6a}$$

$$|\psi_\alpha^{(+)}(E)\rangle = G_\beta^{(+)}(E)V^\beta|\psi_\alpha^{(+)}(E)\rangle, \tag{4.4.6b}$$

respectively, where all $\beta(\neq\alpha)$ are included, has only a trivial solution. Any proper subset of Eqs. (4.4.6a, b) has nontrivial solutions for the omitted channels, but the full set excludes such solutions.

4.5. Scattering Equations for Three or More Noninteracting Clusters in the Initial State

LS-type integral equations can, of course, be written down for scattering processes initiated by a multicluster initial state.[1] In general, they do not have unique solutions, and so they cannot be used for computational purposes except after sufficient truncation. In this section, we present some of these scattering integral equations for three- and four-cluster initial states, and discuss some difficulties that they present. Most of our previous conclusions about nonuniqueness can be generalized to these cases.

As an illustration, let us consider a four-particle system. We denote the n-cluster partitions by $a_n, b_n, c_n, \ldots$, etc. A general three-cluster initial state of four particles, $|\phi_{a_3}\rangle$, has the form

$$|\phi_{a_3}(E)\rangle = |\hat{\phi}_{a_3}\rangle|\mathbf{q}^{(1)}\rangle|\mathbf{q}^{(2)}\rangle, \tag{4.5.1}$$

where $|\hat{\phi}_{a_3}\rangle$ is a two-particle bound state, and $|\mathbf{q}^{(1)}\rangle$ and $|\mathbf{q}^{(2)}\rangle$ are two plane waves that describe the relative motion between the three clusters in the c.m. frame. The wave function describing this scattering process is

$$|\psi_{a_3}^{(+)}(E)\rangle = \lim_{\varepsilon\to 0+} i\varepsilon G(E + i\varepsilon)|\phi_{a_3}(E)\rangle. \tag{4.5.2}$$

The LS-type scattering equations are derived using the resolvent identities

$$G(E + i\varepsilon) = G_{b_n}(E + i\varepsilon) + G_{b_n}(E + i\varepsilon)V^{b_n}G(E + i\varepsilon), \tag{4.5.3}$$

along with Eq. (4.5.2):

$$|\psi_{a_3}^{(+)}\rangle = \lim_{\varepsilon\to 0+} i\varepsilon G_{b_n}(E + i\varepsilon)|\phi_{a_3}\rangle + G_{b_n}^{(+)}(E)V^{b_n}|\psi_{a_3}^{(+)}\rangle, \tag{4.5.4}$$

where in the last term on the right-hand side we have assumed that the limit of the product is the product of the limit.

It remains to evaluate the inhomogeneous term in Eq. (4.5.4); if $n = 4$, the arguments of the preceding section imply that this term vanishes. Hence, the only nonzero possibilities are

$$\lim_{\varepsilon\to 0+} i\varepsilon G_{b_3}(E + i\varepsilon)|\phi_{a_3}(E)\rangle = |\phi_{a_3}(E)\rangle\,\delta_{a_3 b_3}, \tag{4.5.5a}$$

$$\lim_{\varepsilon\to 0+} i\varepsilon G_{b_2}(E + i\varepsilon)|\phi_{a_3}(E)\rangle = |\phi_{a_3}^{(+)}(E)\rangle\,\Delta_{b_2 a_3}, \tag{4.5.5b}$$

where $\Delta_{b_2 a_3}$ is unity if the partition a_3 can be obtained by decomposing one of the clusters of the partition b_2, and is zero otherwise. Here $|\phi_{a_3}^{(+)}\rangle$ is the four-particle subsystem scattering state generated by the interactions internal to b_2, so

$$(E - H_{b_2})|\phi_{a_3}^{(+)}(E)\rangle = 0. \tag{4.5.6}$$

The Eqs. (4.5.4) then become

$$|\psi_{a_3}^{(+)}(E)\rangle = G_0^{(+)}(E)V|\psi_{a_3}^{(+)}(E)\rangle, \tag{4.5.7a}$$

$$|\psi_{a_3}^{(+)}(E)\rangle = \delta_{a_3 b_3}|\phi_{b_3}(E)\rangle + G_{b_3}^{(+)}(E)V^{b_3}|\psi_{a_3}^{(+)}(E)\rangle, \tag{4.5.7b}$$

$$|\psi_{a_3}^{(+)}(E)\rangle = \Delta_{b_2 a_3}|\phi_{a_3}^{(+)}(E)\rangle + G_{b_2}^{(+)}(E)V^{b_2}|\psi_{a_3}^{(+)}(E)\rangle. \tag{4.5.7c}$$

Equations (4.5.7a–c) are true for an N-particle system with a suitable generalization of the definition (4.5.1). For the four-particle system and a given a_3, Eq. (4.5.7a) represents a single homogeneous equation, the Eqs. (4.5.7b) represent a single inhomogeneous and five homogeneous equations, and the Eqs. (4.5.7c) represent three inhomogeneous and four homogeneous equations.

Although $|\psi_{a_3}^{(+)}(E)\rangle$ satisfies Eqs. (4.5.7a–c), these equations do not define this state uniquely because of the existence of homogeneous equations describing other scattering processes. Hence the LS equations (4.5.7a–c) can not be used for numerical calculation unless additional boundary conditions are provided or the many-particle problem is approximated so as to obviate the need for these conditions.

One can construct LS equations for the N-particle system for scattering initiated by an n-cluster state. However, these sets of equations are not expected to have unique solutions except under very special circumstances.

4.6. Discussion

The proof of nonuniqueness in Section 4.3 makes use of certain plausible, but not fully substantiated, mathematical assumptions about the asymptotic behavior of the solution of the multiparticle Schrödinger equation and of the Green function. This behavior has been established in the case of a three-particle system.[1] Studies of solvable three-particle models support the general validity of these assumptions.[2]

Kernel noncompactness and nonuniqueness are different, but not entirely separate, issues. The first issue is a practical one relating to whether the LS equation can be solved numerically, while the second refers to whether the LS equation has solutions that do not satisfy the proper boundary conditions. When the kernel is compact, uniqueness is assured, but a compact kernel is not a necessary condition for uniqueness. For complex energies ($\varepsilon \neq 0$), the homogeneous LS equations are absent, so the multiparticle LS equation has a unique solution. The multicluster channel(s), unlike the open two-cluster rearrangement channels, do not cause any manifest nonuniqueness problems. This is due to the fact that the multicluster channel wave function does not satisfy homogeneous LS equations, in contrast to the rearrangement-channel wave functions. This suggests that, when the rearrangement channels are not present, the LS equation has a unique solution even though it has a noncompact kernel.

The nonuniqueness problems of the wave-function LS equation implies that the corresponding transition-operator LS equations also do not have unique solutions.

There are two alternatives to the LS equation. The first is to develop other scattering integral equations with compact kernels, thus simultaneously circumventing the problems of nonuniqueness and noncompactness. The second is to make approximations sufficient to reduce the N-particle LS equation(s) to equations with compact kernels, that may be effective m-particle equations, with $m < N$.

The second strategy is very frequently used for calculations of dynamical collision problems in atomic, nuclear, and intermediate energy physics. Examples include the coupled-channel or the coupled-reaction-channel formalisms of atomic and nuclear physics, the optical potential formalism for the elastic scattering of two composite objects, the resonating-group method, the distorted-wave Born approximation, various multiple-scattering methods, and the Dirac optical potential approach. Sometimes the errors engendered in making these approximations are compensated phenomenologically in ways not consistent with the complete dynamics. Whether these errors are inconsequential or not, this strategy does not provide any means for making successive improvements in a consistent manner. Sometimes such improvements are difficult to contemplate, such as for 1 GeV proton scattering from ^{208}Pb treated using the Dirac optical potential, but they certainly would be useful in interpreting the physics of such approaches.

The connected-kernel collision integral equations we develop in Ch. 5 provide a dynamical strategy for obtaining an exact solution since they possess compact kernels. They also provide the basis for approximation schemes that can be successively improved upon in a consistent and calculable fashion. By way of comparison, the conventional approximation schemes are based on a single LS equation with a noncompact kernel; uniqueness and the compactness of the kernel are achieved only as the result of the approximation. Consequently, such an approximation scheme cannot be improved upon in its own terms because of its intrinsic mathematical limitations, and *not* its numerical complexity. It is, however, of considerable interest to see how one can recover successful conventional approximation schemes by systematic truncations of a set of connected-kernel equations in order to provide some indication as how to improve on such schemes in a dynamically consistent fashion.

References and Notes (4)

Section 4.1

1. Lippmann (1950).
2. Lippmann (1956).

3. The inclusion of the permutation symmetries resulting from identity among groups of the particles would only obscure the basic issues we wish to consider in this chapter.

Section 4.2

1. Epstein (1957), Foldy (1957).
2. Gerjuoy (1958, 1971).
3. Faddeev (1960, 1965).
4. Glöckle (1973, 1974).
5. Gerjuoy (1985, 1986a, 1987a, b).
6. Adhikari (1980a, 1983b).
7. Glöckle (1970a, b). Glöckle's discussion of the three-particle system was extended to a multiparticle system by Cattapan (1979a).
8. Mukherjee (1978, 1981a, b).
9. Benoist-Gueutal (1986).
10. Sawada (1988).
11. Levin (1984), Gerjuoy (1986a, b; 1987a, b), Adhikari (1980a, 1983b).
12. Lippmann (1950, 1956). For related discussions on the Lippmann identities, see Bencze (1982a), Tobocman (1983a), L'Huillier (1978), Picklesimer (1983a), and Lovitch (1982).

Section 4.3

1. Gerjuoy (1986a).
2. See also Tobocman (1981b).
3. Gerjuoy (1958, 1971).
4. Prugovecki (1971), pp. 482, 485, 624.
5. Lippmann (1950, 1956).
6. Gerjuoy (1985, 1987a, b).
7. Komarov (1980).
8. Sloan (1981), Faddeev (1981).
9. McGuire (1964).
10. Adhikari (1980a).

Section 4.4

1. Ref. 7 in Sec. 4.2.

Section 4.5

1. See, for example, Gerjuoy (1971), Glöckle (1983a), Cattapan (1979a).

Section 4.6

1. Glöckle (1983a).
2. Gerjuoy (1985).

Chapter 5

The Connected-Kernel Approach

5.1. Introduction

The modern era of the quantum theory of collision processes began with Faddeev's resolution of the three-body problem.[1] Collision integral equations similiar to those found by him had appeared previously for zero-range two-body interactions[2] or as a special case of Watson's multiple-scattering theory.[3] However, he first pointed out that these equations were free of the Lippmann–Schwinger nonuniqueness problem. He also first realized the crucial fact that for $N = 3$ a result of the manipulations involved in constructing multiple scattering theories was to remove the Dirac δ-function singularities from the kernels of the LS equations, achieving what are now referred to as *connected-kernel* (C-K) collision integral equations.

Faddeev's work led to an explosion of activity in three-particle collision and bound-state problems in atomic, nuclear, and particle physics.[4] Many alternative sets of three-particle equations were proposed, some much more suitable for practical calculation than the original Faddeev version.[5] The first generation of realistic three-particle calculations appeared in the 1960s for the three-nucleon and meson/two-nucleon systems at low energies.[6] The numerical results were impressive, considering the crude representation of the two-particle interaction that was a common feature of these early works.

The superiority of these results over those obtained previously using other approximate methods left little doubt that the C-K approach was successful in dealing with the full dynamical complexity of multiparticle collisions.

Simultaneously, a surge in interest concerning multiparticle collisions took place in high-energy physics. This was motivated mainly by the spate of multihadronic resonances that were being discovered at that time. The work of Amado and coworkers, Lovelace, and others was especially influential in publicizing the Faddeev approach by demonstrating the practicality of physically reasonable, fully unitary, three-particle calculations.[7]

Van Winter and Weinberg were the first to propose N-particle C-K scattering integral equations for $N > 3$.[8] These two works on N-particle collisions, along with the many other proposals for generalizations of the $N = 3$ Faddeev equations,[3,9] revealed the special simplicity of the $N = 3$ case and a better understanding of the significance of Faddeev's work. His results can be summarized in the following ennumeration of the principal features of the Faddeev equations:

(i) Their homogeneous forms are fully equivalent to the Schrödinger equation for $N = 3$.
(ii) After one iteration they possess connected kernels; consequently, they are effectively Fredholm integral equations of the second kind.[10]
(iii) A physical form of the Fredholm alternative applies: The only nontrivial solutions of the homogeneous form of these equations correspond to three-body bound states.

Most of the alternatives proposed to the original Faddeev equations, and all of the early generalizations of them to $N > 3$, were constructed with the aim of achieving property (ii). A shortcoming of these alternatives and generalizations is that property (i) is not always apparent. This opens the door to so-called *spurious solutions*, the possibility of which was first demonstrated by Federbush[11] for a special case of Weinberg's C-K equation for the full three-body Green function.

N-body scattering integral equations satisfying all three of properties (i)–(iii) were first established by Yakubovskii[12] using a generalization of the original Faddeev strategy. Equivalent equations were later found.[13] Yakubovskii-type equations are a complete and consistent characterization of N-body collisions for which there is a definite, but very complicated, solution algorithm. Namely, the problem is reduced to the solution of multidimensional, coupled, linear integral equations that are effectively of the Fredholm type that, in principle, we know how to solve, quite unlike the LS equations. As such, they represent the theoretical foundation of all of nonrelativistic collision theory. Therefore, no treatment of collision theory is complete without showing how they are derived and discussing their properties. We should note that the general solution algorithm for collision theory also implies a method for solving the N-particle bound-state problem. Bound states are associated with the solutions of the homogeneous form of the Faddeev–Yakubovskii integral equations that exist at energies where the Fredholm determinant vanishes.

On the other hand, we should not be under any illusions as to the immediate practicality of such a set of what we will refer to as a complete, constructive system (C.C.S.) of collision integral equations for $N \geq 3$. (It should always be kept in mind throughout the remainder of this book that the full collection of LS equations for a given N is *not* a C.C.S.) A C.C.S. of equations plays a role in collision theory analogous, in part, to that of the Navier–Stokes equations in describing the flow of fluids. Both form the basis for the phenomena they are intended to describe, but their use without approximation is seldom practical. In one case, the obstacle is the multidimensionality of the coupled integral equations, in the other, the nonlinearity of the coupled partial differential equations. A major difference, of course, is that there is no general solution algorithm for the Navier–Stokes equations.

In practice, a set of fundamental equations in hand can serve as the basis for physically motivated approximations that render them tractable. This may be better realized, so far, in fluid mechanics than it is in collision theory. Several well-understood approximations to the Navier–Stokes equations can be made under clearly stated physical conditions, e.g., inviscid flows, incompressible flows, and so on. The various C.C.S.'s of collision integral equations, in contrast, have not been too enlightening in providing substantiation for any of the well-known approximation techniques of collision theory, nor in providing new ones except, to a limited degree, for $N \leq 4$.

There are a number of reasons for this situation besides the complexity of multiparticle problems. One of them is simply unfamiliarity. It is only at a comparatively late date in the history of the quantum theory of collisions that the *need* for a C.C.S. of scattering integral equations came to be appreciated. Partly because of this, they have been studied sparingly, and the number of investigations that have taken them as a starting point for the development of calculational techniques is perhaps several orders of magnitude less than the number of studies using more conventionally phrased approximation scenarios.

Another reason is that usually one is not interested in the enormous detail in the form of the large number of auxiliary scattering amplitudes that are generated by solving a C.C.S. of scattering integral equations. Typically, only a small subset of the possible variety of occurrences in N-particle collisions is of interest either experimentally or theoretically. Most experiments initiate from carefully prepared two-cluster channels, and are concerned primarily with outgoing channels corresponding to only a few clusters, usually two or three. It is plainly of some advantage to model such processes using dynamical equations in which the channels that are not of interest are treated in such a way that their effects can either be ignored or included only partially as part of a proposed approximation procedure.

Some of the N-particle C-K scattering integral equations that do not satisfy property (i) explicitly have proven more fruitful in suggesting practical approximations as well as recovering conventional approximation schemes. These C-K equations may or may not be C.C.S.'s of equations.[14] If they are not, it is because they possess spurious solutions.[11] Such solutions may be irrelevant once the full dynamics is truncated in the formulation of an approximation scheme, just as the nonuniqueness of the LS equation is irrelevant if one has truncated the equation to such a degree that the solution is unique.

In this chapter, we discuss a variety of sets of C-K scattering integral equations. Some of these sets are C.C.S.'s of equations, others may not be, but all have proven to be of some interest in the theory of collisions. An important omission from our development is the configuration-space approach of Merkuriev *et al.*,[15] which is especially convenient for including the Coulomb interaction. Only a small fraction of the formal analysis or the numerical calculations of multiparticle collisions have been carried out using this method.

5.2. Interactions, Transition Operators, Green Functions, and Wave Functions

In this section, we develop a matrix notation in the partition indices $[a, b, c, \ldots]$ for expressing the fundamental relationships of N-particle collision theory. Partitions, their association with channels, and the relevant labeling notation are introduced in Ch. 1. We recall that $\alpha, \beta, \ldots$, refer to two-cluster partitions, $a = 0$ to the N-cluster partition, and i, j to $(N - 1)$-cluster partitions, i.e., pair indices. Sometimes we also need the notation $a = \underline{1}$ for the single-cluster partition.

5.2a. *Matrix Notation*

Two particular partition-indexed matrices play a central role in multiparticle scattering theory.[1] The first is the anti-Kronecker-delta matrix, $\bar{\delta}$, whose elements are

$$\bar{\delta}_{ab} = 1 - \delta_{ab}. \tag{5.2.1}$$

One verifies that the inverse of $\bar{\delta}$ is given by

$$\bar{\delta}^{-1} = [\bar{\delta} + \underline{\delta}(2 - D)](D - 1)^{-1}, \tag{5.2.2}$$

where D is the matrix dimensionality. In Eq. (5.2.2), $\underline{\delta}$ is the identity matrix, in the partition space, as distinguished from I in the Hilbert space, with components that are the Kronecker deltas, δ_{ab}.

The second matrix is the *sum* $\mathscr{S}$:

$$(\mathscr{S})_{ab} \equiv (\underline{\delta} + \bar{\delta})_{ab} = 1, \tag{5.2.3}$$

whose elements are all unity. It is basically a projection, and so does not possess an inverse, since

$$\mathscr{S}^2 = D\mathscr{S}. \tag{5.2.4}$$

The usefulness of $\mathscr{S}$ is a consequence of the relations

$$(\underline{A}\mathscr{S})_{ac} = \sum_b A_{ab}, \tag{5.2.5a}$$

$$(\mathscr{S}\underline{A})_{ac} = \sum_b A_{bc}, \tag{5.2.5b}$$

which indicate that $\mathscr{S}$ *sums* over the indices of any adjacent multiplicative matrices $\underline{A}$.

Equations (5.2.1–5.2.5) give some examples of our general notational conventions. In partition space, there are matrices and scalars with respect to the partition indexing:

(i) Capital Latin characters such as $\underline{A}$ or $\bar{A}$ refer to matrices with sub- or superscripted components, respectively. A scalar such as D, H, or H_0 has no over- or under-bar.

(ii) The Greek characters $\underline{\delta}$, $\bar{\delta}$, $\underline{\Delta}$, and $\bar{\Delta}$ all refer to special matrices whose elements are either unity or zero.

(iii) The script capital Latin characters, $\mathscr{G}$, $\mathscr{G}^{\iota}$, $\mathscr{S}$, $\mathscr{V}$, and $\mathscr{V}^{\iota}$, all represent specific matrices.

The properties of $\mathscr{S}$ allow a matrix representation of partition-degenerate decompositions such as

$$H = H_a + V^a, \tag{5.2.6}$$

for the complete Hamiltonian H, where

$$V^a \equiv V - V_a \tag{5.2.7}$$

is the residual interaction in the channel corresponding to the partition a and $H_a = H_0 + V_a$ is the channel Hamiltonian. Namely, one can rewrite (5.2.6) as

$$H\mathscr{S} = (\underline{H} + \bar{V})\mathscr{S}, \tag{5.2.8a}$$

$$H\mathscr{S} = \mathscr{S}(\underline{H} + \bar{V}). \tag{5.2.8b}$$

where

$$(\bar{V})_{ab} = V^a \delta_{ab}, \tag{5.2.9a}$$

$$(\underline{V})_{ab} = V_a \delta_{ab}, \tag{5.2.9b}$$

$$(\underline{H})_{ab} = H_a \delta_{ab} = (H_0 \underline{\delta} + \underline{V})_{ab}. \tag{5.2.10}$$

Here, H and H_0 are regarded as scalars in partition space.

The special characteristics of the two limiting partitions, $a = 0, \underline{1}$, should be noted. In order to interpret $H_a = H_0 + V_a$ consistently, we must have $V_0 = V^{\underline{1}} = 0$ and $V_{\underline{1}} = V^0 = V$, along with $H_{\underline{1}} = H$. The partition indices enter into scattering theory in order to label subsystems. So the one-cluster partition, $a = \underline{1}$, is seldom needed and is, by a long-standing and nearly universal convention, omitted from H and $G(z)$, for example, and, except under unusual circumstances, plays no role in our matrix notation.

Partition degeneracy means the possibility of partition-dependent decompositions of quantities such as H or V that themselves have no intrinsic dependence on the partition labels. Partition-degenerate quantities appear often in collision theory, and their matrix representation simplifies and unifies formal manipulations considerably. For example, the partition-degenerate resolvent identities for the full Green function $G(z) = (z - H)^{-1}$ in terms of the subsystem Green functions $G_a(z) = (z - H_a)^{-1}$ are in matrix notation and suppressing all z-dependences,

$$G\mathscr{S} = \underline{G}\mathscr{S} + \underline{G}\bar{V}\mathscr{S}G, \tag{5.2.11a}$$

$$G\mathscr{S} = \mathscr{S}\underline{G} + G\mathscr{S}\bar{V}\underline{G}, \tag{5.2.11b}$$

where

$$(\underline{G})_{ab} = G_a \delta_{ab}, \tag{5.2.12}$$

and, of course, G is a partition-space scalar. We have suppressed the dependence on the complex parameter z, as we often do in what follows. Equations (5.2.11a, b) suggest the following representations of the Green function G:

$$G\mathscr{S} = \underline{G}\mathscr{S} + \underline{G}\underline{T}^{(+)}\underline{G}, \tag{5.2.13a}$$

$$G\mathscr{S} = \mathscr{S}\underline{G} + \underline{G}\underline{T}^{(-)}\underline{G}. \tag{5.2.13b}$$

Equations (5.2.13a, b) are asymmetrical in $\mathscr{S}$, despite the fact that $\mathscr{S}$ and G commute. The symmetrical representation is

$$G\mathscr{S} = \underline{G} + \underline{G}\underline{U}\underline{G}. \tag{5.2.13c}$$

The preceding representations for $(\underline{T}^{(\pm)})_{ab} = T^{(\pm)}_{ab}$ and $(\underline{U})_{ab} = U_{ab}$ are of physical interest for $a, b \neq \underline{1}$ because of the relationship (Section 1.7) of the physical transition amplitudes to the residues of the subsystem poles of G:

$$T(\phi_b, \phi_a) = \langle \phi_b(E) | \vec{G}^{(+)}_b(E)^{-1} G^{(+)}(E) \overleftarrow{G}^{(+)}_a(E)^{-1} | \phi_a(E) \rangle, \tag{5.2.14}$$

which is the on-shell transition amplitude for the process $a \to b$ when the channel eigenstates satisfy

$$H_a | \phi_a(E) \rangle = E | \phi_a(E) \rangle. \tag{5.2.15}$$

The arrow notation in Eq. (5.2.14) designates the right- and left-handed inverses, corresponding to $\leftarrow$ and $\rightarrow$, respectively, described in detail in Section 1.7 and in Ch. 4.

Because of Eq. (5.2.15), the first terms on the right-hand sides of Eqs. (5.2.13a–c) do not contribute to $T(\phi_b, \phi_a)$. Consequently, either $T^{(+)}_{ba}$, $T^{(-)}_{ba}$ or U_{ba} can be taken as *transition operators*, namely, operators whose on-shell matrix elements with respect to $|\phi_a(E)\rangle$ and $|\phi_b(E)\rangle$ yield $T(\phi_b, \phi_a)$. Other possible transition operators will not be related so simply to G, but all will differ from $T^{(\pm)}_{ba}$ and U_{ba} by terms whose $\langle \phi_b | \cdots | \phi_a \rangle$ matrix elements vanish on-shell.

Equations (5.2.13a, b) can be solved for $\underline{T}^{(\pm)}$ to yield, for arbitrary z,

$$\underline{T}^{(+)} = \bar{V}\mathscr{S}G\overleftarrow{\underline{G}}^{-1} \quad \text{(Post Form)}, \tag{5.2.16a}$$

$$\underline{T}^{(-)} = \vec{\underline{G}}^{-1}G\mathscr{S}\bar{V} \quad \text{(Prior Form)}, \tag{5.2.16b}$$

respectively. In Section 1.7, we explained why Eq. (5.2.13c) does not determine a unique set of operators U_{ab} unless the differences between $\vec{G}_c^{-1}$ and $\overleftarrow{G}_c^{-1}$, $c = a, b$, in the operators $[U_{ab} - T^{(\pm)}_{ab}]$ are not relevant for whatever purposes U_{ab} is employed. Unless we indicate otherwise, we assume that this is the case. With this understood we have, from Eq. (5.2.13c),

$$\begin{aligned} \underline{U} &= \bar{\delta}\underline{G}^{-1} + \underline{T}^{(+)} \\ &= \underline{G}^{-1}\bar{\delta} + \underline{T}^{(-)} \end{aligned} \quad \text{(AGS Form)}. \tag{5.2.16c}$$

If we again use Eqs. (5.2.11a, b) in Eqs. (5.2.16a, b), we see that

$$\underline{T}^{(+)} = \bar{V}\mathscr{S} + \bar{V}G\mathscr{S}\bar{V} \tag{5.2.17a}$$

and

$$\underline{T}^{(-)} = \mathscr{S}\bar{V} + \bar{V}G\mathscr{S}\bar{V}, \tag{5.2.17b}$$

respectively.

In index form, Eq. (5.2.17a) becomes

$$T_{ba}^{(+)} = V^b + V^b G V^a. \tag{5.2.18}$$

All three operators (5.2.16a, b, c) are identical for the process $(a = 0) \to (b = 0)$:

$$T_{00}^{(\pm)} = U_{00} = V + VGV. \tag{5.2.19}$$

In terms of the post operator, the breakup process, $\alpha \to 0$, corresponds to

$$T_{0\alpha}^{(+)} = V(1 + GV^{\alpha}) \tag{5.2.20}$$

while, for the two-cluster to two-cluster, $\alpha \to \beta$, processes, we have

$$T_{\beta\alpha}^{(+)} = V^{\beta}(1 + GV^{\alpha}). \tag{5.2.21}$$

5.2b. *Integral Equations for the Transition Operators*

Let us consider the post operators $T_{ba}^{(+)}$. The essential trick for obtaining generic integral equations for these operators involves the rewriting of the potential V^b, which represents the interactions among the different clusters depicted by b, in terms of an array $\mathscr{V}^{ba}$ $(a, b \neq \underline{1})$, of operators called a *channel-coupling scheme*,[2] such that

$$\bar{V}\mathscr{S} = \mathscr{V}\mathscr{S}. \tag{5.2.22}$$

If we call

$$(\mathscr{V}^{t})_{ab} \equiv \mathscr{V}^{ba}, \tag{5.2.23}$$

then

$$\mathscr{S}\bar{V} = \mathscr{S}\mathscr{V}^{t}, \tag{5.2.24}$$

as well.

A special class of $\mathscr{V}$'s incorporates what is called a channel-coupling *array*,[3] $(\underline{W})_{ab} = W_{ab}$,

$$\mathscr{V} = \bar{V}\underline{W}, \tag{5.2.25}$$

where

$$\underline{W}\mathscr{S} = \mathscr{S}. \tag{5.2.26}$$

Equation (5.2.26) is the defining characteristic of a channel-coupling array. All of the various formal collision theories[3] that are based on the introduction of W-arrays just amount to the replacement (5.2.26), or its counterpart with $\underline{W}$ and $\mathscr{S}$ transposed (left W-array), for $\mathscr{S}$ in the matrix expressions of the transition operators and Green functions.

Now, using Eq. (5.2.22) in conjunction with Eqs. (5.2.17a) and (5.2.11a, b), and the identities (1.7.15a, b), we immediately obtain the integral equations

$$\underline{T}^{(+)} = \mathscr{V}\underline{G}\mathscr{S}\overleftarrow{\underline{G}}^{-1} + \mathscr{V}\underline{G}\underline{T}^{(+)}, \tag{5.2.27a}$$

$$\underline{T}^{(+)} = \mathscr{V}\mathscr{S} + \underline{T}^{(+)}\underline{G}\mathscr{V}^{t}, \tag{5.2.27b}$$

$$\underline{T}^{(-)} = \vec{\underline{G}}^{-1}\mathscr{S}\underline{G}\mathscr{V}^{t} + \underline{T}^{(-)}\underline{G}\mathscr{V}^{t}, \tag{5.2.27c}$$

$$\underline{T}^{(-)} = \mathscr{S}\mathscr{V}^{t} + \mathscr{V}\underline{G}\underline{T}^{(-)}. \tag{5.2.27d}$$

Evidently, all possible transition-operator integral equations in W-array theories are special cases of Eqs. (5.2.27a–d). Note that, for any given array $\mathscr{V}$, Eqs. (5.2.27a–d) give rise to other such sets with the substitutions (5.2.26).

With the choice $\mathscr{V} = \bar{V}$, we obtain the LS equations from Eq. (5.2.27a):

$$T_{ba}^{(+)} = V^{b}G_{b}\overleftarrow{G}_{a}^{-1} + V^{b}G_{b}T_{ba}^{(+)}. \tag{5.2.28}$$

5.2c. *Integral Equations for the Green Functions*

Contrasting styles in collision theory can be classified in terms of preferences for using the dynamical equations for transition operators or for wave functions. Each, obviously, is equivalent to the other in terms of physical content, but one of them may be a better vehicle for a particular type of approximation. Actually, the real basis of both lies in the Green function, or resolvent, equations of multiparticle scattering theory.[4]

Consider the Green function matrix $\mathscr{G}$ with elements $(\mathscr{G})_{ab}$, $a, b \neq \underline{1}$, defined by

$$\mathscr{G} = \underline{G} + \underline{G}\mathscr{V}\mathscr{G}, \tag{5.2.29a}$$

$$\mathscr{G} = \underline{G} + \mathscr{G}\mathscr{V}\underline{G}. \tag{5.2.29b}$$

We assume that $(\underline{\delta} - \underline{G}\mathscr{V})^{-1}$ exists, so that Eqs. (5.2.29a, b) are equivalent.

Since a typical potential matrix $\mathscr{V}$ is not symmetric, one can define another, distinct, resolvent $\mathscr{G}^{t}$ as the solution of the equations

$$\mathscr{G}^{t} = \underline{G} + \underline{G}\mathscr{V}^{t}\mathscr{G}^{t}, \tag{5.2.30a}$$

$$\mathscr{G}^{t} = \underline{G} + \mathscr{G}^{t}\mathscr{V}^{t}\underline{G}. \tag{5.2.30b}$$

We note that

$$[\mathscr{G}^t(z)]^\dagger = \mathscr{G}(z^*), \tag{5.2.31}$$

where the adjoint includes a transpose operation with respect to the partition indices, so in component form we have

$$[\mathscr{G}^t_{ab}(z)]^\dagger = \mathscr{G}_{ba}(z^*). \tag{5.2.32}$$

Evidently, if $(\underline{\delta} - \underline{G}\mathscr{V})$ is invertible, so is $(\underline{\delta} - \underline{G}\mathscr{V}^t)$.

The crucial aspect of the analysis is the relationship of $\mathscr{G}$ and $\mathscr{G}^t$ to the full Green function G, which is a scalar in the partition space. Because $(\underline{\delta} - \underline{G}\mathscr{V})^{-1}$ and $(\underline{\delta} - \underline{G}\mathscr{V}^t)^{-1}$ are presumed to exist, these relationships are

$$\mathscr{G}\mathscr{S} = \mathscr{S}G, \tag{5.2.33a}$$

$$\mathscr{S}\mathscr{G}^t = \mathscr{S}G. \tag{5.2.33b}$$

5.2d. *Integral Equations for the Wave Functions*

Wave-function formalisms have proven useful only when the in- and out-states, $|\psi_\alpha^{(\pm)}\rangle$, evolve from a two-cluster eigenstate $|\phi_\alpha\rangle$ of H_α. We confine ourselves to this case.

In terms of our matrix notation we have, in component form and with $z = E + i\varepsilon$, $\lim \varepsilon \to 0+$, assumed for all wave-function expressions below, where E is the energy entering in the various state vectors,

$$|\psi_\alpha^{(+)}\rangle = (\mathscr{S}G\overleftarrow{\underline{G}}^{-1})_{b\alpha}|\phi_\alpha\rangle, \tag{5.2.34a}$$

and

$$\langle\psi_\alpha^{(-)}| = \langle\phi_\alpha|(\overrightarrow{\underline{G}}^{-1}G\mathscr{S})_{\alpha b}, \tag{5.2.34b}$$

for any partition b. Given these expressions, we see the possibility for developing wave-function integral equations using the identities (5.2.33a, b).

In particular, the identity (5.2.33b) in conjunction with Eq. (5.2.34a) suggests that we introduce the auxiliary wave functions ($b \neq \underline{1}$)

$$|\chi_b^{(+)}(\alpha)\rangle \equiv (\mathscr{G}^t\overleftarrow{\underline{G}}^{-1})_{b\alpha}|\phi_\alpha\rangle, \tag{5.2.35}$$

which evidently satisfy

$$|\psi_\alpha^{(+)}\rangle = \sum_b \bar{\delta}_{\underline{1}b}|\chi_b^{(+)}(\alpha)\rangle \tag{5.2.36}$$

and[5]

$$|\chi_b^{(+)}(\alpha)\rangle = \delta_{b\alpha}|\phi_\alpha\rangle + G_b\sum_c \mathscr{V}^{cb}|\chi_c^{(+)}(\alpha)\rangle. \tag{5.2.37}$$

In contrast to this, if we use the identity (5.2.33a) in Eq. (5.2.34a) we obtain, for any $b \neq \underline{1}$,

$$|\psi_\alpha^{(+)}\rangle = (\mathcal{G}\mathcal{S}\overleftarrow{\underline{G}}^{-1})_{b\alpha}|\phi_\alpha\rangle, \tag{5.2.38}$$

from which we deduce that

$$|\psi_\alpha^{(+)}\rangle = \sum_c \mathcal{G}_{bc}\overleftarrow{G}_\alpha^{-1}|\phi_\alpha\rangle. \tag{5.2.39}$$

This suggests that the pertinent auxiliary functions should be, with $b \neq \underline{1}$,

$$|\bar{\chi}_c^b(\alpha)\rangle = \mathcal{G}_{bc}\overleftarrow{G}_\alpha^{-1}|\phi_\alpha\rangle, \tag{5.2.40}$$

so that

$$|\psi_\alpha^{(+)}\rangle = \sum_c \bar{\delta}_{c\underline{1}}|\bar{\chi}_c^b(\alpha)\rangle, \tag{5.2.41}$$

$$|\bar{\chi}_c^b(\alpha)\rangle = \delta_{bc}\, G_b\overleftarrow{G}_\alpha^{-1}|\phi_\alpha\rangle + G_b\sum_d \mathscr{V}^{bd}\,\bar{\delta}_{d\underline{1}}|\bar{\chi}_c^d(\alpha)\rangle, \tag{5.2.42}$$

which is not quite so convenient a set of equations as the set (5.2.37).

The states

$$|\hat{\chi}_b(\alpha)\rangle \equiv (\mathcal{G}\overleftarrow{\underline{G}}^{-1})_{b\alpha}|\phi_\alpha\rangle, \tag{5.2.43}$$

have been considered in the literature.[6] We immediately see from Eq. (5.2.29a) that

$$[\underline{\delta} - \underline{G}\mathscr{V}][\mathcal{G}\overleftarrow{\underline{G}}^{-1} - \mathcal{S}G\overleftarrow{\underline{G}}^{-1}] = -[\underline{G}\,\bar{\delta}\overleftarrow{\underline{G}}^{-1}]. \tag{5.2.44}$$

However, by the Lippmann identity,

$$[\underline{G}\,\bar{\delta}\overleftarrow{\underline{G}}^{-1}]_{b\alpha}|\phi_\alpha> = 0, \tag{5.2.45}$$

for any partition $b \neq \underline{1}$, so we learn from Eq. (5.2.44) that

$$(\mathcal{G}\overleftarrow{\underline{G}}^{-1})_{b\alpha}|\phi_\alpha\rangle = (\mathcal{S}G\overleftarrow{\underline{G}}^{-1})_{b\alpha}|\phi_\alpha\rangle, \tag{5.2.46}$$

and, therefore,

$$|\hat{\chi}_b(\alpha)\rangle = |\psi_\alpha^{(+)}\rangle, \tag{5.2.47}$$

independently of the value of b, so the states (5.2.43) are partition degenerate, in contrast to the states (5.2.35) or (5.2.40), which are said to be "well-labeled" or "true" components.[7]

Interestingly, the seemingly trivial relationship (5.2.47) does have important consequences. From Eqs. (5.2.43) and (5.2.29a) we find

$$|\hat{\chi}_b(\alpha)\rangle = G_b\overleftarrow{G}_\alpha^{-1}|\phi_\alpha\rangle + G_b\sum_c \mathscr{V}^{bc}\,\bar{\delta}_{c\underline{1}}|\hat{\chi}_c(\alpha)\rangle, \tag{5.2.48}$$

which has the appearance of a coupled set of equations, but for a *single* quantity just as the LS equations (1.6.1). The value of equations of the form (5.2.48) is that they can be used for nontrivial iterations.

For example, let us single out α and we label the remaining $N_2 - 1$ two-cluster partitions by $\beta(i)$, $i = 1, \ldots, N_2 - 1$. (Note: $N_2 = 2^{N-1} - 1$ for N particles.) Let

$$\mathscr{V}^{\alpha c} = V^{\alpha}\, \delta_{c, \beta(1)}, \tag{5.2.49a}$$

$$\mathscr{V}^{\beta(i), c} = V^{\beta(i)}\, \delta_{c, \beta(i+1)}. \tag{5.2.49b}$$

Then, after $N_2 - 1$ iterations of Eq. (5.2.48) we obtain, with the identification (5.2.47), the choice $b = \alpha$, and use of the Lippmann identity, the single integral equation,

$$|\psi_{\alpha}^{(+)}\rangle = |\phi_{\alpha}\rangle + G_{\alpha} V^{\alpha} G_{\beta(1)} V^{\beta(1)} \cdots G_{\beta(N_2-1)} V^{\beta(N_2-1)} |\psi_{\alpha}^{(+)}\rangle, \tag{5.2.50}$$

for the wave function. No matter how the interaction potentials are decomposed, the kernel of Eq. (5.2.50) is a sum of terms, each of which contains a product of interactions that connect every one of the two-cluster partitions. Since there are no two groups of particles that are translationally invariant with respect to each other in such products, they must be connected operators. Thus, (5.2.50) is a C-K equation for the wave function $|\psi_{\alpha}^{(+)}\rangle$.

5.3. Three-Particle Equations

In this section, we develop the most frequently employed C-K equations for $N = 3$. It is difficult to overestimate the importance of this case in dynamical collision theory, so we devote a disproportionate amount of attention to its exposition.[1] We ignore spin and any internal symmetries such as isospin in all of our formal development.

In three-body scattering, we have the two-cluster channels labeled by the partitions $a = \alpha(=1, 2, 3)$, where $\alpha = 1$, for example, corresponds to the partition (1)(23), while the three-free-particle channel is $a = 0 = (1)(2)(3)$. These channels may not all be realized physically in a given situation. For instance, in the three-nucleon problem, only those partitions $(i)(jk)$ where (jk) can form a spin-one, isosinglet combination refer to physical channels. The three-cluster channel is not accessible from two-cluster channels at energies below the breakup threshold, but three free particles can always scatter into another state of three free particles.

Whatever the physics of a particular system, all of the possible channel partition labels enter into the indexing of the quantities appearing in three-particle scattering integral equations. Only for $N = 3$ are the numbers of particles and nontrivial partitions (i.e., excluding $a = 0$ and $a = \underline{1}$) equal; this combinatorial accident underlies the special formal simplicity of three-particle scattering theory compared to the description of collisions involving four or more particles.

The three-particle Hamiltonian

$$H = H_0 + V, \tag{5.3.1}$$

decomposes into the total kinetic energy operator H_0 and the complete interaction

$$V \equiv \sum_{\alpha=1}^{3} V_\alpha, \tag{5.3.2}$$

which we suppose is made up solely of pair potentials. (In the $N = 3$ case, the two-cluster indices suffice as pair indices.) The inclusion of three-body forces into the formalism is straightforward, but not trivial.[2] The usual three-particle channel Hamiltonians are

$$H_a = H_0 + V_a, \tag{5.3.3}$$

where $a = \alpha$. It is sometimes useful to extend this definition to the three-cluster partition $a = 0$ by taking $V_0 = 0$, and $V^0 = V$.

We have, for any a,

$$H = H_a + V^a, \tag{5.3.4}$$

where

$$V^a = \sum_b \bar{\delta}_{ab} V_b. \tag{5.3.5}$$

5.3a. Transition-Operator Equations

The LS equations (5.2.28) for $N = 3$ provide a perspective for the C-K equations that we develop. For simplicity, let us take $a = b$ in the Eqs. (5.2.28):

$$T_{aa}^{(+)} = V^a + V^a G_a T_{aa}^{(+)}, \tag{5.3.6}$$

where we have ignored all dependences on the parametric energy. The kernel $V^a G_a$ splits up into pieces that are invariant with respect to the translations of one particle and those that are not:

$$V^a G_a = \sum_b \bar{\delta}_{ab} V_b (G_0 + G_0 t_a G_0), \tag{5.3.7}$$

where

$$t_a = V_a + V_a G_a V_a \tag{5.3.8}$$

is the two-particle transition operator defined on the three-particle space. Thus, the kernel of Eq. (5.3.6) has Dirac δ-function singularities arising from the $\bar{\delta}_{ab} V_b G_0$ terms in Eq. (5.3.7), which takes us outside the class of integral equations for which a standard inversion algorithm can be defined.

Equation (5.3.6), whatever its other shortcomings, is an *exact identity* for $T_{aa}^{(+)}$. All implementable approximation schemes predicated only on Eq. (5.3.6) introduce truncations of the dynamics that either convert it into an approximate *expression* for $T_{aa}^{(+)}$, e.g., the iteration series

$$T_{aa}^{(+)} = V^a + V^a G_a V^a + \cdots, \tag{5.3.9}$$

or yield an approximate *integral equation* for $T_{aa}^{(+)}$, typically for a limited class of its matrix elements, that is of the Fredholm type. In both cases, the troublesome dynamical possibilities are, in effect, either thrown out or neutralized as a part of the approximation scenario. If these discarded degrees of freedom are really unimportant, the approximation may be very successful. We need something more comprehensive if we want to solve the $N = 3$ collision problem. Therefore, we seek realizations of the Eqs. (5.2.27a) other than the LS equations.

With the choice

$$\mathscr{V} = \bar{\delta}\underline{V} \tag{5.3.10}$$

for the potential matrix, the set (5.2.27a) becomes something very different from the LS equations:

$$\underline{T}^{(+)} = \bar{\delta}\underline{V}\underline{G}\mathscr{S}\overleftarrow{\underline{G}}^{-1} + \bar{\delta}\underline{V}\underline{G}\underline{T}^{(+)}. \tag{5.3.11}$$

We see from the matrix form of Eq. (5.3.8), viz.,

$$\underline{t} = \underline{V} + \underline{V}\underline{G}\underline{V}, \tag{5.3.12}$$

where $(\underline{t})_{ab} = t_a \delta_{ab}$, that

$$\underline{V}\underline{G} = \underline{t}G_0, \tag{5.3.13}$$

so

$$\underline{T}^{(+)} = \bar{\delta}\underline{t}G_0\mathscr{S}\overleftarrow{\underline{G}}^{-1} + \bar{\delta}\underline{t}G_0\underline{T}^{(+)}. \tag{5.3.14}$$

The three-particle equations, proposed by Alt, Grassberger, and Sandhas (AGS),[3] can be obtained from Eq. (5.3.14) by using Eq. (5.2.16c) to express $\underline{T}^{(+)}$ in terms of $\underline{U}$:

$$\underline{U} = \bar{\delta}G_0^{-1} + \bar{\delta}\underline{t}G_0\underline{U}, \tag{5.3.15}$$

where we have omitted the arrow on the inverse of G_0 in accord with our understanding (Section 5.2) concerning the meaning of the AGS formalism. The essential features of three-particle scattering can be discussed in terms of $\underline{T}^{(+)}$, $\underline{T}^{(-)}$, or $\underline{U}$. For example, the Lovelace[4] equations have the same kernels as the Eqs. (5.3.15), and are obtained by inserting the potential matrix (5.3.10) into Eqs. (5.2.27b) and (5.2.27d). The more symmetrical $\underline{U}$ will be used in our initial formal development.

The kernel, $\underline{K} = \bar{\delta}\underline{t}G_0$, of Eqs. (5.3.14–5.3.15) has components

$$(\underline{K})_{ab} = \bar{\delta}_{ab}\, t_b G_0. \tag{5.3.16}$$

Since $K_{a0} = 0$, the Eqs. (5.3.15), for example, are closed among the two-cluster partitions, namely,

$$U_{\alpha\beta} = \bar{\delta}_{\alpha\beta}\, G_0^{-1} + \sum_{\gamma=1}^{3} \bar{\delta}_{\alpha\gamma}\, t_\gamma G_0 U_{\gamma\beta}. \tag{5.3.17}$$

The breakup operators $U_{0\beta}$ are determined by the $U_{\alpha\beta}$:

$$U_{0\beta} = G_0^{-1} + \sum_{\gamma=1}^{3} t_\gamma G_0 U_{\gamma\beta}. \tag{5.3.18}$$

An alternative form of the AGS equations is obtained if we start instead with Eq. (5.2.27c),

$$\underline{T}^{(-)} = \vec{\underline{G}}^{-1}\mathscr{S}G_0\underline{t}\,\bar{\delta} + \underline{T}^{(-)}G_0\underline{t}\,\bar{\delta}, \tag{5.3.19}$$

which leads to

$$\underline{U} = \bar{\delta}G_0^{-1} + \underline{U}G_0\underline{t}\bar{\delta}. \tag{5.3.20}$$

We note that Eqs. (5.3.15) and (5.3.20) do not form dual integral equations of the form (1.9.21) unless they act upon states where the directionality of the inverse of G_0 is irrelevant. Consequently, some care must be exercised in assigning too literal of an interpretation of Eqs. (5.3.15) and (5.3.20) as formal LS structures,[5] such as

$$\underline{U} = \underline{V}_{\text{AGS}} + \underline{V}_{\text{AGS}}\underline{G}_0^{\text{AGS}}\underline{U}, \tag{5.3.21a}$$

$$\underline{U} = \underline{V}_{\text{AGS}} + \underline{U}\underline{G}_0^{\text{AGS}}\underline{V}_{\text{AGS}}, \tag{5.3.21b}$$

where

$$\underline{G}_0^{\text{AGS}} \equiv G_0\underline{t}G_0, \tag{5.3.21c}$$

and

$$\underline{V}_{\text{AGS}} \equiv \bar{\delta}G_0^{-1}. \tag{5.3.21d}$$

For many calculations of physical interest, however, $\underline{V}_{\text{AGS}}$ is well defined.

We can now determine the remaining transition operators, $U_{\alpha 0}$ and U_{00}, in terms of the two-cluster operators $U_{\alpha\beta}$:

$$U_{\alpha 0} = G_0^{-1} + \sum_{\gamma=1}^{3} U_{\alpha\gamma} G_0 t_\gamma, \tag{5.3.22}$$

$$U_{00} = \sum_{\gamma=1}^{3} t_\gamma G_0 U_{\gamma 0}, \tag{5.3.23}$$

so

$$U_{00} = \sum_{\gamma=1}^{3} t_\gamma + \sum_{\gamma,\lambda=1}^{3} t_\gamma G_0 U_{\gamma\lambda} G_0 t_\lambda. \tag{5.3.24}$$

It should be pointed out that both $U_{0\beta}$ and $U_{\alpha 0}$ also satisfy integral equations, e.g.,

$$U_{0\alpha} = G_0^{-1} + \sum_{\gamma=1}^{3} U_{0\gamma} G_0 t_\gamma \bar{\delta}_{\gamma\alpha}. \tag{5.3.25}$$

These equations can be used to determine the two-to-three breakup scattering without having to calculate the two-to-two half-off-shell matrix elements of $U_{\alpha\beta}$. The U_{00} operator *does not* satisfy an integral equation such as Eq. (5.3.25) because $t_0 = 0$.

The three-to-three transition amplitude is the plane-wave matrix element

$$\langle \mathbf{k}_1', \mathbf{k}_2', \mathbf{k}_3' | U_{00} | \mathbf{k}_1, \mathbf{k}_2, \mathbf{k}_3 \rangle, \tag{5.3.26}$$

which receives contributions from several different types of physical processes. The distinctive features of multiparticle scattering are reflected in how these processes enter into the structure of the amplitude (5.3.26).

The simplest components of (5.3.26) are the two-particle matrix elements

$$\langle \mathbf{k}_1', \mathbf{k}_2', \mathbf{k}_3' | t_\gamma | \mathbf{k}_1, \mathbf{k}_2, \mathbf{k}_3 \rangle = \delta^{(3)}(\mathbf{K}' - \mathbf{K})\, \delta^{(3)}(\mathbf{k}_\gamma' - \mathbf{k}_\gamma) \langle \mathbf{k}_{\alpha\beta}' | \hat{t}_\gamma | \mathbf{k}_{\alpha\beta} \rangle, \tag{5.3.27}$$

where

$$\mathbf{K} \equiv \sum_{\alpha=1}^{3} \mathbf{k}_\alpha, \tag{5.3.28}$$

$$\mathbf{k}_{\alpha\beta} \equiv \mathbf{k}_\alpha - \mathbf{k}_\beta, \tag{5.3.29}$$

are the total and $\alpha\beta$-relative momenta, respectively. In Eq. (5.3.27), the indices α, β, γ are unequal and in cyclic order; $\hat{t}_\gamma$ is the transition operator on the (relative) two-particle space. The matrix element (5.3.27) is the simplest nontrivial example of a *disconnected process*, and it can be depicted in terms of the disconnected graphs shown in Fig. 1.2, where the disconnected line corresponds to $\delta^{(3)}(\mathbf{k}'_\gamma - \mathbf{k}_\gamma)$, which results from the invariance of $\hat{t}_\gamma$ with respect to translations of particle γ, while the bubble refers to the matrix element of $\hat{t}_\gamma$; the δ-function resulting from overall translational invariance is not graphically represented. This depiction of disconnected processes resulting from translational invariance with respect to subgroups of particles generalizes straightforwardly to systems of arbitrary numbers of particles. Only the number of disconnected possibilities increases as N increases, but the physical interpretation of their occurrence is the same, and their graphical depictions are similar to the $N = 3$ case.

The remaining part of the three-particle to three-particle amplitude is connected and corresponds to the c-labeled bubble in Fig. 1.2. The simplest contribution to that bubble is the double-scattering term,

$$U_{00}^{(2)} = \sum_{\gamma,\lambda=1}^{3} t_\gamma G_0 \bar{\delta}_{\gamma\lambda} t_\lambda, \tag{5.3.30}$$

that arises from the inhomogeneous term $\bar{\delta} G_0^{-1}$ in the AGS equations (5.3.17) and is graphically depicted in Fig. 1.1. This term has the special feature of having pole singularities that correspond to the long-range, on-shell propagation intermediate to successive on-shell, two-particle collisions. Such processes are ubiquitous: They occur, for example, in ordinary laboratory-staged two-particle scattering processes, if the detection process is counted as one of the double-scatterings.

These δ-function and pole singularities reflect an ordering of relative probabilities. Crudely speaking, the disconnected processes are infinitely more probable than the pole contributions, which are, in turn, infinitely more probable than the remainder of the connected process. The representation of the probability of these different processes as a relative ordering of powers of infinities is a legacy of using the unphysical, but convenient, limit of states of well-defined momenta of the three particles that correspond to the limit of infinite space–time volume. The scattering amplitudes calculated with normalizable states would be nonsingular. But doing things this way is usually not practical, it is also generally basis-dependent in ways not identifiable with the typical experimental situation, and, ultimately, it is really not necessary.

Apart from collision detection procedures, isolated three-to-three processes are rarely encountered under laboratory conditions. Nonetheless, they occur frequently in nature. For example, quasi-free collisions of this kind appear strikingly as three-hadron resonances in multihadron final states, which are usually initiated by a two-hadron collision; the $3(^4\text{He}) \to (^{12}\text{C})^*$ reaction contributes to the production of stellar carbon; also, the connected part of the three-to-three amplitude is required for the calculation of the third virial coefficient. Evidently, the satisfaction of unitarity above the three-particle threshold mandates consideration of the three-to-three process.

The structure of the connected part of U_{00} makes it apparent that the kernel of the once-iterated form of the AGS equation (5.3.15) (or Eq. (5.3.17)),

$$\underline{U} = \bar{\delta} G_0^{-1} + \bar{\delta} \underline{t} \bar{\delta} + \underline{K}^2 \underline{U}, \tag{5.3.31}$$

is connected. Similar remarks apply to $\underline{T}^{(+)}$ (*cf*. Eq. (5.3.11)). This achieves our goal of obtaining C-K equations for the three-particle transition operators.

There are other singularities that can appear in U_{00} because of the bound-state poles in the two-particle transition operators $t_\alpha(E)$. Now,

$$t_\alpha(E) = V_\alpha + V_\alpha G_\alpha^{(+)}(E) V_\alpha, \tag{5.3.32}$$

where we can rewrite H_α in the Green function as

$$H_\alpha = K_{\text{tr}}(\alpha) + \hat{H}_\alpha(\text{rel}), \tag{5.3.33}$$

where K_{tr} is the sum of kinetic-energy operators of particle α and the operator representing the (translational) kinetic energy of the c.m. of the other two particles (β, γ), and $\hat{H}_\alpha(\text{rel})$ is the relative Hamiltonian of this pair. Using the completeness of the eigenstates of $\hat{H}_\alpha$, we can decompose $G_\alpha^{(+)}(E)$ into a part that contains all of the contributions from the discrete (normalizable) eigenstates $|\varepsilon(\alpha)\rangle$ of $\hat{H}_\alpha(\text{rel})$:

$$\hat{H}_\alpha |\varepsilon(\alpha)\rangle = \varepsilon(\alpha) |\varepsilon(\alpha)\rangle, \tag{5.3.34}$$

and a part $G_\alpha^{(+)}(E)_c$, that contains only continuum contributions:

$$G_\alpha^{(+)}(E) = G_\alpha^{(+)}(E)_c + \sum \frac{|\varepsilon(\alpha)\rangle\langle\varepsilon(\alpha)|}{E - K_{\text{tr}}(\alpha) - \varepsilon(\alpha) + i0}. \tag{5.3.35}$$

Consequently,

$$t_\alpha(E) = t_\alpha^{\text{c}}(E) + t_\alpha^{\text{B}}(E), \tag{5.3.36}$$

where

$$t_\alpha^{\text{B}}(E) = \sum V_\alpha \left\{ \frac{|\varepsilon(\alpha)\rangle\langle\varepsilon(\alpha)|}{E - K_{\text{tr}}(\alpha) - \varepsilon(\alpha) + i0} \right\} V_\alpha, \tag{5.3.37}$$

and $t_\alpha^{\text{c}}(E)$ is simply the remaining part of $t_\alpha(E)$. The sums in Eqs. (5.3.35) and (5.3.37) are over all the discrete states.

Now $\varepsilon(\alpha) < 0$ for any reasonable two-particle bound state, so there is a range of negative E values for which $[E - \varepsilon(\alpha)] > 0$ and the propagator in Eq. (5.3.37) has the cut structure reflective of scattering. However, since no free-three-particle scattering can occur for $E < 0$, the condition $E = \varepsilon(\alpha)$

corresponds to the threshold for two-cluster scattering of particle α and the bound (β, γ) pair.

The insertion of a complete set of plane-wave eigenstates of $K_{tr}(\alpha)$ converts Eq. (5.3.37) into a sum of integrals. Thus, the singularity structure of $t_\alpha^B(E)$ in the complex E plane is that of cuts beginning at $E = \varepsilon(\alpha) < 0$ generated by the two-body bound-state poles. If the plane-wave matrix elements of $t_\alpha^B(E)$ were continued to unphysical values of the relative energy of the (β, γ) pair, these cuts would appear as poles, just as in the isolated two-particle case. With this in mind, we see from Eq. (5.3.24) that the on-shell, two-cluster transition amplitudes emerge as the residues of the double, two-body, bound-state-pole singularities of the three-to-three transition amplitude. In this regard, we note that if

$$H_\alpha |\phi_\alpha(\tilde{E})\rangle = \tilde{E} |\phi_\alpha(\tilde{E})\rangle, \tag{5.3.38}$$

then

$$G_0^{(+)}(\tilde{E}) V_\alpha |\phi_\alpha(\tilde{E})\rangle = |\phi_\alpha(\tilde{E})\rangle, \tag{5.3.39}$$

which converts the numerators of the pole terms in Eq. (5.3.37) into channel states. (Note that $|\phi_\alpha(\tilde{E})\rangle$ is a product of $|\varepsilon(\alpha)\rangle$ and a plane-wave state of particle α.) The breakup amplitudes are identified with the residues of single, two-body bound-state pole singularities of the connected part of the three-to-three amplitude using Eq. (5.3.22), or a relation similar to it, for $U_{0\alpha}$. These identifications, that again indicate the fundamental role of the three-to-three process, were exploited extensively by Faddeev.[6]

The preceding singularity properties provide motivation for considering integral equations for the connected parts of U_{00}:

$$C_{\alpha\beta} = t_\alpha G_0 U_{\alpha\beta} G_0 t_\beta. \tag{5.3.40}$$

Using our matrix notation, we deduce from Eq. (5.3.15) that

$$\underline{C} = \underline{t} G_0 \bar{\delta} \underline{t} + \underline{t} G_0 \bar{\delta} \underline{C}. \tag{5.3.41}$$

A somewhat more convenient quantity, namely

$$\underline{M} \equiv \underline{t} + \underline{C}, \tag{5.3.42}$$

satisfies

$$M_{\alpha\beta} = t_\alpha \delta_{\alpha\beta} + t_\alpha G_0 \sum_\gamma \bar{\delta}_{\alpha\gamma} M_{\gamma\beta}. \tag{5.3.43}$$

Summing Eq. (5.3.43) over β yields the original Faddeev equations[7]

$$T_\alpha = t_\alpha + t_\alpha G_0 \sum_\gamma \bar{\delta}_{\alpha\gamma} T_\gamma, \tag{5.3.44}$$

where

$$T_\alpha \equiv \sum_\beta M_{\alpha\beta}, \tag{5.3.45}$$

so

$$U_{00} = \sum_{\alpha=1}^{3} T_\alpha. \tag{5.3.46}$$

The kernels of Eqs. (5.3.41), (5.3.43), and (5.3.44) become connected after a single iteration.

If $t_\alpha = (1 - V_\alpha G_0)^{-1} V_\alpha$ is inserted into Eq. (5.3.44), we see that

$$T_\alpha = V_\alpha + V_\alpha G_0 \sum_\gamma T_\gamma, \tag{5.3.47}$$

which when summed over α yields the identity

$$\sum_\alpha T_\alpha = V + VG_0 \sum_\gamma T_\gamma. \tag{5.3.48}$$

This calculation, which also reflects, in reverse order, Faddeev's original algebraic derivation of the Eqs. (5.3.44) *starting* from Eq. (5.3.48), shows that any solution of the Eqs. (5.3.44) yields an operator U_{00} that satisfies the LS equation

$$U_{00} = V + VG_0 U_{00}. \tag{5.3.49}$$

The same calculation holds starting with the homogeneous form of Eq. (5.3.44), and leads again to Eqs. (5.3.47–5.3.49), but with no inhomogeneous term. These last equations show that there are no trivial solutions of the Schrödinger equation ($\Sigma_\alpha T_\alpha = 0$) that correspond to nontrivial solutions ($T_\alpha \neq 0$ for some α) of the homogeneous form of Eq. (5.3.44). Therefore, solutions of the Eqs. (5.3.44) are in one-to-one correspondence with solutions of the Schrödinger equation as embodied in Eq. (5.3.49). By a similar argument, it follows that the AGS equations are also equivalent to the Schrödinger equation. More directly, since the AGS equations are related to the Faddeev equations by means of the inverse operation $\bar{\delta}^{-1}$, both sets of equations have identical physical content.

Although the Eqs. (5.3.44) stand by themselves, alternative derivations do not have the step-by-step reversibility of Faddeev's algebraic derivation. For example, since

$$T_\alpha G_0 = V_\alpha G, \tag{5.3.50}$$

so

$$T_\alpha G_0 = V_\alpha G_\alpha\left(1 + \sum_\beta \bar{\delta}_{\alpha\beta} V_\beta G\right), \tag{5.3.51}$$

from which one readily recovers the Eqs. (5.3.44). The point is that here one can only retrace the derivation by using Eqs. (5.3.47) and (5.3.49) in the intermediate steps, whereas they are precisely what are to be recovered as starting points. In the present instance, this is of little consequence since the final results, the Eqs. (5.3.44), are the same. The different modes of derivation, viz., the reversible algebraic method and other not explicitly reversible approaches that use graph summation and resolvent identities, among other techniques, generally yield inequivalent collision integral equations.

This is not to say that these alternative techniques are not useful. For example, the Eqs. (5.3.44) follow immediately from the basic definition (5.3.47) by graph summation. Let us decompose T_α into its disconnected (d) and connected (c) parts:

$$T_\alpha = [T_\alpha]_{\mathrm{d}} + [T_\alpha]_{\mathrm{c}}. \tag{5.3.52}$$

A three-particle to three-particle *graph* is represented by three horizonal lines corresponding to the particles and an arbitrary number of vertical curly lines, each representing an interaction potential between two particles; internal particle lines represent G_0 propagators (Fig. 5.1). Since T_α consists of all graphs that end on the left with a interaction between the (β, γ)-pair, it follows that

$$[T_\alpha]_{\mathrm{d}} = t_\alpha \tag{5.3.53}$$

and

$$[T_\alpha]_{\mathrm{c}} = t_\alpha G_0 A, \tag{5.3.54}$$

where A is the sum of all graphs that end on the left with a V_γ interaction, $\gamma \neq \alpha$. By the definition of T_γ, we see that

$$A = \sum_{\gamma \neq \alpha} T_\gamma. \tag{5.3.55}$$

The essential idea, which can be generalized to $N > 3$, is the observation that $[T_\alpha]_{\mathrm{c}}$ is a product of a set of *almost-connected* graphs (for $N = 3$, the t_α's)

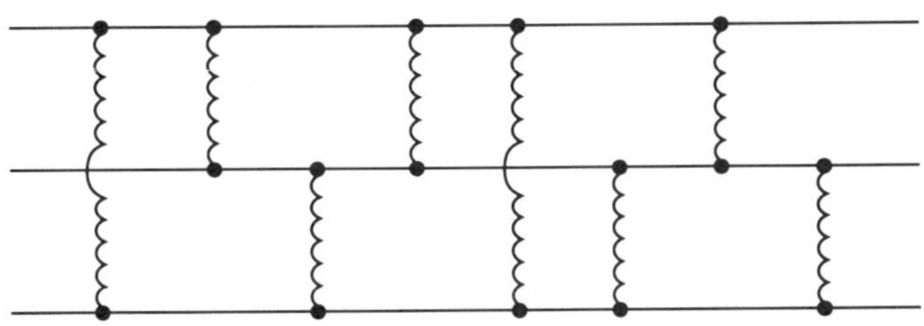

FIG. 5.1. A graph that represents a contribution to the three-to-three particle process.

attached to all graphs that end on the left with exactly that interaction needed to attain full connectivity.

The preceding graphical argument is interesting in several respects. First, after a few extensions of our notation, the argument can be applied to collisions of an arbitrary number of particles. Second, it shows that the Faddeev approach more clearly represents the important physical features of multiparticle scattering than do the LS equations. Third, it shows that the Faddeev equations are the consequence of those physical features rather than of some formal manipulations equivalent to the Fredholm reduction of the LS equations. (This reduction is carried out using the subtraction technique formalism of Section 1.8 with the identifications $K_S \sim V_\alpha G_0 \bar{\delta}_{\alpha\gamma}$ and $K_R \sim V_\alpha G_0 \delta_{\alpha\gamma}$ in Eq. (5.3.47). We see then that $\gamma_R B \sim t_\alpha$, while $\gamma_R K_S$ corresponds to $t_\alpha G_0 \bar{\delta}_{\alpha\gamma}$.) From this perspective, the LS equations appear both physically awkward and mathematically ill-defined.

5.3b. *Subtraction Techniques with Degenerate Kernels*

Let us suppose that the two-particle transition operators decompose into the sum

$$t_\alpha = t_\alpha^{(s)} + t_\alpha^{(r)} \tag{5.3.56}$$

of a piece $t_\alpha^{(s)}$, which we suppose is of finite rank, and a residual part $t_\alpha^{(r)}$, which has an arbitrary structure. We can then affect a subtraction procedure on the AGS equations with the results

$$\underline{U} = \underline{\hat{U}} + \underline{\hat{U}}\underline{\kappa}^{(s)}\underline{U}, \tag{5.3.57a}$$

$$\underline{U} = \underline{\hat{U}} + \underline{U}\underline{\kappa}^{(s)}\underline{\hat{U}}, \tag{5.3.57b}$$

where

$$\underline{\kappa}^{(s)} \equiv G_0 \underline{t}^{(s)} G_0, \tag{5.3.58}$$

and

$$\underline{\hat{U}} = \bar{\delta} G_0^{-1} + \bar{\delta}\underline{t}^{(r)} G_0 \underline{\hat{U}}, \tag{5.3.59a}$$

$$\underline{\hat{U}} = \bar{\delta} G_0^{-1} + \underline{\hat{U}} G_0 \underline{t}^{(r)} \bar{\delta}. \tag{5.3.59b}$$

The subtraction technique (5.3.56–5.3.59) has a number of applications. In the case of interest to us, namely when the restriction of $t_\alpha^{(s)}(z)$ to the two-body Hilbert space has the structure[8]

$$\hat{t}_\alpha^{(s)}(z) = \sum_{n,m} |\alpha, n; z\rangle \tau_{\alpha nm}(z) \overline{\langle \alpha, m; z|}, \tag{5.3.60}$$

the forms of the Eqs. (5.3.57) are not the most convenient ones. This is because we would like, instead of $U_{\beta\alpha}$, an operator whose matrix elements with respect to the states appearing in (5.3.60) are off-shell scattering amplitudes. We keep in mind that the bra and ket vectors appearing in Eq. (5.3.60) are not necessarily duals of each other.

In order to see why this is so, let us note that in terms of the Jacobi variables (Section A.1a), and with the total-momentum Dirac δ-function factored out, Eq. (5.3.27) becomes, with $t^{(s)}$ replacing t,

$$\langle \mathbf{p}'_\alpha, \mathbf{q}'_\alpha | t^{(s)}_\alpha(z) | \mathbf{p}_\alpha, \mathbf{q}_\alpha \rangle = \delta(\mathbf{q}'_\alpha - \mathbf{q}_\alpha) \langle \mathbf{p}'_\alpha | \hat{t}^{(s)}_\alpha(z - \omega_\alpha) | \mathbf{p}_\alpha \rangle. \tag{5.3.61}$$

Thus, from Eq. (5.3.60), we see that

$$t^{(s)}_\alpha(z) = \sum_{n,m} \int d^3 q_\alpha |\alpha, n; z - \omega_\alpha\rangle |\mathbf{q}_\alpha\rangle \tau_{\alpha nm}(z - \omega_\alpha) \times \langle \mathbf{q}_\alpha | \overline{\langle \alpha, m; z - \omega_\alpha |} \tag{5.3.62}$$

and

$$\langle \mathbf{p}'_\alpha | \hat{t}^{(s)}_\alpha(z) | \mathbf{p}_\alpha \rangle = \sum_{n,m} g_{\alpha n}(\mathbf{p}'_\alpha, z) \tau_{\alpha nm}(z) \bar{g}_{\alpha m}(\mathbf{p}_\alpha, z), \tag{5.3.63}$$

where

$$g_{\alpha n}(\mathbf{p}'_\alpha, z) \equiv \langle \mathbf{p}'_\alpha | \alpha, n; z \rangle, \tag{5.3.64a}$$

$$\bar{g}_{\alpha m}(\mathbf{p}_\alpha, z) \equiv \overline{\langle \alpha, n; z} | \mathbf{p}_\alpha \rangle. \tag{5.3.64b}$$

When a contribution to the sum in Eq. (5.3.60), say for $n = m = n_0$, arises because of the factorization of the residue at the bound-state pole of $t_\alpha(z)$, then

$$|\alpha, n_0; z\rangle = V_\alpha |\hat{\phi}_{\alpha, n_0}\rangle, \tag{5.3.65}$$

where $|\hat{\phi}_{\alpha, n_0}\rangle$ is a two-body bound-state in the two-body c.m. frame. Since, by definition,

$$[\varepsilon_{\alpha, n_0} - \hat{v}_\alpha] |\hat{\phi}_{\alpha, n_0}\rangle = V_\alpha |\hat{\phi}_{\alpha, n_0}\rangle, \tag{5.3.66}$$

where $\hat{v}_\alpha$ is the relative kinetic-energy operator, if $E = \varepsilon_{\alpha, n_0} + \omega_\alpha$ and $z = E + i0$, then (*cf*. Eq. (A.5.26))

$$G_0^{(+)}(E)^{-1} |\phi_{\alpha, n_0}(E, \mathbf{q}_\alpha)\rangle = |\alpha, n_0; \mathbf{q}_\alpha\rangle, \tag{5.3.67}$$

where

$$|\phi_{\alpha, n_0}(E, \mathbf{q}_\alpha)\rangle \equiv |\hat{\phi}_{\alpha, n_0}\rangle |\mathbf{q}_\alpha\rangle. \tag{5.3.68}$$

From Eq. (5.3.67), we see that the on-shell bound-state transition amplitudes are

$$\begin{aligned}\langle\beta, m_0; \mathbf{q}'_\beta| G_0^{(+)}(E) U_{\beta\alpha}(E) G_0^{(+)}(E)|\alpha, n_0; \mathbf{q}_\alpha\rangle \\ = \langle\phi_{\beta,m_0}(E, \mathbf{q}'_\beta)| U_{\beta\alpha}(E)|\phi_{\alpha,n_0}(E, \mathbf{q}_\alpha)\rangle.\end{aligned} \tag{5.3.69}$$

This shows that it is easier to exploit the finite-rank structure of $t^{(s)}$ with the operators

$$\underline{F}(z) \equiv G_0(z)\underline{U}(z)G_0(z), \tag{5.3.70}$$

$$\underline{\hat{F}}(z) \equiv G_0(z)\underline{\hat{U}}(z)G_0(z), \tag{5.3.71}$$

that satisfy the integral equations

$$\underline{F}(z) = \underline{\hat{F}}(z) + \underline{\hat{F}}(z)\underline{t}^{(s)}(z)\underline{F}(z), \tag{5.3.72a}$$

$$\underline{F}(z) = \underline{\hat{F}}(z) + \underline{F}(z)\underline{t}^{(s)}(z)\underline{\hat{F}}(z), \tag{5.3.72b}$$

and

$$\underline{\hat{F}}(z) = \zeta(z) + \zeta(z)\underline{t}^{(r)}\underline{\hat{F}}(z), \tag{5.3.73a}$$

$$\underline{\hat{F}}(z) = \zeta(z) + \underline{\hat{F}}(z)\underline{t}^{(r)}\zeta(z), \tag{5.3.73b}$$

respectively, where

$$\zeta(z) \equiv G_0(z)\,\bar{\delta}. \tag{5.3.74}$$

We have taken care at this point to show the z-dependence of all quantities.

Now we call

$$X^{nm}_{\beta\alpha}(\mathbf{q}'_\beta, \mathbf{q}_\alpha) \equiv \langle\mathbf{q}'_\beta|\{\overline{\langle\beta, n; z| F_{\beta\alpha}(z)|\alpha, m; z\rangle}\}|\mathbf{q}_\alpha\rangle, \tag{5.3.75}$$

with a similar quantity $\hat{X}^{nm}_{\beta\alpha}(\mathbf{q}'_\beta, \mathbf{q}_\alpha)$, defined in terms of $\hat{F}_{\beta\alpha}(z)$. Then, we see that Eq. (5.3.72a), e.g., becomes a set of multichannel two-body LS equations:

$$\begin{aligned}X^{nm}_{\beta\alpha}(\mathbf{q}'_\beta, \mathbf{q}_\alpha) &= \hat{X}^{nm}_{\beta\alpha}(\mathbf{q}'_\beta, \mathbf{q}_\alpha) \\ &\quad + \sum_{\gamma,n',m'} \int d^3q''_\gamma\, \hat{X}^{nm'}_{\beta\gamma}(\mathbf{q}'_\beta, \mathbf{q}''_\gamma)\tau_{\gamma m'n'}(z - \omega''_\gamma)X^{n'm}_{\gamma\alpha}(\mathbf{q}''_\gamma, \mathbf{q}_\alpha).\end{aligned} \tag{5.3.76}$$

In the degenerate-kernel limit, $t^{(r)} = 0$,

$$\hat{X}^{nm}_{\beta\alpha}(\mathbf{q}'_\beta, \mathbf{q}_\alpha) = \bar{\delta}_{\beta\alpha} Z^{nm}_{\beta\alpha}(\mathbf{q}'_\beta, \mathbf{q}_\alpha), \tag{5.3.77}$$

where

$$Z_{\beta\alpha}^{nm}(\mathbf{q}'_\beta, \mathbf{q}_\alpha) = \bar{g}_{\beta_n}(\mathbf{p}'_\beta, z) g_{\alpha m}(\mathbf{p}_\alpha, z) D(z; \mathbf{q}'_\beta, \mathbf{q}_\alpha)^{-1}, \tag{5.3.78}$$

$$D = z - \frac{\mathbf{q}_\alpha^2}{2\mu_\beta} - \frac{1}{2\mu_\alpha}\mathbf{q}_\beta'^2 - \frac{\mathbf{q}_\beta \cdot \mathbf{q}_\alpha}{m_\gamma}, \tag{5.3.79}$$

$$\mathbf{p}'_\beta = \mp\left(\mathbf{q}_\alpha + \frac{\mu_\beta}{m_\gamma}\mathbf{q}'_\beta\right), \qquad \mathbf{p}_\alpha = \pm\left(\mathbf{q}'_\beta + \frac{\mu_\alpha}{m_\gamma}\mathbf{q}_\alpha\right), \tag{5.3.80}$$

and $\mu_\alpha = m_\beta m_\gamma (m_\beta + m_\gamma)^{-1}$. In this case, even if there are no two-body bound-states, we see from Eq. (5.3.24) that all one requires is the solution of the Eqs. (5.3.76). The simplest one-term form of Eq. (5.3.76) with the Born term of Eq. (5.3.78) is called the Amado model; it was found by him with a different approach.[8] The homogeneous form of the Amado model was found earlier by Mitra in a study of the three-particle bound state.[8]

Generally, the auxiliary amplitudes $X_{\beta\alpha}^{nm}$ have no physical interpretation even on-shell (if an energy shell *can* be defined), but are introduced to exploit the finite-rank structure of $t^{(s)}$. Lovelace[4] has advocated an extended physical interpretation of these amplitudes when some of the terms in $t^{(s)}$ are taken to correspond to the approximate factorization of t at resonance poles. Thus, conceivably, one could calculate resonance-bound-state or resonance–resonance transition amplitudes. For example, in the three-pion problem, the 3π-to-3π amplitude could be described in terms of $\pi-\rho$-meson amplitudes. This is the basis of a consistent three-particle realization of the isobar model of moderate-energy processes.[9] While these models are, of course, of limited accuracy, they have proven to be useful phenomenological tools.

The inhomogeneous term (5.3.77) corresponds to the three-particle rescattering diagram (Fig. 1.1) with the pairs of initial- and final-state two-particle collisions grouped together as with a bound state or resonance. This description leads to the graphical representation of the set of integral equations (5.3.76) in the degenerate-kernel limit in Fig. 5.2. We see that the various scattering amplitudes can be viewed as infinite iterations of the fundamental rescattering processes represented by the $Z_{\beta\alpha}^{nm}$.

In Section 1.11, we developed the general machinery for incorporating permutation symmetries into integral equations of the sort we have considered in this section. We apply this analysis in Section 5.4f, where the $N = 3$ case emerges as an example. In order to make the present discussion self-contained, we adapt a simple argument due to Lovelace[4] for symmetrizing the Eqs. (5.3.76). We assume that all three particles are identical.

Suppose that the states $|\alpha, n; z\rangle$ are correctly symmetrized, and that we maintain a cyclic ordering of the particle labels specifying the momenta. Thus,

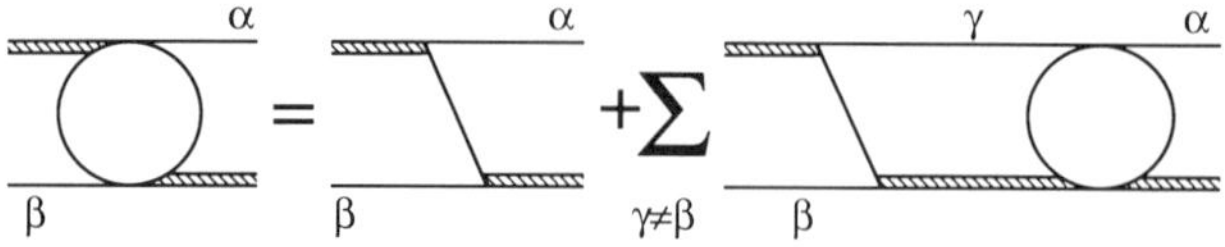

FIG. 5.2. Graphical representation of three-body collision integral equations in the degenerate-kernel limit.

an interchange of the particle labels implicit in $|\alpha, n; z\rangle$ yields $(\pm)|\alpha, n; z\rangle$, where $+ \sim$ Bose and $- \sim$ Fermi statistics. (This is merely the imposition of a "standard" convention for the construction of the three-particle states $|\mathbf{q}_\alpha\rangle|\alpha, n; z\rangle$.)

The consequence of identity is that there are only two distinct $X^{nm}_{\beta\alpha}$ amplitudes for fixed momenta:

$$X^{nm}_{\mathrm{D}}(\mathbf{q}', \mathbf{q}) \equiv X^{nm}_{\alpha\alpha}(\mathbf{q}'_\alpha, \mathbf{q}_\alpha), \tag{5.3.81a}$$

for all $\mathbf{q}' = \mathbf{q}'_\alpha$, $\mathbf{q} = \mathbf{q}_\alpha$, and

$$X^{nm}_{\mathrm{N}}(\mathbf{q}', \mathbf{q}) \equiv X^{nm}_{\beta\alpha}(\mathbf{q}'_\beta, \mathbf{q}_\alpha), \tag{5.3.81b}$$

for all $\mathbf{q}' = \mathbf{q}'_\beta$, $\mathbf{q} = \mathbf{q}_\alpha$, $\beta \neq \alpha$. That is, we have a direct (D) and an exchange (N) amplitude. Since there are two exchange and one direct possibilities, we expect that the complete physical amplitude will either be a sum or a difference of X_{D} and X_{N}. The correct choice is

$$X^{nm}_{sym} \equiv X^{nm}_{\mathrm{D}} + 2X^{nm}_{\mathrm{N}} \tag{5.3.82}$$

for either statistics since the general symmetrized amplitude for the scattering (i $\rightarrow$ f) of a particle (0) from n target particles ($k = 1, \ldots, n$),

$$\langle \mathrm{f}|T|\mathrm{i}\rangle \pm \sum_{k=1}^{n} \langle P_{0k}\mathrm{f}|T|\mathrm{i}\rangle,$$

becomes

$$\langle \mathrm{f}|T|\mathrm{i}\rangle + n\langle \mathrm{f}_{\mathrm{ex}}|T|\mathrm{i}\rangle$$

because of identity and our convention, which imply that $|P_{0k}f\rangle = \pm|f_{\mathrm{ex}}\rangle$ under the exchange $0 \leftrightarrow k$ represented by P_{0k}. With the identification (5.3.82), the set (5.3.76) decouples, yielding a closed set of equations for the amplitude (5.3.82):

$$X^{nm}_{\mathrm{sym}}(\mathbf{q}', \mathbf{q}) = \hat{X}^{nm}_{\mathrm{sym}}(\mathbf{q}', \mathbf{q})$$
$$+ \sum_{n', m'} \int d^3q''\, \hat{X}^{nm'}_{\mathrm{sym}}(\mathbf{q}', \mathbf{q}'')\tau_{m'n'}(z - \omega'')X^{n'm}_{\mathrm{sym}}(\mathbf{q}'', \mathbf{q}), \tag{5.3.83a}$$

where $\hat{X}^{nm}_{\text{sym}}$ is defined analogously to X^{nm}_{sym}, and we have deleted the unnecessary γ index from τ. In the degenerate-kernel limit we have, for spinless bosons,

$$\hat{X}^{nm}_{\text{sym}}(\mathbf{q}', \mathbf{q}) \equiv 2Z^{nm}_{\text{sym}}(\mathbf{q}', \mathbf{q}) = 2\bar{g}_n(\mathbf{p}', z)g_m(\mathbf{p}, z)D(z; \mathbf{q}', \mathbf{q})^{-1}, \quad (5.3.83\text{b})$$

where redundant channel indices have been dropped. The spin and isospin recoupling factors that appear in the generalization of Eq. (5.3.83b) for three arbitrary identical particles are discussed in Section A.5b.

A similar analysis can be applied to develop one-vector-variable integral equations for the breakup amplitudes starting from equations such as (5.3.25). We note that quadrature rules such as Eq. (5.3.22) make direct use of the X amplitudes only when $\underline{t}^{(\text{r})} = 0$.

For general t_α, the AGS equations (5.3.17) are integral equations in two vector variables, as are Eqs. (5.3.73a, b) for nonseparable $t_\alpha^{(\text{r})}$. For example, when referring to a collision initiated in a two-cluster channel α, the Eqs. (5.3.17) lead to a set of integral equations for the matrix elements ${}_\beta\langle \mathbf{p}'_\beta, \mathbf{q}'_\beta | U_{\beta\alpha} | \phi_\alpha \rangle$, where we keep in mind that each set of Jacobi coordinates $(\mathbf{p}'_\beta, \mathbf{q}'_\beta)$ defines a distinct basis $\{|\mathbf{p}'_\beta, \mathbf{q}'_\beta\rangle_\beta\}$ for each β (*cf*. Section A.1a). In Section A.5b, it is indicated how the AGS, and similar two-vector-variable three-particle equations, reduce to two-dimensional, multichannel integral equations.

The preceding general situation is in marked contrast to the integral equations (5.3.76) in one vector variable, which reduce (*cf*. Section A.5b) to a set of one-dimensional integral equations after partial-wave analysis. When $\underline{t}^{(\text{r})} = 0$, the entire computational problem reduces to the solution of a set of one-dimensional, two-body, multichannel integral equations. The "potentials" in these equations are energy-dependent and possess a singularity structure that depends on the initial and final momenta as well as the parametric energy.

The reduction to the relatively simple Eqs. (5.3.76) depends, of course, on the finite-rank structure of $\underline{t}^{(\text{s})}$. However, it is a consequence of the special coupling scheme that distinguishes the AGS–Faddeev equations that it is possible to exploit this structure so efficiently. Indeed, it is only for these types of C-K equations that exact $N = 3$ scattering calculations have been performed.

Other C-K, $N = 3$ collision integral equations have been proposed. A class of these that describes the two-to-two and two-to-three processes simultaneously can be obtained from the AGS equations.[6,10,11] The set due to Karlsson and Zeiger (KZ)[11] has the attractive feature of requiring the two-body input in terms of only the two-body bound-state wave functions and the half-off-shell two-body transition matrices for nonnegative parametric energies. Some proposals for using the KZ equations to construct systematic approximation techniques have achieved good convergence for the case of the Amado model.[12] No major calculational projects have yet been launched using KZ equations despite their exceptional features. One can conjecture that as the availability of supercomputing power becomes more common, these features, particularly when one wishes to compare the off-shell characteristics of different potentials, will motivate more realistic KZ-based calculations.

A number of other C-K, $N = 3$ equations have been proposed. Of the infinite number of possibilities, we mention only two other types that have attracted more than passing attention.

The Weinberg equation for the full Green function G is a single integral equation whose kernel is already connected with an inhomogeneous term that has no connected part. Although this equation implicitly incorporates the Faddeev coupling, the final form of the kernel is such that finite-rank two-body t-matrices produce no significant simplification. The Weinberg equation (for any N) has proven useful only as an auxiliary theoretical device.[13]

On the other hand, the Kouri–Levin–Tobocman (KLT)[14] equations, which are derived in Section 5.4, were originally proposed as a practical calculational alternative to the usual C-K collision integral equations. One of the main reasons for this initial optimism is the unusual coupling scheme that allows the subsystem input to be expressed in terms of the channel interactions V_α and the channel Green functions. Moreover, the same coupling idea works for arbitrary N, which yields integral equations for the physical transition operators of stunning simplicity compared with other proposals. Despite sustained analysis of their properties since their inception, they have yet to yield an exact calculation for $N = 3$. One of the major reasons for this is that the coupling scheme, despite its usefulness for approximations to the spectral representation of the channel Green functions, does not lend itself to the exploitation of simplifications in the two-body input.

5.3c. Other Subtraction Techniques

We next review some other alternatives to the splitting (5.3.56), relevant to the subtraction technique (5.3.57–5.3.59), other than to take $\underline{t}^{(r)} = 0$, with $\underline{t}^{(s)}$ a finite-rank representation of $\underline{t}$. This last choice leads to Eqs. (5.3.76–5.3.78), in which the three-particle dynamics is treated exactly, but if $\underline{t}$ is not actually of finite rank, the two-body system is approximated. It is perhaps worthwhile to examine approaches that achieve the opposite effect.

In the long run, these alternative methods may prove valuable for understanding how to approximate multiparticle dynamics. Such approximations are now inevitable when more than three particles are involved, or when the three-body system of interest is a subsystem of a more complex colliding system of particles. It seems plausible that progress in multiparticle collision theory will be enhanced by finding and understanding the three-body counterparts of the rich variety of techniques and approximation methods that we have at our disposal in the much simpler two-body problem.[15] Compared to the amount of work that has been expended on two-particle scattering in this regard, relatively little has been done for $N = 3$.

One simple generalization of Eqs. (5.3.76–5.3.78) deserves mention. Suppose that the residual t-matrix has only one component $\bar{t}_\gamma$ so

$$t_\alpha^{(r)} = \delta_{\alpha\gamma} \bar{t}_\gamma, \tag{5.3.84}$$

where γ is fixed. Then the *exact solution* for $\hat{U}_{\beta\alpha}$ is

$$\hat{U}_{\beta\alpha} = \bar{\delta}_{\beta\alpha} G_0^{-1} + \bar{\delta}_{\beta\gamma} \bar{t}_\gamma \bar{\delta}_{\gamma\alpha}. \tag{5.3.85}$$

A possible use for this is the three-nucleon problem where a cut-off Coulomb force between two protons in a ppn configuration is included in addition to a separable nuclear interaction.[16] This result is relevant to any three-body configuration in which a particular two-particle subsystem is distinguished in some way.

A splitting (5.3.56) that interchanges the roles of $\underline{t}^{(s)}$ and $\underline{t}^{(r)}$ realizes the "two-potential" type of subtraction technique.[17] One now solves Eqs. (5.3.59a, b) (or Eqs. (5.3.73a, b)) exactly, while Eqs. (5.3.57a, b) (or Eqs. (5.3.72a, b)) are handled perturbatively, e.g.,

$$\underline{U} \simeq \underline{\hat{U}} + \underline{\hat{U}}\underline{K}^{(s)}\underline{\hat{U}}. \tag{5.3.86}$$

This technique was used by Pieper[18] to include the effects of the D- and P-wave components of the nucleon–nucleon (NN) interaction in a perturbative fashion in order to calculate elastic nucleon–deuteron (Nd) scattering. Pieper used an exact three-body calculation that included only the NN S-wave components. The results yielded new predictions for the various spin observables in Nd scattering, and were in reasonable agreement with what experimental data existed, as well as with a subsequent exact calculation.[19]

Another application of the general subtraction technique has been to obtain a quasi-unitary form of the so-called impulse approximation[20–22] and a three-body version of the optical potential.[23,24] The essential idea here is to take $\underline{t}^{(s)}$ as all or part of the separable contributions within $\underline{t}$ that arise from the two-body bound-state pole terms. If $\underline{t}^{(s)}$ is taken as the complete bound-state contribution to $\underline{t}$, then Eqs. (5.3.57–5.3.59) represent a three-body, and possibly coupled-channels (for more than one $\underline{t}^{(s)}$), realization of the optical-model formalism.[23]

The Finkel–Rosenberg–Sloan[20,21] idea is to identify $\underline{t}^{(s)}$ as the Dirac δ-function part of the bound-state portion of $\underline{t}$ for every channel in which there is a two-particle bound state. The dynamical approximation then consists in taking

$$\hat{U} \simeq \bar{\delta}G_0^{-1} + \bar{\delta}\underline{t}^{(r)}\bar{\delta}. \qquad (5.3.87)$$

Because of the specific structure of $\underline{t}^{(s)}$ in this instance, Eqs. (5.3.57a, b) have Heitler-like kernels; they reduce to a set of algebraic relations after partial-wave decomposition in the case of the scattering from a bound state. The resultant transition matrices include the impulse approximation in a manner consistent with unitarity,[25] and so we call the approximation (5.3.87) the first-order unitary model (UM-1).

Calculations using the UM-1 were the first to indicate that the nucleon polarization in elastic Nd scattering from 14 MeV up to 40 MeV is influenced mainly by the P-wave components of the NN force, while the D-wave plays little role.[26] This conclusion was substantiated by much more refined calculations[18,19] and it has some bearing upon the influence of off-shell NN effects, particularly in the 14 MeV range of Nd scattering, since at these energies the NN P-wave phase shifts are very small.

One can construct a generalization of the UM-1 that satisfies three-body unitarity exactly. Here, the $\underline{t}^{(s)}$, $\underline{t}^{(r)}$ splittings are the same as in the UM-1. The difference arises from a more complete dynamical treatment of $\hat{U}$ that is related to a three-body $\hat{K}$-matrix formalism.[27] In this formalism, $t_\alpha^{(r)}$ is chosen just as in the UM-1, and then it is related to a Hermitian two-body operator k_α by

$$k_\alpha(E) = t_\alpha^{(r)}(E) + i\pi t_\alpha^{(r)}(E)\delta(E - H_0)k_\alpha(E), \qquad (5.3.88)$$

for parametric energy $z = E + i0$, so for $E < 0$, $k_\alpha(E) = t_\alpha^{(r)}(E)$. We note that $k_\alpha(E)$ is not the usual two-particle $\hat{K}$-matrix introduced in Ch. 2. Nevertheless, it can be shown[28] that $k_\alpha(E)$ has real-axis poles at any resonant two-body energies, and that a principal-value prescription is to be employed when integrating over these poles when the $k_\alpha(E)$ are used in the three-body integral equations.

The various proposed unitary three-particle formalisms differ sufficiently in appearance so that what seem like the same low-order approximations are in fact different, as has been borne out in calculations.[29] The reasons for this are understood, and amount to inequivalent input.[27]

The unitary $\hat{K}$-matrix formalism for three-particle scattering was completed by the development of C-K Heitler equations.[30] An alternative $N = 3$, $\hat{K}$-matrix formalism[31] requires very complicated constraints on the $\hat{K}$-matrix elements in order to satisfy unitarity exactly.[32] Since this formalism is just the $N = 3$ version of a $\hat{K}$-matrix technique for arbitrary N, the viability of this alternative method as a unitary model is open to question.[32,33] At present, we are not aware of any candidates for a unitary $\hat{K}$-matrix formalism for $N > 3$ that can be used like the $N = 2$ and $N = 3$ versions.

The calculations[29] of Nd scattering using an exactly unitary generalization of the UM-1, which we call UM_{ex}, were somewhat disappointing in that only qualitative agreement with the exact results[34] were obtained. The use of the UM_{ex} for three-pion scattering was disastrous.[35] The source of the difficulty is that the UM_{ex} is equivalent to approximating the $\hat{K}$-matrix by its disconnected parts.[36] This results in a multilation of the three-body dynamics that destroys much of the characteristic analyticity properties and singularity structure of the three-particle amplitude.[35–37] For example, the double-scattering pole in the three-to-three amplitude becomes replaced by its Dirac δ-function part. This results in a manifest violation of the known threshold behavior of this amplitude[38] and leads to spurious resonances.[35,37] The latter enhancements result from the movement onto the physical sheet of certain ("Peierls") unphysical-sheet singularities whose influence on the physical sheet is normally suppressed.[37]

While the minimal[36] $\hat{K}$-matrix model that is equivalent to the UM_{ex} is certainly inadequate for dealing with three-pion resonant behavior, it has a number of other characteristics that are worth pointing out. It yields a set of on-shell integral equations for the 3π-to-3π transition amplitudes in which the only input consists of the on-shell two-pion amplitudes.[39] The fact that one requires no further details about two-pion dynamics, either in the form of analyticity properties or off-shell behavior, constitutes the principal phenomenological appeal of the model,[35,36] as well as its main shortcoming. The model illustrates some of the subtleties of multiparticle dynamics.

The preceding example suggests that an approximation scenario that preserves the essential structure of the three-particle amplitudes may be preferable to strategies that do not. Some positive support for this is provided by the calculation of Nd scattering by Bencze and Doleschall.[40] Here, the Doleschall[19] NN transition amplitudes are used with $\underline{t}^{(s)}$ identified with the exact S- and D-wave parts of the NN interaction, while $\underline{t}^{(r)}$ is the P-wave part and is treated in first-order perturbation theory (*cf.* Eq. (5.3.87)). Good agreement with the exact calculation[19] was achieved. This strategy was generalized and refined to yield a fully unitary, both off-shell as well as on-shell, three-particle perturbation theory that preserves the important physical characteristics of the three-particle scattering amplitudes.[41] In particular, the correct threshold behaviors are necessarily reproduced and the location and residues of the physical-region double-scattering poles are the same as for the exact amplitudes.

The crucial feature of this structure-invariant perturbation theory[41] is that within the context of the AGS subtraction technique $\underline{t}^{(s)}$ is identified as the "essential" part $\underline{t}^{e}$ of the exact two-particle transition operator. The required $\underline{t}^{e}(z)$ is Hermitian analytic and contains all of the bound-state poles of $\underline{t}(z)$, and except within some arbitrarily small regions of the poles, is bounded for all negative z and on the positive z cut. Moreover, $\underline{t}^{e}(z)$ must be continuous across the negative z axis, and such that, on the two-body Hilbert space, $\langle \mathbf{p}|\hat{\underline{t}}^{(r)}|\mathbf{p}'\rangle = 0$ for z on the positive-energy cut and for either $|\mathbf{p}|$ or $|\mathbf{p}'|$ equal to $|z|^{1/2}$. It then follows that $\underline{t}^{(r)}(z)$ is continuous across the entire real z axis and $\underline{t}^{e}$ satisfies the two-body off-shell unitarity condition. Finally, in order to make the model practical, it is assumed that for real z the projection of both $\theta(z)\underline{t}^{e}(z)$ and $\theta(-z)\underline{t}^{e}(z)$ onto states of well-defined angular momentum are finite-rank operators, although not necessarily of the same rank. It is straightforward to show, using the methods of Section 2.4, that such an "essential" $\underline{t}^{e}$ exists and it is at most of rank three in any given partial wave. If the half-shell functions are nonsingular, $\underline{t}^{e}$ can be identified with $\underline{t}_{KN}$ given by Eq. (2.4.49). In structure-invariant perturbation theory the integral equation for $\hat{\underline{U}}$ is nonsingular.

Another off-shell three-body unitary formalism has been proposed by Carew and Rosenberg[42] based on a splitting of the two-body transition operators similar to the structure-invariant one just described. The form of the three-particle subtraction procedure is also similar, yet different than the AGS type. The role of $\hat{\underline{U}}$ is instead played by an "effective potential" (EP). The most interesting new feature that arises is that the equations that determine the EP can be put into variational form. This leads to a systematic procedure for generating exactly unitary approximations to three-particle scattering that are *variational* in nature.

Variational methods, as we have seen in Ch. 3, sometimes provide a viable alternative to the integral equations approach we have chosen to emphasize for multiparticle collisions. Although there have been important calculations[18,43] carried out in the past using this technique, it has been used far less frequently than direct integral equation methods. One of the principal reasons for this is the necessity of the careful specification of the asymptotic behavior of the three-particle scattering wave functions.[44] Consequently, the effectiveness of multiparticle variational principles for scattering is much less well understood than in the two-body case.

To conclude this subsection, we consider some results that may have some bearing upon the use of off-shell three-particle amplitudes in collision integral equations for four or more particles. Compared to the extensive analysis of the representations of the two-body off-shell amplitudes, very little is known about the off-shell properties of multichannel amplitudes.

Karlsson[45] has shown how the use of three-body analogues of the representations (2.4.42) and (3.3.23) allows one to obtain a three-particle version of the Kowalski–Noyes[46] reduction technique. The three-particle counterpart of the two-particle half-shell function f_s also satisfies a nonsingular integral equation. As in the two-particle case, this method provides a means of generating exactly unitary approximations that, with an appropriate choice[47] of two-particle input, are structure-invariant. Karlsson shows via the Yakubovskii equations that for suitable amplitudes the representation (2.4.42) has an N-particle generalization.

A simple modification[48] of the subtraction technique introduced by Karlsson[45] enhances the resemblance to the two-particle off-shell problem even further. In the absence of three-body forces, it is shown that a three-body f_s function exists, and enters into the kernel of a phase-space-type integral equation for the on-shell, three-particle amplitudes. After partial-wave decomposition, this integral equation reduces to a set of one-dimensional integral equations with finite domains of integration. Again a canonical representation analogous to Eq. (2.4.42) exists manifesting the off-shell unitary properties of the three-particle amplitude; these properties are, of course, crucial when such amplitudes enter as input into $N > 3$ collision integral equations.

Unfortunately, these factorization properties seem not to be realized in such a transparent fashion when many-particle forces are included.[48] Further research into the structural properties of multiparticle off-shell amplitudes would appear to be a prerequisite to progress in the calculation of collision problems in which these amplitudes occur as input.

5.3d. *Wave-Function Equations*

We write out the explicit $N = 3$ forms of the integral equations (5.2.37) for the "well-labeled" or "true" components $|\chi_b^{(+)}(\alpha)\rangle$ of the wave function initiated from a two-cluster channel with the potential matrix (5.3.10). In this case, only the two-cluster components are relevant ($|\chi_0^{(+)}(\alpha)\rangle = 0$), and these satisfy

$$|\chi_\beta^{(+)}(\alpha)\rangle = \delta_{\beta\alpha}|\phi_\alpha\rangle + G_\beta^{(+)}(E)V_\beta \sum_\gamma \bar{\delta}_{\beta\gamma}|\chi_\gamma^{(+)}(\alpha)\rangle, \tag{5.3.89}$$

where we have explicitly indicated the parametric energy in the Green functions but have suppressed it elsewhere. The combinations $G_\beta^{(+)}(E)V_\beta$ can, of course, be replaced by $G_0^{(+)}(E)t_\beta$. The Eqs. (5.3.89) have been frequently employed as a means of investigating various conventional approximation methods of atomic and nuclear reaction theory in a simple three-body setting. Just how this is done, and what it implies is discussed in further detail in Sections 9.4–9.5.

Three-body bound states correspond to solutions $|\chi_\beta\rangle$ of the homogeneous form of the Eqs. (5.3.89) for energies below the lowest scattering threshold:

$$|\chi_\beta\rangle = G_0(E)t_\beta \sum_\gamma \bar{\delta}_{\beta\gamma}|\chi_\gamma\rangle, \tag{5.3.90}$$

where the Green function is now nonsingular. The complete three-body bound-state wave function $|\psi_{3\mathrm{B}}\rangle$ is given by

$$|\psi_{3\mathrm{B}}\rangle = \sum_{\beta=1}^{3} |\chi_\beta\rangle. \tag{5.3.91}$$

5.3e. *Relativistic Generalizations*

The extension of the Faddeev approach to hadronic physics including relativistic effects was an early objective of three-particle collision theory.[4,49] Early calculations[50] concentrated upon the three-meson system, and the isobar models[4] designed to treat them were brought to a high state of phenomenological usefulness.[9] The meson–NN system,[51] and to a much smaller degree, the 3N system,[52] have been the focus of interest.

The meson–NN system includes the possibility of absorption, and so two-baryon channels generally have to be factored into the dynamical equations rather than serving merely as input. This has proven to be a major complication in the pion–NN problem, where it appears that the treatment of absorption is the major differentiating feature among extant πd calculations.[51] On the other hand, the relativistic Faddeev approach seems to work well for systems such as moderate-energy K^+d scattering where absorption is not possible.[51]

None of the relativistic three-body Faddeev formalisms is well founded in terms of a fundamental field theory. The situation is much like that already described in Section 2.7 for the two-body problem. The derivations of the collision integral equations are by heuristic methods designed to preserve unitarity, and that are apparently consistent with the overall requirements of relativity. The practical forms of these equations do not differ enough from their nonrelativistic counterparts to merit a detailed elaboration. Therefore, except for a brief review in Ch. 7 and some general remarks in Ch. 9, these models are not relevant to our subsequent discussion.

5.4. *N*-Particle Equations. I

This category covers a lot of ground. The exclusion of $N = \infty$ (nuclear matter, for example) and $N \leq 3$ helps somewhat, but still the possibilities are immense. What we have in mind is the extension of the Faddeev point of view to the collisions of nonrelativistic systems of finite numbers (≥ 4) of particles interacting through short-ranged potentials. We take this to include the deduction of few-body models of many-particle scattering.

In this section, we consider C-K equations of the form (5.2.27a–d) that are generated by a channel-coupling scheme (CC-type). The more fundamental Yakubovskii type (Y-type) is considered in Section 5.5.

The CC-type equations can be obtained in many cases from those of the Y-type by means of irreversible operations. This introduces the possibility of (spurious) solutions of the CC-type equations that do not correspond to physics.[1] This possibility is of little practical consequence when approximations are necessary, as they always are for more than very few particles. Moreover, it appears that the CC-type theories can provide a useful bridge between the full *N*-particle collision theory that a Y-type set of equations represents and the conventional models for collisions based upon the truncation of incomplete sets of LS equations. Because of these applications, which are facilitated by their relatively simple structure, we first discuss the CC-type theories on their own terms, rather than as a consequence of the Y-type theories.

Equations of the form (5.2.27a–d) are called *partition-labeled* in reference to the finite-dimensional matrix structure in the partition indices that they possess. Partitions are an unavoidable aspect of multiparticle scattering theory, and the matrix notation introduced in Section 5.2 is a very convenient way to handle this complication. The nomenclature and combinatorial results we require are developed in the Appendix. For most of this section, we suppress the dependence of the Green functions and transition operators on the parametric energy.

We learned in Section 5.2 that the objective of the CC-type approach leading to Eqs. (5.2.27a–d), viz.,

$$\underline{T}^{(+)} = \mathscr{V}\,\underline{G}\,\mathscr{S}\,\overleftarrow{\underline{G}}^{-1} + \mathscr{V}\,\underline{G}\,\underline{T}^{(+)}, \tag{5.4.1}$$

is to find a potential distribution matrix $\mathscr{V}$ such that $(\mathscr{V}\underline{G})^m$ is a connected operator for some integer $m \geq 1$. The wide variety of extant C-K partition-labeled equations[2] reflects the possibilities for choosing potential-distribution matrices that satisfy the connectivity requirement. The $\mathscr{V}$ matrices have often been found by indirect means: a class of C-K equations is found and the appropriate $\mathscr{V}$ is then inferred. An exception to this occurred in the case of the Kouri–Levin–Tobocman (KLT) equations,[3] where a particular $\mathscr{V}$ was chosen for reasons other than connectivity, but was subsequently shown to satisfy the connectivity requirement. The only known constraint, in addition to Eq. (5.2.26), is that all two-cluster partitions α be included. Namely, for every α, there is at least one partition b such that $\mathscr{V}^{b\alpha} \neq 0$.[4] The coupling scheme is said to be *minimally coupled* if, for any two-cluster partition β, we have $\mathscr{V}^{\beta a} = 0$ when $n_a \neq 2$.

5.4a. Rosenberg Equations

We follow an approach that begins with a generalization of the graphical method introduced in Section 5.3 for $N = 3$. We assume only pair interactions.

The relevance and power of graphical methods in nonrelativistic many-particle scattering theory was demonstrated by Weinberg,[5] and then applied by Rosenberg[6] to obtain C-K equations for the pair-labeled transition operators

$$M_{ij} = V_i \delta_{ij} + V_i G V_j. \tag{5.4.2}$$

Here V_i is the interaction between the pair of particles labeled by the index $i = 1, \ldots, N(N-1)/2$. We need consider only the operators

$$\begin{aligned} T_i &\equiv \sum_j M_{ij} \\ &= V_i[1 + GV]. \end{aligned} \tag{5.4.3}$$

An *N-particle scattering graph* is a representation via lines and vertices of a product of pair potentials and a generic Green function g, of the form

$$V_i g V_j g \cdots g V_k. \tag{5.4.4}$$

An example of a graphical representation of a product of the form (5.4.4) is depicted in Fig. 5.3. The *intrinsic connectivity structure* of the graph is defined solely by the pairings affected by the potentials in (5.4.4). Here, we will take $g = G_0 = (z - H_0)^{-1}$, the free Green function, but other choices are possible.[7] Then, T_i is simply the collection of all N-particle graphs that end on the left

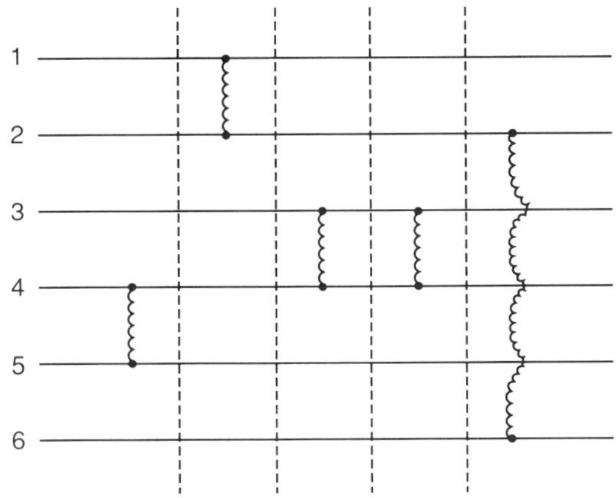

FIG. 5.3. $N = 6$ graph representing the product $V_{45}gV_{12}gV_{34}gV_{34}gV_{26}$. This graph has an intrinsic connectivity structure corresponding to the two-cluster partition (126) (345) (from Kowalski (1983b)).

with the pair interaction V_i. Note that we do not require that the naive sum of the analytical representations of these graphs converge to T_i. In the end, the correct "summation" of these graphs is affected by the integral equations we obtain.

Our definition of the intrinsic connectivity of a graph means that to each graph there is a unique partition of the N-particle system determined by the potentials in the expression (5.4.4). If an n_b-cluster graph corresponds in this way to a partition b, the graph is said to have *g-referenced connectivity b*. When $g = G_0$, one then encounters the customary *string connectivity*,[7] in which the n_b clusters of the graph are translationally invariant with respect to one another. This was the notion of connectivity used in Section 5.3, and it is what we mean by "connectivity" when we use the term without a qualifier. Generally, a graph with intrinsic connectivity b may decompose into a finite number of terms, some of which have string connectivities corresponding to fewer than n_b clusters; this is not possible for a graph with string connectivity b.

The operator T_i possesses the connectivity decomposition

$$T_i = \sum_{a \in P} [T_i]_a, \tag{5.4.5}$$

where P is the set of all partitions of the N particles, and $[T_i]_a$ is the a-connected part of T_i, i.e., the sum of all a-connected graphs that end with the pair i on the left. Connectivity expansions, along with the associated problem of determining the operators $[T_i]_a$, are analyzed in detail in the Appendix. Only the fully-connected piece $[T_i]_{\underline{1}}$ of T_i involves the full N-particle dynamics: the lower-connectivity components $[T_i]_a$, $a \neq \underline{1}$, are calculable in terms of dynamics involving less than N particles.

The crucial step in deriving C-K equations for the T_i consists in the observation that any graph that contributes to $[T_i]_{\underline{1}}$ has the unique form $L(\gamma)R(r)$ of a product of a two-cluster graph $L(\gamma)$ and a graph $R(r)$ that ends on the left with a pair potential external to γ, so that the partition r is not contained in γ. An example of this is shown in Fig. 5.4. We say that a partition b is *not contained in*, a partition a, denoted $a \not\supset b$, if b cannot be obtained by breaking the clusters of a; if b can be obtained from a in this way, we say that b is *contained in a*, and denote this by $a \supset b$. Then, using the graphical interpretation of T_i implied by Eq. (5.4.3), we see that

$$[T_i]_{\underline{1}} = \sum_j \left\{ \sum_\gamma [T_i]_\gamma \bar{\Delta}_{\gamma j} \right\} G_0 T_j, \tag{5.4.6}$$

where the sums are over all two-cluster and pair partitions γ and j, respectively, and $\bar{\Delta}_{\gamma j} = 1$ if $\gamma \not\supset j$, and is equal to zero otherwise. Inserting the expression (5.4.6) into the expansion (5.4.5) gives us the Rosenberg equations[8]

$$T_i = \sum_c \bar{\delta}_{n_c,1} [T_i]_c + \sum_j \left\{ \sum_\gamma [T_i]_\gamma \bar{\Delta}_{\gamma j} \right\} G_0 T_j. \tag{5.4.7}$$

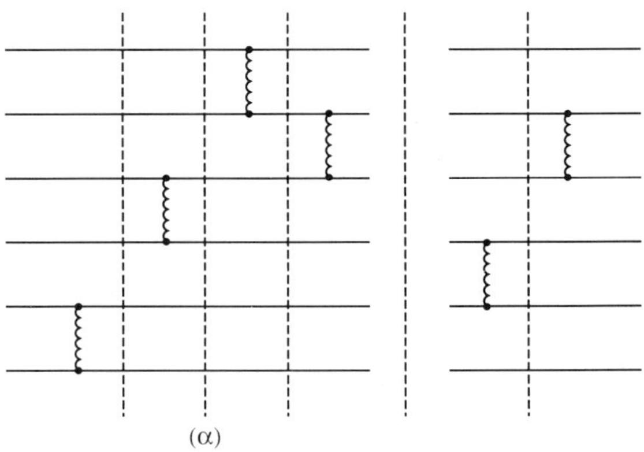

FIG. 5.4. Unique separation of a connected graph into an almost-connected graph on the left and a graph which has its leftmost interaction external to α (from Kowalski (1982c)).

Since products such as $[T_i]_\gamma \bar{\Delta}_{\gamma j} G_0 [T_j]_\beta$ are fully connected, the kernels of the Eqs. (5.4.7) become connected after a single iteration. For $N = 3$, the Eqs. (5.4.7) are identical to the Faddeev equations (5.3.44).

Given the solutions T_i of the coupled equations (5.4.7), one can calculate any physical transition operator. For example, we have the following relationship to the post form (5.2.16a):

$$T_{ba}^{(+)} = \sum_j \bar{\Delta}_{bj} T_j G_0 \overleftarrow{G}_a^{-1}, \tag{5.4.8}$$

since

$$V^b = \sum_j \bar{\Delta}_{bj} V_j. \tag{5.4.9a}$$

The last equation generalizes Eq. (5.3.5) to N bodies, and is a consequence of the decompositions

$$\begin{aligned} V &= \sum_i V_i \\ &= V_a + V^a, \end{aligned} \tag{5.4.9b}$$

since the channel interaction V_a has the form

$$V_a = \sum_j \Delta_{aj} V_j, \tag{5.4.9c}$$

where $\Delta_{aj} = 1 - \bar{\Delta}_{aj} = 1$ when $a \supseteq j$, and is zero otherwise. Further properties of the matrices $\underline{\Delta}$ and $\bar{\Delta}$ are collected in Section A.5a.

5.4b. *Bencze–Redish–Sloan Equations*

An interesting challenge is to obtain integral equations for the transition operators $T_{ba}^{(+)}$ from the Rosenberg equations. From the form of Eq. (5.4.8), it appears that only the operators $T_{\gamma a}^{(+)}$ would appear in the kernels of the equations one would obtain by multiplying Eq. (5.4.7) by $\bar{\Delta}_{bi} G_0 \overleftarrow{G}_a^{-1}$ on the right and summing over all i:

$$T_{ba}^{(+)} = \sum_c \bar{\delta}_{n_c,1} W^{b0}(c) G_0 \overleftarrow{G}_a^{-1} + \sum_\gamma W^{b0}(\gamma) G_0 T_{\gamma a}^{(+)}, \tag{5.4.10}$$

where

$$W^{b0}(c) \equiv \sum_i \bar{\Delta}_{bi} [T_i]_c. \tag{5.4.11}$$

Since the operators $W^{b0}(\gamma) G_0 W^{\gamma 0}(\alpha)$ are connected, the Eqs. (5.4.10) have connected kernels after a single iteration. This is the simplest argument leading to the closed system of C-K equations for the two-cluster indexed operators $T^{(+)}_{\beta\alpha}$, first found by Bencze, Redish, and Sloan (BRS):[9]

$$T^{(+)}_{\beta\alpha} = \sum_c \bar{\delta}_{n_c,1} W^{\beta 0}(c) G_0 \overleftarrow{G}_\alpha^{-1} + \sum_\gamma W^{\beta 0}(\gamma) G_0 T^{(+)}_{\gamma\alpha}. \qquad (5.4.12)$$

An instructive exercise is to show that the BRS equations reduce to the Faddeev–Lovelace equations (5.3.14) when $N = 3$, and so represent their N-particle generalization. Here the "pair" partitions i are the same as the two-cluster partitions α.

Generalizations, called *sigma-set equations*, of the BRS equations to include operators $T^{(+)}_{c\alpha}$, where $n_c > 2$, in the kernel are possible, but rest on the existence of left-hand inverses of the $\bar{\Delta}$ matrices.[10] Given these inverses, one can invert the BRS equations to recover the Rosenberg equations thus, as in the $N = 3$ case where $\bar{\delta} = \bar{\Delta}$, establishing the one-to-one correspondence of the two. Other generalizations involving intrinsic connectivities other than the string type, multiparticle forces, and different subsystem operators have been obtained;[11] these extensions include a C-K version of Watson multiple-scattering theory.[12]

The original BRS equations have been studied by a number of investigators in connection with various possible applications. In order to understand this work, we will have to develop the content of Eqs. (5.4.10–5.4.12) in greater detail.

The Eqs. (5.4.10), of course, are straightforward consequences of the Rosenberg equations. There have been two major advances beyond this. The principal one was to identify the potential-distribution matrix, which we call $\mathscr{V}_C$, that generates Eqs. (5.4.10–5.4.12).[13] The other had to do with the investigation and simplification of the inhomogeneous terms in the Eqs. (5.4.12), in particular.

The identification of $\mathscr{V}_C$ follows from the properties (Section A.5a) of the matrix $\underline{\Delta}$, one of the most important of which is that it possesses an inverse. Since $\Delta_{\underline{1}b} = 1$ for any b, we obtain, from $(\underline{\Delta}^{-1}\underline{\Delta})_{\underline{1}b} = \delta_{\underline{1}b}$, the sum rule

$$\sum_a C_a \bar{\delta}_{a\underline{1}} \Delta_{ab} = \bar{\delta}_{\underline{1}b}, \qquad (5.4.13)$$

where[13,14] $C_a \equiv -(\underline{\Delta}^{-1})_{\underline{1}a}$, $a \neq \underline{1}$, and is equal to $(-1)^{n_a}(n_a - 1)!$ for a partition with n_a clusters. Henceforth, unless explicitly indicated to the contrary, we exclude the $n_{\underline{1}} = 1$ partition from our matrix expressions. With this understood, the sum rule (5.4.13) can be rewritten in matrix form as

$$\mathscr{S}\underline{C}\underline{\Delta} = \mathscr{S}, \qquad (5.4.14a)$$

so

$$\underline{\Delta}^t \underline{C}\mathscr{S} = \mathscr{S}, \qquad (5.4.14b)$$

where $(\underline{C})_{ab} = C_a \delta_{ab}$ and $(\underline{\Delta}^t)_{ab} = \Delta_{ba}$. We see then that

$$\underline{W}_{\text{BRS}} \equiv \underline{\Delta}^t \underline{C} \tag{5.4.15}$$

is a channel-coupling array (*cf*. Eq. (5.2.26)).

Next, we note that we can rewrite Eq. (5.4.9a) in matrix form as

$$\bar{V}\mathscr{S} = \bar{\Delta}\underline{P}_{N-1}\underline{V}\mathscr{S}, \tag{5.4.16}$$

where $(\underline{P}_{N-1})_{ab} = \delta_{n_a, N-1}\,\delta_{ab}$ is the projector on the $(N-1)$-cluster partitions labeling the pair interactions. Thus, we can identify an infinity of channel-coupling schemes that have the form

$$\mathscr{V} = \bar{\Delta}\underline{P}_{N-1}\underline{V}\underline{W}, \tag{5.4.17}$$

where $\underline{W}\mathscr{S} = \mathscr{S}$ is an arbitrary channel-coupling array. Obviously, not all of the arrays, e.g., $\underline{W} = \underline{\delta}$, are interesting, for we want $\mathscr{V}$ to generate a C-K equation. As we have signaled, the array that leads to the BRS equations is $\underline{W}_{\text{BRS}}$, and we next verify this.

Inserting $\underline{W}_{\text{BRS}}$ into Eq. (5.4.17), we find

$$\mathscr{V}_{\text{C}} \equiv \bar{V}_{\text{M}}\underline{C}, \tag{5.4.18}$$

where

$$\bar{V}_{\text{M}} \equiv \bar{\Delta}\underline{P}_{N-1}\underline{V}\underline{\Delta}^t, \tag{5.4.19}$$

defines the "mixed" (M) interaction matrix. (For $N = 3$, $\bar{\Delta}_{\alpha i}\Delta_{\beta i} = \bar{\delta}_{\alpha\beta}\Delta_{\beta i}$.) The kernel of Eq. (5.4.1) now becomes

$$\underline{K}_{\text{C}} \equiv \mathscr{V}_{\text{C}}\underline{G}. \tag{5.4.20}$$

The crucial thing is to establish that some power of $\underline{K}_{\text{C}}$ is connected.[15]

In order to prove this, we note that

$$\underline{K}_{\text{C}} = \underline{\tau}_{\text{M}}\underline{C}G_0, \tag{5.4.21}$$

where

$$\underline{\tau}_{\text{M}} \equiv \bar{V}_{\text{M}}(\underline{\delta} + \underline{G}\underline{V}). \tag{5.4.22}$$

According to Eqs. (A.5.18–A.5.19), any operator, and $\underline{\tau}_{\text{M}}$ in particular, possesses a connectivity expansion:

$$\underline{\tau}_{\text{M}} = \underline{M}\,\underline{\Delta}^t, \tag{5.4.23}$$

where

$$(\underline{M})_{bd} = W^{b0}(d) \tag{5.4.24}$$

is the d-connected part of $(\underline{\tau}_{\mathrm{M}})_{b1} = V^b(I + GV)$, is identical to the quantity defined by Eq. (5.4.11). (Unfortunately, both the operators (5.4.24) and the channel-coupling arrays are conventionally denoted by W; however, their matrix representations and the context of their use are distinct). Thus,

$$\underline{K}_{\mathrm{C}} = \underline{M} G_0 \underline{W}_{\mathbf{BRS}}. \tag{5.4.25}$$

The unique property of $\underline{W}_{\mathbf{BRS}}$, which distinguishes it from other channel-coupling arrays, is that is also satisfies (*cf.* Eq. (A.5.16))

$$\underline{W}_{\mathbf{BRS}} \bar{\Delta} P_{N-1} = \underline{P}_2 \bar{\Delta} \underline{P}_{N-1}, \tag{5.4.26}$$

where $(\underline{P}_2)_{ba} = \delta_{ba}\,\delta_{n_a,2}$ is the projector onto the two-cluster partitions. Since $\underline{M}$ is proportional to $\bar{\Delta}\underline{P}_{N-1}$, we immediately see that

$$\underline{W}_{\mathbf{BRS}} \underline{K}_{\mathrm{C}} = \underline{P}_2 \underline{K}_{\mathrm{C}}. \tag{5.4.27}$$

Combining Eqs. (5.4.25) and (5.4.27), we obtain[15]

$$(\underline{K}_{\mathrm{C}})^2 = \underline{M} G_0 \underline{P}_2 \underline{K}_{\mathrm{C}}. \tag{5.4.28}$$

In component form, this becomes

$$[(\underline{K}_{\mathrm{C}})^2]_{ba} = \sum_{\gamma, d} W^{b0}(\gamma) G_0 W^{\gamma 0}(d) G_0 (\underline{W}_{\mathbf{BRS}})_{da}, \tag{5.4.29}$$

which a is sum of connected operators. Thus, $\mathscr{V}_{\mathrm{C}}$ leads to a C-K equation of the form (5.4.1).

It remains to show how the C-K equation can be re-expressed in the form (5.4.10). This is accomplished by using the resolvent equation,

$$\underline{R}_{\mathrm{C}} = \underline{K}_{\mathrm{C}} + \underline{K}_{\mathrm{C}} \underline{R}_{\mathrm{C}}, \tag{5.4.30a}$$

to show that

$$\underline{W}_{\mathbf{BRS}} \underline{R}_{\mathrm{C}} = \underline{P}_2 \underline{R}_{\mathrm{C}}. \tag{5.4.30b}$$

Equation (5.4.30a) then becomes

$$\underline{R}_{\mathrm{C}} = \underline{K}_{\mathrm{C}} + \underline{M} G_0 \underline{P}_2 \underline{R}_{\mathrm{C}}. \tag{5.4.30c}$$

The relevant form of Eq. (5.4.1),

$$\underline{T}^{(+)} = \mathscr{V}_{\mathrm{C}} \underline{G} \mathscr{S} \overleftarrow{G}^{-1} + \underline{K}_{\mathrm{C}} \underline{T}^{(+)}, \tag{5.4.31a}$$

whose solution is

$$\underline{T}^{(+)} = \{\underline{\delta} + \underline{R}_{\mathrm{C}}\} \mathscr{V}_{\mathrm{C}} \underline{G} \mathscr{S} \overleftarrow{\underline{G}}^{-1}, \tag{5.4.31b}$$

can then be rewritten as

$$\underline{T}^{(+)} = \{\underline{\delta} + \underline{M} G_0 [\underline{W}_{\mathbf{BRS}} - \underline{P}_2]\} \mathscr{V}_{\mathrm{C}} \underline{G} \mathscr{S} \overleftarrow{G}^{-1} + \underline{M} G_0 \underline{P}_2 \underline{T}^{(+)}. \tag{5.4.32}$$

Using the identity (5.4.26), we see that this last equation reduces to

$$\underline{T}^{(+)} = \underline{M}G_0\mathscr{S}\overleftarrow{G}^{-1} + \underline{M}G_0\underline{P}_2\underline{T}^{(+)}, \tag{5.4.33}$$

i.e., Eq. (5.4.10) in matrix form.

We can employ Eqs. (5.4.10) and (5.4.33) as quadrature rules for the operators $T_{b\alpha}^{(+)}$, where $n_b > 2$. Thus, for studying two-cluster initiated reactions, the minimally coupled form (5.4.12) suffices. Since $T_{\beta\alpha}^{(+)}$ operates on $|\phi_\alpha\rangle$ when we calculate physical matrix elements, it is also sufficient to suppose that the Eqs. (5.4.12) are half-on-shell. The relevant inhomogeneous term then has the form

$$B_{\beta\alpha}|\phi_\alpha\rangle \equiv \sum_d (\mathscr{V}_C)^{\beta d}\,\bar{\delta}_{d\underline{1}}\,G_d\overleftarrow{G}_\alpha^{-1}|\phi_\alpha\rangle, \tag{5.4.34}$$

which, after using the Lippmann identity, simplifies to

$$B_{\beta\alpha}|\phi_\alpha\rangle = V_\alpha^\beta|\phi_\alpha\rangle, \tag{5.4.35}$$

where $V_\alpha^\beta = (\bar{V}_M)_{\beta\alpha}$. The justification of the use of the Lippmann identity in the inhomogeneous term of an integral equation is a matter of some subtlety.[13, 16] Its usage in connection with the integral equation (5.4.33) is proper because $\underline{M}G_0\underline{P}_2\mathscr{V}_C$ is connected, so that in none of the higher-order iterations of the inhomogeneous term in Eq. (5.4.33) is there an operator that can cancel the vanishing of $G_d\overleftarrow{G}_\alpha^{-1}|\phi_\alpha\rangle$ for $d \neq \alpha$.

Although we have stated and used the Lippmann identity a number of times in Chs. 1, 4 and in present chapter, only at this point is it appropriate to state the general form[17] of this identity:

$$G_d^{(+)}(E)\overleftarrow{G}_a^{-1}|\phi_a(E)\rangle = \delta_{da}|\phi_a(E)\rangle, \tag{5.4.36}$$

for $a \supseteq d$ and $2 \le n_a \le N-1$. When $a = \alpha$ is a two-cluster partition, then Eq. (5.4.36) is true for all $d \neq \underline{1}$.

We see then that, in order to calculate all two-cluster-initiated transition amplitudes, we need consider only the transition amplitudes $\underline{T}$ that satisfy

$$\underline{T} = \bar{V}_M + \underline{M}G_0\underline{P}_2\underline{T}. \tag{5.4.37}$$

The restriction of this set to two-cluster partitions yields what might be called the final form of the BRS equations:

$$T_{\beta\alpha} = V_\alpha^\beta + \sum_\gamma W^{\beta 0}(\gamma)G_0T_{\gamma\alpha}, \tag{5.4.38}$$

where $T_{\beta\alpha}$ is half-on-shell equivalent to $T_{\beta\alpha}^{(+)}$.

In what may be viewed[9] as the off-shell transformation that led to Eqs. (5.4.37–5.4.38), some remarkable characteristics have been acquired by the new set of transition operators. For example, all of the disconnected structure of $T_{\beta\alpha}$ resides in V_{α}^{β}. This has many unusual consequences, including the fact that in Born approximation $T_{\alpha\alpha}$ vanishes. The major implication of this is that much of the intuition for constructing approximations that is based on the off-shell extensions $T_{\beta\alpha}^{(\pm)}$ has to be developed anew in dealing with $T_{\beta\alpha}$ because most of the physics resides in the highly structured kernels $W^{\beta 0}(\gamma)G_0$. This means new strategies are required for approximation scenarios rather than implying some intrinsic shortcoming of this approach.[18]

How can one avoid arriving at the form (5.4.38) of the BRS? One way is to introduce an auxiliary operator

$$\underline{\tau}_{\mathrm{u}} \equiv \bar{V}_{\mathrm{u}} + \bar{V}\mathscr{S}G\bar{V}, \tag{5.4.39}$$

where

$$\bar{V}_{\mathrm{u}} = \bar{\Delta}\underline{P}_{N-1}\underline{V}\bar{\Delta}^{t}, \tag{5.4.40}$$

This move is motivated by the fact that, since $\bar{\Delta} + \underline{\Delta} = \mathscr{S}$, we have $\bar{V}\mathscr{S} = \bar{V}_{\mathrm{M}} + \bar{V}_{\mathrm{u}}$ and so

$$\underline{T}^{(+)} = \bar{V}_{\mathrm{M}} + \underline{\tau}_{\mathrm{u}}. \tag{5.4.41}$$

We indicate in Section A.5a how to show that $\tau_{\mathrm{u}}^{\beta\alpha} = (\underline{\tau}_{\mathrm{u}})_{\beta\alpha}$ satisfies integral equations with the same kernels as the BRS equations (5.4.38), but with drastically different inhomogeneous terms:

$$\tau_{\mathrm{u}}^{\beta\alpha} = \sum_{m=2}^{N-1} \sum_{d(m)} W^{\beta\alpha}[d(m)] + \sum_{\gamma} W^{\beta 0}(\gamma)G_0\tau_{\mathrm{u}}^{\gamma\alpha}. \tag{5.4.42}$$

Here,

$$W^{\beta\alpha}[d(m)] = \sum_{ij} \bar{\Delta}_{\beta i} M_{ij}[d(m)]\, \bar{\Delta}_{\alpha j}, \tag{5.4.43}$$

where $M_{ij}[d(m)]$ is the connected part of the operator (5.4.2) and the partition $d(m)$ has m clusters. We now see that the conventional Born approximation $T_{\beta\alpha}^{(+)} \simeq V^{\beta}$ is obtained from Eq. (5.4.41) if we keep only the Born approximations to the $d(N-1)$-connected terms of the inhomogeneous part of Eq. (5.4.42).

The principal advantage of the set (5.4.41–5.4.42), which have been referred to as the PLUMS equations,[19] is that is works within a conventional off-shell extension. This leads to a relatively familiar operator structure and framework for the statement of approximations. This is crucial for transition-operator statements of approximation scenarios that are dependent in their form on the off-shell extension chosen, as the Born-approximation example indicated.

Off-shell extensions play a less explicit role in wave-function formalisms. Therefore, it is interesting to see what options present themselves in this regard using $\mathscr{V}_{\mathrm{C}}$. First, we consider the Eqs. (5.2.37), which in this case become

$$|\chi_b^{(+)}(\alpha)\rangle = \delta_{b\alpha}|\phi_\alpha\rangle + G_b \sum_d (\mathscr{V}_{\mathrm{C}}^t)^{bd} |\chi_d^{(+)}(\alpha)\rangle. \tag{5.4.44}$$

The kernel of these equations in matrix form is

$$\underline{K}_{\mathrm{C}}^t \equiv \underline{G}\mathscr{V}_{\mathrm{C}}^t. \tag{5.4.45}$$

The proof that $(\underline{K}_{\mathrm{C}}^t)^2$ is connected follows in the same way as for $(\underline{K}_{\mathrm{C}})^2$, except now $\underline{W}_{\mathrm{BRS}}^t$ functions as a left-handed channel-coupling array (*cf.* Eq. (5.4.14a)). Thus, the Eqs. (5.4.44) are C-K equations for the wave-function components $|\chi_b^{(+)}(\alpha)\rangle$. However, in marked contrast to the transition operator BRS equations, we obtain no restriction of the sum in Eq. (5.4.44) to only two-cluster partitions.

Wave-function equations of this C-K type can be obtained[16] by using the Eqs. (5.4.37) as quadrature rules for the components $T_{0\alpha}$. This is relevant because $|\psi_\alpha^{(+)}\rangle = G_0 T_{0\alpha}|\phi_\alpha\rangle$. Then, we obtain

$$|\psi_\alpha^{(+)}\rangle = \sum_\gamma |\psi_\alpha^{(\gamma)}\rangle, \tag{5.4.46}$$

where

$$|\psi_\alpha^{(\gamma)}\rangle \equiv \delta_{\gamma\alpha}|\phi_\alpha\rangle + G_0 W^{00}(\gamma) G_0 T_{\gamma\alpha}|\phi_\alpha\rangle. \tag{5.4.47}$$

We then infer, using the identity

$$T_{\gamma\alpha}|\phi_\alpha\rangle = \vec{G}_\gamma^{-1}[|\psi_\alpha^{(+)}\rangle - \delta_{\gamma\alpha}|\phi_\alpha\rangle], \tag{5.4.48}$$

the set of minimally coupled equations[16]

$$|\psi_\alpha^{(\gamma)}\rangle = \delta_{\gamma\alpha}|\phi_\alpha\rangle + g_\gamma v_\gamma \sum_\lambda \bar{\delta}_{\gamma\lambda}|\psi_\alpha^{(\lambda)}\rangle, \tag{5.4.49}$$

with

$$v_\gamma \equiv W^{00}(\gamma) G_0 \vec{G}_\gamma^{-1}, \tag{5.4.50}$$

and

$$g_\gamma = (z - H_0 - v_\gamma)^{-1}. \tag{5.4.51}$$

5.4c. *Kouri–Levin–Tobocman Equations*

The Kouri–Levin–Tobocman (KLT)[3,20] collision integral equations are by far the simplest C-K equations of the form (5.4.1) in their derivation and their final form. We will review some of their properties.

Some aspects of KLT theory were anticipated by Avishai,[21] Glöckle,[22] and Lippmann.[23] However, perhaps the most important feature introduced by KLT in connection with transition-operator equations[24] was the exploitation of the idea of Baer and Kouri[3] that the interactions V_b did not have to be written as the usual simple, unity-weighted sum of pair interactions. This is the concept realized in Eqs. (5.2.22–5.2.26). The motivation for doing this was that it provides the opportunity for generating infinite classes of collision integral equations with alternative distributions of the potential. Some of these equations may be more useful than others for specific purposes.

The contribution of KLT was to find a particular distribution W that leads to C-K equations. Specifically, KLT chose special W-matrices called channel-permuting arrays (CPA) that are equal to the identity except on the subspace of two-cluster partitions. A typical CPA has elements

$$W_{ab} = \delta_{ab}\,\bar{\delta}_{n_a,2}, \tag{5.4.52a}$$

$$W_{\alpha_i,\beta} = \delta_{\hat{\alpha}_{i+1},\beta}, \tag{5.4.52b}$$

where

$$\hat{\alpha}_{i+1} = \alpha_{i+1}\,\bar{\delta}_{i,N_2} + \alpha_1\,\delta_{i,N_2}, \tag{5.4.52c}$$

and the set $\{\alpha_i;\, i = 1, \ldots, N_2\}$ represents an enumeration of the N_2 two-cluster partitions. Thus, on the two-cluster subspace a CPA is an $N_2 \times N_2$ matrix such that each column and each row has only one nonvanishing element, and that element never lies on the diagonal. The $(N_2 - 1)!$ permutations of the rows, for example, which do not place an element on the diagonal generate the set of all $N_2 \times N_2$ channel-permuting arrays. It is clear that Eq. (5.2.26) is satisfied.

The set of all possible KLT equations is given by Eqs. (5.2.27a–d), to within nondynamical off-shell transformations, such as using the AGS operators or simplifying via the Lippmann identity, with

$$\mathscr{V}_{\mathrm{KLT}} = \bar{V}\underline{W}_{\mathrm{CPA}}. \tag{5.4.53}$$

We have, for example, using Eqs. (5.4.52a–c) to define $\underline{W}_{\mathrm{CPA}}$, and restricting oneselves to two-cluster partitions,

$$T^{(+)}_{\alpha_i\beta} = V^{\alpha_i}G_{\hat{\alpha}_{i+1}}\overleftarrow{G}_{\beta}^{-1} + V^{\alpha_i}G_{\hat{\alpha}_{i+1}}T^{(+)}_{\hat{\alpha}_{i+1}\beta}. \tag{5.4.54}$$

By using the Lippmann identity in the usual way, we see that the operators $T_{\alpha_i\beta}$ that satisfy

$$T_{\alpha_i\beta} = V^{\alpha_i}\,\delta_{\hat{\alpha}_{i+1}\beta} + V^{\alpha_i}G_{\hat{\alpha}_{i+1}}T_{\hat{\alpha}_{i+1}\beta}, \tag{5.4.55}$$

are physically equivalent to the $T^{(+)}_{\alpha_i\beta}$. If we now make $N_2 - 1$ iterations of either Eq. (5.4.54) or Eq. (5.4.55), we obtain a single equation for $T^{(+)}_{\alpha_i\beta}$ or $T_{\alpha_i\beta}$, respectively, with a connected kernel. This kernel has a structure very similar to that of the kernel of Eq. (5.2.50) and both are connected for the same reason.

The Eqs. (5.4.55) are of the form that are usually called KLT equations. They have a number of properties that contrast with other C-K equations. First, their form and the argument establishing the C-K property stay the same whether or not there are many-body forces included in the interactions V^b. Thus, we can say Eqs. (5.4.54–5.4.55) are "invariantly" minimally coupled in that they involve only two-cluster partitions independently of the possible multiparticle structure of the interactions. The KLT equations are distinguished among the known partition-labeled equations in this respect. For example, the BRS equations lose their minimal coupling property once multiparticle interactions are included.[25] This property of invariant minimal coupling has been shown to be an important advantage of the KLT formalism for constructing few-body Hamiltonian models of nuclear reactions.[26]

The simplicity of the Eqs. (5.4.55) is also attractive as a starting point for so-called bound-state approximations. In such schemes, only the contributions of the cluster-states $|\phi_\alpha\rangle$ are retained in the spectral representation of the Green functions G_α. Actually, since one will obtain C-K equations after making this approximation no matter what sort of W array is chosen in Eq. (5.4.53), the significance of the C-K property of the KLT equations is unclear. These points are discussed with reference to calculational performance in Section 9.4.

Other features of the KLT equations have limited its usefulness. For example, the very simplicity of the kernel and driving terms of the Eqs. (5.4.55) means that it is difficult to incorporate approximations, and the simplifications that result from them, to the subsystem dynamics. No calculations, even for $N = 3$, have been carried out to date with these equations with all of the multiparticle degrees of freedom taken into account. In Section 5.4f, it is shown that permutation symmetry also presents a problem.

5.4d. *Other Partition-Labeled Equations*

A number of collision integral equations have been proposed for $N \geq 4$ that are basically different realizations of the Rosenberg–Weinberg approach. Those that were established only for $N = 4$ have not played a role in subsequent developments.[27]

Two sets of N-particle integral equations were developed by Weyers.[28] One set (W-1) shares with the Rosenberg equations the property of having $N(N-1)/2$ constituents, but for the *physical operators* $T_{ib}^{(-)}$, where i labels a pair and b an arbitrary (fixed) partition. These equations are derived using what appears to be the first implicit use of the channel-coupling-scheme idea, and for this alone they merit special mention. The W-1 equations, however, are different from the Rosenberg equations even when the later are re-expressed in terms of the $T_{ib}^{(-)}$. For example, their kernels contain fully connected, as well as disconnected pieces, unlike the two-cluster structure of the Rosenberg kernels. The kernels of the W-1 equations become fully connected upon iteration.

Weyers has also proposed a coupling scheme incorporating all of the two-cluster partitions that yield a set (W-2) of only N equations for the physical transition operators $T_{\alpha_1 b}^{(-)}$, b fixed, where α_1 refers to two-cluster partitions for which one of the clusters has only a single particle, as would be relevant for elastic scattering from a composite target. These equations seem to be equivalent to what one would obtain by using Bencze's decoupling trick[29] to express the Rosenberg-equivalent equations for $T_{ib}^{(-)}$ in terms of $T_{\alpha_1 b}^{(-)}$. The W-2 and W-1 kernels have similar properties, and despite different clustering structures, involve no more complicated input than the Rosenberg equations.

Connected-kernel equations for the full Green function are interesting for several reasons. One is historical importance: The first C-K equations for $N > 3$ that were proposed, independently, by van Winter[30] and Weinberg[5] are of this type. From a modern perspective, these and similar equations can be obtained in an efficient manner from the iterated form of (*cf*. Eqs. (5.2.11a) and (5.2.22))

$$G\mathscr{S} = \underline{G}\mathscr{S} + \underline{G}\mathscr{V}G\mathscr{S}, \tag{5.4.56}$$

namely

$$G\mathscr{S} = \underline{G}[I + \underline{A}(p)]\mathscr{S} + \underline{G}(\mathscr{V}\underline{G})^p\bar{V}G\mathscr{S}, \tag{5.4.57}$$

which is obtained from Eq. (5.4.56) after p iterations, and where

$$\underline{A}(p) \equiv \sum_{n=1}^{p} (\mathscr{V}\underline{G})^n. \tag{5.4.58}$$

If $\mathscr{V} = \mathscr{V}_C$, the kernel of Eq. (5.4.57) is connected for $p = 1$ and, as pointed out by Bencze and Tandy,[31] the resultant equation for $[G\mathscr{S}]_{0b}$ is exactly the van Winter–Weinberg equation for G. (Note that $[\mathscr{V}_C\underline{G}]_{0b} = V_b C_b G_b$.)

Another version of Eq. (5.4.57) with a completely connected kernel is obtained with $\mathscr{V}_{KLT}$.The $[G\mathscr{S}]_{0b}$ component is then equivalent to the C-K equation proposed by Avishai.[21]

We observe that any coupling scheme that might be called[32] "simply connected," namely there is an integer p such that $(\mathscr{V}_C \underline{G})^p \bar{V}$ is fully connected but $\underline{A}(p)$ is not, will yield the (same) van Winter–Weinberg equation in the same manner as for $\mathscr{V}_C$. Note that $\mathscr{V}_{KLT}$ is not simply connected, so that the Avishai C-K equation for G is different from the van Winter–Weinberg equation.

5.4e. *Spurious Solutions*

Let us consider a generic C-K collision integral equation for the transition operators of the form

$$T = B + KT, \tag{5.4.59}$$

where any possible discrete-index matrix structure is left implicit. We suppose that K is a Fredholm kernel. The constrained Fredholm alternative (C.F.A.) is the situation where all nontrivial solutions $|\chi\rangle$ of the homogeneous form of Eq. (5.4.59),

$$|\chi\rangle = K|\chi\rangle, \tag{5.4.60}$$

are in one-to-one correspondence with the bound-state solutions of the Schrödinger equation. A (set of) C-K integral equation(s) that does *not* satisfy a C.F.A. is said to possess *spurious* solutions. The Faddeev–Yakubovskii equations satisfy a C.F.A., so that the possibility of spurious solutions is meaningful only for the types of equations considered in this section so far.

Shortly after the van Winter–Weinberg equations were proposed, Federbush[33] found spurious solutions of them for $N = 3$ in a special, unphysical, model. In this model, one particle is fixed, while the other two are unconstrained, but interact only with the fixed particle, via separable interactions, and not with each other. The N-particle version of the model has been given by Chandler[34] along with sufficient conditions for spurious solutions for several equations of the type (5.4.59).

The possibility of spurious solutions of Eq. (5.4.59), e.g., was related by Newton[35] to the existence of the factorization property

$$(I - K) = A(z - H), \tag{5.4.61}$$

where z is the relevant parametric energy. That is, if there is a vector $|\chi\rangle$ such that $(z - H)|\chi\rangle \neq 0$, but $A(z - H)|\chi\rangle = 0$, then $|\chi\rangle$ is a spurious solution.

This factorization is related to the use of multipliers to reduce the LS equations to a C-K form.[36] As pointed out by Blankenbecler and Sugar,[37] if $(I - K)$ cannot be inverted by Fredholm methods, as in the LS case where K

has the form $V^{\alpha}G_{\alpha}$, then, provided L^{-1} and $[R(I - K)L]^{-1}$ exist, a solution of Eq. (5.4.59) is $T = L[R(I - K)L]^{-1}RB$. The general possibilities for spurious solutions arising from multipliers can be appreciated from the case of the van Winter–Weinberg equation for $N = 3$:

$$G = [R(I - G_0 V)]^{-1} R G_0, \tag{5.4.62}$$

where the multiplier is

$$R = 1 + \sum_{\alpha=1}^{3} G_0 t_{\alpha}. \tag{5.4.63}$$

There is a possibility of a spurious solution if there are zero-eigenvalue eigenvectors of $R(I - G_0 V)$ that are not solutions of the Schrödinger equation.

The Green-function and transition-operator equations such as Eqs. (5.4.57) and (5.4.1), respectively, have a matrix structure in the partition indices. Let $\underline{K}$ denote a generic matrix kernel for either type of C-K equation, where we now explicitly denote the matrix structure by an underbar. Suppose we have, as a generalization of Eq. (5.4.61), the factorization property

$$(\underline{I} - \underline{K}) = (\underline{I} - \underline{K}_1)(\underline{I} - \underline{K}_2), \tag{5.4.64}$$

where either $(\underline{I} - \underline{K}_1)$ or $(\underline{I} - \underline{K}_2)$ has eigenvectors with zero eigenvalues that are in one-to-one correspondence with solutions of the Schrödinger equation. Then, spurious solutions correspond to vectors $|\underline{\chi}_{\mathrm{sp}}\rangle$ such that $(\underline{I} - \underline{K})|\underline{\chi}_{\mathrm{sp}}\rangle = 0$, but $(\underline{I} - \underline{K}_j)|\underline{\chi}_{\mathrm{sp}}\rangle \neq 0$, where $\underline{K}_j$ is the Schrödinger-related kernel. Explicit factorizations (5.4.64), for both $j = 1$ and $j = 2$, have been constructed for three-particle channel-coupling array wave-function and transition-matrix equations, and for the four-particle Rosenberg and Sloan equations.[38]

For the wave-function counterpart of Eq. (5.4.59),

$$|\underline{\chi}\rangle = |\underline{\phi}\rangle + \underline{K}'|\underline{\chi}\rangle, \tag{5.4.65a}$$

the condition for spurious solutions is

$$|\underline{\chi}_{\mathrm{sp}}\rangle = \underline{K}'|\underline{\chi}_{\mathrm{sp}}\rangle. \tag{5.4.65b}$$

Often the solution of Eq. (5.4.65a) is constrained so that the Schrödinger wave function $|\psi\rangle$ is $\{\mathscr{S}|\underline{\chi}\rangle\}_0$ where is 0 is the N-cluster partition. For many of the channel-coupling schemes of interest, all solutions of Eq. (5.4.65b) are ruled out except those for which $\mathscr{S}|\underline{\chi}_{\mathrm{sp}}\rangle = 0$.[34,38]

When multipliers are applied to obtain an equation such as (5.4.59), spurious solutions are only relevant with respect to the solution of the homogeneous form (5.4.60) or, equivalently, to the discrete spectrum of the multiplier R in the case that $L = 1$.[38] In solving the scattering problem, the effect of the multipler will cancel out, but one could encounter numerical instabilities if one of the discrete spurious solutions appears close to the scattering energy of interest.[38]

Spurious solutions can arise either from the multipliers that are introduced to invert the non-Fredholm LS equations or, equivalently, from the multiplications that are needed to obtain C-K equations of the form (5.4.1) from the Yakubovskii equations. All claims to date that they *cannot* exist in the BRS or KLT equations are false. It appears that they are not a concern for models of physical interest, and even if they did occur their effect can, at least in principle, be canceled out. However, if they exist and reflect a legitimate inadequacy in the formulation of the collision problem, then the C.F.A. property can always be restored by building in additional constraints.[34] It is also possible that their presence would be neutralized in the course of approximation, as happens for certain approximations to the LS equations in connection with their nonunique solution characteristics.[38] Nonetheless, the question of whether spurious solutions will affect practical physical applications of arbitrary C-K equations still remains somewhat obscure.[1]

5.4f. *Inclusion of Particle Identity*

We use the formalism of Section 1.11 to incorporate permutation symmetry and the associated statistics into integral equations of the form (5.4.1) in such a way as to preserve the C-K properties. We mean by "symmetrization" either Bose or Fermi statistics. For simplicity, we assume that all particles are identical.

We recall from Section 1.11 that the action of a permutation P on a partition a is another partition denoted as $P(a)$. A permutation cannot change the number of clusters in a partition. Also, if $a \supset b$ $[a \not\supset b]$, then $P(a) \supset P(b)$ $[P(a) \not\supset P(b)]$, so

$$n_{P(a)} = n_a, \tag{5.4.66a}$$

$$\Delta_{ab} = \Delta_{P(a),P(b)}, \tag{5.4.66b}$$

$$\bar{\Delta}_{ab} = \bar{\Delta}_{P(a),P(b)}. \tag{5.4.66c}$$

For any $P \in S$, where S is the group of permutations, the pair potentials V_i are label transforming (L.T.):

$$U(P)V_iU(P)^\dagger = V_{P(i)}. \tag{5.4.67}$$

This result coupled with Eqs. (5.4.66a–c) implies that all of the usual channel interactions are also L.T.:

$$U(P)V_aU(P)^\dagger = V_{P(a)}, \tag{5.4.68a}$$

$$U(P)V^aU(P)^\dagger = V^{P(a)}, \tag{5.4.68b}$$

$$U(P)V_b^aU(P)^\dagger = V_{P(b)}^{P(a)}. \tag{5.4.68c}$$

Therefore, each of the transition operators $T_{ba}^{(\pm)}$, U_{ba} is also L.T.

The next step is to investigate the L.T. property of the potential matrix $\mathscr{V}$. Now because

$$U(\underline{P})(\underline{W}_{\mathrm{CPA}})^{ab}U(P)^\dagger = (\underline{W}_{\mathrm{CPA}})^{ab}, \tag{5.4.69a}$$

but

$$(\underline{W}_{\mathrm{CPA}})^{ab} \neq (\underline{W}_{\mathrm{CPA}})^{P(a),P(b)}, \tag{5.4.69b}$$

we see that $\mathscr{V}_{\mathrm{KLT}}$ is *not* label transforming. Thus, one cannot apply that part of Section 1.11 that depends upon the L.T. property to obtain C-K versions of the KLT equations for the symmetrized transition operators.

There have been different methods that have been proposed to deal with this symmetrization problem. The first[39] collapses the KLT equations by successive substitution into a set involving only the operators $T_{\beta_{\mathrm{D}},\alpha}^{(+)}$, where β_{D} refers to the set of physically distinguishable two-cluster partitions. The resultant equations are summed over α with exchange-equivalent partitions weighted in the appropriate manner. This yields a set of C-K equations for the symmetrized transition operators, but with kernels and inhomogeneous terms more complicated than for the original equations. The other method[40] is similar to the Glöckle[41] approach outlined in Section 1.11; it yields C-K equations for the symmetrized transition operators, but with no reduction in the number of equations as compared to the unsymmetrized case.

On the other hand, because of Eq. (5.4.66a), we have $C_a = C_{P(a)}$, and so by Eq. (5.4.68c) we conclude that the potential matrix $\mathscr{V}_{\mathrm{C}}$ is L.T.:

$$\mathscr{V}_{\mathrm{C}}^{ab} = \mathscr{V}_{\mathrm{C}}^{P(a),P(b)}. \tag{5.4.70}$$

Since G_a is L.T., we then infer that $\underline{K}_{\mathrm{C}}$, as given by Eq. (5.4.20), is also L.T. Since, in contrast to $\underline{W}_{\mathrm{CPA}}$, we have from Eq. (5.4.15) that $\underline{W}_{\mathrm{BRS}}$ is L.T., our result for $\underline{K}_{\mathrm{C}}$ and Eq. (5.4.25) implies that $M_{bd} = W^{b0}(d)$ is L.T. as well.

The proof of this last statement is not trivial and serves as a good example of symmetrization technology. By definition (Eq. (A.5.22)), $W^{ba}(d)$ is the d-connected part of the operator $V^{ba} + V^bGV_a$. It then follows that ($d \neq \underline{1}$)

$$W^{ba}(d) = [t_d^{ba}]_d, \tag{5.4.71}$$

where, for an arbitrary partition f, $[t_f^{ba}]_d$ is the d-connected part of the L.T. operator

$$t_f^{ba} \equiv V_f^{ba} + V_f^b G_f V_f^a, \tag{5.4.72}$$

and

$$V_f^{ba} \equiv \sum_i \bar{\Delta}_{bi} \bar{\Delta}_{ai} V_i \Delta_{fi}. \tag{5.4.73}$$

We then infer that

$$W^{ba}(d) = \sum_f (\Delta^{-1})_{df} \bar{\delta}_{f\underline{1}} t_f^{ba}. \tag{5.4.74}$$

Since $\underline{\Delta}^{-1}\underline{\Delta} = I$, $(\Delta^{-1})_{df}$ is L.T., and so $W_{ba}(d)$ is L.T. This calculation illustrates that all connectivity structures are preserved under permutations.

The preceding results can now be used to write down C-K, symmetrized versions of the PLUMS equations (5.4.42) in terms of the class-labeled operators $\tilde{\tau}_{\mathrm{u}}^{\bar{\beta}\bar{\alpha}}$:

$$\tilde{\tau}_{\mathrm{u}}^{\bar{\beta}\bar{\alpha}} = \tilde{W}_{\mathrm{MS}}^{\bar{\beta}\bar{\alpha}} + \sum_{\bar{\gamma}} \tilde{K}^{\bar{\beta}\bar{\gamma}} \tilde{\tau}_{\mathrm{u}}^{\bar{\gamma}\bar{\alpha}}. \tag{5.4.75}$$

Here, $W_{\mathrm{MS}}^{\beta\gamma}$ is defined as the inhomogeneous term of Eq. (5.4.42) and $\tilde{W}_{\mathrm{MS}}^{\bar{\beta}\bar{\alpha}}$ is its symmetrized form defined by Eq. (1.11.49). The symmetrized kernel $\tilde{K}^{\bar{\beta}\bar{\gamma}}$ and $\tilde{\tau}_{\mathrm{u}}^{\bar{\beta}\bar{\alpha}}$ are defined analogously, where $K^{\beta\gamma} \equiv W^{\beta 0}(\gamma) G_0$. The sum in Eq. (5.4.75) is over the equivalance classes $\bar{\gamma}$ of the two-cluster partitions.

For three identical particles, there is just one equivalence class and the Eqs. (5.4.75) reduce to a single equation, which, of course, is consistent with our treatment of identity in Section 5.3. It is instructive to make the connection between the two treatments more explicit by considering the three-particle AGS equations (5.3.17). Let us choose $\alpha = 1$ as our canonical partition. The symmetrized AGS operator

$$U_{\mathrm{sym}} = R_1[\bar{U}(P_{11})U_{11} + \bar{U}(P_{12})U_{21} + \bar{U}(P_{13})U_{31}], \tag{5.4.76}$$

where R_1 is the symmetrizer on the particles 2 and 3, and $\bar{U}(P_{\alpha\beta})$ is the unitary permutation operator corresponding to the permutation $P_{\alpha\beta}$, including the permutation-parity factor. If we take matrix elements of Eq. (5.4.76) with respect to states constructed in the same "standard" fashion as those that entered into the argument summarized in Eqs. (5.3.81–5.3.83), we obtain a form similar to (5.3.82). In carrying out this calculation it is instructive to note that, e.g., $U_{31} = \bar{U}(P_{23}) U_{21} \bar{U}(P_{23})^{\dagger}$.

The inhomogeneous terms, $\bar{\delta}_{\alpha\beta} G_0^{-1}$, of the Eqs. (5.3.17) have the symmetrized form $\mathcal{N}_1 G_0^{-1}$, where

$$\mathcal{N}_1 \equiv R_1[\bar{U}(P_{12}) + \bar{U}(P_{13})]. \tag{5.4.77}$$

The matrix elements of $\mathcal{N}_1$ correspond to the exchange integrals that appear in a variety of approaches to identical-particle scattering.

Finally the symmetrized kernel is

$$\tilde{K}_1 \equiv \mathcal{N}_1 t_1 G_0. \tag{5.4.78}$$

Confidence in the meaning of the various exchange operators can be gained by verifying that $(\tilde{K}_1)^2$ is a connected operator.

As another example, we consider a set of auxiliary integral equations that may or may not be minimally coupled. Namely, for an arbitrary potential matrix $\mathscr{V}$ and a projector $\underline{Q}(\alpha)$ that is defined below, we consider the operators $\Gamma_{ba}(\alpha) \equiv [\underline{\Gamma}(\alpha)]_{ba}$ that satisfy

$$\underline{\Gamma}(\alpha) = \mathscr{V} + \mathscr{V}\underline{Q}(\alpha)\underline{G}\underline{\Gamma}(\alpha). \tag{5.4.79a}$$

The only properties of $\underline{Q}(\alpha)$ that we really need at the moment are that $[\underline{Q}(\alpha)]_{ba}$ is L.T. with respect to all three partition indices and that $[\underline{Q}(\alpha), \underline{G}] = \underline{0}$. This last implies that Eq. (5.4.79a) couples only $\underline{Q}(\alpha)\underline{\Gamma}(\alpha)\underline{Q}(\alpha)$ as does

$$\underline{\Gamma}(\alpha) = \mathscr{V} + \underline{\Gamma}(\alpha)\underline{Q}(\alpha)\underline{G}\mathscr{V}. \tag{5.4.79b}$$

Typically, we have

$$[\underline{Q}(\alpha)]_{ba} = \delta_{ba} - P_\alpha \delta_{ba} \delta_{\alpha a}, \tag{5.4.79c}$$

where P_α is a projector on a particular set of cluster states $|\phi_\alpha\rangle$, and for this reason we refer to $\underline{\Gamma}(\alpha)$ as a dynamical optical potential, after its original use.[42] Similar quantities $\underline{\Gamma}^t(\alpha)$, generated by $\mathscr{V}^t$, e.g.,

$$\underline{\Gamma}^t(\alpha) = \mathscr{V}^t + \mathscr{V}^t\underline{Q}(\alpha)\underline{G}\underline{\Gamma}^t(\alpha), \tag{5.4.80}$$

enter into a theory of the optical potential that was motivated by taking a C-K wave-function formalism as a starting point.[40]

Applying the prescriptions of Section 1.11, one finds that Eq. (5.4.79a), for example, assumes the symmetrized form

$$\Gamma_{\bar{b}\bar{a}}(\bar{\alpha}) = \mathscr{V}_{\bar{b}\bar{a}} + \sum_{\bar{c}} \mathscr{V}_{\bar{b}\bar{c}} Q_{\bar{c}}(\bar{\alpha}) G_{\bar{c}} \Gamma_{\bar{c}\bar{b}}(\bar{\alpha}). \tag{5.4.81}$$

A similar set of equations follows from the Eqs. (5.4.80). Note that $(\mathscr{V}^t)_{\bar{b}\bar{a}} \neq \mathscr{V}_{\bar{a}\bar{b}}$ in general, even though $(\mathscr{V}^t)_{ab} = (\mathscr{V})_{ba}$. Also note that the Eqs. (5.4.81) no longer have a matrix structure with respect to the partition indices. Rather, the symmetrized, dynamical optical potential operator $\Gamma_{\bar{b}\bar{a}}(\bar{\alpha})$ is indexed by the equivalence classes $\bar{a}$, $\bar{b}$, and $\bar{\alpha}$.

The symmetrized forms of both the PLUMS and the dynamical optical potential equations are important for some of the applications considered in

Ch. 9. They were chosen as examples in this subsection with this in mind. This does not, of course, exhaust viable alternatives. We have not discussed here the path to antisymmetrized dynamical equations, such as (5.4.81), by means of a wave-function formalism, although this is considered in Ch. 9.

Also, the implementation of permutation symmetries in time-dependent N-particle scattering theory[43] has not been discussed. Our primary reason for this omission at this point is that the fundamental dynamical equations are not of the C-K type. This type of theory is reviewed in Section 9.5a.

5.5. *N*-Particle Equations. II

The class of collision integral equations (Y-type) first obtained by Yakubovskii[1] is the only one of the C-K variety where complete equivalence to the Schrödinger equation is explicit. This is a result of the reversibility of the operations used in deriving these equations. The price that seems to have to be paid for the guaranteed absence of spurious solutions is the necessity for the most detailed specification of how the wave functions and transition amplitudes are built up from the pair interactions compared to other types of N-body equations. Our objective in this section is to derive in detail for $N = 4$ the Y-type equations, in a way that we hope clears through the thicket of notation in order to emphasize the elementary idea of Fredholm reduction upon which such equations are based.

The detailed labeling of amplitudes requires additional combinatorial technology involving *chains of partitions*. Unfortunately, the unfamiliarity of this notation seems to have limited the understanding of the general Y-type equations, as compared to the $N = 3$, Faddeev equations, for example, for which the familiar pair indices suffice. Consequently, the body of literature concerned with the properties of these equations and their applications is anomalously small. It is possible that, with increased familiarity, Y-type collision integral equations will not only provide new insights into many-particle collision processes, but also present opportunities for constructing more sophisticated approximation scenarios. We hope that the present development will encourage further research that realizes these expectations.

We refer to any C-K equations as Y-type equations if their derivation from the Schrödinger equation plus boundary conditions is reversible at every step. Such equations have been derived for components of wave functions, transition operators, and various auxiliary quantities such as resolvents, by four different groups of investigators.

Faddeev ($N = 3$) and Yakubovskii (all N) were, of course, the first to obtain what we call the FY version of Y-type of equations. Their strategy was based on the repeated application of the Faddeev idea for removing the disconnected pieces from the kernel of the LS equation. This was done by introducing a progression of auxiliary operators, $T^{a_{N-1}, a_{N-2}, \ldots a_k}$, labeled by a nested set of partitions, $a_{N-1} \subset a_{N-2} \subset \cdots \subset a_k$ called a *chain.* At each stage in the progression, a reversible inversion is performed whose effect is to produce a kernel of higher connectivity. This chain of step-by-step Fredholm reductions ends at $k = 2$ with a kernel whose $(N - 1)^{\text{th}}$ power is connected.

Shortly after Yakubovskii's work, Alt, Grassberger, and Sandhas (AGS)[2], proposed an inductive procedure, that we refer to as the method of *Lippmann–Schwinger analogies*, for obtaining C-K integral equations for the AGS off-shell extension of the transition operators. The AGS approach has the advantages of simplicity in derivation and of being expressed in terms of quantities closely related to experiment, two features not prominent in the original Yakubovskii treatment. However, originally it was not evident that the results of the LS analogies induction were equivalent to the Yakubovskii equations. Also, it was not obvious that the new chain-partition-indexed AGS operators were related to the physical transition operators for arbitrary N. Both of these shortcomings were removed by Karlsson and Zieger (KZ),[3] who showed for, arbitrary N, that the FY and AGS equations are identical in content in the sense of yielding integral equations with the same kernels. Morever, KZ showed that the AGS operators are directly related to the transition operators for elastic and rearrangement scattering initiated from two-cluster partitions. They also identified the operators that correspond to complete and partial breakup, and then found a new, third set of Y-type equations for the full set of two-cluster to arbitrary-cluster processes.

Since the work of KZ most of the formal developments on Y-type equations have been generated by Vanzani *et al.*[4] These investigators have introduced several innovations that have simplified a number of aspects of the subject.[5]

Many of the intricacies of the $N > 3$ collision problem are already present at the $N = 4$ level. In order to avoid the notational and combinatorial problems that accompany the scattering problem for arbitrary N, we first discuss the $N = 4$ case in some detail. This is the only case, besides $N = 3$, where calculations using dynamical C-K equations have been carried out for molecular and nuclear systems.[6]

Let us consider the transition operator for the scattering of N-particles into N-particles:

$$T_{00} = V + VG_0 T_{00}, \tag{5.5.1}$$

where V is the full N-body interaction

$$V = \sum_i V_i. \tag{5.5.2}$$

Here, as in Ch. 5, we have denoted the $(N-1)$-cluster partitions a_{N-1} by Latin indices $i, j, \ldots$. Also, since $a = 0$ corresponds to N free particles, we have written $T_{00} = T_{00}^{(\pm)} = U_{00}$. In Eq. (5.5.1), and for the remainder of this chapter, we have suppressed all dependencies on the parametric energy. Then, as in the $N = 3$ case, let us introduce the auxiliary amplitudes

$$T_i = V_i + V_i G_0 T_{00}, \tag{5.5.3}$$

so that

$$T_{00} = \sum_i T_i. \tag{5.5.4}$$

Since the operators (5.4.2) satisfy

$$M_{ij} = V_i \delta_{ij} + V_i G_0 \sum_k M_{kj}, \tag{5.5.5}$$

we see that

$$T_i = \sum_j M_{ij}. \tag{5.5.6}$$

Equations (5.5.3–5.5.5) for T_i and M_{ij} are of the non-C-K type.

The Fredholm reduction that consists in separating out the disconnected pieces of the kernel of Eq. (5.5.5), for example, begins with the isolation of the "diagonal" components M_{ij}, namely,

$$(1 - V_i G_0) M_{ij} = V_i \delta_{ij} + V_i G_0 \sum_k \bar{\delta}_{ik} M_{kj}. \tag{5.5.7}$$

If we operate on the left of Eq. (5.5.7) with $(1 - V_i G_0)^{-1}$, we obtain

$$M_{ij} = t_i \delta_{ij} + t_i G_0 \sum_k \bar{\delta}_{ik} M_{kj}, \tag{5.5.8}$$

where t_i is the two-body transition operator. Upon iteration, the kernel of Eq. (5.5.8) remains disconnected except for $N = 3$.

5.5a. *Four-Particle Scattering*

For $N = 4$, the Eqs. (5.5.8) are not a set of C-K equations. The FY idea for obtaining such a set is to go through the procedure starting from Eq. (5.5.3) and ending with Eq. (5.5.8) once more, but with restrictions on the pair indices i that label the three-cluster partitions. The relevant combinatorial consideration here is that the three-cluster partitions are distinguished by the two-cluster partitions, $\alpha, \beta, \sigma, \ldots$, they are contained within. With this in mind, let us define the auxiliary amplitudes

$$Y_{ij}^{\sigma} \equiv t_i \Delta_{\sigma i} G_0 \sum_k \bar{\delta}_{ik} \Delta_{\sigma k} M_{kj}. \tag{5.5.9}$$

Our subsequent manipulations are simplified if we denote

$$\bar{\delta}_{ij}(\sigma) \equiv \bar{\delta}_{ij} \Delta_{\sigma i} \Delta_{\sigma j}, \tag{5.5.10a}$$

$$\delta_{ij}(\sigma) \equiv \delta_{ij} \Delta_{\sigma i}. \tag{5.5.10b}$$

Since, if $i \neq j$, there is a unique two-cluster partition σ such that $\sigma \supset i, \sigma \supset j$, we have the identity

$$\sum_{\sigma} \bar{\delta}_{ij}(\sigma) = \bar{\delta}_{ij}. \tag{5.5.10c}$$

Then, using a matrix notation in the pair indices, we can rewrite the Eqs. (5.5.9) in the form

$$\underline{Y}^{\sigma} = \underline{t}\,\underline{\delta}(\sigma)\, G_0\, \bar{\delta}(\sigma)\, \underline{M}. \tag{5.5.11}$$

We infer from $\underline{\delta}(\sigma)\,\bar{\delta}(\sigma) = \bar{\delta}(\sigma)$ and Eq. (5.5.10c) that

$$\sum_{\sigma} \underline{Y}^{\sigma} = \underline{t}\,\bar{\delta} G_0 \underline{M} = \underline{M} - \underline{t}. \tag{5.5.12}$$

Thus, the Y_{ij}^{σ} are linearly related to T_{00}, and so are suitable auxiliary transition operators.

Combining Eqs. (5.5.11–5.5.12), we obtain

$$[I - \underline{t} G_0\, \bar{\delta}(\sigma)]\, \underline{Y}^{\sigma} = \underline{t} G_0\, \bar{\delta}(\sigma)\, \underline{t} + \underline{t} G_0\, \bar{\delta}(\sigma) \sum_{\tau} \bar{D}^{\sigma\tau}\, \underline{\delta}(\tau)\, \underline{Y}^{\tau}, \tag{5.5.13}$$

where the $\bar{D}^{\sigma\tau}$ are two-cluster-indexed matrices with components

$$\bar{D}_{ij}^{\sigma\tau} = \delta_{ij}(1 - D_{ij}^{\sigma\tau})\, \Delta_{\sigma i} \Delta_{\tau j}, \tag{5.5.14a}$$

$$D_{ij}^{\sigma\tau} = \delta_{ij}\, \delta^{\sigma\tau}\, \Delta_{\sigma i}. \tag{5.5.14b}$$

The form of Eq. (5.5.13) suggests that we introduce the subsystem operators

$$\underline{M}^{\sigma} \equiv [\underline{I} - \underline{t} G_0\, \bar{\delta}(\sigma)]^{-1} \underline{t}\, \underline{\delta}(\sigma), \tag{5.5.15}$$

which satisfy the integral equations

$$\underline{M}^{\sigma} = \underline{t}\, \underline{\delta}(\sigma) + \underline{t} G_0\, \bar{\delta}(\sigma)\, \underline{M}^{\sigma}, \tag{5.5.16a}$$

$$= \underline{t}\, \underline{\delta}(\sigma) + \underline{M}^{\sigma} G_0\, \bar{\delta}(\sigma)\, \underline{t}. \tag{5.5.16b}$$

For any of the four two-cluster partitions such as $\sigma = (1)(234)$, $\underline{M}^{\sigma}$ is a three-body subsystem transition operator. On the other hand, the three partitions (12)(34), (13)(24), and (14)(23) correspond to M^{σ}-operators that represent a concatenation of two-body processes. With the identification (5.5.15), we can then convert Eq. (5.5.13) into the following set of coupled integral equations:

$$\underline{Y}^\sigma = \underline{M}^\sigma G_0\,\bar{\delta}(\sigma)\,\underline{t} + \underline{M}^\sigma G_0\,\bar{\delta}(\sigma)\sum_\tau \bar{\underline{D}}^{\sigma\tau}\,\underline{Y}^\tau, \tag{5.5.17}$$

where we keep in mind that $\underline{M}^\sigma = \delta(\sigma)\,\underline{M}^\sigma$ and $\underline{Y}^\sigma = \underline{\delta}(\sigma)\,\underline{Y}^\sigma$. The Eqs. (5.5.17), except for a difference in the inhomogeneous terms, are the Yakubovskii equations for $N = 4$.

In assessing the connectivity properties of the kernels of the Eqs. (5.5.17), we note that $\underline{M}^\sigma$ has to be regarded as possessing the connectivity properties of its *least* connected part, namely $\underline{t}\,\underline{\delta}(\sigma)$. Upon noting this, it follows that after two iterations, the Eqs. (5.5.17) become a set of C-K integral equations. The inhomogeneous term in Eq. (5.5.17) is just $\underline{M}^\sigma$ minus its two-particle disconnected pieces. Consequently, from the Eqs. (5.5.17), we see that $\underline{Y}^\sigma$ is a sum of two-cluster disconnected pieces plus the fully connected kernel term.

With no restrictions on i and j, there would be 252 independent operators Y_{ij}^σ; however, the restrictions $\sigma \supset i, j$ imply that only 48 are independent. Moreover, since the index j plays a dummy role in the Eqs. (5.5.17) for Y_{ij}^σ, this set actually consists of "only" 18 equations. This interdependence of the two-and three-cluster indices is a harbinger of the role that chains of partitions play in Y-type equations.

The operators Y_{ij}^σ are not the most useful for developing other aspects of $N = 4$ collision theory. To motivate the introduction of operators that are more convenient, it is fruitful to examine the structure of the four-particles-to-four-particles operator T_{00} because through its singularity structure it embodies the most symmetrical representation of the operators corresponding to collisions with clustered initial and/or final states. One finds that

$$T_{00} = \sum_i t_i + \sum_{\sigma,\tau}\sum_{i,j} M_{ij}^{\sigma\tau}, \tag{5.5.18}$$

where, in matrix notation with respect to the pair indices,

$$\underline{M}^{\sigma\tau} \equiv \underline{t}\,\bar{\delta}(\sigma)\,G_0\,\underline{D}^{\sigma\tau}\underline{t} + \underline{t}G_0\,\bar{\delta}(\sigma)\,\underline{M}\,\bar{\delta}(\tau)\,G_0\underline{t}, \tag{5.5.19}$$

$(\underline{M})_{ij} \equiv M_{ij}$, and the M_{ij} are defined by the Eqs. (5.5.8).

Next we note that

$$\underline{M}^{\sigma\tau} = \underline{t}\,\bar{\delta}(\sigma)\,G_0\,\underline{D}^{\sigma\tau}\underline{t} + \underline{Y}^\sigma\,\bar{\delta}(\tau)\,G_0\underline{t}. \tag{5.5.20}$$

The last term on the right-hand side of this equation can be expressed in terms of the $\underline{M}^{\sigma\tau}$ by using the Eqs. (5.5.17), so one finds that

$$\underline{M}^{\sigma\tau} = \underline{M}^\sigma\,\bar{\delta}(\sigma)\,G_0\underline{t}\,\underline{D}^{\sigma\tau} + \underline{M}^\sigma G_0\,\bar{\delta}(\sigma)\sum_\gamma \bar{\underline{D}}^{\sigma\gamma}\,\underline{M}^{\gamma\tau}, \tag{5.5.21a}$$

from which we also infer that

$$\underline{M}^{\sigma\tau} = \underline{t}\,\bar{\delta}(\sigma)\,G_0\underline{M}^\sigma\,\underline{D}^{\sigma\tau} + \sum_\gamma \underline{M}^{\sigma\tau}\,\bar{\underline{D}}^{\gamma\tau}\,\bar{\delta}(\tau)\,G_0\underline{M}^\sigma. \tag{5.5.21b}$$

We see that $\underline{M}^{\sigma\tau}$ is precisely the sort of symmetrical object that is most useful for analyzing the structure of collisions. In this regard, let us note that, by Eqs. (5.5.16a, b), the inhomogeneous terms in Eqs. (5.5.21a, b) are identical. Also, for ease of manipulation, Eqs. (5.5.21a, b) can be compressed further by means of a matrix notation in the two-cluster indices as well. If we define the supermatrices

$$\mathscr{M} \equiv (\underline{M}^{\sigma\tau}), \tag{5.5.22a}$$

$$\mathscr{K} \equiv (\underline{M}^{\sigma}\,\bar{\delta}(\sigma)\,\underline{D}^{\sigma\tau}), \tag{5.5.22b}$$

$$\tilde{t} \equiv (\underline{t}\,\underline{\delta}(\sigma)\,\underline{D}^{\sigma\tau}), \tag{5.5.22c}$$

then Eqs. (5.5.21a, b) become

$$\mathscr{M} = \mathscr{K}G_0\tilde{t} + \mathscr{K}G_0\bar{D}\mathscr{M}, \tag{5.5.23a}$$

$$\mathscr{M} = \tilde{t}G_0\mathscr{K} + \mathscr{M}\bar{D}G_0\mathscr{K}, \tag{5.5.23b}$$

where we keep in mind that $\mathscr{K}G_0\tilde{t} = \tilde{t}G_0\mathscr{K}$.

When $N \geq 4$, the integral relations satisfied by the pair-indexed AGS operators $(\underline{U})_{ij} = U_{ij}$ namely,

$$\underline{U} = \bar{\delta}G_0^{-1} + \bar{\delta}tG_0\underline{U}, \tag{5.5.24a}$$

$$= \bar{\delta}G_0^{-1} + \underline{U}G_0\underline{t}\,\bar{\delta}, \tag{5.5.24b}$$

do not have kernels that become connected after any number of iterations. We have omitted the arrows on the inverse Green functions in Eqs. (5.5.24a, b). This is with the understanding that the dual forms of Eqs. (5.5.24a, b), as well as the subsequent manipulations that involve converting these equations, are to be used only under circumstances where the directionality of the inverse is irrelevant. The next thing to do is to proceed with the process of Fredholm reduction. The strategy of LS analogies proposed by AGS is to place Eqs. (5.5.24a, b) in LS form, and execute the same formal steps that one uses in passing from the LS equation to the Faddeev equations for $N = 3$. One repeats the process until the Fredholm reduction is complete.

The first step is to cast Eqs. (5.5.24a, b) into LS form by means of the identifications

$$\tilde{V} \equiv \bar{\delta}G_0^{-1}, \tag{5.5.25a}$$

$$\tilde{G}_0 \equiv G_0\underline{t}G_0, \tag{5.5.25b}$$

so that Eqs. (5.5.24a, b) become

$$\tilde{T} = \tilde{V} + \tilde{V}\tilde{G}_0\tilde{T} \tag{5.5.26a}$$

$$= \tilde{V} + \tilde{T}\tilde{G}_0\tilde{V}, \tag{5.5.26b}$$

where we have written $\tilde{T} \equiv \underline{U}$ to emphasize the analogy with the LS equation.

The "full" Green function can be defined by

$$\tilde{G} \equiv G_0 \underline{M} G_0, \tag{5.5.27}$$

where $(\underline{M})_{ij} = M_{ij}$ satisfies the Eqs. (5.5.8) so that, in the pair-indexed notation, we have

$$\underline{M} = \underline{t} + \underline{t}G_0\,\bar{\delta}\underline{M}, \tag{5.5.28a}$$

$$\underline{M} = \underline{t} + \underline{M}\,\bar{\delta}G_0\underline{t}. \tag{5.5.28b}$$

Then,

$$\tilde{G} = \tilde{G}_0 + \tilde{G}_0\tilde{V}\tilde{G}, \tag{5.5.29a}$$

$$\tilde{G} = \tilde{G}_0 + \tilde{G}\tilde{V}\tilde{G}_0 \tag{5.5.29b}$$

are the resolvent identities, and the identification

$$\tilde{G} = \tilde{G}_0 + \tilde{G}_0\,\tilde{T}\tilde{G}_0 \tag{5.5.29c}$$

leads to the expression

$$\tilde{T} = \tilde{V} + \tilde{V}\tilde{G}\tilde{V}. \tag{5.5.30}$$

The relationship between $\tilde{T}$ and $\underline{M}$ implicit in Eq. (5.5.30), along with what we learned from the Yakubovskii equations, suggests the first step in the Fredholm reduction, namely the identification of an appropriate set of auxiliary operators. These operators are represented by a supermatrix $\tilde{T}^{(1)}$ whose components $U_{ij}^{\sigma\tau}$ are indexed by the $(N-2)\to(N-1)$ chains of partitions,

$$(\tilde{T}^{(1)})_{ij}^{\sigma\tau} \equiv U_{ij}^{\sigma\tau}, \tag{5.5.31}$$

in addition to the pair indices i,j. The next step is to determine the equations satisfied by $\tilde{T}^{(1)}$.

To this end we note that

$$\tilde{V} = \sum_{\sigma} \tilde{V}^{\sigma}, \tag{5.5.32}$$

where

$$\tilde{V}^{\sigma} \equiv \bar{\delta}(\sigma)\,G_0^{-1}. \tag{5.5.33}$$

If we define

$$\tilde{M}^{\sigma\tau} \equiv \underline{D}^{\sigma\tau}\tilde{V}^{\sigma} + \tilde{V}^{\sigma}\tilde{G}\tilde{V}^{\tau}, \tag{5.5.34}$$

we find that

$$\tilde{M}^{\sigma\tau} = \underline{U}^{\sigma}\underline{D}^{\sigma\tau} + \underline{U}^{\sigma}\tilde{G}_0\sum_{\gamma}\bar{D}^{\sigma\gamma}\tilde{M}^{\gamma\tau}, \tag{5.5.35a}$$

$$\tilde{M}^{\sigma\tau} = \underline{U}^{\sigma}\underline{D}^{\sigma\tau} + \sum_{\gamma}\tilde{M}^{\sigma\gamma}\bar{D}^{\gamma\tau}\tilde{G}_0\underline{U}^{\tau}, \tag{5.5.35b}$$

where

$$\underline{U}^{\sigma} = \tilde{V}^{\sigma} + \tilde{V}^{\sigma}\tilde{G}_0\underline{U}^{\sigma} \tag{5.5.36a}$$

$$= \tilde{V}^{\sigma} + \underline{U}^{\sigma}\tilde{G}_0\tilde{V}^{\sigma}. \tag{5.5.36b}$$

The pair-indexed operators U_{ij}^{σ} refer to disconnected two-cluster systems, and can be determined from the C-K equations (5.5.36a, b), which are analogous to Eqs. (5.5.16a, b). In fact,

$$\underline{U}^{\sigma} = \tilde{V}^{\sigma} + \bar{\delta}(\sigma)\,\underline{M}^{\sigma}\bar{\delta}(\sigma). \tag{5.5.37}$$

Next, corresponding to the identifications (5.5.25a, b) and (5.5.27), we define

$$\tilde{G}_0^{\sigma\tau} \equiv \tilde{G}_0\underline{U}^{\sigma}\underline{D}^{\sigma\tau}\tilde{G}_0, \tag{5.5.38}$$

$$\tilde{G}^{\sigma\tau} \equiv \tilde{G}_0\tilde{M}^{\sigma\tau}\tilde{G}_0. \tag{5.5.39}$$

From Eqs. (5.5.19), (5.5.27), and (5.5.34), it follows that

$$\tilde{G}^{\sigma\tau} = G_0\underline{M}^{\sigma\tau}G_0. \tag{5.5.40}$$

We also infer from Eqs. (5.5.35a, b) the new resolvent identities

$$\mathcal{G} = \mathcal{G}_0 + \mathcal{G}_0\mathcal{V}^{(1)}\mathcal{G}, \tag{5.5.41a}$$

$$\mathcal{G} = \mathcal{G}_0 + \mathcal{G}\mathcal{V}^{(1)}\mathcal{G}_0, \tag{5.5.41b}$$

where the Green-function supermatrices are given by

$$(\mathcal{G})^{\sigma\tau} \equiv \tilde{G}^{\sigma\tau}, \tag{5.5.42a}$$

$$(\mathcal{G}_0)^{\sigma\tau} \equiv \tilde{G}_0^{\sigma\tau}, \tag{5.5.42b}$$

$$\mathcal{V}^{(1)} \equiv \bar{D}^{\sigma\tau}\tilde{G}_0^{-1}. \tag{5.5.42c}$$

The crucial identification at this stage of the LS-analogy prescription is the next set of auxiliary transition operators by means of

$$\mathcal{G} \equiv \mathcal{G}_0 + \mathcal{G}_0\tilde{T}^{(1)}\mathcal{G}_0, \tag{5.5.43}$$

from which it follows, with the aid of Eqs. (5.5.41a, b), that

$$\tilde{T}^{(1)} = \mathscr{V}^{(1)} + \mathscr{V}^{(1)}\mathscr{G}_0\tilde{T}^{(1)} \tag{5.5.44a}$$

$$= \mathscr{V}^{(1)} + \tilde{T}^{(1)}\mathscr{G}_0\mathscr{V}^{(1)}, \tag{5.5.44b}$$

which are the AGS equations for $N = 4$. It is instructive to write out these equations explicitly:

$$U_{ij}^{\sigma\tau} = \bar{\delta}^{\sigma\tau}(G_0 t_i G_0)^{-1}\,\delta_{ij}\,\Delta_{\sigma i}\,\Delta_{\tau j} + \sum_{\gamma,k}\bar{\delta}^{\sigma\gamma}\,\Delta_{\sigma i}\,\Delta_{\gamma i}\,U_{ik}^{\gamma}G_0 t_k G_0\,\Delta_{\gamma k}\,U_{kj}^{\gamma\tau}\,\Delta_{\tau j}. \tag{5.5.45}$$

Further properties and simplifications of these equations are taken up in Ch. 8, where calculations based upon them are reviewed.

5.5b. N(>4)-Particle Scattering

In this subsection, we confront the nontrivial problem of how to obtain Y-type equations for arbitrary N. The essential obstacle here is combinatoral, and for a complete treatment it is a formidable one. We describe what appear to be the essential aspects of the problem, and review recent progress towards the clarification of Y-type, N-particle collison theories and the development of the associated combinatoric technology.

We saw for $N = 4$ that a Y-type set of integral equations involves a considerably more detailed indexing of the auxiliary transition operators than is the case with the partition-labeled formalisms. In fact, we will see that it is the *most* detailed nonredundant indexing.

This indexing is in terms of a chain of partitions $\alpha(a_k)$, which is a sequential fragmentation, denoted by α, of a k-cluster partition a_k of N objects ending in its decomposition into N-clusters:

$$\alpha(a_k) \equiv \left\{a_m | m \geq k, \prod_{i=k}^{N-1} \Delta_{a_i a_{i+1}} = 1\right\}, \tag{5.5.46}$$

where we use the standard notation for a set of objects along with the constraints that define it. When $k = 2$, we have what is called a *complete* chain of partitions, which can be represented by an ordered tree graph in which there is a progressive joining of branches. The $k = N$ partition is not referred to in any chain because it is contained in any partition; the $k = 1$ partition contains all partitions, so that its role is implicit in a complete chain.

The relevance of such sets of nested partitions to the analysis of many-particle systems arises from the fact that every connected Weinberg graph can be identified with a complete chain of partitions. More generally, every graph of connectivity a_k can be identified with a chain of partitions $\alpha(a_k)$. Evidently, a labeling of graphs by chains of partitions provides a much more detailed classification of graphs than just by their connectivity, and corresponds to the breaking of different successive connections between particle lines as one follows the N-particle lines through the graph. The vastly increased indexing options that appear with chain-indexed transition operators produces a rich variety of combinatorial possibilities and with this a more complicated formalism to organize these possibilities. The chain-labeling introduces a structure that may appear remote from the usual intuitive pictures of reaction mechanisms.

It was only fairly recently that the numbers of coupled Y-type equations in various situations were determined.[7] These numbers are quite large compared to equations labeled by partitions. For example, for five distinguishable particles the number $Y(N)$ of Y-type equations jumps from $Y(4) = 18$ to $Y(5) = 180$, and to $Y(8) \simeq 1.6 \times 10^6$ for $N = 8$. Of course, in most instances, the number of dynamically active species of particles is very small. When there are permutation symmetries, the figure of merit is the number of equivalence classes of chains. The solution of this enumeration problem was given in a notably elegant fashion by Karlsson,[7] who found a significant reduction in the numbers of equations as a result of identity; for N identical particles, he found that $Y(4) = 2$, $Y(5) = 5$, and $Y(8) = 272$, which are from one to four orders of magnitude smaller than the case of distinguishable particles.

A transparent way of illustrating how chains enter into collision theory is by graphical means.[8] Let us consider the operator

$$[T]_{a_i}^{\alpha(a_{i+1})} \equiv \left(\prod_{m=i}^{N-1} V_{a_m}^{a_{m+1}} G_{a_m} \right) \overleftarrow{G}_0^{-1}, \tag{5.5.47}$$

corresponding to a chain of partitions, $\alpha(a_{i+1})$. In Eq. (5.5.47), the product is understood as beginning on the left with $V_{a_{N-1}}$ and ending on the right with $G_{a_{i+1}}$. We recall that $V_b^a = (\bar{V}_M)_{ab}$, as per Eq. (5.4.19), and $V_b^{a_N} = V_b$. The index a_i on the left-hand side of Eq. (5.5.47) signifies that the right-hand side is a_i-connected; this is established using the definition of V_b^a and the observation that the Green functions are irrelevant to the connectivity of each of the graphs to which $[T]_{a_i}^{\alpha(a_{i+1})}$ corresponds.

The organization (5.5.47) corresponds to the fact that an a_i-connected graph ends on the right with some succession of pair interactions, V_j, $a_i \supset j$, after the last interaction V_k, $a_i \supset k$, $a_{i+1} \not\supset k$. We note that $G_{a_i}\overleftarrow{G}_0^{-1} = 1 + G_0 T_{a_i}$, where $T_{a_i} = V_{a_i} + V_{a_i} G_{a_i} V_{a_i}$.

Now, $[T]_{a_i}^{\alpha(a_{i+1})}$ consists of a sum of graphs of the type described. On the other hand, it is clear that each Weinberg graph in the perturbative representation of T_{00} is contained in one and only one collection of graphs $[T]_{a_i}^{\alpha(a_{i+1})}$. Therefore, we have

$$T_{00} = \sum_{i=1}^{N} \sum_{a_i} \sum_{\alpha(a_{i+1})} [T]_{a_i}^{\alpha(a_{i+1})}, \tag{5.5.48}$$

where the sum over $\alpha(a_{i+1})$, which enumerates all chains beginning with the partition a_{i+1}, includes a sum over a_{i+1} for fixed i.

Next, we define

$$T^{b_{N-1}} \equiv \sum_{i=1}^{N} \sum_{b_i} \sum_{b_{i+1}} \cdots \sum_{b_{N-2}} [T]_{b_i}^{\alpha(b_{i+1})}, \tag{5.5.49}$$

in which the sum over chains $\alpha(b_{i+1})$ is restricted to those that end with a particular pair partition labeled by b_{N-1}, so that $T^{b_{N-1}}$ consists of all graphs ending on the left with the interaction $V_{b_{N-1}}$. By the almost-connected idea, we see that

$$[T]_{a_1}^{\alpha(a_2)} = [T]_{a_2}^{\alpha(a_3)} G_0 \sum_{b_{N-1}} \bar{\Delta}_{a_2 b_{N-1}} T^{b_{N-1}}, \tag{5.5.50}$$

where $\alpha(a_2) \supset \alpha(a_3)$, namely, the two chains are identical except for a_2 and $a_2 \supset a_3$. From Eq. (5.5.49), we see that the Eqs. (5.5.50) are integral equations for $[T]_{a_1}^{\alpha(a_2)}$ in terms of subsystem amplitudes. They can be transformed into equations for the operator

$$T^{\alpha(a_2)} \equiv \sum_{i=1}^{N} \sum_{a_i} [T]_{a_i}^{\alpha(a_2)}, \tag{5.5.51}$$

by adding the disconnected parts of each side of Eq. (5.5.50).

The Eqs. (5.5.50) are archetypal Y-type equations. They can be used to obtain collision integral equations with less detailed characterizations of the auxiliary transition operators.[8] Combinatorially, this classification is associated with an ordered tree. Summation over ordered trees that differ in the order of the switching on of the interactions in different branches yields the Omnes[9] N-particle equations. On the other hand, a summation over all trees with fixed first interactions gives the Rosenberg[10] equations. Finally, summation of Eq. (5.5.50) over all the partitions, $a_3, \ldots, a_{N-2}$, contained in $\alpha(a_2)$ gives rise to the Narodetskii–Yakubovskii[8] equations.

The graphical approach is useful in that it provides some insight as to how the chains of partitions come into the picture. However, it fails to provide the assurance that there are no spurious solutions. This can only be provided by the algebraic Fredholm reduction by means of reversible operations. Let us outline the logic of the approach in the general N-particle context. One begins with the decomposition of T_{00} into operators

$$T^{a_{N-1}} = V_{N-1} + V_{N-1} G_0 T_{00}. \tag{5.5.52}$$

Then, the diagonal portion of Eq. (5.5.52) is inverted, yielding a set of integral equations that have disconnected kernels for $N > 3$. One further decomposes into auxiliary operators $T^{a_{N-1},a_{N-2}}, \ldots, T^{a_{N-1},\ldots,a_2}$ inverting the diagonal parts at each step, where $a_2 \supset a_3 \supset \cdots \supset a_{N-1}$. Only after the chain is extended to a_2 does one obtain C-K equations.

Neither the representation of the successive reduction or the proof of the C-K property are easy to show for arbitrary N. The original proofs[1–4] of both the Fredholm reduction and the C-K character used either direct inductive procedures or the method of LS analogies, which in its original form is a plausibility argument rather than a rigorous proof. When carried out in full detail, the combinatorics lead to complex decomposition and recurrence relations.

A significant step towards simplifying things has occurred recently.[5,11] An important technical advance involves the decomposition of partition-labeled operators such as V_a, H_a, G_a into chain-labeled components.[5] This has then been exploited to develop chain-labeled C-K resolvent equations again representing the scattering in its most finely decomposed form, and from which all of the various scattering equations can be derived.[11] These chain-labeled resolvent structures appear to represent the true primordial LS structure underlying multiparticle scattering theory. The formalism of Cattapan and Vanzani,[11] while too complicated to be presented here in full detail, circumvents the complicated recurrence relations that appear in the Yakubovskii–Karlsson–Zieger[1,2] treatments and justifies the inductive method of LS analogies advocated by Sandhas.[2] This work seems to represent an advance in improving the manipulative facility of Y-type collision theories. As such, it represents an opportunity for exploring further this most general class of collision integral equations.

Another important simplification of the AGS form of the Y-type collision theory was obtained by Haberzettl.[12] He found a partial decoupling of the AGS equations so that the only transition operators that appear are labeled by those two-cluster partitions that correspond to a definite split $(n, N - n)$ of the N particles. This is achieved by means of effective interactions that contain the effects of the suppressed channels, using the fact that some of the original integral equations function as quadrature rules. For identical particles, a reduction to a single equation is obtained. We show how this decoupling works for $N = 4$ in Ch. 8.

Undoubtedly, there is much to be learned about Y-type collision theories. At present their complexity is daunting. There is, however, a fascinating gap in our knowledge in that it is not understood how so much of the intricate clustering patterns that lead to that complexity can be ignored when dealing with typical few-cluster collision problems involving all or nearly all identical particles. Phenomenological tools of the practical collision theory trade such as "Pauli-blocking" behavior have yet to be recovered from Y-type theories. These, among many more questions, remain challenges for the future.

Appendix (5)

A.5a. Partition Combinatorics

We present a self-contained account of the techniques that are needed in the text to deal with the clustering possibilities for systems of more than two particles.[1] The underlying mathematical basis for this material consists in combinatorics, especially as it relates to partitions and lattice theory.[2]

The clustering ideas that we utilize appear in a number of areas besides collision theory, such as statistical mechanics and field theory. Often the same ideas have been rediscovered a number of times in different contexts without recognition of their common basis. When this has happened in multiparticle collision theory, it usually has been accompanied by an entirely new set of notational conventions that frustrate comparisons among different papers. Our notation has been chosen to facilitate the matrix constructions in partition space that allow simple and manipulatively useful statements of the integral equations of N-particle collision theory.

A partition of a set X of N objects is a grouping of these objects into K $(1 \leq K \leq N)$ nonempty, disjoint equivalence classes. For the case of N particles, to which we confine ourselves, the equivalence classes are called *clusters*. The Latin letters $a, b, c, \ldots$ denote partitions of X into $n_a, n_b, n_c, \ldots$ clusters, except for the N-cluster and the 1-cluster partitions which are denoted by 0 and $\underline{1}$, respectively. A partition a with m clusters is denoted as $a(m)$. We denote by $L(N)$ the set of all partitions of the N particles.

A partition a is said to *contain* another partition b, denoted $a \supset b$, if each cluster of b is wholly contained within some cluster of a ($a \supseteq b$ denotes the possibility of equality). We introduce a matrix $\underline{\Delta}$ with elements

$$(\underline{\Delta})_{ab} = 1, \qquad a \supseteq b, \qquad \text{(A.5.1)}$$
$$= 0, \qquad \text{otherwise.}$$

Also the matrix $\bar{\Delta}$,

$$\bar{\Delta} \equiv \mathscr{S} - \underline{\Delta}, \qquad \text{(A.5.2)}$$

has nonvanishing elements $\bar{\Delta}_{ab}$ equal to unity only when $a \not\supseteq b$, where $(\mathscr{S})_{ab} = 1$, for all a and b. Most of the important properties of the matrix $\underline{\Delta}$, and thus $\bar{\Delta}$, are related to the fact that, for every $a \in L(N)$, there exist the unique elements 0 and $\underline{1}$ such that $\underline{1} \supseteq a \supseteq 0$.

One of the most important properties of $\underline{\Delta}$ is that it possesses an inverse $\underline{\Delta}^{-1}$. The proof consists in showing that det $\underline{\Delta} = 1$. Let us order the rows and columns of the $\underline{\Delta}$-matrix by $a(n) > b(m)$, whenever $n > m$. With this ordering, $\underline{\Delta}$ is upper triangular and has diagonal elements all equal to unity. Therefore, det $\underline{\Delta} = 1$. This method of proof indicates that, in fact, $\underline{\Delta}^{-1}$ exists on any subset $L(N)_R \subset L(N)$ such that, for each $a \in L(N)_R$, there are unique partitions 0_R and 1_R for which $0_R \subset a \subset 1_R$. Any subset of $L(N)$ not including 0 and $\underline{1}$ can be bracketed in this way. The inverse $\underline{\Delta}^{-1}$ also exists on the extended sets which include either, or both of, 0_R and 1_R. In lattice theory,[2] $\underline{\Delta}^{-1}$ is referred to as the Möbius function for the partial ordering defined by $\supseteq$.

Partition-labeled decompositions of an operator A that have the form

$$A = \sum_a [A]_a \qquad \text{(A.5.3)}$$

occur frequently in collision theory and many-particle physics in general. They represent what we refer to generically as a *cluster decomposition*. For example, $[A]_a$ may be the a-connected part of A. Associated with the sum (A.5.3) are two restrictive decompositions defined by

$$A_a \equiv \sum_b \Delta_{ab} [A]_b, \qquad \text{(A.5.4)}$$

$$A^a = \sum_b \bar{\Delta}_{ab} [A]_b. \qquad \text{(A.5.5)}$$

From (A.5.4), we have

$$[A]_a = \sum_b (\underline{\Delta}^{-1})_{ab} A_b, \qquad \text{(A.5.6)}$$

which corresponds to the representation of the cumulants in terms of moments of a distribution.[1]

The identity $(\underline{\Delta}^{-1}\underline{\Delta})_{\underline{1}b} = \delta_{\underline{1}b}$ implies the sum rule

$$\sum_a C_a \bar{\delta}_{a\underline{1}} \Delta_{ab} = \bar{\delta}_{\underline{1}b}, \tag{A.5.7}$$

where

$$C_a \equiv -(\underline{\Delta}^{-1})_{\underline{1}a}, \qquad a \neq \underline{1}. \tag{A.5.8}$$

Now the Stirling number of the second kind S_n^m is equal to the number of distinct m-cluster partitions that can be made from n distinguishable objects:

$$S_{n_b}^m = \sum_a \delta_{mn_a} \Delta_{ab}, \tag{A.5.9}$$

for any b. The standard properties[3] of S_n^m can be used to show that

$$\sum_{m=2}^{N} (-1)^m (m-1)! S_k^m = 1, \qquad k \geq 2. \tag{A.5.10}$$

Comparing the sum rule (A.5.10) with Eq. (A.5.7), we see that

$$C_a = (-1)^{n_a}(n_a - 1)!, \qquad a \neq \underline{1}. \tag{A.5.11}$$

Next, let i denote a $(N-1)$-cluster partition corresponding to a pair index. A direct calculation yields

$$\sum_a \delta_{n_a m} \Delta_{ab} \Delta_{ai} \bar{\Delta}_{bi} = \bar{\Delta}_{bi} S_{n_b - 1}^m, \qquad n_b \geq 2, \tag{A.5.12}$$

so that

$$\sum_a \delta_{n_a m} \Delta_{ab} \bar{\Delta}_{ai} = \sum_a \delta_{n_a m} \Delta_{ab} \bar{\Delta}_{ai} \bar{\Delta}_{bi}$$

$$= (S_{n_b}^m - S_{n_b - 1}^m) \bar{\Delta}_{bi}, \qquad n_b \geq 2. \tag{A.5.13}$$

Since, from Eq. (A.5.10), we have with $c_m = (-1)^m (m-1)!$ that

$$\sum_{m=2}^{N} c_m (S_k^m - S_{k-1}^m) = \delta_{k2}, \qquad k \geq 2, \tag{A.5.14}$$

we conclude from Eq. (A.5.13) that

$$\sum_{m=2}^{N} c_m \sum_a \delta_{mn_a} \Delta_{ab} \bar{\Delta}_{ai} = \delta_{n_b 2} \bar{\Delta}_{bi}, \tag{A.5.15}$$

or, in matrix form,

$$\underline{\Delta}^t \underline{C}\, \bar{\Delta} \underline{P}_{N-1} = \underline{P}_2 \bar{\Delta} \underline{P}_{N-1}, \tag{A.5.16}$$

where $\underline{P}_2$ is the projector on two-cluster partitions and $(C)_{ab} = C_a \delta_{ab}$.

Next, let us define, using the notation of Sections 5.2 and 5.4:

$$(\underline{\tau}_{\mathrm{M}})_{ba} \equiv V_a^b + V_a^b G_a V_a. \tag{A.5.17}$$

Consider the cluster decomposition

$$(\underline{\tau}_{\mathrm{M}})_{b\underline{1}} = \sum_d M_{bd}, \tag{A.5.18}$$

where $M_{bd} \equiv [(\underline{\tau}_{\mathrm{M}})_{b\underline{1}}]_d$ is the d-connected part of $(\underline{\tau}_{\mathrm{M}})_{b\underline{1}}$. We see that

$$(\underline{\tau}_{\mathrm{M}})_{ba} = \sum_b M_{bd}\,\Delta_{ad}. \tag{A.5.19}$$

The full interpretation of the BRS formalism of Section 5.4 depends upon the identities we establish next. The connectivity expansion of the pair-indexed operator

$$M_{ij} = V_i \delta_{ij} + V_i G V_j \tag{A.5.20}$$

can be written as

$$M_{ij} = \sum_a M_{ij}(a), \tag{A.5.21}$$

where $M_{ij}(a)$ is the a-connected part of M_{ij}. Then, we define

$$W^{ba}(d) \equiv \sum_{ij} \bar{\Delta}_{bi}\, M_{ij}(d)\, \bar{\Delta}_{aj}. \tag{A.5.22}$$

One finds after a lengthy, but straightforward, calculation that[4]

$$\sum_d{}' W^{ba}(d) = \sum_d{}' W^{b0}(d) G_0 \overleftarrow{G}_a^{-1} + \sum_\gamma W^{b0}(\gamma) G_0 V_a^\gamma - V_a^b, \tag{A.5.23}$$

where the prime on the sum means $2 \leq n_d \leq N-1$, and we recall that $n_\gamma = 2$. The potentials are $(\bar{V}_{\mathrm{M}})_{ba}$ (*cf*. Eq. (5.4.19)).

One can derive the Eqs. (5.4.42) by making the substitution (5.4.41) into Eq. (5.4.32), and then using the identity (A.5.23).

Half-on-shell we deduce, using the explicit structure of $W^{b0}(d)$ and the Lippmann identity, that

$$W^{b0}(d) G_0 \overleftarrow{G}_\alpha^{-1} |\phi_\alpha\rangle = \delta_{d\alpha} V_\alpha^b |\phi_\alpha\rangle. \tag{A.5.24}$$

Thus, from the identity (A.5.23), we infer that half-on-shell

$$\sum_d{}' W^{b\alpha}(d) |\phi_\alpha\rangle = \sum_\gamma W^{b0}(\gamma) G_0 V_\alpha^\gamma |\phi_\alpha\rangle, \tag{A.5.25}$$

which reveals that the multiple-scattering structure and the remainder of the usual Born approximation are contained in the *second* Born term of the off-shell transformed BRS equations (5.4.38).

In conclusion, we remark that the techniques that we have briefly outlined here have, in more elaborate forms, been applied to the analysis of multifermion Bethe–Salpeter equations in relativistic field theory[5] and to relativistic N-particle quantum-mechanical collision theory.[6]

A.5b. Partial-Wave Decomposition of Three-Body Equations

In order to affect a numerical solution of a set of three-body collision integral equations, one must first carry out a partial-wave decomposition (p.w.d.). The basis for doing this derives from angular momentum conservation and the finite ranges of the interparticle interactions. In this subsection, we outline the essential features of such decompositions.

The simplest three-body p.w.d. problem involves the quasi-two-body equations (5.3.76), but even in this instance the details are too lengthy to be fully reproduced here.[7,8] The relevant states (we suppress z),

$$|\alpha, n; \mathbf{q}_\alpha\rangle \equiv |\alpha, n; z\rangle|\mathbf{q}_\alpha\rangle, \tag{A.5.26}$$

are specified in the three-body c.m. by the spectator (α) particle's momentum $\mathbf{q}_\alpha$ and internal quantum numbers (e.g., parity, spin, spin-projection, isospin, etc.), along with the quantum numbers of the other (interacting) pair of particles that are labeled collectively by n. Only the spin j_α ($\bar{j}_\alpha$) and the spin projection m_α ($\bar{m}_\alpha$) of the spectator (interacting pair) will be relevant for the p.w.d. Thus, for the labeling of the internal quantum numbers of the states (A.5.26), it suffices to replace (α, n) in Eq. (A.5.26) by the set $(\bar{m}_\alpha, m_\alpha, n_\alpha)$, where n_α now denotes the remaining internal quantum numbers, including $\bar{j}_\alpha$ and j_α.

To progress any further, it is necessary to specify a channel-spin coupling scheme by first coupling the pair spin j_α to the spectator spin S_α (with a z-projection σ_α). This is then coupled to the angular momentum of the spectator with respect to the pair c.m., L, to yield the total angular momentum J. Then (*cf*. Eq. (5.3.75)), we have the p.w.d.

$$\begin{aligned} X^{n'n}_{\beta\alpha}(\mathbf{q}'_\beta, \mathbf{q}_\alpha) &= \langle \mathbf{q}'_\beta; \bar{m}'_\beta, m'_\beta, n'_\beta | F_{\beta\alpha} | \mathbf{q}_\alpha; \bar{m}_\alpha, m_\alpha, n_\alpha \rangle \\ &\equiv \sum \langle \bar{j}'_\beta, \bar{m}'_\beta; j'_\beta, m'_\beta | S'_\beta, \sigma'_\beta \rangle \langle L'_\beta, M'_\beta; S'_\beta, \sigma'_\beta | JM \rangle Y^{M'_\beta}_{L'_\beta}(\tilde{\mathbf{q}}'_\beta, \tilde{\zeta}_z) \\ &\quad \times \langle L'_\beta, S'_\beta, n_\beta; q'_\beta | X^J | L_\alpha, S_\alpha, n_\alpha; q_\alpha \rangle Y^{M_\alpha}_{L_\alpha}(\tilde{\mathbf{q}}_\alpha, \tilde{\zeta}_z)^* \\ &\quad \times \langle L_\alpha, M_\alpha; S_\alpha, \sigma_\alpha | JM \rangle \langle \bar{j}_\alpha \bar{m}_\alpha; j_\alpha, m_\alpha | S_\alpha, \sigma_\alpha \rangle, \end{aligned} \tag{A.5.27}$$

as the definition of the M-independent partial-wave amplitude $\langle \cdots | X^J | \cdots \rangle$. The sums in Eq. (A.5.27) are over the repeated discrete indices, and we use standard definitions for the Clebsch–Gordon coefficients and the spherical harmonics.[8,9] The axis of quantization is defined by $\tilde{\zeta}_z$. A p.w.d. similar to Eq. (A.5.27) also holds for $\hat{X}^{n'n}_{\beta\alpha}(\mathbf{q}'_\beta, \mathbf{q}_\alpha)$.

Finally, we need to specify the appropriate components of the two-particle τ-functions that appear in the kernels of the Eqs. (5.3.76). Evidently, these τ-functions will be diagonal in the internal indices and, as a result of their independence of the spin projections $(\bar{m}_\gamma, m_\gamma)$, can be replaced by functions τ_{N_γ} labeled only by the set $N_\gamma = (L_\gamma, S_\gamma, n_\gamma)$. After using the orthogonality relations for the vector-coupling coefficients and the Y^M_L, one finds that the Eqs. (5.3.76) reduce to the one-dimensional partial-wave forms:

$$X^J_{N'_\beta, N_\alpha}(q'_\beta, q_\alpha; z) = \hat{X}^J_{N'_\beta, N_\alpha}(q'_\beta, q_\alpha; z) + \sum_{N_\gamma, \gamma} \int_0^\infty dq''_\gamma q''^2_\gamma \hat{X}^J_{N'_\beta, N_\gamma}(q'_\beta, q''_\gamma; z)\tau_{N_\gamma}(z - \omega''_\gamma) X^J_{N_\gamma, N_\alpha}(q''_\gamma, q_\alpha; z), \quad \text{(A.5.28)}$$

where

$$X^J_{N'_\beta, N_\alpha}(q'_\beta, q_\alpha; z) \equiv \langle N'_\beta; q'_\beta | X^J(z) | N_\alpha; q_\alpha \rangle, \quad \text{(A.5.29)}$$

with a corresponding definition for $\hat{X}^J_{N'_\beta, N_\alpha}(q'_\beta, q_\alpha; z)$.

In the degenerate-kernel limit, $\hat{X}^{nm}_{\beta\alpha}(\mathbf{q}'_\beta, \mathbf{q}_\alpha)$ is given by Eqs. (5.3.77–5.3.80). The determination of the partial-wave amplitudes $\hat{X}^J_{N'_\beta, N_\alpha}(q'_\beta, q_\alpha; z)$ in the case of arbitrary spin and angular momentum of the interaction pair is carried out in great detail in the Appendix of Ref. 8.

In the case of equations where explicit three-particle states enter in, such as the equations for the breakup amplitudes, or the Faddeev equations when the two-particle t-matrices are not of finite rank, the p.w.d. is somewhat more complex. There are two coupling schemes that have been proposed, the symmetrical method of Omnes[10] and the more popular asymmetrical coupling used by Ahmadzadeh and Tjon.[11] We make no explicit use of either technique in our development; full descriptions of both methods are readily available and lead to two-dimensional versions of the Eqs. (A.5.28).[7]

References and Notes (5)

Section 5.1

1. Faddeev's original work (Faddeev (1960–1962)) on the three-particle scattering problem is developed in comprehensive detail in his monograph on the subject (Faddeev (1965)).
2. Skornyakov (1956).
3. The chronology of these ideas is an interesting and not widely appreciated part of the history of scattering theory. The idea of decomposing an N-particle collision

operator into a sum of auxiliary operators that are labeled by each of the N particles was exploited earlier by Watson (1953, 1956). (See also the discussion in Watson (1967), pp. 53–58.) The objective of this work was entirely different than Faddeev's and the accidental coincidence of the equations for $N = 3$ is reflective of the fact that only in this case does one obtain connected-kernel (C-K) equations. However, Rosenberg (1965) showed how the Watson–Faddeev approach could be extended to N-particle scattering to obtain C-K equations by use of the graphical method of Weinberg (1964). As pointed out by Narodetskii (1980), the works by Mitra (1965b), Takahashi (1965, 1966). Alessandrini (1966), and Weyers (1966) all arrive at C-K equations with much the same content as the Rosenberg equations. Rosenberg's ideas were also used implicitly to obtain C-K equations for the physical transition operators by Bencze (1973a, 1975) and Redish (1974a). The Watson–Rosenberg approach was pursued again by Picklesimer (1982a, 1983a) and Kowalski (1983a) to obtain a new class of C-K equations that resembled the Watson equations. Finally, a simple graphical derivation of all these Rosenberg-related results was given by Kowalski (1983b), and generalized to include multiparticle forces by Picklesimer (1983b).

4. The magnitude of this "explosion" is reflected, in part, by the size of our bibliography.

5. Some examples of alternative sets of C-K equations for $N = 3$ are given in Lovelace (1964a, b), Alt (1967), Osborn (1971), Kouri (1974a, b), and Karlsson (1975, 1977a). See also Schmid (1974) and Glöckle (1983a).

6. Much of this work is reviewed in Ch. 7.

7. Amado (1963, 1969, 1970b), Lovelace (1964a, b), Aaron (1964–1966), Freedman (1966).

8. van Winter (1964), Weinberg (1964).

9. The "classical" period from 1960 to 1968 is reviewed in Faddeev (1970).

10. The fact, Faddeev (1965) showed that the fifth powers of the kernels of his equations are compact on a certain Banach space, and this suffices for the essential Fredholm properties to hold true.

11. Federbush (1966).

12. Yakubovskii (1967).

13. Alt (1972), Karlsson (1974a, b), Riahi (1969), Vanzani (1971, 1976). See Narodetskii (1980) for a review.

14. To date, there are no proofs that any of these equations are free from spurious solutions or, on the other hand, that they possess them except under unphysical circumstances.

15. Merkuriev (1976, 1984). See also Glöckle (1988).

Section 5.2

1. Kowalski (1969, 1977b).

2. The general concept of channel-coupling schemes is systematically exploited in Bencze (1977a). The usefulness of this device for discussions of uniqueness was

illustrated by Sandhas (1976, 1978). Special cases and applications are contained in Vanzani (1971, 1976), Bencze (1973a), Kouri (1974a, b; 1975a, b), Redish (1974a), and Kowalski (1977b).

3. Baer (1973) and Kouri (1974a, b; 1975a, b), Goldflam (1980c). The *W*-matrices are sometimes called partition-permuting arrays.
4. Sandhas (1978), Kowalski (1978b, c; 1984), Polyzou (1979a, b).
5. Bencze (1977a).
6. Sandhas (1976, 1978), Levin (1980).
7. L'Huillier (1978), Levin (1980).

Section 5.3

1. Treatments of three-particle scattering with varying degrees of completeness and attention to practical methods of solution are given in Faddeev (1965), Watson (1967), Schmid (1974), Joachain (1975), Thomas (1977), and Newton (1982).
2. The modification of three-body scattering theory to include three-body forces is considered in Kowalski (1973a, 1976b, 1977a) and Glöckle (1983b).
3. Alt (1967).
4. Lovelace (1964a, b).
5. Sandhas (1974, 1976, 1978, 1986).
6. Faddeev (1965).
7. Faddeev (1960–1962).
8. The simplifications in three-particle scattering that result when the two-particle interactions are of finite rank have been known for a long time, *cf.* Mitra (1962). The simplicity of the Amado model (Amado (1963)) results from the fact that it is equivalent to a nonrelativistic three-particle problem with this type of interaction. Further references and a persuasive rationale for using such interactions (see, however, Bollé (1970)) are contained in Lovelace (1964a, b). For comprehensive reviews, see Amado (1969), Mitra (1969), Fuda (1971), and Duck (1968).
9. Aaron (1977).
10. Osborn (1971).
11. Karlsson (1975). For the extension to $N = 4$, see Karlsson (1977a).
12. Kholyavin (1977), Eyre (1979).
13. Hunziker (1977).
14. Kouri (1974a, b; 1975a, b), Tobocman (1974a, b; 1975a–c).
15. See Chs. 2, 3 and the references cited therein.
16. Alessandrini (1968).
17. Sloan (1972c), Kowalski (1972a).
18. Pieper (1971, 1972b).
19. Doleschall (1973).
20. Sloan (1968a, 1969).
21. Finkel (1968).

22. The impulse approximation is discussed in detail in Section 9.3.
23. Grassberger (1969). See also Alt (1967).
24. The optical potential is discussed in Sections 1.8 and 9.2.
25. Kowalski (1969).
26. Pieper (1972a).
27. Kowalski (1972b). This work was shown (Kowalski (1974b)) to be equivalent, for identical inputs, to the unitary formalisms of Cahill (1972) and Sasakawa (1973); for a concise derivation, see Karlsson (1978).
28. Kowalski (1972b).
29. Tandy (1972), Sharma (1973).
30. Kowalski (1972d).
31. Kouri (1976).
32. Kowalski (1977a).
33. Kouri (1974a).
34. Aaron (1965).
35. Ascoli (1975), Lock (1976a).
36. This point is made in a formalism-independent manner in Kowalski (1973b), where the UM_{ex} for the three-to-three amplitude when there are no two-body bound states is said to generate "minimal" scattering amplitudes. Evidently, "sub-minimal" would have been a more accurate appellation.
37. Aitchison (1968, 1975).
38. Amado (1970a).
39. These integral equations are analyzed in detail in Lock (1975, 1976b) and Kowalski (1976a).
40. Bencze (1973b).
41. Kowalski (1973c, 1977c).
42. Carew (1973).
43. See, e.g., Pieper (1970a, b).
44. Some useful sources on the derivation of three-body scattering variational principles and their utilization include Rosenberg (1968), Lieber (1972), Carew (1973), Merkuriev (1973), and Hahn (1976). For a somewhat different approach to multiparticle variational principles, see Goldflam (1981).
45. Karlsson (1977b).
46. Kowalski (1965b), Noyes (1965). See Chs. 2 and 3 for other relevant references.
47. Karlsson (1977b) is not too specific about the nature of the two-body off-shell functions for negative energies. There are a number of plausible options that avoid introducing branch cuts in the f_s functions for negative energies and these are correctly pointed out by Karlsson.
48. Kowalski (1977c).
49. Alessandrini (1965), Freedman (1966).
50. Basdevant (1966), Mennessier (1972).
51. The Faddeev approach to the meson–NN problem is concisely reviewed in Garcilazo (1988a, b). See also Eisenberg (1980).

52. Rupp (1988) treats the $3N$-bound-state problem using a Bethe–Salpeter version of the Faddeev equations. Relativistic N-particle quantum-mechanical collision theories are developed in Coester (1982) and Mutze (1984).

Section 5.4

1. A very clear discussion of these points is contained in Narodetskii (1980).
2. See the reviews of N-particle scattering by Vanzani (1978a), Kowalski (1978c, 1984), Levin (1980), Narodetskii (1980, 1981), and Cattapan (1985a, 1986).
3. Baer (1973), Kouri (1974a, b; 1975a, b), Tobocman (1974a, b), Goldflam (1980a, c).
4. A direct proof is contained in Kowalski (1977d). Consistency arguments leading to the same result were given earlier by Benoist-Gueutal (1975) and Bencze (1977a).
5. Weinberg (1964).
6. Rosenberg (1965).
7. See Kowalski (1981d, 1982c, 1983b) and Picklesimer (1983a). The irrelevance of the Green functions in what might be taken to be the definition of minimal connectivity (the g's cannot decrease it) was appreciated in association with sets of (quasi-) Lippmann–Schwinger equations: Avishai (1971), Tobocman (1974a, b), Kouri (1975a, b), Kowalski (1978c).
8. The present derivation of the Rosenberg (1965) equations closely follows Kowalski (1982c). These equations have been rederived in many different ways: Redish (1974a), Narodetskii (1980), Picklesimer (1982b, 1983a), Kowalski (1983b).
9. The BRS equations were originally obtained in three different ways by Sloan (1972b) for $N = 4$, and by Bencze (1973a) and Redish (1974a) for arbitrary N. Subsequently, they have been derived in still another fashion by using a channel-coupling scheme, *cf.* L'Huillier (1976), Kowalski (1977b), and Benoist-Gueutal (1978).
10. This possibility was suggested in Picklesimer (1982b, 1983a). The mathematical problem entailed in obtaining sigma-set equations was stated and solved for a number of cases in Kowalski (1982c).
11. See Picklesimer (1982b, 1983a, b) and Kowalski (1982c, 1983a).
12. This was first developed in Picklesimer (1983a). The inhomogenous and kernel terms of these C-K Watson equations involve N-particle operators, rather than true subsystem operators, for which no solution algorithm is known. See Kowalski (1982c, 1983a).
13. The subscript C refers to *combinatorial*. It is an alternative to a more unwieldly acronym that represents the discoverers of this potential matrix: Benoist-Gueutal, L'Huillier, Redish, and Tandy (BLRT). See L'Huillier (1976) and Benoist-Gueutal (1978).
14. The numbers C_a occur repeatedly in combinatorial arguments. For example, they appear as coefficients of a sum rule satisfied by the Stirling numbers of the second kind (Abramowitz (1965)), and arose on the N-particle scattering scene in the calculations carried out by Weinberg (1964).
15. The proof is first given in Kowalski (1977b). In the work cited in Ref. 13, the BRS

equations are derived without explicitly addressing the question of whether some power of $\underline{K}_C$ is connected.

16. L'Huillier (1978).
17. See Benoist-Gueutal (1975).
18. A notable attempt to develop such strategies is contained in Polyzou (1979a, b).
19. The acronym refers to partition-labeled and multiple-scattering equations with the substitution of the German form of the conjunction. See Goldflam (1980a) and Kowalski (1979a).
20. This acronym is sometimes extended to include Baer (1973). We use the KLT abbreviation to refer to those developments specifically concerned with C-K properties.
21. Avishai (1971).
22. Glöckle (1970a, b).
23. Lippmann (1956).
24. Similar ideas for wave-function equations had been considered earlier by Hahn (1968, 1969, 1972). The relationship of the reduction method of Hahn to that of the channel-coupling array is explored in Hahn (1974, 1977).
25. See, e.g., Polyzou (1979a, b) and Goldflam (1980a).
26. Goldflam (1980c).
27. Mitra (1965b), Takahashi (1965, 1966), Alessandrini (1966).
28. Weyers (1966).
29. Bencze (1976).
30. Van Winter (1964).
31. Bencze (1977a).
32. Kowalski (1978b).
33. Federbush (1966).
34. Chandler (1978b, 1981a). Some of this work is reviewed in Kowalski (1978c, 1984), which contains references to other literature on the topic.
35. Newton (1967).
36. Kazaks (1970), Adhikari (1979b), Vanzani (1978b).
37. Blankenbecler (1964).
38. Adhikari (1979b, c).
39. Goldflam (1978, 1979). The method in Tobocman (1975c) leads to non-C-K equations.
40. Adhikari (1984a).
41. Glöckle (1970a, b).
42. Kowalski (1981a, 1986).
43. We follow Bencze (1982b) in the use of the term "time-dependent."

Section 5.5

1. Yakubovskii (1967).
2. Alt (1972). In its original form, Grassberger (1967), the AGS method yields C-K

equations only after approximations to the dynamics. Sandhas (1974–1976, 1978, 1986) has given a number of detailed expositions of this method.
3. Karlsson (1974a, b).
4. Vanzani (1971, 1976, 1978b), Cattapan (1979b, c), Lovitch (1979). For recent developments and extensive referencing to prior work, see Cattapan (1983a, 1985a, b; 1986).
5. See also Benoist-Gueutal (1982).
6. A description of this work along with relevant references is given in Ch. 8.
7. Bencze (1977c) and Karlsson (1982).
8. Narodetskii (1971, 1980).
9. Omnes (1968).
10. Rosenberg (1965).
11. Cattapan (1984, 1985a, b).
12. Haberzettl (1981a, b).

Section A.5

1. Reviews from which we draw much of our material and which should be consulted for further developments and more detailed citations are Kowalski (1981d) and Benoist-Gueutal (1982). Some important early work remains unpublished, L'Huillier (1976), Polyzou (1979a); parts of these papers are contained in Benoist-Gueutal (1978) and Polyzou (1979b, 1980), respectively.
2. Useful mathematical references are Szasz (1963), Rota (1964), Abramowitz (1965), Birkhoff (1967), Donellan (1968), and Andrews (1976).
3. See Abramowitz (1965), p. 825, Kowalski (1977b).
4. Kowalski (1979a).
5. Kowalski (1979c).
6. Coester (1982), Mutze (1984).
7. The full details that accompany the various three-body p.w.d.s are contained in the comprehensive developments of Watson (1967), Schmid (1974), Afnan (1977b), Glöckle (1983a), Micha (1985), and references cited therein.
8. See, in particular, Afnan (1977b).
9. Edmonds (1957).
10. Omnes (1964).
11. Ahmadzadeh (1965).

Chapter 6

Singularity Structure of Multiparticle Amplitudes

6.1. Introduction

One cannot usually solve multiparticle-scattering problems exactly. Instead, one makes use of simple, tractable models of atomic and subatomic collisions that involve many particles. Nevertheless, the singularity structure of the exact multiparticle amplitudes is of practical interest. This structure provides guidance in constructing models of physical processes, because the singularities are often crucial to the reproduction of experimental results. Some examples concern the positions and natures of the final-state interaction peaks in two-to-three breakup processes, and the low-temperature behavior of the cluster-coefficients in quantum statistical mechanics. This structure also leads to various dispersion relations for the scattering amplitudes, provides constraints on approximation and numerical schemes, and is essential to the explanation of unusual multiparticle phenomena such as the Efimov and the Thomas effects. In this chapter, we consider those aspects of the singularity structure of multiparticle scattering amplitudes that have been of some practical value in describing collision processes.

Our discussion is based on physically plausible arguments and conjectures rather than on rigorous mathematical analysis such as in Refs. 1. The present development is at a level similar to that in Refs. 2.

Given the complexity of multiparticle scattering amplitudes as well as their number, even for a two-cluster initial state, it is not always clear which amplitudes are most amenable to study or which singularies are of greatest relevance. Also, except for the simplest two-body processes, the optimum choice of energy/momentum variables for expressing the singularity structure is usually not obvious. Therefore, we only consider the simplest cases that have been investigated rather attempt a general approach.

There are essentially two types of singularities in the off-shell amplitudes that satisfy some sort of scattering integral equation. One type we call *dynamical singularities* because they are associated with the general dynamical constraints imposed by unitarity. The other we refer to as *potential singularities* in that they arise from the "potential" driving terms in the scattering integral equations, and depend upon the particular nature of the interparticle interactions. Both types reflect important features of the scattering process.

6.2. Singularities of the Two-Particle Amplitudes

In this section, we consider the threshold behavior of the partial-wave off-shell t-matrix. We also review the analytic behaviors of certain on-shell amplitudes in the complex-energy plane, which lead to their representation by dispersion relations.

6.2a. Threshold Behavior of the Off-Shell Amplitude

We first discuss the low-energy behavior of the two-body off-shell elastic scattering amplitude as a function of the two-particle c.m. energy E, through terms of order $E^{3/2}$.[1] We assume the potential is short-ranged and, for the sake of simplicity, central.

The partial-wave expansion of the off-shell two-body t-matrix is (*cf.* Eq. (2.4.31))

$$\langle \mathbf{k}|t(E)|\mathbf{k}'\rangle = \frac{2}{\pi}\sum_{L,M} Y_L^M(\tilde{\mathbf{k}},\tilde{\zeta})Y_L^M(\tilde{\mathbf{k}}',\tilde{\zeta})^* t_L(k,k';E), \tag{6.2.1}$$

where

$$t_L(k,k',E) = \int d^3x\, d^3x'\, j_L(kr)\langle \mathbf{x}|t(E)|\mathbf{x}'\rangle j_L(k'r'), \tag{6.2.2}$$

in units $\hbar = m = 1$, where m is the mass of each of the particles. We are interested in the behavior of $t_L(k,k',E)$ for small values of k, k', and E. If we just keep the terms through $E^{3/2}$, only the $L = 0, 1$ amplitudes contribute to the sum (6.2.1).

At low energies, a power series expansion of $t_0(k, k, E)$, $k^2 = E$, in $\sqrt{E}$ is appropriate because of the $\sqrt{E}$ behavior arising from the unitarity cut; the lowest-order term is the scattering length a. In the off-shell case, if we make a small k, k' expansion using Eq. (6.2.2), we see that only even powers of k, k' appear for $L = 0$. So

$$t_0(k, k'; E) = a + if\sqrt{E} + bE + c(k + k'^2) + idE^{3/2} + \cdots, \tag{6.2.3}$$

where

$$a = \lim_{k, k' \to 0} t_0(k, k', E = 0) \tag{6.2.4}$$

is the S-wave scattering length, while b, c, d, f are undetermined parameters.[1] The effective-range expansion,

$$k \cot \delta = -\frac{1}{a} + \frac{1}{2} r_0 k^2 + \cdots, \tag{6.2.5}$$

is associated with the unitary parameterization of the on-shell t-matrix:

$$t_0(k, k, k^2) = -\frac{1}{k \cot \delta - ik}. \tag{6.2.6}$$

Consistency between the on-shell version of Eq. (6.2.3) and Eqs. (6.2.5–6.2.6) leads us to the identifications

$$f = -a^2, \tag{6.2.7}$$

$$b + 2c = \tfrac{1}{2} a^2 (r_0 - 2a), \tag{6.2.8}$$

$$d = a^3 (a - r_0). \tag{6.2.9}$$

We see that f and d are defined in terms of a and r_0, but only the combination $(b + 2c)$ is constrained by Eq. (6.2.8). Even without knowing b and c separately, expansion (6.2.3) is useful for studying the analytic properties of connected three-particle amplitudes. For the on-shell, two-particle t-matrix only the combination $(b + 2c)$ is relevant.

The threshold expansion of the $L = 1$ amplitude is

$$t_1(k, k', E) = kk'(a_1^3 + ib_1\sqrt{E} + \cdots), \tag{6.2.10}$$

which follows from Eq. (6.2.2) for $L = 1$. Here,

$$a_1^3 = \lim_{k, k' \to 0} (kk')^{-1} t_1(k, k', E = 0), \tag{6.2.11}$$

where a_1 is called the P-wave scattering length and b_1 is an expansion parameter. The coefficients of the terms in the parentheses in Eq. (6.2.10) can be determined from the effective-range expansion for $L = 1$ and/or from the knowledge of the potential.

6.2b. *Forward Dispersion Relations (FDR)*

This subject has been treated in detail in many books, so it is only necessary to quote the essential results of this analysis.[2]

The forward-scattering amplitude at energy $E = k^2$,

$$f(E) \equiv \langle \mathbf{k}|t(E)|\mathbf{k}\rangle = \langle \mathbf{k}|V|k\rangle + \langle \mathbf{k}|VG^{(+)}(E)V|\mathbf{k}\rangle, \qquad (6.2.12)$$

is an analytic function in the complex-E plane, except for dynamical singularities arising from the unitarity cut that extends along the real positive-energy axis and the bound-state poles.[2-4] It is expected that if V is a short-ranged, local potential, for which $\langle \mathbf{k}|V|\mathbf{k}\rangle$ is independent of energy, then $f(E)$ has no other singularities in the complex-E plane. Klein and Zemach[3] have shown that $\hat{f}(E) \equiv [f(E) - \langle \mathbf{k}|V|\mathbf{k}\rangle]$ has this singularity structure and tends to zero as $|E| \to \infty$.

Cauchy's theorem with the contour C of Fig. 6.1a implies that

$$\hat{f}(z) = \frac{1}{2\pi i}\oint_C \frac{\hat{f}(z')}{z' - z}\,dz', \qquad (6.2.13)$$

where z is the complex-energy parameter. Then, using the fact that $\hat{f}(z)$ is Hermitian-analytic $[\hat{f}(z^*) = \hat{f}^*(z)]$, we find that the contributions of the unitarity cut and the bound-state poles yield

$$\hat{f}(z) = \sum_i \frac{r_i}{z + B_i} + \frac{1}{\pi}\int_0^\infty \frac{\operatorname{Im}\hat{f}(E')}{E' - z}\,dE'. \qquad (6.2.14)$$

Here the B_i's are the binding energies and the r_i's are the corresponding (real) residues.

In the limit of real energies, taking the real part of Eq. (6.2.14) leads to the *forward dispersion relation* (FDR)

$$\operatorname{Re} f(E) = \langle \mathbf{k}|V|\mathbf{k}\rangle + \sum_i \frac{r_i}{E + B_i} + \frac{P}{\pi}\int_0^\infty \frac{\operatorname{Im} f(E')\,dE'}{E' - E}, \qquad (6.2.15)$$

where P denotes the principal-value prescription and Re (Im) denotes real (imaginary) part. From Eq. (6.2.12) and the spectral analyses of Chs. 1 and 2, we see that

$$r_i = \lim_{k^2 \to -B_i} |\langle \mathbf{k}|V|B_i\rangle|^2, \qquad (6.2.16)$$

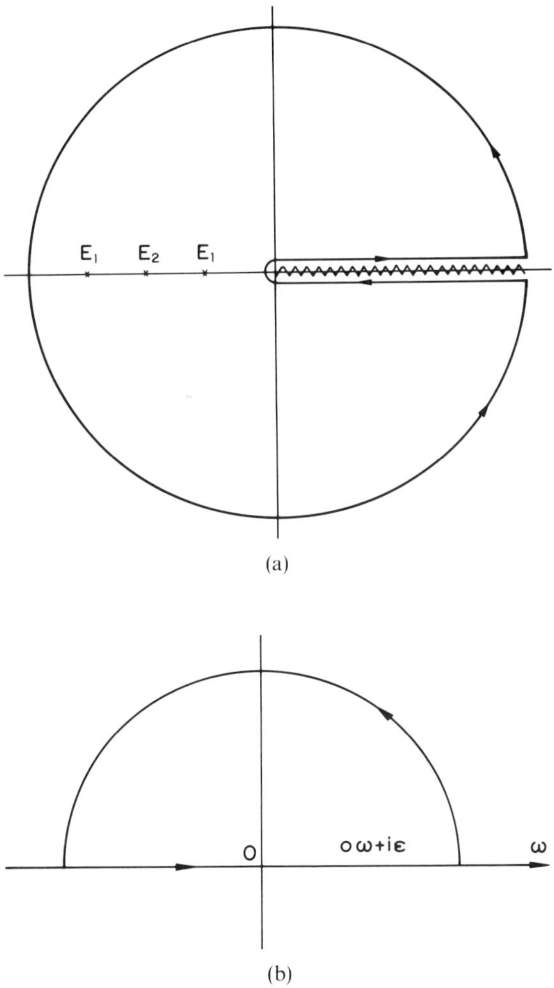

FIG. 6.1. (a) Contour used to derive Eqs. (6.2.13) and (6.2.14); the crosses denote the bound states. (b) Contour used to derive Eq. (6.2.21).

where $|B_i\rangle$ is the bound-state wave function at energy $-B_i$. Also, one has the optical theorem (*cf*. Eq. (A.1.22)):

$$\operatorname{Im} f(E) = \frac{k}{4\pi}\,\sigma_{\mathrm{T}}(E). \tag{6.2.17}$$

Equation (6.2.15) provides an important check on both theory and experiment since the first term on the right-hand side is model dependent, whereas the remaining terms can be evaluated approximately from experiment. A sum-rule for the S-wave scattering length results by taking $E = 0$ in Eq. (6.2.15):

$$a = \langle \mathbf{k}|V|\mathbf{k}\rangle + \sum_i \frac{r_i}{B_i} + \frac{P}{\pi}\int_0^\infty \frac{\operatorname{Im} f(E')\,dE'}{E'}. \tag{6.2.18}$$

The on-shell partial-wave t-matrix elements for a local potential have potential cuts along the real negative-energy ($E < 0$) axis, but these singularities are canceled in the sum over partial waves leading to $f(E)$. Usually, the potential between two objects, at least one of which is composite, is nonlocal in nature.[5] Hence, in many physically interesting problems, the forward-scattering amplitude does not have the analyticity properties required for the FDR (6.2.15) to hold, and it also has intrinsic potential singularities along the $E < 0$ axis that cannot be eliminated even by subtracting the Born term.[6,7] Attempts have been made to rewrite FDR's for such processes by incorporating the left-hand singularities in some way; we give a brief account of this work.

6.2b.1. Electron–Atom Collisions

FDR's of the form (6.2.15) for electron–atom collisions were conjectured by Gerjuoy and Krall[4] for the electron–hydrogen and electron–helium atomic systems. Of course, because of the spins of the electrons, they had to construct amplitudes with appropriate symmetry by taking proper linear combinations of direct and exchange amplitudes. Later, it was pointed out[6] that even in the zero-energy limit, the Gerjuoy–Krall FDR is in contradiction with experiment. It is now generally agreed that the potential singularities of the exchange amplitude on the $E < 0$ axis are responsible for the breakdown of these FDR's; a number of these singularities have been identified.[6] Plausibility arguments are usually given for the absence of such singularities in the *direct* amplitude, but no proof is known. Though the absence of such singularities in the direct amplitude alone does not validate the Gerjuoy–Krall FDR's for the electron–atom system, it does suggest the existence of such dispersion relations for the positron–atom system, where the exchange amplitudes are absent.

6.2b.2. Hadron–Nucleus Collisions

Ericson and Locher[7] have presented a formalism for writing FDR's for hadron–nucleus systems including the potential singularities. They use relativistic kinematics for the collision of a projectile of mass m, lab energy ω, and lab kinetic energy $(\omega - m)$ with a target of mass M. They consider the following function of the scattering amplitude $f(\omega)$:

$$F(\omega) \equiv \frac{1}{\omega - \omega_0}[f(\omega) - f(\omega_0)], \tag{6.2.19}$$

where ω_0 is a fixed energy. This combination tends to zero sufficiently rapidly at infinity, so that in the integral (6.2.13) the contribution of the contour at infinity can be neglected. Then, we have

$$\operatorname{Re} F(\omega) = \frac{1}{\pi} P \int_{-\infty}^{\infty} \frac{\operatorname{Im} F(\omega')\, d\omega'}{\omega' - \omega}, \tag{6.2.20}$$

where we have ignored the contributions of any bound-state poles and where the integral now extends over the right-hand and left-hand cuts as in Fig. 6.1b. Equations (6.2.19–6.2.20) together lead to

$$\operatorname{Re} f(\omega) = \operatorname{Re} f(\omega_0) + \frac{(\omega - \omega_0)}{\pi} P \int_{-\infty}^{\infty} \frac{\operatorname{Im} f(x)}{(x - \omega_0)(x - \omega)} dx. \tag{6.2.21}$$

This result can be obtained from the formal dispersion relation

$$\operatorname{Re} f(\omega) = \frac{1}{\pi} P \int_{-\infty}^{\infty} \frac{\operatorname{Im} f(x)}{x - \omega} dx, \tag{6.2.22}$$

which may have convergence difficulties since $f(\omega)$ does not usually tend to zero at infinity. Nevertheless, Eq. (6.2.21) is a valid FDR that is referred to as the once-subtracted version of Eq. (6.2.22), because it can be derived by subtracting $\operatorname{Re} f(\omega_0)$ from $\operatorname{Re} f(\omega)$. (The FDR (6.2.15) is a special case of Eq. (6.2.21) with $\omega_0 = \infty$ and the identification $\operatorname{Re} f(\omega_0) = \langle \mathbf{k} | V | \mathbf{k} \rangle$.)

In order to do practical calculations with the FDR (6.2.21), one must express the integral below the scattering threshold in terms of some experimentally known quantity and/or make a model for it. This is usually achieved by utilizing crossing symmetry that relates the forward-scattering amplitude for the hadron–nucleus system to that of the antihadron–nucleus system by[7]

$$f(-\omega) = \bar{f}^*(\omega), \tag{6.2.23}$$

where $\bar{f}$ is the scattering amplitude for the antihadron on the same target. It should be noted that $\operatorname{Im} f(-\omega) = -\operatorname{Im} \bar{f}(\omega)$. In particular, this relation connects the direct channel for $\omega < 0$ to the crossed channel for $\omega > 0$. In addition, one needs models for other left-hand singularities. We consider applications of this formalism to the pion–nucleus and nucleon–nucleus systems.

6.2b.2A. Pion–Nucleus Collisions. For pion–nucleon scattering the singularity structure in the unphysical region is very simple. Namely, there are only nucleon poles in the direct and crossed channels.[7,8]

The unphysical region for pion–nucleus scattering, on the other hand, has many poles associated with nuclear excited states, and a cut below the threshold for scattering arising because of pion absorption. The poles correspond to the ground and excited states of a nucleus differing from the target nucleus (X) only by the charge of the pion. The pole terms in the direct and crossed channels are given by

$$\frac{2r_i}{\omega - \omega_i} \quad \text{and} \quad \frac{-2r_i}{\omega + \omega_i}, \tag{6.2.24}$$

respectively, where $\pm\omega_i$ are the pole positions that are determined kinematically. The real numbers r_i are the dimensionless residues.

If we neglect the Coulomb interaction, the amplitudes

$$f^{(\pm)}(\omega) = \tfrac{1}{2}[f_{\pi^+X}(\omega) \pm f_{\pi^-X}(\omega)] \tag{6.2.25}$$

have the crossing properties

$$f^{(\pm)}(-\omega) = \pm f^{(\pm)*}(\omega). \tag{6.2.26}$$

So, noting that the conventional FDR's can be written as

$$\operatorname{Re} f_{\pi^\pm X}(\omega) = \pm\sum_i \frac{2r_i}{\omega \mp \omega_i} + \frac{1}{\pi} P \int_{-\infty}^{\infty} \frac{\operatorname{Im} f_{\pi^\pm X}(\omega')\, d\omega'}{\omega' - \omega}, \tag{6.2.27}$$

we see that the amplitudes $f^{(\pm)}(\omega)$ satisfy

$$\operatorname{Re} f^{(+)}(\omega) = \sum_i \frac{2\omega_i r_i}{\omega^2 - \omega_i^2} + \frac{1}{\pi} P \int_{-\infty}^{\infty} \frac{\operatorname{Im} f^{(+)}(\omega')}{\omega' - \omega}\, d\omega', \tag{6.2.28a}$$

$$\operatorname{Re} f^{(-)}(\omega) = \sum_i \frac{2\omega r_i}{\omega^2 - \omega_i^2} + \frac{1}{\pi} P \int_{-\infty}^{\infty} \frac{\operatorname{Im} f^{(-)}(\omega')}{\omega' - \omega}\, d\omega'. \tag{6.2.28b}$$

Crossing symmetry restricts these integrations to positive ω:

$$\operatorname{Re} f^{(+)}(\omega) = \sum_i \frac{2\omega_i r_i}{\omega^2 - \omega_i^2} + \frac{2}{\pi} P \int_0^{\infty} \frac{\omega'\, d\omega' \operatorname{Im} f^{(+)}(\omega')}{\omega'^2 - \omega^2}, \tag{6.2.29}$$

$$\operatorname{Re} f^{(-)}(\omega) = \sum_i \frac{2\omega r_i}{\omega^2 - \omega_i^2} + \frac{2\omega}{\pi} P \int_0^{\infty} \frac{d\omega' \operatorname{Im} f^{(-)}(\omega')}{\omega'^2 - \omega^2}. \tag{6.2.30}$$

The integral in Eq. (6.2.30) is convergent since $\mathrm{Im}\, f^{(-)}(\omega) \sim [\sigma_{\pi^+ X}(\omega) - \sigma_{\pi^- X}(\omega)]$ tends to zero sufficiently rapidly as $\omega \to \infty$ (Pomerankchuk theorem), where σ represents the total cross section. This is not true for the integral in Eq. (6.2.29), and so we make a subtraction at the lowest scattering threshold $\omega = m$, to obtain instead

$$\mathrm{Re}\, f^{(+)}(\omega) = \mathrm{Re}\, f^{(+)}(m) + \sum_i \frac{2\omega_i r_i}{\omega^2 - \omega_i^2} \frac{k^2}{k_i^2} + \frac{2k^2}{\pi} P \int_0^\infty \frac{\omega'\, d\omega'\, \mathrm{Im}\, f^{(+)}(\omega')}{k'^2(\omega'^2 - \omega^2)}, \tag{6.2.31}$$

where we have used $k^2 = \omega^2 - m^2$, and $k'^2 = \omega'^2 - m^2$.

It has been estimated that the effect of the pion-absorption cuts can often be neglected. Ericson and Locher have proposed approximate methods for evaluating the integrals and pole terms in Eqs. (6.2.30–6.2.31).[7]

Allowing for the approximate and nondynamical nature of the treatment, these FDR's yield very good results for both $f^{(+)}$ and $f^{(-)}$. The targets ^{2}H, ^{4}He, and ^{12}C were used for $f^{(+)}$, while for $f^{(-)}$ the targets ^{9}Be and ^{12}C were considered.[7]

6.2b.2B. Nucleon–Nucleus Collisions. In this case, the unsubtracted FDR is[7]

$$\mathrm{Re}\, f(\omega) = \sum_i \frac{r_i}{\omega - \omega_i} + \frac{1}{\pi}\left[P\int_m^\infty + \int_{-\infty}^{\omega_b}\right] \frac{\mathrm{Im}\, f(\omega')}{\omega' - \omega}\, d\omega', \tag{6.2.32}$$

where the integral extends over the scattering (unitarity) cut from $\omega' = m$ to $\omega' = \infty$, and over the left-hand cut from $\omega' = -\infty$ to $\omega' = \omega_b$. From crossing, we have $\mathrm{Im}\, f(-\omega) = -\mathrm{Im}\, \bar{f}(\omega)$, where $\bar{f}(\omega)$ is antinucleon–nucleus amplitude, so we can rewrite Eq. (6.2.32) as

$$\begin{aligned}\mathrm{Re}\, f(\omega) = {} & \sum_i \frac{r_i}{\omega - \omega_i} + \frac{1}{\pi} P \int_m^\infty \mathrm{Im}\, f(\omega') \left[\frac{1}{\omega' - \omega} + \frac{1}{\omega' + \omega}\right] d\omega' \\ & + \frac{1}{\pi} \int_{-\omega_b}^{m} \frac{\mathrm{Im}\, \bar{f}(\omega')}{\omega' + \omega}\, d\omega' \\ & + \frac{1}{\pi} \int_m^\infty \frac{[\mathrm{Im}\, \bar{f}(\omega') - \mathrm{Im}\, f(\omega')]}{\omega' + \omega}\, d\omega',\end{aligned} \tag{6.2.33}$$

where the argument of $\mathrm{Im}\, \bar{f}$ is the energy in the exchange channel. If we use the optical theorem, perform a subtraction at threshold, and introduce the kinetic energy, $E \equiv \omega - m$, and the momentum, $k^2 \equiv \omega^2 - m^2$, variables, we obtain

$$\operatorname{Re} f(E) = f(0) + \sum_i \frac{Er_i}{(E - E_i)E_i} + \frac{k^2}{2\pi^2} P \int_0^\infty \frac{\sigma(k')\,dk'}{k'^2 - k^2}$$
$$+ \frac{E}{\pi} \int_{-2m}^{E_b} \frac{\operatorname{Im} f(E')\,dE'}{(E' - E)E'}$$
$$- \frac{E}{4\pi^2} \int_0^\infty k'\,dE' \frac{[\bar{\sigma}(E') - \sigma(E')]}{(E' + 2m)(E' + E + 2m)}. \tag{6.2.34}$$

Since $[\bar{\sigma}(E) - \sigma(E)] \to 0$ for $E \to \infty$ by the Pomeranchuk theorem, the last integral in Eq. (6.2.34) is convergent. Here, we do not use amplitudes such as $f^{(\pm)}$ because relatively little is known about antinucleon–nucleus scattering.

Let us describe the dominant singularities of $f(E)$ for the nucleon–deuteron (nd) system.[9] It is convenient to describe these singularities in the complex-E plane. Here, E is the three-body c.m. energy that is related to the wave number k in the c.m. frame by

$$E + \alpha^2 = \tfrac{3}{4}k^2, \tag{6.2.35}$$

where α^2 is deuteron binding energy in units $\hbar^2 = m = 1$, and m is the nucleon mass. In this case, $f(E)$ has unitarity cuts associated with the thresholds for elastic scattering and breakup starting at $E = -\alpha^2$ and $E = 0$ that extend along the entire $E > 0$ axis; it also has the triton bound-state pole at $E = -8.5$ MeV. The one-nucleon-exchange Born term produces a pole in $f(E)$ at $E = -4\alpha^2/3$. In the on-shell partial-wave amplitude, this pole gets smeared into a cut along the $E < 0$ axis between $E = -4\alpha^2/3$ and $E = -4\alpha^2$. Multi-nucleon exchanges contribute to cuts starting from the left of the one-nucleon-exchange pole. Two-nucleon exchange corresponds to a left-hand cut starting at $E = -4\alpha^2$, while exchanges of more than two nucleons contribute to cuts further removed from the physical region. Physically, multiple exchanges of pions or a combination of pions and nucleons are possible (one-pion exchange is not allowed by isospin conservation); the positions of the associated cuts can be calculated by applying the Landau rules to the related Feynman diagrams. In addition, one has a cut for elastic scattering in the crossed channel of antinucleon–deuteron ($\bar{\text{n}}$d) scattering; this appears as a left-hand cut that is usually neglected in low-energy treatments. The discontinuities across the elastic scattering cuts for the nd and $\bar{\text{n}}$d systems are given by the optical theorem, and are implicit in the FDR (6.2.34). The effect of other exchange cuts in nd scattering are included via appropriate models. Similar considerations apply for other nucleon–nucleus systems.

The cut structure of $f(E)$ is complicated enough so that approximations have to be made in order to apply the FDR (6.2.34). Usually, at low energies, only the effects of the singularities near $E = 0$ are taken into account. It has been argued that the last term involving an integral over $(\bar{\sigma} - \sigma)$ in Eq. (6.2.34) can be neglected; the real part of $f(E)$ has been calculated in this approximation for the n-^{4}He and n-^{12}C systems with results that compare reasonably well with experiment.[7]

The FDR's allow one to calculate the residues at the bound-state poles,[10] which are, in turn, related to the asymptotic normalization parameter (ANP) of the bound-state wave function. The ANP, or the related vertex constant, can be determined from a theoretical (model) calculation of the wave function and also from an extrapolation of the scattering data to the bound-state pole, for example, via an effective-range expansion.[10] It is generally accepted that the ANP is as important in characterizing the nuclear bound state as the binding energy.[11]

Overall, the results obtained for the pion–nucleus and the nucleon–nucleus systems demonstrate the utility of FDR's for connecting certain experimentally measured and theoretically calculated quantities.

6.2c. *Partial-Wave Dispersion Relations (PWDR)*

In this subsection, we write down a dispersion relation[12,13] for the nonrelativistic on-shell partial-wave t-matrix $t(E)$, which is given by the partial-wave form of Eq. (6.2.12) ($E = k^2$):

$$t(k^2) \equiv t(k, k; E) = V(k, k) + [VG^{(+)}V](k, k; E), \tag{6.2.36}$$

where we suppress the angular momentum label. The amplitude $t(k^2)$ has the same dynamical singularity structure as Eq. (6.2.12), with a unitarity cut starting at the lowest scattering threshold and extending along the whole $E > 0$ axis, along with bound-state poles for $E < 0$. Otherwise, the singularity structures of $t(k^2)$ and $f(k^2)$ of Eq. (6.2.12) are very different. For local potentials, $\langle \mathbf{k}|V|\mathbf{k}\rangle$ is an energy-independent constant, whereas $V(k, k)$ is an energy-dependent, singular function. If the potential $V(r)$ has a tail that smoothly goes to zero for large r, $V(k, k)$ has a branch cut which extends from $E = E_C$ to $E = -\infty$, where $E_C < 0$ is a model-dependent constant. The potential $V(k, k)$ also contributes to the other branch cuts in $t(k^2)$ that are generated via the last term of Eq. (6.2.36) for $E_C > E > -\infty$. For real energies, $t(k^2)$ is parametrized in terms of a real phase shift δ as

$$t(k^2) = -k^{-1}e^{i\delta}\sin\delta, \tag{6.2.37}$$

so

$$\operatorname{Im} t^{-1}(k^2) = k, \qquad k^2 \geq 0, \tag{6.2.38a}$$

$$\operatorname{Im} t(k^2) = -k|t(k^2)|^2, \qquad k^2 \geq 0. \tag{6.2.38b}$$

Since $t(k^2)$ is Hermitian-analytic, $\operatorname{Im} t(k^2)$ is the discontinuity of $t(k^2)$ across the unitarity cut. The discontinuity across the left-hand cut depends on the details of the potential and is not easily calculable.

We can represent $t(k^2)$ in the so-called N/D form[12,13]

$$t(k^2) = \frac{N(k^2)}{D(k^2)}. \tag{6.2.39}$$

We choose $N(k^2)$ to have the singularities of $t(k^2)$ arising from the potential V in Eq. (6.2.36), e.g., the singularities lying on the $E < 0$ axis between $E = E_C$ and $E = -\infty$. Then, for $E > E_C$, $N(k^2)$ is real. The function $D(k^2)$ is taken to have the unitarity cut between $E = 0$ and $E = \infty$; at the bound-state poles of $t(k^2)$, $D(k^2)$ is assumed to be zero. For $E < 0$, $D(k^2)$ is required to be real. In the limit of large $|E|$, $N(k^2)$ and $[D(k^2) - 1]$ are assumed to vanish at least as fast as $1/\sqrt{E}$. These properties enable us to use Cauchy's theorem to express $N(k^2)$ and $[D(k^2) - 1]$ as integrals over the left- and right-hand cuts, respectively:[12,13]

$$D(k^2) = \frac{1}{\pi}\int_0^\infty \frac{\operatorname{Im} D(k'^2)}{k'^2 - k^2 - i0}\,dk'^2, \tag{6.2.40}$$

$$N(k^2) = \frac{1}{\pi}\int_{-\infty}^{E_C} \frac{\operatorname{Im} N(k'^2)}{k'^2 - k^2 - i0}\,dk'^2. \tag{6.2.41}$$

On the right-hand cut, Eqs. (6.2.37) and (6.2.39) lead to

$$\operatorname{Im} D(k^2) = kN(k^2), \qquad E > 0, \tag{6.2.42}$$

whereas on the left-hand cut we have

$$\operatorname{Im} N(k^2) = [\operatorname{Im} t(k^2)]D(k^2), \qquad E < E_C. \tag{6.2.43}$$

Using Eq. (6.2.42) in the Eq. (6.2.40) and normalizing $D(0) = 1$ by a subtraction, we obtain[12,13]

$$D(k^2) = 1 + \frac{k^2}{\pi}\int_0^\infty \frac{N(k'^2)}{k'(k'^2 - k^2 - i0)}\,dk'^2. \tag{6.2.44}$$

Equations (6.2.41) and (6.2.44), when supplemented by Eq. (6.2.43) and a reasonable approximation for $\operatorname{Im} t(k^2)$ for $E < E_C$, provide the starting point for a calculational scheme. In general, drastic approximations are needed in order to construct $\operatorname{Im} N(k^2)$ on the left-hand cut.

If the potential V is known, the set (6.2.41) and (6.2.44) do not present any advantage over the exact solution of the LS equation. However, this set is the starting point for phenomenological descriptions of two-particle and multiparticle systems in atomic, nuclear, and particle physics.[14–18]

6.2c.1. *Application to the S-Wave Spin-Doublet Neutron–Deuteron (nd) System*

As an illustration of the power of PWDR's we consider an application to the trinucleon system, which is the only multiparticle system for which accurate dynamical calculations exist. If appropriate assumptions are made about the N and D functions on the right- and left-hand cuts, the PWDR yields results in close agreement with the exact solution of the problem.

We take the S-wave spin-doublet elastic nd amplitude to be given by Eq. (6.2.37), where now δ denotes the nd phase shift. The wave number k, the total three-particle c.m. energy E, and the deuteron binding energy α^2 are related by Eq. (6.2.35). The amplitude $t(k^2)$ has right-hand unitarity cuts due to the two- and three-body thresholds starting, respectively, at $k^2 = 0$ and $E = 0$. It also has a left-hand cut, due to the exchange of nucleons and mesons, starting at $k^2 = b = -4\alpha^2/9$ and extending to $-\infty$. The N/D equations for this problem have been solved numerically by different workers, even including the three-particle unitarity cut, and they have yielded results in agreement with the exact dynamical solution.[9,16] Instead of reviewing these results in detail, we next consider a simpler, approximate model of nd scattering.[15]

In our treatment, we ignore the three-particle unitarity cut and take δ to be real for all $E > 0$. These are reasonable assumptions for low energies. Substituting Eq. (6.2.41) into Eq. (6.2.44) and recalling Eq. (6.2.43), we find

$$D(k^2) = 1 + \frac{ik}{\pi}\int_{-\infty}^{b} dk'^2 \frac{D(k'^2)\,\mathrm{Im}\,t(k'^2)}{k'(k + k')}. \tag{6.2.45}$$

We further approximate Eq. (6.2.45) in order to obtain an analytic solution that yields the correct ^{3}H ground and virtual-state energies when it predicts the experimental nd doublet scattering-length $\bar{a}$.

The integral term in Eq. (6.2.45) over the left-hand cut is approximated as follows. We treat the one-nucleon exchange part of $\mathrm{Im}\,t(k^2)$ exactly, i.e., we replace $t(k^2)$ by the Born term $B(k^2)$ of the Amado model. The neglected portion of the left-hand cut is expanded as a power series in k, which we suppose converges at low enough energies. The integral over the one-nucleon exchange cut is written as a contour integral C around this cut in the counterclockwise sense. If we retain only one term in the power series, we can rewrite Eq. (6.2.45) as[15]

$$D(k^2) = 1 - ic_1 k - \frac{k}{2\pi} \oint_C \frac{dk'^2 D(k'^2) B(k'^2)}{k'(k + k')}. \tag{6.2.46}$$

The replacement of the one-nucleon exchange cut by a pole at $k^2 = -L_0^2$ with residue d',

$$B(k'^2) = \frac{d'}{k'^2 + L_0^2}, \tag{6.2.47}$$

allows us to perform the contour integration to get

$$D(k^2) = 1 - ic_1 k + dk/(k + iL_0), \tag{6.2.48}$$

where $d \equiv -d'D(-L_0^2)/L_0$.[18] Equation (6.2.48) is the approximate analytic solution of the N/D equation that we parametrize for our study of the trinucleon system.

Now from Eqs. (6.2.37), (6.2.39), and (6.2.42) we have, for $k^2 > 0$,

$$k \cot \delta = -\frac{\operatorname{Re} D(k^2)}{N(k^2)} = -k \frac{\operatorname{Re} D(k^2)}{\operatorname{Im} D(k^2)}, \tag{6.2.49}$$

which can be used to eliminate the parameter d in terms of the scattering-length a:

$$-a = c_1 + d/L_0. \tag{6.2.50}$$

With this relation, Eq. (6.2.48) takes the form

$$D(k^2) = 1 + iak - i\frac{(a + c_1)k^2}{k + iL_0}. \tag{6.2.51}$$

We parametrize Eq. (6.2.51) by imposing the condition that it should produce the experimental ^{3}H ground- and virtual-state energies for the experimental scattering-length $a \equiv \bar{a} = 0.65$ fm. The ^{3}H ground and virtual states correspond to zeros of $D(k^2)$ for

$$k = iL = \pm i[4(\varepsilon - \varepsilon_2)/3]^{1/2}, \tag{6.2.52}$$

where ε and ε_2 are the triton and deuteron binding energies, respectively. The parameter L is positive (negative) for a ground (virtual) state.

Let L_B and L_V represent the L's corresponding to the experimental ^{3}H ground- and virtual-state energies, respectively, for the experimental scattering length $\bar{a}$. Then, we have, from Eq. (6.2.51),

$$1 - \bar{a}L_B + L_B^2(c_1 + \bar{a})/(L_B + L_0) = 0, \tag{6.2.53a}$$

and

$$1 + \bar{a}|L_V| + L_V^2(c_1 + \bar{a})/(L_0 - |L_V|) = 0, \tag{6.2.53b}$$

which yield

$$L_0 = L_B|L_V|[L_B|L_V|\bar{a} + L_B - |L_V|]^{-1}, \tag{6.2.54}$$

and

$$c_1 = -[L_B|L_V|\bar{a} + L_B - |L_V|]^{-1}. \tag{6.2.55}$$

Using Eqs. (6.2.49) and (6.2.51) we can now obtain $k \cot \delta$. The condition for having a ^{3}H ground or virtual state at $k = iL$ is given by $D(-L^2) = 0$, or

$$1 - aL + (c_1 + a)L^2/(L + L_0) = 0, \tag{6.2.56}$$

which, with Eqs. (6.2.54–6.2.55), we can rewrite as

$$aL = [|L_V|L_B + L(\bar{a}L_B|L_V| + L_B - |L_V|) - L^2]/(L_B|L_V|). \tag{6.2.57}$$

The parameters a and L refer to the approximate analytic solution of the N/D equations, and give the scattering length and the binding energy of the nd system. By construction, if $a = \bar{a}$ and $L = L_B$ or $= -|L_V|$, Eq. (6.2.57) is trivially satisfied. The approximate solution of the N/D problem of the form (6.2.57) suggests that a and L should be correlated in a definite way; we discuss this in Ch. 7.

6.2d. *Other Applications of Dispersion Relations*

With some simplifying assumptions about the unphysical and unitarity cuts, the N/D equations (6.2.40–6.2.41) can be solved numerically.[14–20] Many such calculations take into consideration the effect of inelasticity (breakup) using the techniques of Ref. 16. Given the positions and the residues of the bound-state poles, the N/D equations can be used to calculate the N and D functions, and hence the scattering observables. Conversely, given the experimental scattering phase-shifts, or the related D-function, these equations yield the residues at the bound-state poles or the ANP of the bound-state wave functions. In the second approach, the residues appear without any crude extrapolation of the scattering data to the bound-state pole as in earlier methods; it, in effect, provides a well-defined extrapolation of that data.

Both the FDR's and the PWDR's have proven to be very useful for the calculation of the bound-state pole residues, or the related vertex constants, or the ANP's.[10] Different FDR's have been used by various groups for these purposes.[7,10,20] Sometimes the effect of the Coulomb interaction has been included approximately. Extensive studies of the ANP have been made for various light nuclei by Ericson, Locher, Plattner, and their coworkers. Also, Kok and Rinat have used PWDR's to investigate the ANP and the related vertex constants for dpn, tnd, ^{3}Hedp, α^3Hen, αtp, ^{6}Liαd, ^{8}Be$\alpha\alpha$, etc. Locher and Mizutani[10] review the calculations of various groups, which are, however, only in qualitative agreement with each other.

6.3. Singularities of the Three-Particle Amplitudes

Amplitudes that refer to scattering with three particles in the initial and/or final states are intrinsically much more complicated than the two-particle amplitudes. They involve more momentum and energy variables and, of course, describe various types of physical processes.

The singularity structure of multiparticle amplitudes can be deduced from the relevant connected-kernel integral equations, but in practice one relies on other methods. One approach is to study the multiple-scattering (MS) series, ignoring questions of convergence, by which one can identify the dynamical, potential, and kinematic singularities. The dynamical singularities result from the unitarity of the S-matrix, and do not depend on the detailed properties of the interparticle interactions. The potential singularities are a consequence of these interactions, and kinematical singularities occur for certain values of initial and final momenta of the interacting particles. However, the potential and kinematic singularities can get mixed up and, except in the lowest-order terms of the MS series, it is not easy to disentangle them. We are mainly concerned with the dynamical singularities, which are analyzed by means of unitarity.

6.3a. Singularities of the Three-to-Three Scattering Amplitude

Rubin, Sugar, and Tiktopoulos (RST)[1] made a detailed study of the MS series for the three-to-three process assuming Yukawa pair interactions. The singularities in the three-to-three scattering amplitude were identified and classified as follows.

(i) *Dynamical singularities.* These consist of the unitarity cuts extending from $E = 0$ to ∞, where E is the three-body c.m. energy, and from $E = -B_i$ to ∞, where B_i is the binding energy of each of the two-body bound states. The first cut arises from three-body unitarity and the second from two-body unitarity. In addition, one has poles at the three-body bound-state energies.

(ii) *Potential singularities.* There are many branch cuts along parts of the negative-energy axis from $E = -\infty$ to some energy $E_C < 0$ that depends on the interaction potential.

(iii) *Anomalous cuts.* These cuts may run from $E = -\infty$ to 0; they are present only for incident or final particle momenta lying in a subdomain of the phase space defined by overall energy–momentum conservation. Another type of cut lies between $E = -B_i$ and $E = -A_i$ ($A_i > B_i$), where B_i is the binding energy of the two-body system and A_i is a function of the incident and final momenta of the particles. For certain values of the momenta, cuts of this kind disappear into the unphysical region through the scattering thresholds.

The RST analysis does not rely on the convergence of the MS series, but only on the assumption that the singularities present in individual terms of this series are also present in the full amplitude. This is a special case of the Feynman-graph analysis of the singularities of the scattering amplitude.[2] We next use this method to determine the low-energy behavior of the three-to-three amplitude for the scattering of three particles of mass m.

We know that the low-energy two-body scattering amplitude is a constant related to the S-wave scattering length. As the three-body c.m. energy E goes to zero, Amado and Rubin[3,4] have shown that the connected three-to-three scattering amplitude T_C has the divergent behavior

$$T_C \sim aE^{-1} + bE^{-1/2} + c\ln E + O(1), \tag{6.3.1}$$

where the constants a, b, and c are determined by the low-energy, on-shell, two-body, S-wave scattering amplitude, and certain kinematical factors. These constants may contain kinematic singularities.

The terms in T_C that diverge as $E \to 0$ arise only from connected parts of the first three terms of the connected MS series; typical examples are represented in Figs. 6.2(a–c). Such amplitudes have three free particles with initial (final)-state momenta $\mathbf{k}_i$ ($\mathbf{k}'_i$), $i = 1, 2, 3$, so

$$2E = k_1^2 + k_2^2 + k_3^2 = k_1'^2 + k_2'^2 + k_3'^2, \tag{6.3.2}$$

again in units $\hbar = m = 1$. For the study of the limit $E \to 0$, it is convenient to introduce new variables

$$\mathbf{k}_i = \sqrt{E}\,\mathbf{y}_i, \qquad \mathbf{k}'_i = \sqrt{E}\,\mathbf{y}'_i. \tag{6.3.3}$$

The contribution of the double-scattering process depicted in Fig. 6.2(a) to T_C is given by

$$\frac{\langle \frac{1}{2}\sqrt{E}(\mathbf{y}_1 - \mathbf{y}_2)|t_3[E(1 - \frac{3}{4}y_3^2)]|\sqrt{E}(\mathbf{y}'_1 + \frac{1}{2}\mathbf{y}_3)\rangle}{E[1 - \frac{1}{2}y_1'^2 - \frac{1}{2}y_3^2 - \frac{1}{2}(\mathbf{y}'_1 + \mathbf{y}_3)^2]}$$
$$\times \langle\sqrt{E}(\mathbf{y}_3 + \tfrac{1}{2}\mathbf{y}'_1)|t_1[E(1 - \tfrac{3}{4}y_1'^2)]|\tfrac{1}{2}\sqrt{E}(\mathbf{y}'_3 - \mathbf{y}'_2)\rangle, \tag{6.3.4}$$

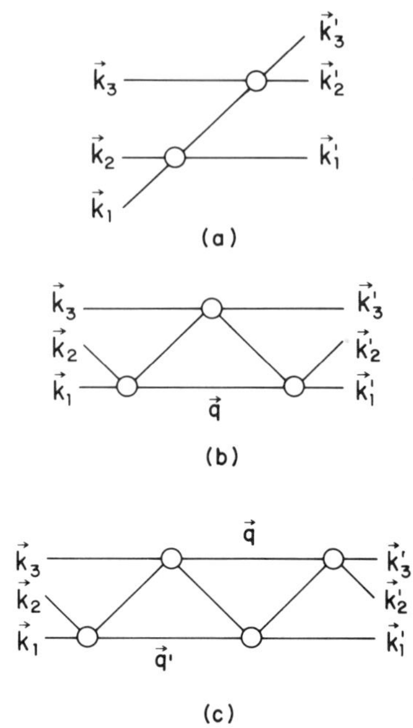

FIG. 6.2. Typical (a) second-, (b) third-, and (c) fourth-order scattering graphs in the c.m. frame. The circles indicate two-body t-matrices (from Adhikari (1973)).

where $\langle \mathbf{p}|t_i(E)|\mathbf{q}\rangle$ is the off-shell two-body t-matrix at energy E for scattering of pair jk ($i \neq j \neq k \neq i$) from relative momentum $\mathbf{q}$ to relative momentum $\mathbf{p}$. When $E \to 0$ we have, as in Section 6.2a,

$$\langle \mathbf{p}|t_i(E)|\mathbf{q}\rangle \to \left[a_i + i(-a_i^2)\sqrt{E}\right]/(2\pi^2), \tag{6.3.5}$$

where a_i is an S-wave scattering length. Using this in Eq. (6.3.4), we find that the double-scattering term in the zero-energy limit has the form

$$\frac{1}{2\pi^4}\left\{\frac{a_3 a_1}{E[2 - y_1'^2 - y_3^2 - (\mathbf{y}_1' + \mathbf{y}_3)^2]} - \frac{i a_3 a_1\left[a_3\sqrt{1 - \frac{3}{4}y_3^2} + a_1\sqrt{1 - \frac{3}{4}y_1'^2}\right]}{\sqrt{E}[2 - y_1'^2 - y_3^2 - (\mathbf{y}_1' + \mathbf{y}_3)^2]}\right\} + 0(1), \tag{6.3.6}$$

which is divergent as $E \to 0$ in the manner of the terms with coefficients a and b in Eq. (6.3.1). We see that a and b have rescattering (kinematical) singularities when the denominators in the expression (6.3.6) vanish. Evidently, a and b of Eq. (6.3.1) receive contributions from the six permutations of particles 123 in

Fig. 6.2. Similar remarks apply to the contributions of the higher-order terms in Fig. 6.2, which we consider next.

The contribution of the triple-scattering diagram of Fig. 6.2 to T_C is represented schematically by[3,4]

$$\frac{4}{\sqrt{E}}\int\frac{d^3x\,t_3t_1t_3}{[2-y_3^2-x^2-(\mathbf{y}_3+\mathbf{x})^2][2-y_3'^2-x^2-(\mathbf{y}'_3+\mathbf{x})^2]},\tag{6.3.7}$$

where $\mathbf{q}\equiv\mathbf{x}\sqrt{E}$ is the loop momentum; we have suppressed the momentum and energy variables of the two-body t-matrices. As $E\to 0$, the leading term in (6.3.7) diverges like $1/\sqrt{E}$:

$$\frac{a_3a_1a_3}{2\pi^6\sqrt{E}}\int\frac{d^3x}{[2-y_3^2-x^2-(\mathbf{y}_3+\mathbf{x})^2][2-y_3'^2-(\mathbf{y}'_3+\mathbf{x})^2-x^2]}.\tag{6.3.8}$$

The expression (6.3.7) has another part which diverges as $\ln E$ when $E\to 0$, and therefore contributes to the coefficient c in Eq. (6.3.1). This is obtained when we expand t_1 in the integral (6.3.7) and only keep the term proportional to $\sqrt{E}$:

$$\frac{-ia_3a_1^2a_3}{2\pi^6}\int_0^{x=\Lambda/\sqrt{E}}\frac{d^3x(1-\frac{3}{4}x^2)^{1/2}}{[2-y_3^2-x^2-(\mathbf{y}_3+\mathbf{x})^2][2-y_3'^2-x^2-(\mathbf{y}'_3+\mathbf{x})^2]}.\tag{6.3.9}$$

The integration in (6.3.7) extends over all of momentum space, but here a momentum cut off, $q=\Lambda$, is needed for convergence. The integral (6.3.9) diverges like $\ln E$ as $E\to 0$. Its divergent part is[4]

$$-i2a_3a_1^2a_3\pi^{-5}\int^{\Lambda/\sqrt{E}}\frac{x^2\,dx(\sqrt{3}/2)ix}{4x^4}=-a_3a_1^2a_3\frac{\sqrt{3}}{8\pi^5}\ln E.\tag{6.3.10}$$

The fourth-order MS diagram also has a part that diverges as $\ln E$. The diagram of Fig. 6.2(c) has the representation

$$\int^{\Lambda/\sqrt{E}}\int^{\Lambda/\sqrt{E}}\frac{8\,d^3x\,d^3z\,t_3t_1t_3t_1}{[2-y_3^2-x^2-(\mathbf{y}_3+\mathbf{x})^2][2-x^2-z^2-(\mathbf{x}+\mathbf{z})^2]}$$
$$\times[2-z^2-y_1'^2-(\mathbf{z}+\mathbf{y}'_1)^2]^{-1},\tag{6.3.11}$$

where now the loop momenta are $\mathbf{q}'=\mathbf{z}\sqrt{E}$ and $\mathbf{q}=\mathbf{z}\sqrt{E}$. If we approximate the two-body t-matrices by their S-wave scattering lengths, the integral (6.3.11) diverges and one has to introduce cut offs in the upper limits of x and z. By dimensional analysis we see that the integral (6.3.11) diverges as $\ln E$ as $E\to 0$; the calculation of the coefficient of the $\ln E$ term is straightforward.[4]

Also by dimensional analysis, one can show that all MS terms higher than the fourth order are finite as $E \to 0$. From the Faddeev equations, it follows that the sum of these higher-order terms is finite as $E \to 0$ so long as there is no three-body bound state at $E = 0$; such a state would simply be included as part of the a-term in Eq. (6.3.1).[4]

The preceding conclusions are clearly independent of the convergence properties of the MS series. These results also remain essentially unchanged in the unequal mass case; the form (6.3.1) remains the same but with different coefficients.

The same ideas can be extended to study the low-energy behavior of the connected N-body scattering amplitude ($N > 3$). For example, if we consider the connected MS series, it has been shown that the connected four-body amplitude, in the limit of vanishing four-body c.m. energy E, diverges as[4]

$$AE^{-2} + BE^{-3/2} + CE^{-1} + DE^{-1}\ln E + FE^{-1/2} + GE^{-1/2}\ln E + H\ln E + I\ln^2 E + O(1).$$

The coefficients of the divergent terms of this expansion are calculable in terms of the low-energy behaviors of the two- and three-body scattering amplitudes. The connected N-body scattering amplitude ($N > 4$) also diverges as $E \to 0$.[4]

The low-energy expansion of the connected three-to-three amplitude has a number of interesting applications. For example, the expansion (6.3.1) can be used to parametrize the three-body decay amplitudes.[5] It can also be applied to study the low-temperature behavior of the quantum-cluster coefficients in statistical mechanics.[6–9]

Dashen, Ma, and Bernstein (DMB)[8] express the grand canonical potential in terms of the S-matrix. They obtain expressions for the nth order quantum-cluster coefficient b_n in terms of energy integrals of the traces of the S-matrix weighted by the Boltzman factor, $\exp(-\beta E)$, with $\beta = 1/kT$, where T is the temperature and k the Boltzman constant. For low T, these integrals are dominated by the low-energy behavior of the scattering amplitudes. The result of DMB can be written as[9]

$$b_n - b_n^{(0)} = -\frac{n^{3/2}}{\pi}\int dE\, e^{-\beta E}\frac{\partial}{\partial E}\operatorname{Re}\{\operatorname{Tr}_n A \arctan[\pi\,\delta(E - H_0)\hat{K}(E)]\}_c, \tag{6.3.12}$$

where b_n and $b_n^{(0)}$ are the nth-order cluster coefficients with and without interactions, respectively; Tr_n means the trace in the n-body space; the symbol A means that the trace is supposed to be taken with respect to properly symmetrized or antisymmetrized states; and $\hat{K}(E)$ is the $\hat{K}$-matrix. The subscript c means that only connected terms are to be considered.

We first apply the expression (6.3.12) to the case $n = 2$. At low energies, the $\hat{K}$-matrix reduces to the S-wave scattering length, and we need keep only $\arctan x \sim x$. Then Eq. (6.3.12) yields the familiar low-T result

$$b_2 - b_2^{(0)} = -2a/\lambda + O(1/\lambda^3),$$

where λ is the thermal wavelength.

For $n = 3$, the low-energy behavior of the $\hat{K}$-matrix has the same form (6.3.1) as the t-matrix. The leading low-T behavior of b_3 comes from the linear term of the arctangent and the first term of the expansion (6.3.1). Essentially on the basis of dimensional arguments, one finds[9]

$$b_3 \sim O(a^2/\lambda^2) + O(a^3/\lambda^3) + \cdots.$$

Prior to the work of Ref. 9, it was believed that b_n should contain only positive and negative powers of λ.[10] However, the $\ln E$ term in Eq. (6.3.1), when substituted in Eq. (6.3.12), yields a term with the nonanalytic behavior $(a/\lambda)^4 \ln(a/\lambda)$. This term is missed without a correct treatment of the three-body problem.

Gibson used these results to calculate the low-T expansion of the third-cluster coefficient for three-dimensional gases obeying Boltzman, Bose–Einstein, and Fermi–Dirac statistics.[6] He did not use the expression (6.3.12) explicitly, but treated the MS series properly in order to arrive at the correct result. Later he extended his analysis to the one- and two-dimensional quantum-mechanical gases. Using similar ideas, one can also find the high-T behavior of the quantum-cluster coefficients.[7]

6.3b. *Low-Energy Behavior of the Two-to-Two and Two-to-Three Scattering Amplitudes*

We next study the behavior of the two-to-two and two-to-three amplitudes of the three-particle system as the c.m. energy E tends to zero. An analysis similar to that of the last section reveals that both types of amplitudes approach constants in this limit.

In order to see this, we consider again the first few terms of the MS expansion, as shown in Fig. 6.3(a) and 6.3(b), representing the lowest-order two-to-two and two-to-three scattering amplitudes, respectively. Apart from the vertex functions and two-body t-matrix elements, both of which tend to constants as $E \to 0$, the two terms shown in Figs. 6.3(a) and 6.3(b) contain the

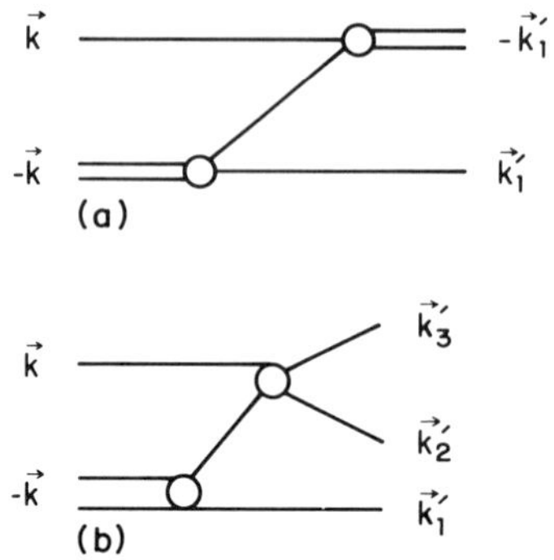

FIG. 6.3. Typical two-to-two and two-to-three amplitudes of lowest order.

energy denominator

$$[(E - \tfrac{1}{2}k^2 - \tfrac{1}{2}k_1'^2 - \tfrac{1}{2}(\mathbf{k} + \mathbf{k}_1')^2]^{-1}, \tag{6.3.13}$$

as in Eq. (6.3.4). For the two-to-two scattering amplitude, the energy-conservation condition is now

$$E + \alpha^2 = \tfrac{3}{4}k^2 = \tfrac{3}{4}k_1'^2, \tag{6.3.14a}$$

where α^2 is the two-particle binding energy while, for the two-to-three scattering amplitude, we have

$$\begin{aligned} E + \alpha^2 &= \tfrac{3}{4}k^2, \\ 2E &= k_1'^2 + k_2'^2 + k_3'^2. \end{aligned} \tag{6.3.14b}$$

With conditions (6.3.14a, b) we see that, unlike with Eq. (6.3.4), the energy denominator (6.3.13) does not necessarily vanish as $E \to 0$, because k^2 tends to a nonvanishing constant unless α^2 is zero. Hence the amplitudes of Figs. 6.3(a) and 6.3(b) do not diverge as $E \to 0$, nor do the higher-order diagrams.

Although they are finite, these amplitudes have a branch-cut singularity at $E = 0$ arising from the scattering threshold.[2] For example, the two-body t-matrix is finite at $E = 0$, but it does have a square-root branch point at this energy.

Next, let us consider[11] the singularity structure of the two-to-two amplitude of the three-particle system at $E = 0$, which consists of a branch point at $E = 0$ with an associated branch cut running from $E = 0$ to $E = \infty$. The discontinuity across this branch cut is obtained from the unitarity relation depicted in Fig. 6.4, where the vertical lines represent the Dirac δ-function part

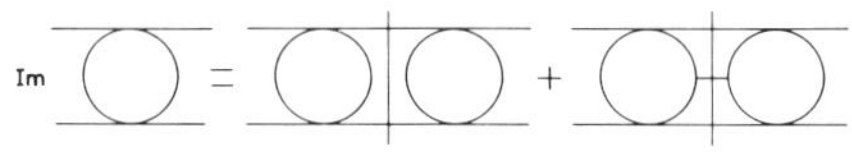

FIG. 6.4. Diagrammatic representation of the unitary relation satisfied by the two-to-two amplitude of the three-body system.

of the Green function corresponding to on-shell propagation. The discontinuity across the $E = 0$ cut is then given by[11]

$$\text{Disc}\, T_{2\to 2} \sim \prod_{i=1}^{3} \int d^3k_i\, \delta(E - \tfrac{1}{2}k_1^2 - \tfrac{1}{2}k_2^2 - \tfrac{1}{2}k_3^2) \times \delta^{(3)}(\mathbf{k}_1 + \mathbf{k}_2 + \mathbf{k}_3)\,|T_{2\to 3}|^2, \tag{6.3.15}$$

to within irrelevant factors. Here, Disc denotes the contribution of the last term in Fig. 6.4 with the three-body intermediate state. The Dirac δ-functions represent energy–momentum conservation. Equation (6.3.15) reduces to

$$\text{Disc}\, T_{2\to 2} \sim \int k^5\, dk\, d\Omega\, \delta(E - k^2)\,|T_{2\to 3}|^2, \tag{6.3.16}$$

where $d\Omega$ represents the angular integration subject to $\mathbf{k}_1 + \mathbf{k}_2 + \mathbf{k}_3 = 0$, and k is the hyperradius in momentum space. If we assume that $T_{2\to 3}$ is constant at the three-body threshold, the energy-dependent part of Disc $T_{2\to 2}$ in the neighborhood of $E = 0$ is given by

$$\text{Disc}\, T_{2\to 2} \sim E^2. \tag{6.3.17}$$

This implies a singularity of the form[11]

$$\text{Sing}\, T^{(3)}_{2\to 2} \sim E^2 \ln(-E), \tag{6.3.18}$$

at the three-body threshold, which is finite at $E = 0$. The superscript (3) in Eq. (6.3.18) denotes the three-body threshold.

The singularity (6.3.18) of the amplitude T_{nd} has been observed at the onset of three-body breakup in the trinucleon system. The nd S-wave, spin-doublet, effective-range function does not depend smoothly on E at the three-body breakup threshold, but has an anomalous behavior suggestive of a singularity.[12,13] Conventional parametrizations of this function fail to reproduce this anomalous behavior. The mere existence of the singularity, however, does not necessarily explain the anomalous behavior. For the anomalous behavior to be visible, the coefficient of the singular term has to be sizable.[13]

Fishbane and Noble[14] make explicit use of a singular term of the form (6.3.18) to parametrize the ^{2}H–^{4}He elastic scattering effective-range function near the deuteron breakup threshold. They obtain a good fit for the S-wave phase shift in this region.

The same idea can be extended to study the singular part of $T_{2\to 2}$, namely Sing $T^{(N)}_{2\to 2}$, at the N-body threshold. The results are, for N odd,[11]

$$\text{Sing } T^{(N)}_{2\to 2} \sim E^{(3N-5)/2} \ln(-E), \tag{6.3.19}$$

and, for N even,

$$\text{Sing } T^{(N)}_{2\to 2} \sim E^{(3N-5)/2}. \tag{6.3.20}$$

The preceding considerations can be used to find the singular part of the two-to-three scattering amplitude $T_{2\to 3}$ at threshold. Again one starts by calculating the discontinuity across the branch cut by using the unitarity relation given diagrammatically in Fig. 6.5. The summation on the right-hand side of this relation is over all three final-state pairs i, which, as we see later, contribute not to the singularity in the total c.m. energy E, but only to the singularities in the sub-energy. The first two terms on the right-hand side of Fig. 6.5 contribute the the singularity in E. The discontinuity we are interested in arises from the second term on the right-hand side of Fig. 6.5, and is given by

$$\text{Disc } T_{2\to 3} \sim \prod_{i=1}^{3} \int d^3k_i\, \delta(E - \tfrac{1}{2}k_1^2 - \tfrac{1}{2}k_2^2 - \tfrac{1}{2}k_3^2) \times \delta^{(3)}(\mathbf{k}_1 + \mathbf{k}_2 + \mathbf{k}_3)\, T_{2\to 3} T^*_{3\to 3}. \tag{6.3.21}$$

This term gives rise to the singularity in E at $E = 0$. The first term on the right hand side of Fig. 6.5 represents the singularity in E at the two-body threshold. We have seen in the last section that, as $E \to 0$, the amplitude $T_{3\to 3}$ is singular and divergent as given by Eq. (6.3.1), but $T_{2\to 3}$ is finite in this limit. The phase-space integral, as before, yields $E^2 \ln(-E)$ for the singular part of $T_{2\to 3}$. Hence, the leading singularities of (6.3.21) are given by the product of $E^2 \ln(-E)$ of the phase-space integral with the singularities of $T_{3\to 3}$ given by Eq. (6.3.1):

$$\text{Sing } T_{2\to 3} \sim \ln(-E)[aE + bE^{3/2} + cE^2 \ln(-E) + \cdots]. \tag{6.3.22}$$

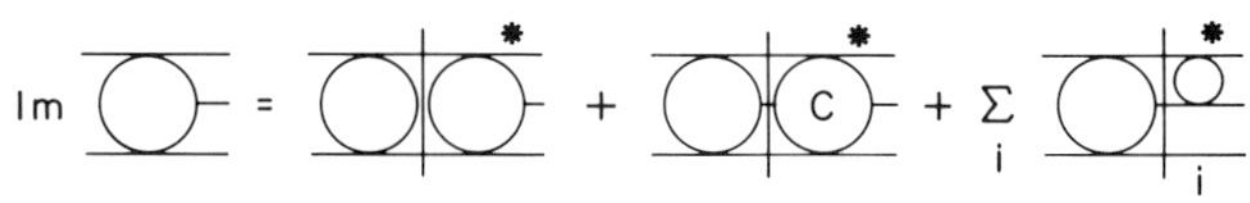

FIG. 6.5. Diagrammatic representation of the unitary relation satisfied by the two-to-three amplitude.

This expression, which is finite at $E = 0$, is the same as for the one-to-three decay amplitude obtained by Amado, Freeman, and Rubin[15] using the MS series.

6.3c. *Subenergy Singularities in the Two-to-Three Scattering Amplitude*

The singularities in the subenergy variables in the breakup amplitude $T_{2\to 3}$ are essential for providing a correct description of final-state interactions.[16] Usually, in breakup processes, one measures the spectrum of a single particle. For example, let us consider the process shown in Fig. 6.6, where three distinguishable particles 1, 2, 3 are produced and we suppose only the spectrum of the particle 1 is measured. Then, one requires the proper singularity structure of this amplitude as a function of the energy of the particle 1 in order to make a correct parametrization. The nature of this singularity can be determined from unitarity (Fig. 6.5). Since the first two terms on the right-hand side of the unitarity relation contribute only to the singularity in E, we confine ourselves to the last term in Fig. 6.5.

The amplitude $T_{2\to 3}$ can be decomposed into a sum of terms depending on which pair interacts last, *viz.*,

$$T_{2\to 3} = \sum_i T_i t_i, \tag{6.3.23}$$

where t_i is the on-shell two-body scattering amplitude of the j-k pair ($i \neq j \neq k \neq i$) and T_i is its coefficient in $T_{2\to 3}$. This is the standard isobar, or Faddeev, or MS decomposition.[17] Using the unitarity relation we can show that T_i has a singular dependence on the $j - k$ relative momentum q_i. The relevant unitarity relation for $T_{2\to 3}$ that represents this singularity is given schematically by[18]

$$\text{Disc}\, T_{2\to 3} = \sum_i T_{2\to 3}\rho_{2i} t_i^*, \tag{6.3.24}$$

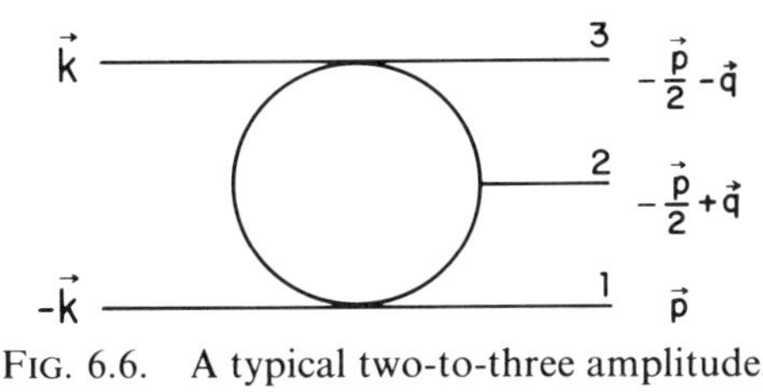

FIG. 6.6. A typical two-to-three amplitude.

where ρ_{2i} represents the usual phase-space integral involving the δ-function. Substituting Eq. (6.3.23) in Eq. (6.3.24), we get

$$\sum_i T_i \operatorname{Disc} t_i + \sum_i (\operatorname{Disc} T_i) t_i^* = \sum_i T_i t_i \rho_{2i} t_i^* + \sum_{j \neq i} T_j t_j \rho_{2i} t_i^*. \tag{6.3.25}$$

The first terms on the left- and right-hand sides of this equation cancel by two body unitarity ($\operatorname{Disc} t_i = t_i \rho_{2i} t_i^*$), and with the phase-space integral ρ_{2i} written explicitly, we find

$$\operatorname{Disc} T_i \sim \sum_{j \neq i} \int d^3 q_i \, \delta(E - \tfrac{3}{4} p_i^2 - q_i^2) \, T_j t_j. \tag{6.3.26}$$

We have used the momentum variables of Fig. 6.6 and suppressed the momentum variables of the amplitudes T and t.

The $\mathbf{q}_i$-integral in Eq. (6.3.26) yields a square-root discontinuity of the form $(E - \frac{3}{4} p_i^2)^{1/2} \equiv \bar{q}_i$. If we assume that the remaining quantities of this equation remain finite when the argument of the δ-function vanishes, the singular part of T_i is given by

$$\operatorname{Sing} T_i \sim \bar{q}_i \sum_{j \neq i} T_j t_j. \tag{6.3.27}$$

Hence T_i has a singular part that is proportional to the non-i part of T. In partial-wave form, this equation still involves an integral over a range of q_i values, and so represents a set of coupled integral conditions involving the T's.

Many phenomenological analyses of three-body final states assume that the i-term in Eq. (6.3.23) carries all the information about the final-state interaction of the pair i, and so suppose that

$$T_{2 \to 3} = - T_i e^{i\delta_i} \frac{\sin \delta_i}{q_i} + B_i, \tag{6.3.28}$$

where

$$B_i \equiv \sum_{j \neq i} T_j t_j \tag{6.3.29}$$

is assumed to be a slowly-varying function of the relative momentum of pair i.[19] This implies that only the i-term in the decomposition (6.3.23) carries the elastic scattering phase of the pair i, while the remaining terms contribute to a slowly-varying background. The result (6.3.27) shows that this line of reasoning is not generally true.

Equation (6.3.27) suggests the parametrization

$$T_i = A_i - i q_i B_i. \tag{6.3.30}$$

Then, Eq. (6.3.28) takes the form

$$T_{2\to 3} = -A\frac{e^{i\delta}\sin\delta}{q} + Be^{i\delta}\cos\delta, \tag{6.3.31}$$

where we have dropped the i subscript for simplicity. This shows that the entire amplitude carries the phase δ coherently.[16,19]

To illustrate this, let us consider the half-shell form of the LS equation (2.5.12a), where since the half-shell t-matrix carries the scattering phase, it might be incorrectly thought that only the kernel term of the LS equation carries this phase. This equation can be rewritten symbolically as

$$t = [V - i\pi V\,\delta(E - H_0)t] + \{VPG_0t\}, \tag{6.3.32}$$

where P denotes the principal-value prescription. Because of the δ-function, the t-matrix in the square-bracket is on-shell. Using the explicit partial-wave forms, we see that the square-bracket term carries the elastic scattering phase. The curly-bracket term also carries the elastic scattering phase. Hence the two terms of the half-shell LS equation (6.3.32) add up coherently to produce the overall elastic scattering phase.[16]

The singularity of Eq. (6.3.27) lies at the boundary of the physical region because the allowed values of p^2 are from 0 to $(4/3)E$. It has been stressed by Aitchison, Amado *et al.* that this singularity leads to rapid variations in the final-state amplitude and is crucial to the phenomenological treatment of the final-state interaction without three-body dynamical equations.[16,18–21] Calculations based upon such dynamical equations, of course, automatically build in this singularity.

The preceding discussion applies to general multibody final states.[22] In special situations, the neglect of the term $-q_iB_i$ in Eq. (6.3.30) may lead to reasonable results. This is true for a special class of nuclear reactions involving projectile fragmentation ($a \to b + x$) that are identified by the detection of the fragment near its grazing angle, when its spectrum is found to peak near its grazing velocity. The inclusive reaction is described by[23]

$$\begin{aligned} a + A &\to b + x + A^* \\ &\to b + X, \end{aligned} \tag{6.3.33}$$

indicating that only b is detected. Although a strong interaction between b and the target A is possible, it is expected that b plays a spectator, or passive, role in the reaction involving the collision of the fragment x with A when its inclusive spectrum peaks near the beam velocity. Consequently, the sum (6.3.23) can be truncated to a single term and T_i taken to be slowly varying. Such a treatment of inclusive reactions has been very successful, but sometimes it has been applied beyond the domain of its applicability.[23,24]

The three-particle, off-shell, two-to-two scattering amplitude T has the singularity (6.3.27). This can be seen from the zero-range form of the Amado-model equation that we write schematically as[21,25,26]

$$\langle \mathbf{p}|T|\mathbf{k}\rangle = \langle \mathbf{p}|B|\mathbf{k}\rangle + \int^{\Lambda} d^3q \, \langle \mathbf{p}|B|\mathbf{q}\rangle \tau(E - \tfrac{3}{4}q^2)\langle \mathbf{q}|T|\mathbf{k}\rangle. \quad (6.3.34)$$

We have set the two-particle interaction form-factors equal to unity; we are interested in a dynamical singularity, and these form-factors contribute to only the potential singularities. Also, $\mathbf{k}$ is the on-shell momentum at energy E, τ is the two-body propagator, and Λ is a cut off to ensure convergence at $q = \infty$. Equation (6.3.34) implies that T must have a branch cut that is obtained by setting the energy-denominator $(E - p^2 - q^2 - \mathbf{p}\cdot\mathbf{q})$ of the nucleon-exchange term $\langle \mathbf{p}|B|\mathbf{q}\rangle$ equal to its Dirac δ-function part. The discontinuity across this branch cut is given by[26]

$$\text{Disc}\,\langle \mathbf{p}|T|\mathbf{k}\rangle \sim \int d^3q \, \tau(E - \tfrac{3}{4}q^2)\,\delta(E - q^2 - p^2 - \mathbf{p}\cdot\mathbf{q})\langle \mathbf{q}|T|\mathbf{k}\rangle, \quad (6.3.35)$$

which is depicted in Fig. 6.4. If we make a partial-wave decomposition

$$\langle \mathbf{p}|T|\mathbf{k}\rangle = \sum_{L,M} Y_L^M(\tilde{\mathbf{p}}, \bar{\zeta})T_L(p,k)Y_L^M(\tilde{\mathbf{k}}, \tilde{\zeta})^*, \quad (6.3.36)$$

we find that Eq. (6 .3.35) reduces to

$$\text{Disc}\, T_L(p,k) \sim \int q^2\,dq\,\tau(E - \tfrac{3}{4}q^2)\int_{-1}^{+1} dz\, P_L(z)\,\delta(E - q^2 - p^2 - pqz)\,T_L(q,k). \quad (6.3.37)$$

The singularity implicit in (6.3.37) is explicated if we make the change of variables $p^2 = x$ and $q^2 = y$, to get

$$\text{Disc}\, T_L(E,x) \sim \frac{1}{\sqrt{x}}\int_{y_-}^{y_+} dy\,\tau\left(E - \frac{3}{4}y\right)P_L\left(\frac{E - x - y}{\sqrt{xy}}\right)T_L(E,y), \quad (6.3.38)$$

where $y_\pm = E - x/2 \pm [x(E - \tfrac{3}{4}x)]^{1/2}$, $T_L(E,x) \equiv T_L(p,k)$, and $T_L(E,y) \equiv T_L(q,k)$. From Eq. (6.3.38), it is obvious that, for x near zero, y values near E contribute; while for x near $(4/3)E$, y values near $(1/3)E$ contribute. Since as x approaches $(4/3)E$, the integration region shrinks to zero, Disc $T_L(E,x)$ behaves like $[E - (3/4)x]^{1/2}$ in this region. This means that $T_L(E,x)$ has a square-root branch-point behavior $[E - (3/4)x]^{1/2}$ at the boundary of the physical region. This branch cut runs from $E - 3x/4 = 0$ to $E - 3x/4 = \infty$, or from $x = 4E/3$ to $-\infty$, just as we would expect for a two-body scattering cut in the three-body space. The physical region is given by $0 \leq x \leq 4E/3$. The role of this branch point was first emphasized by Amado and collaborators, who demonstrated its importance in the analyses of final states, especially in the three-pion and the nd systems.[18,26,27]

The presence of the singularity implicit in Eq. (6.3.27) or (6.3.38) can also be demonstrated by other means, for example, with the Faddeev equations or the MS series. Nevertheless, it is not easy to incorporate this singularity properly into phenomenological analyses of few-body final states.[20,26] The inclusion of condition (6.3.27) in such analyses without violating the constraints of analyticity calls for a full dynamical three-body treatment of the problem. This casts doubt over analyses of the inclusive spectrum in reactions of the type (6.3.33) and some analyses of few-hadron final states as well.[26,27]

6.4. Efimov and Thomas Effects

In this section, we study a very peculiar behavior of the three-body amplitude as the strength of the two-body potential is progressively increased. It has been shown by Efimov that when this is done so as to bind the first S-wave, zero-energy, two-body bound state (the *Efimov limit*), the number of three-body bound states increases to infinity. This surprising result is called the *Efimov effect*.[1–6] When the two-body potential is strengthened further to obtain a nonzero two-body binding energy, the number of three-body bound states is finite. This unusual discontinuous effect does not depend on the details of the two-body interaction so long as it is *short ranged*. To summarize: In the Efimov limit, the three-body amplitude has an accumulation of an infinite number of bound-state poles at $E = 0$. Consequently, the kernel of the three-body scattering integral equation in the Efimov limit has an infinite number of eigenvalues greater than unity at $E = 0$ (see Eq. (3.4.4)), and it is noncompact so the dynamical equations we developed for the three-particle system do not hold in this limit.[3] The Efimov effect has been the subject of many theoretical and numerical studies, which we review in some detail.[5–7]

Specifically, Efimov showed that if three nonrelativistic, identical bosons interact via a short-range, attractive two-particle interaction $-\lambda V(r)$, characterized by a range parameter r_0, then in the limit as $\lambda \to \lambda_c$, where λ_c is the strength needed to support the first zero-energy two-particle bound state, the number N of three-particle S-wave bound states tends to infinity as

$$N \sim \pi^{-1} \ln(|a|/r_0), \tag{6.4.1}$$

where a is the two-particle scattering length that tends to infinity as $\lambda \to \lambda_c$.

In a different context, Thomas[8] showed that when three nonrelativistic identical bosons interact via a two-particle interaction $-\lambda V(r)$ of range r_0 and fixed two-particle binding energy, then as $r_0 \to 0$, the binding energy B of the three-particle system increases beyond any limit: $B \to \infty$. However, as $r_0 \to 0$, the parameter λ has to be increased so that the binding energy of the two-particle system remains fixed (the *Thomas limit*). This phenomena is known as the *Thomas effect*.

These two apparently different effects are related to the divergence of the trace of the kernel of the three-particle scattering equation in two different limits. In fact, we will see that, as $r_0 \to 0$ in the Thomas limit, the three-particle system is expected to have an infinite number of bound states with an accumulation point at infinite binding energy, instead of $E = 0$ as in the Efimov effect.

The original demonstration of Efimov was carried out in hyperspherical coordinate space (hyperspace). He demonstrated that the effective three-particle interaction develops a long-range R^{-2} tail, where R is the hyperradius, in the Efimov limit that is responsible for the infinite number of three-body bound states. Instead of using this technique, we follow the approach of Amado and Noble,[3] and its generalization by Adhikari *et al.*,[4] based on the three-body Faddeev equations. There are other approaches to the problem, such as the Born–Oppenheimer model[5] of two heavy and one light particles, which we do not consider.

We first try to make plausible the appearance of an infinite number of Efimov states, then we show how can these states disappear as the potential strength is increased. For simplicity, we consider a system of three identical bosons, though most of our considerations are more general. For identical bosons, the Faddeev equations reduce to a single equation and, as we are interested in bound states, we need consider only its homogeneous, S-wave form, which is, schematically,

$$|\psi_n\rangle = K(E, \lambda)|\psi_n\rangle, \tag{6.4.2}$$

where n labels the different eigenfunctions. The kernel K is a function of the potential strength λ and the three-body c.m. energy $E \leq 0$.

In general, the eigenvalue equation for K is

$$K(E, \lambda)|\phi_n\rangle = \eta_n|\phi_n\rangle, \tag{6.4.3}$$

where both the eigenvalues η_n and eigenfunctions $|\phi_n\rangle$ are functions of E and λ. This eigenvalue problem allows us to relate the divergences of the traces, $\operatorname{Tr} K^m$, where m is any positive integer, with the appearances of the Efimov and the Thomas effects. Whenever $\eta_n = 1$, the eigenfunction $|\phi_n\rangle$ coincides with

$|\psi_n\rangle$. For $E < 0$ and fixed λ, η_n is monotonic in E, hence $\eta_n = 1$ at most once for $-\infty < E < 0$. Therefore, if an infinite number of bound states, each with $E < -B$, appears for some given λ, then an infinite number of η_n's must exceed unity for $E = -B$. Thus, $\mathrm{Tr}\, K^m = \sum_n \eta_n^m$ diverges at $E = -B$. If the traces of powers of K diverge in both the Efimov and the Thomas limits, this is an indication that the kernel is noncompact in these limits, and the Fredholm method no longer applies.

The trace is representation independent. It is convenient, however, to study the trace of the kernel of the Faddeev equations in the momentum representation. This was done in detail by Amado and Noble, and we quote their results in what follows.[3] The divergence of $\mathrm{Tr}\, K^m$ comes from the low-energy, low-momentum region, where it is valid to use the effective-range expansion for the two-body t-matrix. Of course, then one has to cut off the momentum-space integrals at a large momentum Λ in order to achieve the convergence normally provided by the two-body form-factors. This corresponds to the zero-range version of the three-body equation (6.4.2). which in S-wave form is

$$\psi_n(p) = \int_0^\Lambda q^2\, dq \frac{1}{pq} \ln\left(\frac{p^2 + q^2 + pq - E}{p^2 + q^2 - pq - E}\right) \frac{1}{1/a + \sqrt{\frac{3}{4}p^2 - E}}\, \psi_n(q), \quad (6.4.4)$$

where the parameter Λ is related to the range r_0 of the two-body interaction by $\Lambda = r_0^{-1}$. Now the Thomas and Efimov limits correspond to taking $r_0 = \Lambda^{-1} \to 0$, and $|a| \to \infty$, respectively.

The trace of the kernel in Eq. (6.4.4) is given by

$$\mathrm{Tr}\, K \sim \int_0^\Lambda dp \ln\left(\frac{\beta^2 + 3p^2}{\beta^2 + p^2}\right) \frac{1}{\alpha + \sqrt{\frac{3}{4}p^2 + \beta^2}}, \quad (6.4.5)$$

where $\beta^2 = -E$ and $\alpha = 1/a$. The divergent part of this integral is independent of the sign of α, and hence of a. On dimensional grounds, one sees that for $\beta^2 = -E = 0$, $\mathrm{Tr}\, K$ diverges at the lower limit ($p \to 0$) as $\ln|\alpha|$ when $|\alpha| \to 0$, and diverges at the upper limit ($p \to \Lambda$) as $\ln \Lambda$ when $\Lambda \to \infty$. The first divergence corresponds to the Efimov limit and the second to the Thomas limit. These two divergences have essentially the same mathematical origin and one implies the other.

In order to see this more clearly, let us rewrite the expression (6.4.5) as ($y \equiv p/\alpha$)

$$\mathrm{Tr}\, K \sim \int_0^{\Lambda/\alpha} dy \ln\left[\frac{\beta^2/\alpha^2 + 3y^2}{\beta^2/\alpha^2 + y^2}\right]\left[1 + \left(\frac{3}{4}y^2 + \frac{\beta^2}{\alpha^2}\right)^{1/2}\right]^{-1}. \quad (6.4.6)$$

Now this form of Tr K is finite at the lower limit $y \to 0$, but may diverge at the upper limit when $(\Lambda/\alpha) \to \infty$. Hence, both the Efimov and the Thomas divergences are related to the large-y divergence of Eq. (6.4.6).[4]

The essential divergence properties of Eq. (6.4.6) can be summarized as follows. For $\beta = 0$, the expression (6.4.6) diverges as $\ln(\Lambda/\alpha)$ if either $\alpha \to 0$ or $\Lambda \to \infty$; for $\beta = \beta_0 > 0$, it diverges as $\ln \Lambda$ as $\Lambda \to \infty$ and remains finite as $\alpha \to 0$. The first divergence means that in both the Efimov and the Thomas limits the three-body system has an infinite number of bound states. The second divergence means that, in the Thomas limit, these infinite number of bound states all have binding energies greater than β_0^2. As this is true for any β_0, however large, we conclude that an infinite number of Thomas states accumulate at infinite binding energy.[4] The fact that for $\beta = \beta_0$ the expression (6.4.6) is finite as $\alpha \to 0$ means that the infinity of Efimov states accumulates at zero energy.

The divergence of Tr K at $E = -B$, however, is only a necessary condition for the appearance of an infinite number of bound states of binding energy greater than B. In order that the infinite number of bound states really appear, one requires that Tr K^m also diverges as $\ln(\Lambda/\alpha)$ for all m, such that the infinite number of eigenvalues of K accumulate at a value greater than unity. Amado and Noble find that this is indeed the case. By considering Tr K^2 they show that, in the Efimov limit,

$$\frac{\operatorname{Tr} K^2}{\operatorname{Tr} K} \to \text{constant} > 1, \tag{6.4.7}$$

from which they argue that an infinite number of eigenvalues of K accumulate at a value greater than unity. The same argument is readily extended to the Thomas limit in order to prove the existence of an infinite number of Thomas bound states.

Apart from this similarity in the mathematical description of the Efimov and Thomas effects, there is a physical relationship between the two. In the Efimov effect, $|a| \to \infty$ and Λ is finite, whereas for the Thomas effect the reverse is true, but in each $|a| \gg 1/\Lambda$ and so $\Lambda|a| \to \infty$ covers both cases. We call the $\ln(|a|\Lambda)$ behavior of Tr K for $\Lambda|a| \to \infty$ the Efimov–Thomas divergence.[4]

The Thomas effect is only of pedagogic interest, since it only occurs in the zero-range limit of the two-body interaction. The Efimov effect, on the other hand, is relevant to physical quantum-mechanical three-body systems as it is quite possible to have a two-particle interaction that is near the Efimov limit and so one, in practice, may have a finite number of quasi-Efimov states. In such a situation, the properties of the three-body system are governed by the fact that the Efimov limit is nearby. This is our motivation for the following elaboration of the Efimov effect.

We first examine how the number of Efimov bound states can decrease with increasing coupling. It follows from Feynman's theorem that, if λ is the potential strength, then

$$\frac{\partial B_2}{\partial \lambda} > 0, \qquad \frac{\partial B_3}{\partial \lambda} > 0, \tag{6.4.8}$$

where B_i is the binding energy of the i-body system. This means that both the two-body and three-body binding energies increase with increasing coupling. Then how can the number of three-body bound states decrease? It cannot, so long as the lowest scattering threshold remains fixed in the process. Strange things may happen when one has moving thresholds in the multiparticle system.[9] As λ increases, B_2 increases too, and it does so at a rate greater than the rate of increase of B_3. (We should remember that $B_3 > B_2$ in order for the three-body bound state to be stable against decay into a two-body bound state and a third free particle.) What happens, then, is that with increasing coupling the lowest scattering threshold moves faster to the left than the three-body bound states and overruns the three-body bound states, which then move into the second Riemann sheet R_2 associated with the square-root singularity of the lowest (two-body) scattering threshold. This happens if

$$\frac{\partial B_2}{\partial \lambda} > \frac{\partial B_3}{\partial \lambda} > 0, \tag{6.4.9}$$

which is consistent with Eq. (6.4.8). Eventually, the infinity of Efimov states become virtual states as they move onto R_2.[6]

The appearance and disappearance of the Efimov states is illustrated in Fig. 6.7, where we represent the motions of the bound-state poles in the complex energy plane. In Fig. 6.7(a), we consider the situation $\lambda < \lambda_c$, where λ_c is the strength corresponding to the Efimov limit; the three-body cut starts at $E = 0$ and the crosses represent a finite number of three-body bound states. In Fig. 6.7(c), for $\lambda = \lambda_c$, we have an infinite number of three-body bound states. In Fig. 6.7(e), $\lambda > \lambda_c$, and we have a nonzero binding energy of the two-body system that generates the two-body cut and forces the infinity of Efimov bound states to become virtual states on R_2.

We have demonstrated the possibility of the Efimov effect in the case of three identical bosons. Depending on the strength of the two-body interaction, the number of S-wave three-body bound states may tend to infinity. Certain special features of the Efimov states are worth emphasizing:[2]

(i) These states disappear when the strength of the two-body interaction is increased.

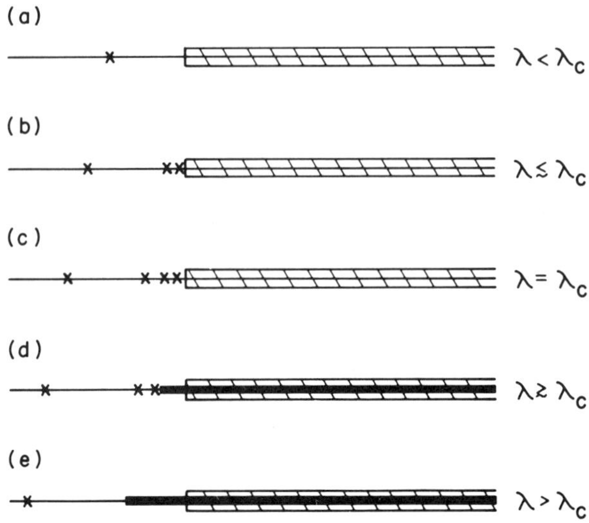

FIG. 6.7. The three-body bound states (denoted by crosses) and the scattering cuts (denoted by hatched lines) in the complex three-body c.m. energy plane for various values of two-body coupling constant λ. In the Efimov limit, $\lambda = \lambda_c$, an infinite number of three-body bound states appear. As the coupling constant increases ($\lambda > \lambda_c$), the moving two-body cut overtakes the three-body bound-state poles and forces them to become virtual states in the unphysical sheet associated with this cut.

(ii) These states are loosely bound and have sizes (root-mean-square radii) that extend well beyond the range of the two-body interaction.
(iii) The binding energies of excited Efimov states show a scaling property. They are linearly spaced on a logarithmic scale of energy.
(iv) As the strength of the two-body interaction is increased, these states are run over by the two-body scattering cut, move into the unphysical sheet associated with this cut, and become virtual states.

The Efimov effect can occur in many asymmetric situations, e.g., nonidentical particles and particles with spin. This is especially interesting since many of the three-body systems we encounter in nature are constituted of such particles. In particular, we demonstrate in the next chapter that the behavior of the trinucleon system seems to be governed largely by the fact that the nucleon–nucleon interaction is weak enough to have a loosely bound deuteron on the nuclear scale, but strong enough so that no actual Efimov states with properties (i)–(iv) appear. Consequently, much of the behavior of the low-energy trinucleon system can be understood by using an effective nd interaction whose form is implied by an approximation to the Efimov effect.

Efimov has shown that in the Efimov limit the long-range part of the effective interaction, $V_{eff}(R)$, in hyperspace develops a R^{-2} behavior that is responsible for the infinite number of bound states. Here, R is the hyperradius of the system defined by[1,2]

$$R^2 = \tfrac{2}{3}[r_{12}^2 + r_{23}^2 + r_{31}^2], \tag{6.4.10}$$

where r_{ij} is the separation between particles i and j. For small R, the potential $V_{eff}(R)$ does not have the R^{-2} behavior and it usually has a finite negative value. Though the specific R^{-2} behavior only emerges from a mathematical treatment in a model, as has been performed by Efimov, the appearance of an attractive effective interaction, well beyond the range r_0 of two-body interactions, in a symmetric state with total angular momentum zero, can be intuitively and qualitatively understood by considering the one-particle-exchange effective interaction of this system. Because of the large size of the two-body system in the Efimov limit, particle i can simultaneously be bound to both the particles j and k $(i \neq j \neq k \neq i)$ provided that the separation between particles j and k does not roughly exceed the sum of root-mean-square radii of the i-j and/or i-k subsystems. More rigorously, in the identical-particle case, the effective long-range interaction has been shown to be cut off at around $R = R_0 \sim |a|$. In the Efimov limit, $|a| \gg r_0$, hence, the range R_0 of the effective interaction could be much greater than r_0. Also the effective interaction saturates at distances smaller than r_0. The preceding discussion also holds in the case where the two-body system has a low-energy virtual state rather than a bound state. The potential $V_{eff}(R)$ has a R^{-2} behavior for distances $r_0 \ll R \ll |a|$.[1,2]

Fonseca *et al.*[5] have considered the Efimov effect in a three-body system constituted of two heavy particles and a light particle, using the Born–Oppenheimer approximation, in the limit where the light–heavy system has the first zero-energy bound state. They employed a separable light–heavy interaction with Yamaguchi form-factors, and then derived the effective heavy–heavy interaction V_{hh}, arising due to the exchange of the light particle between the two heavy particles. They found that V_{hh} has a R_{hh}^{-2} behavior at intermediate distances for $|a| \gg R_{hh} \gg r_0$, where a is the light–heavy scattering length, r_0 is the range of the light–heavy interaction, and R_{hh} is the heavy–heavy separation. In the Efimov limit, $|a| \to \infty$, so the R_{hh}^{-2} tail extends to infinity and has the correct coefficient in order to give rise to an infinite number of Efimov states. This is the first study to demonstrate that a long-range R_{hh}^{-2} interaction in physical configuration space, rather than hyperspace, is responsible for the Efimov effect.

In the treatment of the low-energy trinucleon problem in the next chapter, we use plausibility arguments in order to conjecture a nucleon–deuteron effective interaction that has a long-range r^{-2} tail, where r is the nd separation. Such an interaction appears to be the essential feature for a coherent understanding the trinucleon problem both at positive and negative energies.

Next, we consider the possible influences of various factors on the Efimov effect. No rigorous estimates have been made concerning the magnitude of the effect of these factors; the arguments are basically qualitative.[2]

(a) *$L \neq 0$ states.* The higher angular momentum three-boson states do not exhibit the Efimov effect because the centrifugal repulsion associated with $L \neq 0$ states dominates the attractive R^{-2} effective interaction at large distances.

(b) *Three-body force.* A short-range three-body force should have no influence on the Efimov effect, since such a force does not modify the long-range part of the effective interaction.

(c) *Excited zero-energy two-body states.* Such states do not lead to the Efimov effect. The existence of a finite-energy two-body ground state prohibits the appearance of an infinite number of zero-energy three-body bound states.

(d) *Charges.* The repulsive Coulomb interaction between any two particles of the three-body system precludes the Efimov effect, since at large R the long-range Coulomb repulsion dominates the R^{-2} attraction.

(e) *Spin one-half fermions.* The effective interaction in this case has the R^{-2} form, but with a modified strength. This will make the effect less pronounced in the case of three identical spin one-half fermions with a weaker, but still attractive, R^{-2} interaction in the S-wave spin-doublet state. In the S-wave spin-quartet state, the effective interaction is repulsive and the Efimov effect is absent. Both these facts are a consequence of the Pauli exclusion principle.

(f) *Unequal mass particles.* The Efimov effect also appears for unequal mass particles, but the number of three-body bound states away from the Efimov limit depends on the masses of the particles.

Search for Efimov States in Nature. It is certainly of interest to search for evidence of Efimov-like states in nature. In the limit of zero deuteron binding energy, the 3N system has excited states of the Efimov type, and these have been studied numerically. However, for the actual deuteron binding energy of 2.2 MeV, this system has no excited bound state and hence has no Efimov type state. However, the ^{3}H nucleus has an excited virtual state of the Efimov type, which turns into a real bound state if the nucleon–nucleon spin-triplet interaction strength is reduced.[6] The first 0^+ excited state of ^{12}C has been suggested [2] as an Efimov state, but numerical studies contradict this.[10]

Since the Efimov effect is expected to be most enhanced in chargeless and spinless three-particle systems, it is unlikely that a nuclear system will exhibit this effect. On the other hand, many atomic and molecular systems are composed of chargeless bosons, and are most likely candidates for the Efimov effect. For example, an Efimov state has been predicted in the ^{4}He trimer. This state has been investigated numerically by several workers. The ^{4}He dimer is at best very weakly bound; consequently, the ^{4}He trimer has one excited state. This excited state is of the Efimov type, and it disappears when the ^{4}He–^{4}He interaction is slightly strengthened.[11]

There is another type of system that is likely to exhibit the Efimov effect. Namely, a negative ion of diatomic molecules that is composed of an electron and two neutral atoms, whenever the electron–atom system has a virtual or bound state of almost zero energy.

Since the pioneering work by Efimov, there have been many studies of this effect both from a general mathematical point of view and also in simple analytic models. The latter studies are generally relevant because the essential features of the Efimov effect are model independent. The Efimov effect has been looked for in other systems, e.g., for the case of four particles in three dimensions[12] and in the three-particle system in two dimensions.[4,13] In both cases, it has been demonstrated that there is no possibility for the appearance of an infinite number of Efimov states. It is now generally accepted that the Efimov effect remains an anomaly of the three-dimensional three-body system.

References and Notes (6)

Section 6.1

1. Faddeev (1965), Amrein (1977), Reed (1979), Derezínski (1987).
2. Goldberger (1964), Eden (1966), Watson (1967), Schmid (1974), Joachain (1975), Thomas (1977), Glöckle (1983a), Newton (1982).

Section 6.2

1. Adhikari (1973).
2. Ref. 2 of Sec. 6.1.
3. Klein (1958a, b).
4. Gerjuoy (1960, 1962), Krall (1960).
5. Feshbach (1958, 1962), Wildermuth (1977), Rawitscher (1987).
6. Byron, Jr. (1975), Blum (1977), Tip (1977), Peccei (1977), Gerjuoy (1978).
7. Ericson (1970).
8. For a discussion of the pion–nucleon problem, see Hamilton (1963).

9. Barton (1969), Fuda (1976).
10. Locher (1978).
11. Kim (1974), Amado (1981), Ericson (1985), Weller (1988).
12. Noyes (1959), Chew (1960). For early applications of the N/D method, see Hamilton (1963).
13. Frautschi (1963), Dodd (1977)
14. Kok (1970), Rinat (1972).
15. Adhikari (1983a, 1984b).
16. Frye (1963), Fuda (1980).
17. Blankenbecler (1959b), Avishai (1969), Kok (1970), Rinat (1971, 1972), Stelbovics (1974), Stingl (1974), Whiting (1976), Girard (1979a, b).
18. Pagels (1965).
19. Bower (1972), Greben (1976a, b; 1978).
20. Locher (1970, 1972, 1975), Viollier (1973), Plattner (1973, 1975, 1977), Peccei (1977), Bornard (1978).

Section 6.3

1. Rubin (1966, 1967a, b).
2. Eden (1966).
3. Amado (1970a).
4. Adhikari (1973).
5. Amado (1971b).
6. Gibson (1972, 1983, 1987).
7. Bollé (1981), Gibson (1986).
8. Dashen (1969–1971).
9. Adhikari (1971).
10. Lee (1959), Larsen (1970), Mascheroni (1970).
11. Graves-Morris (1967).
12. Efimov (1981).
13. Tkachenko (1981), Bhasin (1982), Shapoval (1987).
14. Fishbane (1967).
15. Amado (1971a).
16. Amado (1967), Cahill (1974).
17. Faddeev (1960), Watson (1956).
18. Aaron (1973).
19. Adhikari (1974a, b).
20. Aitchison (1965, 1975).
21. Amado (1974).
22. Adhikari (1978a, b).
23. Baur (1976, 1984), Kerman (1979), Hussein (1985), Austern (1987).
24. Herndon (1975a, b).

25. Skornyakov (1956), Noyes (1982).
26. Amado (1975).
27. Aaron (1975).

Section 6.4

1. Efimov (1970a, b; 1973).
2. Efimov (1976, 1981).
3. Amado (1971a, 1972).
4. Adhikari (1988c).
5. Fonseca (1979), Ouchinnikov (1979).
6. Adhikari (1972, 1982a, b), Doolen (1974), Frederico (1988c).
7. Stelbovics (1972), Antunes (1976).
8. Thomas (1935).
9. Simon (1970, 1977).
10. Huber (1978).
11. Lim (1977), Nakaichi-Maeda (1982), Huber (1985), Cornelius (1986).
12. Amado (1973), Adhikari (1981a), Karner (1987).
13. Lim (1980a, b).

Chapter 7

Solution Methods and Techniques: Three-Particle Scattering

7.1. Introduction

In Ch. 3, we have introduced some methods for obtaining numerical solutions of the integral equations for potential scattering. We now move on to deal with the far more difficult question of seeking practical methods for solving multiparticle scattering integral equations with compact kernels. It is useful to point out the major causes for this additional difficulty. We also report on numerical results obtained for various physical three-particle systems.

The most obvious of the preceding difficulties is the increasingly multivariable character of the scattering integral equations as the number N of particles increases, reflecting a growth in the active degrees of freedom that in three-dimensional space is $3N$. This manifests itself as increases in the dimensionality and number of integral equations.

Also, as N increases, the number of possible channels grows as the number of partitions of N, i.e., exponentially. Each of the distinct partitions correspond to possible subsystems that are reflected in the singularity structure of the kernels of the scattering integral equations that, for $N > 2$, is far more complex than for potential scattering. Thus, much more powerful methods are needed to supplement the techniques developed in Ch. 3.

After the c.m. motion is separated out, an N-particle problem involves $N - 1$ vector momenta; after all possible partial-wave projections are affected, $N - 1$ independent scalar momentum variables still remain. So N-particle scattering leads to integral equations with a minimum dimensionality of $N - 1$; numerical methods for their solution entail discretizations of $(N - 1)$-dimensional momentum integrals over semi-infinite domains. At present, essentially exact numerical solutions for $N > 4$ are not practical. Up to now, only the momentum-space, three-nucleon scattering problem has been solved exactly by Kloet and Tjon, by Takemiya, and, more recently, by Glöckle and coworkers[1] for a realistic nucleon–nucleon interaction entailing two-dimensional integral equations. Typically, approximations are introduced to reduce the dimensionality to unity. We remark that all so-called conventional approximation methods,[2] since they are not based on integral equations with compact kernels, involve not only the reduction of the dimensionality to unity, but also the approximation of noncompact kernels by compact ones.

In this chapter, we shall describe some techniques that have proven to be extremely useful for solving the three-particle scattering equations approximately. The recent nucleon–deuteron (Nd) scattering calculations by Glöckle *et al.*[1] may make the techniques of this chapter less attractive for obtaining very accurate solutions. Nevertheless, these techniques are very useful when approximate solutions are called for, for example, in the case of the four-body problem or for other three-particle systems where either the dynamics or the interactions are not well known. The physical systems in this later category are meson–deuteron systems and many-body systems in atomic and nuclear physics, where a three-body model often allows an excellent description of the actual state of affairs. The pion–deuteron, atom–diatom and deuteron–nucleus systems are examples of such systems where a three-body model has been successfully employed.

The input to the N-particle Faddeev–Yakubovskii[3] equations is expressed in terms of subsystem transition matrices. If M-body subsystem transition matrices are replaced by M-connected finite-rank forms for all $M < N$, such equations can be reduced after partial-wave analysis to a finite set of coupled one-dimensional equations.[4] This is a drastic approximation, in general, but it does take into account important features of the multiparticle dynamics. A somewhat more refined approach would be to take into account that for $N > 2$ these subtransition matrices are disconnected, and a program of finite-rank approximations should respect this structure. In this regard, it should be pointed out that there are N-particle equations, such as the Weinberg and the channel-coupling array equations,[5] that do not reduce to one-dimensional equations after making finite-rank approximations to the subsystem amplitudes. Such sets of equations are not considered in the present discussion.

The example of primary interest to us in this chapter, and the one that has been the most actively pursued, is the three-particle scattering problem. In this case, the most commonly used set of equations due to Faddeev is expressed entirely in terms of two-particle transition operators. These operators, as we know from Ch. 3, can be approximated systematically by finite-rank forms. The results of such approximations are finite sets of one-dimensional integral equations labeled by discrete angular momentum indices.

When there are bound states in the two-body subsystems, these equations can be rewritten as a closed set of equations for the off-shell two-cluster amplitudes. The breakup and three-to-three amplitudes can be obtained by quadrature. One obtains the same form of equations when no bound states exist in some or in all of the two-particle channels, such as in three-neutron scattering. Then, only the three-to-three amplitude has physical significance; it is obtained by quadrature as before. Regardless, the basic set of equations has the noteworthy feature of preserving all aspects of the multiparticle dynamics. The price for this dynamical integrity are the singularities that appear in the kernels of these equations above the three-particle threshold. These so-called moving-point singularities are not of the type that can be treated using the methods of Section 3.3, which apply only to fixed-point singularities. We usually need the contour-rotation method and a real-axis interpolation method.

Such sophisticated techniques are not required in more conventional approaches to scattering such as the resonating-group, coupled-reaction-channel and close-coupling methods.[2] The reason for this is that there all multiparticle states are assumed to be dominated by two-cluster configurations. For example, in the six-nucleon system with a deuteron–alpha (dα) initial state, it is assumed that the system channels preferentially into dα and ^{3}H–^{3}He final states. Configurations such as d-d-d are either ignored or accounted for in a phenomenological manner. The net effect is a set of coupled two-particle dynamical scattering equations.

Besides describing the techniques for solving three-particle equations, we also report detailed numerical results for various physical three-body systems, such as the trinucleon, alpha–deuteron, pion–deuteron, and atom–diatom systems. These calculations represent the major triumph of the method of connected-kernel equations; many experimental results were explained satisfactorily by them for the first time. In the case of the trinucleon system, we discuss the results of these calculations for the bound state, the Nd scattering channel, and the photodisintegration of the trinucleon system because they appear to be very closely related.

7.2. Numerical Methods

7.2a. Contour Rotation Method for Three-Particle Scattering

In addition to the fixed-point singularities of the two-particle LS equations studied in Ch. 3, a multiparticle integral equation also involves moving-point singularities in its kernel. In the present section, we discuss handling such singularities using the technique of contour rotation. The basic idea of this method is simple, although its implementation can become quite involved. We confine ourselves to the illustrative example generated by the nonrelativistic three-particle problem with separable two-particle interactions; in principle, the method is applicable in more complicated cases.[1]

For separable S-wave two-particle interactions, and after a partial-wave projection, the basic three-identical-particle scattering integral equations have the form

$$X_L(q,q';E) = 2Z_L(q,q';E) + 8\pi\int_0^\infty q''^2\,dq''\,Z_L(q,q'';E)\tau(E-3q''^2/4)X_L(q'',q';E), \quad (7.2.1)$$

where

$$Z_L(q,q';E) = \frac{1}{2}\int_{-1}^{1} du\,P_L(u)\frac{g(q\cdot q'-\frac{3}{4}q^2)g(q\cdot q'-\frac{3}{4}q'^2)}{E+i0-q\cdot q'}, \quad (7.2.2)$$

with

$$q\cdot q' \equiv q^2+q'^2+qq'u. \quad (7.2.3a)$$

Here, we are taking $\hbar = m = 1$, where m is the mass of each particle. Also,

$$[\tau(E)]^{-1} = \frac{1}{\lambda} - 4\pi\int_0^\infty q^2\,dq\frac{g^2(q)}{E-q^2+i0}, \quad (7.2.3b)$$

where the two-particle interaction has the separable S-wave form

$$\langle p|V|q\rangle = \lambda g(p)g(q). \quad (7.2.4)$$

Let us investigate the singularity structure of the function $Z_L(q,q'';E)$-$\tau(E-3q''^2/4)$ that enters into the kernel of Eq. (7.2.1). The dynamical singularities of the two-particle t-matrix are contained in the τ-function. The two-body bound-state pole occurs at a unique value of q'' defined by $(E - 3q''^2/4) = -B$, where B is the two-particle binding-energy, while the branch point corresponding to the two-particle unitary cut is at a value of q'' such that $E - 3q''^2/4 = 0$.

The remaining singularities of the kernel arise from the branch cuts of Z_L. In order to identify these singularities, we rewrite Eq. (7.2.2) for $q' = q''$ as

$$Z_L(q, q''; E) = \frac{1}{2qq''} \int_{-1}^{1} du \frac{P_L(u)}{v - u} g \cdot g, \tag{7.2.5a}$$

where

$$v = (E + i0 - q''^2 - q^2)/qq'', \tag{7.2.5b}$$

and we have suppressed the arguments of the form-factors, the g's, since they do not contribute to the cut structure. The branch cuts of Z_L arise from the possibility of the vanishing of the denominator of the integrand in Eq. (7.2.5a). This can only occur if $|v| \leq 1$, so $v = \pm 1$ are branch points of the cut running between these two points. This cut is logarithmic in structure.

The $v = \pm 1$ branch points are each realized for fixed q and E by two different values of q'', which indicates *two* branch cuts. One cut (I) has branch points in the complex q'' plane at

$$(q'')_{\mathrm{I}} = \frac{1}{2} q \left(\mp 1 + \left[\frac{4E + i0}{q^2} - 3 \right]^{1/2} \right\}, \tag{7.2.6a}$$

while the other (II) has branch points at

$$(q'')_{\mathrm{II}} = \frac{1}{2} q \left\{ \mp 1 - \left[\frac{4E + i0}{q^2} - 3 \right]^{1/2} \right\}. \tag{7.2.6b}$$

Even though E is fixed throughout a calculation, because of the preceding q-dependence, the positions of the branch points depend upon the choice of integration mesh.

The branch cuts of Z_L are defined by $v = u$ or

$$E + i0 - q \cdot q'' = 0, \tag{7.2.7a}$$

which when solved for q'' yields

$$q'' = -\frac{1}{2} uq \pm \left[E + i0 + \frac{1}{4}(u^2 - 4)q^2 \right]^{1/2}. \tag{7.2.7b}$$

Equations (7.2.6a, b) are recovered in the limiting cases $u = \pm 1$ corresponding to the ends of the branch cuts.

The branch cuts are depicted in Figs. 7.1a–c for the three different domains: (a) $q^2 < E$, (b) $q^2 > (4/3)E$, and (c) $E < q^2 < (4/3)E$, respectively. In the domain (a), we see from Eq. (7.2.7b) that q'' is real for $|u| < 1$, and the two different branch cuts are depicted in Fig. 7.1a. The termini denoted as $\pm\pm$ and $\pm\mp$ correspond to the four possibilities allowed by Eqs. (7.2.6a, b). For instance, $\mp\pm$ represent the points $\mp q/2 \pm (E - 3q^2/4)^{1/2}$.

When $q^2 > (4/3)E$, the square roots in Eqs. (7.2.6a, b) are purely imaginary, while $-uq/2$, of course, is purely real. We see then from Eq. (7.2.7b) that

$$(\mathrm{Re}\, q'')^2 + (\mathrm{Im}\, q'')^2 = (q^2 - E) > 0, \tag{7.2.8}$$

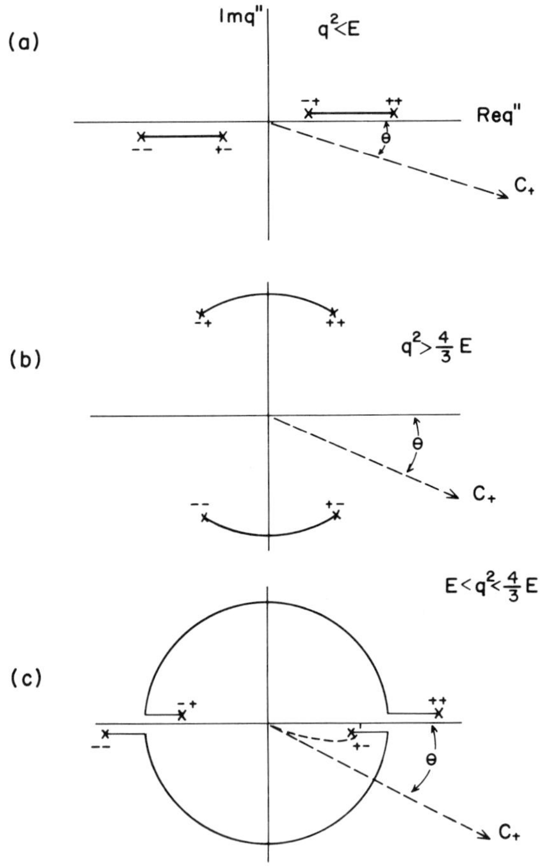

FIG. 7.1. Contour rotation used for Eq. (7.2.11).

which is a circle of radius $(q^2 - E) > 0$ in the q''-plane. The branch cuts are arcs of this circle with end points defined by Eqs. (7.2.6a, b) as shown in Fig. 7.1b. In the limiting case $q^2 = (4/3)E$, the square roots in Eqs. (7.2.6a, b) vanish and the two arcs join to complete a circle.

Finally, when $(4/3)E > q^2 > E$, we have a combination of the structure realized in the other two domains depending upon the value of u. The termini (7.2.6a, b) are clearly all real, but for small enough $|u|$ the argument of the square root in Eq. (7.2.7b) becomes negative, so that Eq. (7.2.8) is satisfied and the trajectory of the cut is along a circle as shown in Fig. 7.1c. In the limiting case $q^2 = E$, the arcs collapse onto the real line and the termini $+-$ and $-+$ coincide.

Let us write Eq. (7.2.1) more compactly as

$$X(q, q'; E) = 2Z(q, q'; E) + \int_0^\infty dq''\, K(q, q''; E)X(q'', q'; E), \qquad (7.2.9)$$

where the form of the kernel $K(q, q''; E)$, whose singularity structure we have studied in detail for real and positive q, q'', is obvious from Eq. (7.2.1). If the integrand of Eq. (7.2.9) is analytic in some region R_+ of the complex q''-plane ($\equiv Q''$) that includes the positive-real axis, then

$$\int_0^\infty dq''\, K(q, q''; E)X(q'', q'; E) = \int_{C_+} dQ''\, K(q, Q''; E)X(Q'', q'; E), \qquad (7.2.10)$$

where C_+ lies wholly within R_+. Given the necessary analyticity properties, the replacement (7.2.10) is trivial. What is not trivial is to suppose that the location of the singularities of $X(Q'', q'; E)$ are the same as those of $Z(q, q'; E)$ and $K(q, Q''; E)$. Then, for example, we can take C_+ as the dotted line in Fig. 7.1a, if $q^2 < E$, and where the contribution from the arc at infinity is ignored.

In order to exploit a continuation such as (7.2.10) for solving the integral equation (7.2.9), we must allow q to be complex, viz., Q. Also, the singularity structure of the kernel K must be such that we thereby obtain the integral equation on the contour, which we refer to generically as C_+, given by

$$X(Q, q'; E) = 2Z(Q, q'; E) + \int dQ''\, K(Q, Q''; E)X(Q'', q'; E), \qquad (7.2.11)$$

where both Q and Q'' are on C_+. The important thing is to justify this continuation. It turns out that it is possible to choose the contour in such a way that, with E real, no singularities of $K(Q, Q''; E)$ in Q or Q'' of the sort discussed previously are crossed.

Let us discuss this in a little more detail. First, we take C_+ as in Fig. 7.1a, so that Q and Q'' on the contour are given by

$$Q = Re^{-i\theta}, \tag{7.2.12a}$$

$$Q'' = R''e^{-i\theta}, \tag{7.2.12b}$$

respectively. Then, we recall that K is simply the product of the functions τ and Z. The poles and cuts of $\tau(E - 3Q''^2/4)$ occur on the real axis. Also, the positions of the singularities of $Z(Q, Q''; E)$ are defined by the complex generalizations of Eqs. (7.2.7a, b), namely

$$E - Q \cdot Q'' = 0 \tag{7.2.13a}$$

and

$$R'' = -uR/2 \pm [Ee^{2i\theta} + \tfrac{1}{4}(u^2 - 4)R^2]^{1/2}, \tag{7.2.13b}$$

where we have used Eqs. (7.2.12a, b). Evidently, Eq. (7.2.13b) cannot be satisfied if $0 < \theta < 2\pi$.

Now, for real q, q'', the two form-factors occurring in Eq. (7.2.5a) are irrelevant to the singularity structure of Z_L. They are quite relevant, however, when we continue to complex values Q and Q''. For then the form-factors have poles in the complex momentum plane. These poles will generate singularities in Z_L under two circumstances. One of the poles may be positioned at one of the end points of the u-integration in Eq. (7.2.5a), which generates what is called an endpoint singularity. Also two poles may approach ("pinch") the integration contour from either side, closing off the possibility of evasion.

A much more subtle problem involves the possibility of the generation of singularities in the solution X that are not apparent from the structure of the kernel and inhomogeneous terms. This question, along with the possibility of form-factor-generated singularities of Z_L referred to in the preceding paragraph, have been investigated in great detail by Brayshaw.[2] The net result is that there are no hidden singularity problems that enter into the contour-rotation technique provided $\theta(> 0)$ is small enough and the contours are chosen as in Fig. 7.1.

The contour depicted in Fig. 7.1c requires more careful specification. In this instance, the contour begins from the origin and progresses along the real axis up to the branch point $+-$. Then, C_+ is taken around this point, crossing over into the second Riemann sheet, and it then progresses back to the origin, and then radially outward as indicated. C_+ remains on the second sheet until it crosses the circular branch cut, upon which it returns to the first Riemann sheet.

In summary, Eq. (7.2.11) is solved on the half-rotated contour. The equation itself with Q continued back to real values can be used as a quadrature rule to calculate $X(q, q'; E)$. We remark that a serious drawback of this technique is the necessity of the knowledge of the analytic properties of the solution as well as the kernel and inhomogeneous terms of the integral equation. Under certain circumstances, these properties cannot be assessed beforehand, which makes the method inapplicable except possibly as an approximation. Then, there is recourse to the method we discuss next.

7.2b. Real-Axis Integration Using Interpolation Functions

Except under very special circumstances, the contour integration technique is difficult to implement because the analytic properties of the input (form-factors) and the solution itself are difficult to determine, or simply not known. An integration technique that confines itself to the real-momentum axis and uses interpolation functions with known analytic forms would therefore be of considerable practical value. Such a method was first suggested by Sohre and Ziegelmann and variants were successfully employed in different situations by Doleschall, and by Larson and Hetherington.[3]

The real-axis integration approach is based upon a relatively simple idea. The equation of interest is (7.2.9) with q' and E fixed at some definite values, so that we need to consider an equation of the general form

$$X(q) = Z(q) + \int dq''\, K(q, q'')X(q''). \tag{7.2.14}$$

Now let $S_i(q)$, $i = 1, \ldots, N$, be a set of linearly independent interpolation functions. The essential strategy is to suppose that Z, K, and therefore X can be approximated, as closely as one chooses, by expansions of the form

$$Z(q) \simeq \sum_{i=1}^{N} Z_i S_i(q), \tag{7.2.15a}$$

$$K(q, q'') \simeq \sum_{i=1}^{N} K_i(q'') S_i(q), \tag{7.2.15b}$$

$$X(q) \simeq \sum_{i=1}^{N} X_i S_i(q), \tag{7.2.15c}$$

where Z_i, $K_i(q'')$, and X_i are suitably chosen coefficients: $Z_i = Z(q_i)$, $K_i(q'') = K(q_i, q'')$, and $X_i = X(q_i)$, where q_i, $i = 1, \ldots, N$, are a suitably chosen set of points in the integration domain of Eq. (7.2.14). In addition, if $S_i(q)$ satisfies

$S_i(q_j) = \delta_{ij}$, then the approximate expansions (7.2.15a–c) are exactly satisfied at the points q_i, $i = 1, \ldots, N$. The points q_i can usually be taken as the integration quadrature points. Inserting Eqs. (7.2.15a–c) into Eq. (7.2.14), we obtain the set of algebraic equations

$$X_i = Z_i + \sum_{j=1}^{N} I_{ij} X_j, \qquad (7.2.16)$$

where

$$I_{ij} = \int dq''\, K_i(q'') S_j(q''). \qquad (7.2.17)$$

Once the integrals (7.2.17) have been calculated, the solution of Eq. (7.2.16) can be accomplished by routine procedures. The approximate continuous functions $X(q)$ are then determined from Eq. (7.2.15c) using these X_i.

The central problem is the selection of the function $S_i(q)$ and the points q_i. Larson and Hetherington[3] use the ansatz

$$S_i(q) = \prod_{j \neq i}^{N} \frac{(q - q_j)}{(q_i - q_j)}. \qquad (7.2.18)$$

The points q_i are usually taken as the integration quadrature points. It is claimed that with this form of $S_i(q)$ the singularity structures of the kernel and the Born terms are adequately represented. Computations with this choice are more stable and accurate than comparable contour rotation calculations.

Of course, Eq. (7.2.18) is not a unique choice for the functions $S_i(q)$, nor do Eqs. (7.2.15a–c) provide a unique way of doing the interpolation. Many variants of Eqs. (7.2.15a–c) and (7.2.18) have been suggested in the literature. The interpolation points q_i can be different from the integration quadrature points.

It is claimed that functions such as given by Eq. (7.2.18) do not provide the best interpolation, and that a better choice is provided by spline functions.[4] The advantage of using them as a basis for solving Faddeev equations has been recognized by several authors.[5,6] Usually, a piecewise, local interpolation is preferred to an interpolation over the whole region as in Eqs. (7.2.15a–c). This allows considerable flexibility in fitting complicated structure into the solution function. B-splines provide a convenient basis for solving two- and three-body scattering integral equations.[5] The cubic spline yields good results for solving the Faddeev equations above the breakup threshold. Product integration rules based on splines can also be used to treat integral equations with singular kernels. Eyre *et al.*[5] have established the usefulness of spline functions for solving Faddeev equations with separable NN potentials. Glöckle and coworkers[6] apply this technique to realistic trinucleon problems.

7.2c. *Direct Integration*

Another possibility is to construct the Neumann series of the original Faddeev equations by performing integrations over known functions—the kernel and the Born terms. The (divergent) Neumann series is then summed by using the technique of Padé approximants. This can also be applied to the two-dimensional form of the Faddeev equation with local two-body interactions.[7]

7.2d. *Numerical Results*

The preceding subsections provide a way of transforming the one-variable, momentum-space Faddeev equation with finite-rank two-body potentials to a set of discrete algebraic linear equations. These equations can then be solved by the techniques of Ch. 3, such as the iteration–subtraction, the continued-fraction, and the Padé approximant methods, or by direct matrix inversion.

Without finite-rank two-body interactions, one has a two-variable momentum-space integral equation even after partial-wave projection. Usually such equations are difficult to deal with, especially above the three-body breakup threshold, and have rarely been solved.[1] An exception is the work by Glöckle and coworkers, who use spline functions and real-axis integration for solving the Faddeev equations to calculate Nd elastic scattering and breakup observables with the meson-theoretic Paris and Bonn potentials.[6]

Doleschall[3] employs the method of real-axis integration using interpolation functions to solve the Faddeev equation with low-rank separable Yamaguchi-type NN potentials in many partial waves; he obtains impressive results for both breakup and polarization observables. The Graz–Osaka collaboration uses low-rank separable expansions with analytic form factors that are found by the Ernst–Shakin–Thaler technique, also with the Bonn and Paris potentials; the Faddeev equations are solved by the contour-rotation method in order to calculate Nd elastic scattering observables.[8]

There are not many comparative studies of the efficiency of different available numerical methods. Combinations of various numerical techniques are used by different groups. It is not easy to assess their comparative numerical efficiency.

The Padé approximant and the iteration–subtraction methods have been compared for the Amado–Lovelace–Mitra model with identical separable NN potentials.[9] We see from Table 7.1 that the convergence obtained with the iteration–subtraction method is extremely rapid, as with the NN-model study of Ch. 3 (Table 3.1), and the Padé technique producing a much slower but, nevertheless, acceptable rate of convergence.

TABLE 7.1. The scattering lengths for the S-wave spin-doublet and spin-quartet neutron–deuteron systems in the Amado model for different iterations N of the Neumann series. The results are taken from Brady (1972) and Tomio (1981).

	Spin Doublet		Spin Quartet	
N	Padé	Iteration Subtraction	Padé	Iteration Subtraction
2	7.08	5.52257	6.642	6.28340
4	2.55	−2.50914	6.321	6.32783
6	−0.68	−1.03804	6.317	6.31792
8	−1.01	−1.03574	6.317	6.31792
10	−1.04	−1.03541		6.31792

Larson and Hetherington[3] compare the method of real-axis integration using interpolation functions with the method of contour integration for the Amado–Lovelace–Mitra model with the same inputs and find that the first method is more accurate.

7.3. Results for the Three-Nucleon System

7.3a. *General Considerations*

A large number of three-nucleon (3N) calculations have been performed since the pioneering work of Amado, Mitra, Sitenko, and Kharchenko, and their collaborators in the early sixties.[1] The dynamical model of the 3N system based on the Faddeev equations has been the principal vehicle for calculations of the 3N bound state and scattering from low to moderate energies utilizing a variety of nucleon–nucleon (NN) pair interactions and 3N interactions.[2,3]

In this section, we review and interpret the results obtained in these calculations since they represent a major success of the connected-kernel approach to multiparticle collisions. Low to moderate energies refer to the energy domain where the nonrelativistic nucleons-only Schrödinger equation provides a good description of the Nd system; typically, "low" means an incident nucleon energy E_N of up to 20–30 MeV, and "moderate" refers to an E_N of up to 50–100 MeV. A "realistic" NN interaction usually has a one-pion exchange tail and fits experimental NN phase-shifts for all partial waves up to about $E_N = 250$ MeV.

The 3N observables can be broadly classified into two categories—central and noncentral (N-C) observables. The central observables are those which can be reasonably explained in a simple, nucleons-only, nonrelativistic 3N dynamical model by introducing only an S-wave NN $^3S_1$, $^1S_0$ potential, e.g., the 3N binding energies, the root-mean-square (r.m.s.) bound-state radii, and the scattering cross section.[4,5] The 3N N-C observables are appreciable only if the NN N-C potential is included, e.g., the asymptotic D-state to S-state normalization ratio of the 3N bound-state wave function, and the deuteron and nucleon analyzing powers of Nd scattering.

Although we are primarily concerned with scattering calculations, in the 3N case we also review the results for the bound-state observables in some detail. This is because the 3N bound-state and scattering results seem to be correlated, so that a consideration of both is important for a complete understanding.[4–6]

Since the pioneering calculations,[1] there has been much work involving more realistic phenomenological NN potentials.[7] The more recent calculations are at least an order of magnitude more precise than the earlier ones because of increased computer capacity. Computations using meson-theoretic NN and 3N potentials of the bound-state trinucleon observables have been performed by the Los Alamos–Iowa collaboration (LAIC) and by the Sendai group.[8] Similar calculations were repeated by a second Los Alamos group.[9]

Nd scattering calculations with meson-theoretic NN potentials have been carried out by the Graz–Osaka collaboration to determine the elastic Nd cross sections, and polarization observables.[10] The Bochum group repeated the Graz–Osaka calculation, and extended it to the breakup spectrum for various final-state configurations and breakup polarization observables.[11] Neither of these two calculations has incorporated a 3N potential. The LAIC also produced accurate numerical results for low-energy S-wave spin-doublet ($^1H-^2H$) and ($n-^2H$) scattering using meson-theoretic NN *and* 3N potentials.[12]

It is important to note that the LAIC solved the coupled-integro-differential form of the Faddeev equations in configuration space, whereas all other groups employed momentum-space integral equations.

The preceding calculations have been crucial in explaining the properties of the 3N system.[8,10,13] Although great effort has been expended in this work, comparatively little has been done to interpret the results obtained or to extract meaningful physics from them.

A certain general trend can be observed in the results of all these calculations, which is termed *universality*. We mean by this the apparent independence of the low-energy 3N observables from the "details" of the short-range behavior of the NN interaction.[4] The nature of these details will be made clear shortly. This independence was first noted by Brayshaw[6] in his study of Nd elastic scattering and breakup. Universality has been emphasized by Efimov, studied by Adhikari, Efimov, and collaborators, and seems to be essential to the interpretation of the general features of 3N calculations.[4,5,14–17] The following discussion of universality is intended to provide a global interpretation of the dynamical calculations of most of the 3N observables, *not* to substitute for these calculations.

7.3b. Universality in the Trinucleon System

From the discussion of Section 6.4, we expect that the spin-doublet S-wave 3N system is dominated by an effective long-range attractive r^{-2} interaction that is responsible for the Efimov effect[18] in the limit of zero two-nucleon binding. Here, r is the hyperspherical radius of the 3N system,[18] or the relative separation between the nucleon and the c.m. of the deuteron.[14] In the limit of large deuteron binding on nuclear scale ($\gtrsim 8$ MeV), the preceding long-range interaction should be transformed into a short-range interaction. The physical deuteron, of course, has intermediate binding ($\simeq 2$ MeV), and so the long-range r^{-2} tail of the interaction is expected to be cut off at distances greater than some value R_0. Moreover, for distances smaller than about the range of nuclear forces r_0, the r^{-2} behavior is cut off and the effective interaction is essentially constant as in the Saxon–Woods optical potential. Such interactions, which are shown in Fig. 7.2 with a range R_0 of the order of 5 fm, have been successfully used in a numerical modeling of the Nd system.[14]

The long-range tail of an effective Nd interaction only reflects the fact that the deuteron is loosely bound. Its short-range component, however, depends on some average properties of the NN interaction. If these average properties are not modified, the 3N system is expected to have the same low-energy behavior. If these average properties are changed while holding the two-nucleon observables fixed, this corresponds to altering the interior part ($r < r_0$) of the effective interaction. This is equivalent to changing the off-shell behavior of the NN interaction and/or introducing a 3N force, but maintaining the same NN observables.[4,14]

The effective long-range r^{-2} interaction occurs only in the spin-doublet ($S = \frac{1}{2}$) S-wave ($L = 0$) state, and is expected to be strongly attractive in nature and therefore sensitive to the NN interaction. The spin-quartet ($S = \frac{3}{2}$) and the $L > 0$ partial waves of the spin-doublet ($S = \frac{1}{2}$) states are repulsive or very weakly attractive because of the Pauli exclusion principle and the centrifugal barrier for $L > 0$, and so are expected to be insensitive to the NN

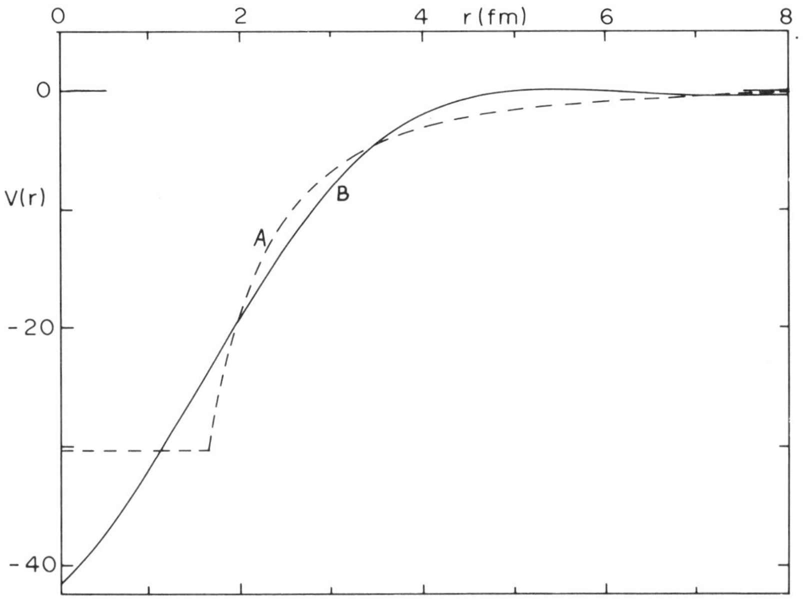

FIG. 7.2. Two Nd effective interactions in MeV with ranges of the order of 5 fm (from Tomio (1987)).

potential. Universality implies that the S-wave spin-doublet 3N observables are sensitive to the changes in NN potentials in a way we discuss below.

In order to determine the 3N wave function, one needs the NN potential at short distances or, equivalently, the NN off-shell behavior. According to universality, the S-wave, spin-doublet 3N system is dominated by the r^{-2} effective interaction at intermediate distances, $r_0 < r < R_0$. For $r < r_0$, this interaction can approximately be parametrized by a single depth parameter.[5] Hence, the relevant parts of the NN potential can also be approximately parametrized by a single parameter, in addition to the low-energy NN properties, in a dynamical 3N calculation. Fixing this parameter means one fixes either the ^{3}H binding-energy, $B(^3\text{H})$, or the spin-doublet S-wave neutron-deuteron (nd) scattering length, $^2a_{\text{nd}}$. Thus, two different dynamical 3N models will yield identical results for *all* central 3N observables provided that they yield identical $B(^3\text{H})$ or $^2a_{\text{nd}}$. *This is what is meant by the independence of 3N calculations from details of the short-range behavior of the NN interaction.*

The existence of a universal effective interaction in the 3N problem is quite similar to the effective interactions that appear in the NN system, in conjunction with the "shape-independent" approximation for low-energy NN parameters in accord with the effective-range expansion (2.4.61a). The effective interaction of the 3N system also implies a shape-independent approximation for the low-energy, S-wave, spin-doublet 3N observables that is implicit in the effective-range-type expansion of Section 6.2; thus, these 3N observables can be parametrized by a small number of parameters. For central 3N observables, this number is, to a first approximation, unity.[19]

In the NN system, which appears in the neutron–proton (np), neutron–neutron (nn), and proton–proton (pp) configurations, the charge of the proton does not destroy the validity of the shape-independent approximation discussed above, although the NN observables do depend on the charge state. Analogously, in the Nd system, the Coulomb repulsion in the proton–deuteron (pd) case does not destroy the idea of universality, though the pd and the nd observables are distinct. Universality in the pd system implies that all 3N models that yield the same ^{3}He binding energy $B(^3\text{He})$ should yield the same pd spin-doublet scattering length $^2a_{\text{pd}}$, and other low-energy central pd observables.[12,20]

7.3b.1. Low-Energy Three-Nucleon Correlations

According to universality, if two 3N dynamical models yield the same values for $B(^3\text{H})$ or $^2a_{\text{nd}}$, they should yield identical low-energy 3N observables. This has been verified in numerical studies of the 3N system.[4,8] A strong correlation between $^2a_{\text{nd}}$ and $B(^3\text{H})$ calculated in the same model was first observed by Phillips.[4] Girard and Fuda[21] found that the S-wave asymptotic normalization parameter (ANP) of ^{3}H, $C_s(^3\text{H})$, is correlated either with $B(^3\text{H})$ or $^2a_{\text{nd}}$. Brayshaw[6] showed, using his boundary-condition model, that even the low-energy nd-breakup results are strongly correlated with $^2a_{\text{nd}}$. A correlation has also been observed between the r.m.s. radius of ^{3}H, $R(^3\text{H})$, and $B(^3\text{H})$.[22] The inclusion of the Coulomb interaction in the pd system does not destroy these correlations.[12,20] Correlations have also been established between $^2a_{\text{pd}}$ and $B(^3\text{He})$; $R(^3\text{He})$ and $B(^3\text{He})$; and $C_s(^3\text{He})$ and $B(^3\text{He})$.[8,12,20,22,23]

7.3b.2. Universality in the Presence of NN Non-Central (N-C) Forces

The 3N N-C observables vanish in the absence of N-C NN potentials. First, we consider those 3N N-C observables that can be adequately described by including *only* a NN tensor potential; an example of such an observable is the triton D-state to S-state asymptotic normalization ratio η^t. In addition, there are other low-energy 3N N-C observables that require both the NN tensor potential and $L \geq 1$ components of NN interaction for a satisfactory description, such as the deuteron and nucleon analyzing powers in Nd scattering.[5] Universality also holds for these observables provided we introduce an extra NN observable to "measure" the D-state of the deuteron[24] or, equivalently, to reflect the strength of the NN tensor force. The deuteron D-state probability is not a model-independent observable. Amado and Friar have suggested[25] that the deuteron asymptotic D-state to S-state ratio η^d be regarded as a measure of the deuteron D-state.

According to universality, if different 3N dynamical models utilize NN potentials fitted to identical low-energy on-shell NN observables, including η^d, they will yield identical low-energy 3N N-C observables provided that they predict identical values of $B(^3H)$. Hence, the general idea of universality is easily extended in the presence of the N-C NN potentials, if we include η^d as one of the fundamental on-shell NN observables. This has been demonstrated in detail for η^t.[15]

In contrast to η^t, the deuteron and nucleon analyzing powers in low-energy Nd scattering are not only sensitive to the NN tensor potential, but are particularly sensitive to the $L \geq 1$ parts of the NN interaction. In this case, universality and the correlations among the low-energy 3N observables seem to break down unless these partial waves are kept constant; then, the discussion of universality becomes more subtle (see Section 7.3c.2B).

Finally, the NN polarization-transfer parameters in Nd scattering and other second-order polarization observables, such as spin correlations, are reasonably explained with only S-wave NN potentials, although small deviations of the calculated and the experimental results can only be explained by including a NN tensor potential and $L \geq 1$ NN interactions.[16,26] Therefore, these 3N observables are predominantly central with a small N-C admixture.

Other 3N observables that have been calculated are the different 3N form-factors—both electric and magnetic.[8] However, they certainly cannot be considered low-energy 3N observables because they are very sensitive to the NN interaction at short distances, and so are subject to corrections due to the composite structure of the nucleon and meson-exchange currents. These effects require a fully relativistic treatment, especially at large momentum transfers.

7.3b.3 Consequences of Universality

Universality has serious implications for the interpretation of those 3N calculations that are intended to distinguish NN interaction models.[5,14–16] The representation of the NN interaction by a potential is uncertain at short distances. The short-range behavior of the NN potential is phenomenologically introduced in meson-theoretic models of the NN interaction at large and intermediate ($r \gtrsim 1$ fm) distances. This uncertainty in the short-range behavior of the NN interaction can be characterized as an off-shell effect. The potential models of 3N interactions, used in 3N dynamical calculations, are also uncertain at short distances, and this behavior is usually chosen arbitrarily. One of the objectives of performing 3N calculations with different sets of "realistic" NN and 3N potentials is to be able to distinguish various 3N models, so that one can learn about both the NN off-shell effects and the short-range 3N interaction. One anticipates that at low energies the sensitivity of 3N calculations to different NN potential models can essentially be parametrized by one 3N observable: $B(^3\mathrm{H})$ or ${}^2a_{\mathrm{nd}}$ (in addition to the usual set of low-energy NN observables). Hence, it may be difficult to fix both the NN off-shell effects and the short-range 3N interaction by performing complicated 3N calculations, since different admixtures of 3N and on-shell equivalent NN potentials will produce identical low-energy 3N results if the 3N models correspond to the same $B(^3\mathrm{H})$.

7.3c. *Numerical Results*

Modern 3N calculations are very sophisticated and may require up to 100 Megabytes of memory on supercomputers. We will not describe all of the technicalities of such calculations; nevertheless, it is relevant to clarify certain aspects of the word "channel" as used in trinucleon calculations.[27] The triton of total spin $\frac{1}{2}$ consists of infinite number of spin-angular momentum states both for the interacting pair and the spectator, which are denoted, respectively, in spectroscopic notation by ${}^{2S+1}L_J$ and l_j. So far calculations have been performed with 34 states of Table 7.2, which we call a 34-channel calculation. If a triton calculation is performed by taking the first 3 (5,18,26,34) states of this table into account, it is called a 3 (5,18,26,34) channel calculation.

7.3c.1. Central Trinucleon Observables

7.3c.1A. Triton Binding Energy. Of all the 3N observables, $B(^3\mathrm{H})$ is most precisely measured (8.482 MeV $\equiv B(^3\mathrm{H})_{\mathrm{exp}}$) and theoretically studied. Calculations of $B(^3\mathrm{H})$ have been performed both in coordinate and momentum space using a variety of numerical methods.[8,9,27–30] They are very reliable in the sense that the values obtained by different groups agree with each other to within less than one percent.

Table 7.3 lists the results of the converged calculations, using realistic NN potentials, of Ishikawa *et al.* (Sendai),[27] Friar *et al.* (LAIC)[29] and

TABLE 7.2. Thirty-four channels in the triton.

Channel number	Interacting pair $^{2S+1}L_J$	Spectator l_j	Channel number	Interacting pair $^{2S+1}L_J$	Spectator l_j
1	$^1S_0$	$s_{1/2}$	18	$^3F_2$	$f_{5/2}$
2	$^3S_1$	$s_{1/2}$	19	$^1F_3$	$f_{5/2}$
3	$^3D_1$	$s_{1/2}$	20	$^1F_3$	$f_{7/2}$
4	$^3S_1$	$d_{3/2}$	21	$^3D_3$	$d_{5/2}$
5	$^3D_1$	$d_{3/2}$	22	$^3G_3$	$d_{5/2}$
6	$^3P_0$	$p_{1/2}$	23	$^3D_3$	$g_{7/2}$
7	$^1P_1$	$p_{1/2}$	24	$^3G_3$	$g_{7/2}$
8	$^1P_1$	$p_{3/2}$	25	$^3F_3$	$f_{5/2}$
9	$^3P_1$	$p_{1/2}$	26	$^3F_3$	$f_{7/2}$
10	$^3P_1$	$p_{3/2}$	27	$^3G_4$	$g_{7/2}$
11	$^3D_2$	$d_{3/2}$	28	$^3G_4$	$g_{9/2}$
12	$^3D_2$	$d_{5/2}$	29	$^1G_4$	$g_{7/2}$
13	$^1D_2$	$d_{3/2}$	30	$^1G_4$	$g_{9/2}$
14	$^1D_2$	$d_{5/2}$	31	$^3F_4$	$f_{7/2}$
15	$^3P_2$	$p_{3/2}$	32	$^3H_4$	$f_{7/2}$
16	$^3F_2$	$p_{3/2}$	33	$^3F_4$	$h_{9/2}$
17	$^3P_2$	$f_{5/2}$	34	$^3H_4$	$h_{9/2}$

TABLE 7.3. Results from 34-channel calculations of $B(^3H)$ (in MeV) with and without TM, BR, and U 3N potentials with $\Lambda_0 = 800$ MeV.

NN potential	NO 3N potential	BR 3N potential	TM 3N potential	U 3N potential
RSC	7.35*	8.89*	8.86*	8.70*
AV	7.68**,7.67*	9.22*	9.29**,9.36*	8.99*
PARIS	7.64**		9.18**	
BONN	8.35***,8.33**			
TRS	7.35**,7.56*		9.71**	

* LAIC (Ref. 29); ** Sendai (Ref. 27); *** Los Alamos II (Ref. 9).

Brandenburg *et al.* (Los Alamos II).[9] Except for the Bonn potential, the triton is underbound by about 1 MeV, which may reflect the omission of a 3N force. The Bonn potential nearly reproduces $B(^3H)_{exp}$ without such a force, but it does not reproduce the NN phase-shifts as well as other NN potentials.[27]

The values for $B(^3H)$ obtained, including three versions of two-pion-exchange 3N potentials[3] known as the Tucson–Melbourne (TM), the Brazil

(BR), and the Urbana (U) 3N potentials, are also listed in Table 7.3. These 3N potentials have similar long-distance behaviors, but they differ somewhat at short distances. The uncertainty in short-distance behavior of the 3N potential is usually handled phenomenologically in terms of an unknown parameter Λ_0 in the πNN form-factor. The inclusion of the 3N potential increases $B(^3\text{H})$ by about 1–2 MeV, so the final, converged result overbinds the triton. Qualitatively, this is encouraging in that the modification is in the right direction. For each NN potential, it is possible to find a value for Λ_0 that yields $B(^3\text{H})_{\text{exp}}$.[27]

In conclusion, most of the realistic NN potentials, except the Bonn potential, underbind the triton by about 1 MeV, and 3N potentials are expected to produce additional binding. Convergence has been achieved with respect to partial waves of the NN potentials. According to universality, the results for $B(^3\text{H})$ are not expected to distinguish between an off-shell variation of the NN potential and a change in the 3N potential.

7.3c.1B. Spin-Doublet Nucleon–Deuteron Scattering Length. The *scattering* observable of the nd system that has been most thoroughly studied is the scattering length $^2a_{\text{nd}}$. Its currently accepted experimental value[31] is $(^2a_{\text{nd}})_{\text{exp}} \equiv 0.65 \pm 0.04$ fm. Numerical calculations have been performed with different NN and 3N potentials. The correlation between $^2a_{\text{nd}}$ and $B(^3\text{H})$ was the first observed in the 3N system and was used by Phillips[4] to *predict* the value of $(^2a_{\text{nd}})_{\text{exp}}$ consistent with $B(^3\text{H})_{\text{exp}}$.

More recently, a similar correlation has been conjectured by the LAIC between $B(^3\text{He})$ and $^2a_{\text{pd}}$. On this basis, they predicted $^2a_{\text{pd}}$ to be in the vicinity of 0 fm, in contradiction with the experiment, namely, $(^2a_{\text{pd}})_{\text{exp}} \gtrsim 2$ fm.[12]

The $B(^3\text{H})-{}^2a_{\text{nd}}$ correlation is exhibited in Fig. 7.3, where each point refers to a single calculation.[19] Consistent with the discussion of universality, explanations of this correlation have been proposed in a two-body model using the dispersion-theoretic (N/D) approach,[19] in a minimal three-body model using only on-shell NN-input,[32] and, finally, in an effective Nd-potential model exploiting the r^{-2} one-nucleon exchange tail.[14] The full curve in Fig. 7.3 represents the N/D calculation using Eq. (6.2.57). This is *not* a dynamical model, but yet it reproduces the results of such 3N calculations very well.

In Fig. 7.4, calculated results for the 3N binding energy versus Nd scattering-length correlation are shown. This represents the work of the LAIC[12] using realistic NN and 3N potentials both for the nd and the pd systems, and that of a Brazilian group using the effective Nd-potential models of Fig. 7.2[14] parametrized to reproduce $B(^3\text{H})_{\text{exp}}$, $B(^3\text{He})_{\text{exp}}$, and $(^2a_{\text{nd}})_{\text{exp}}$. The actual

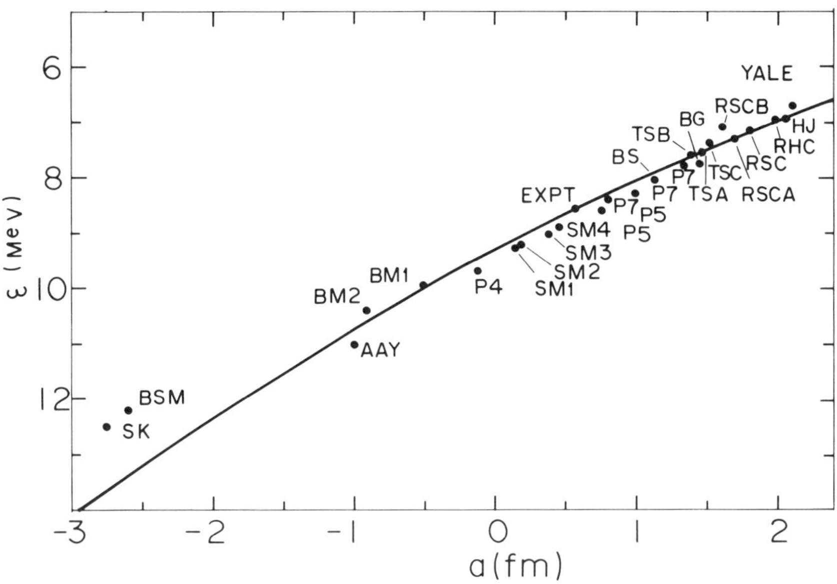

FIG. 7.3. The ${}^2a_{nd} - B({}^3H)$ correlation. The full line is the result of an effective-range theory for the Nd system (see Section 6.2c) and the points are results of realistic calculations (from Adhikari (1983a)).

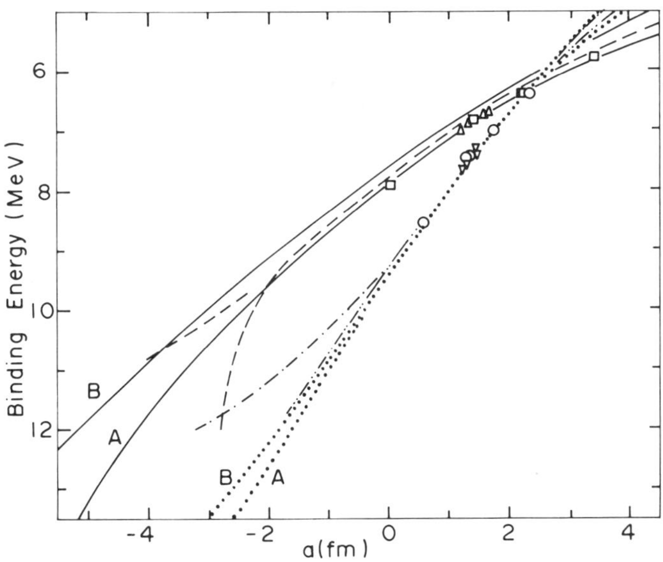

FIG. 7.4. Binding-energy/scattering-length plot for the trinucleon ground state. The points are from the calculations of Friar (1984a) [FR] and Chen (1986a) [CH]: circles (n–d), Ref. 7; squares (p–d), Ref. 7; triangles (p–d), Ref. 8; inverted triangles (n–d), Ref. 8; dashed line (p–d), plot of Refs. 7 and 8; dashed–dotted and dashed–dotted–dotted lines (n–d), plot of Refs. 7 and 8; solid lines and dotted lines are the p–d and n–d plots of the interactions A and B of Fig. 7.1, respectively (from Tomio (1987)).

$B(^3\mathrm{H})-{}^2a_{\mathrm{nd}}$ and $B(^3\mathrm{He})-{}^2a_{\mathrm{pd}}$ correlations are well simulated in this model by varying the depth of the effective Nd potential.

It was entirely unexpected that an Nd effective-potential model, which does not explicitly use three-particle dynamics, could simulate the results of fully dynamical 3N calculations, for example, the $B(^3\mathrm{H})-{}^2a_{\mathrm{nd}}$ correlation. The fact that the results of Nd scattering can be described so well in such simple effective-potential models indicates that the usual dynamical 3N calculations conceal some elementary features of the 3N system.

7.3c.1C. Root-Mean-Square Radii of ^{3}H and ^{3}He. Another fundamental property of the trinucleon bound state is its r.m.s. radius. Calculations of $R(^3\mathrm{H})$ and $R(^3\mathrm{He})$ appear to be correlated with $B(^3\mathrm{H})$ and $B(^3\mathrm{He})$, respectively. Again the correlation is universal in that it does not distinguish between different 3N models.[29]

7.3c.1D. S-State Asymptotic Normalization Parameter C_s. The S-state ANP's of ^{3}H and ^{3}He seem to be correlated with $B(^3\mathrm{H})$ and $B(^3\mathrm{He})$, respectively. From this correlation, Friar *et al.*[33] predict $C_s = 1.85 \pm 0.02$ for both ^{3}H for ^{3}He. It is to be recalled that the result for ^{3}He should be consistent only with $B(^3\mathrm{He})$ and not with $B(^3\mathrm{H})$. The correlated behavior between C_s and trinucleon binding energy, or equivalently that between C_s and the Nd scattering length, was first studied by Girard and Fuda,[21] who used it to predict $C_s(^3\mathrm{H}) = 1.82$. Petrov[14] finds $C_s(^3\mathrm{H}) = 1.86$ with an effective-Nd-potential model.

7.3c.1E. Electromagnetic Effects on the Trinucleon Bound State. This effect requires a more complete discussion. Naively, one may just add the pp Coulomb potential to the purely nuclear 3N Hamiltonian. The Coulomb force is the dominant electromagnetic effect, but there are other features that cannot be ignored. Brandenburg *et al.*[34] have discussed some of these, including charge-symmetry breaking, the finite size of nucleons, and the p–n mass difference. Although the effect of the static Coulomb force is expected to be consistent with universality, the various small corrections can lead to its breakdown. The estimates of Brandenburg *et al.* of the contributions of these corrections to the $^3\mathrm{H}-{}^3\mathrm{He}$ binding energy difference, $\Delta E_{\mathrm{total}}$, are shown in Table 7.4. The positive and negative signs of the various effects contributing to $\Delta E_{\mathrm{total}}$ demonstrate their compensating behavior. The reproduction in Fig. 7.4 of the $B(^3\mathrm{He})-{}^2a_{\mathrm{pd}}$ correlation using an Nd effective interaction only demonstrates that our discussion of universality is valid in the presence of the static pp repulsion. In Fig. 7.5, we represent a plot of

TABLE 7.4. Charge-asymmetric contributions to the $^3H-^3He$ binding energy difference.

	ΔE (keV)
Static Coulomb	687
Charge-symmetry breaking forces	59
Finite-size effects	−39
p−n mass difference in kinetic energy	12
Second-order perturbation corrections	−6
Other electromagnetic effects	29
Total (theory)	742
Experiment	764

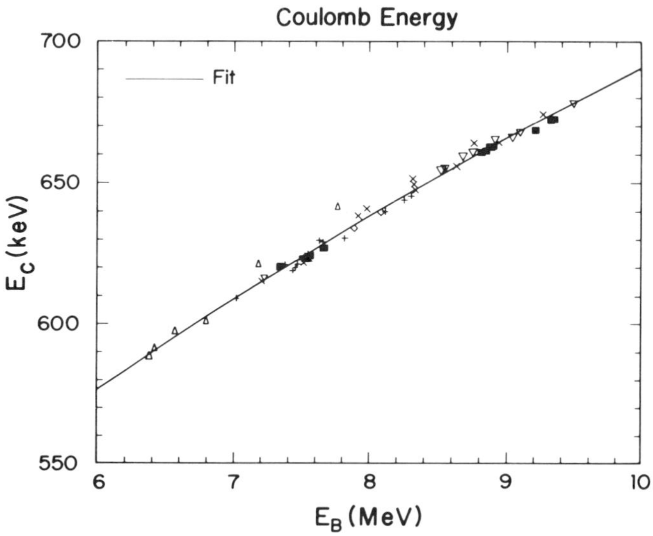

FIG. 7.5. The ^3He Coulomb energy in first-order perturbation theory for diverse trinucleon models plotted versus trinucleon binding energy (from Friar (1987b)).

static Coulomb energy versus the 3N binding energy.[35] Again we have a correlation of narrow width.

7.3c.1F. Observables of Nd Scattering. The Nd scattering observables are expressed in terms of the scattering phase shifts. These phase shifts follow a general trend as required by universality. Since we are mostly concerned here with a central 3N observable, we consider a simple Nd model involving only $^1S_0$ and $^3S_1$ NN potentials. Arvieux and Bruinsma *et al.*[36] have shown with separable two-body interactions that once these potentials are constrained to reproduce low-energy NN properties such as $B(^2H)$, the NN scattering lengths, and effective ranges, the low-energy Nd phase-shifts, with the exception of $^2\delta_{L=0}$, are insensitive to the form-factors of the NN potentials. A similiar conclusion also holds for local NN potentials.[16] Hence the sensitivity of a low-energy Nd-scattering observable on the on- and off-shell behaviors of the NN potentials is contained only in $^2\delta_{L=0}$, besides that information already included in the low-energy NN observables.

Frederico and Goldman[37] observed that even above the breakup threshold the real part of $^2\delta_{L=0}$ is strongly correlated with $^2a_{\mathrm{nd}}$, whereas its imaginary part is essentially model independent. Earlier parametrizations of $^2\delta_{L=0}$ as a function of $^2a_{\mathrm{nd}}$ also support the existence of a $^2\delta_{L=0}-{}^2a_{\mathrm{nd}}$ correlation.[19]

Hence, all of the low-energy central Nd scattering observables, such as the differential, total, and even breakup cross sections, will be correctly predicted by a 3N model calculation provided that it reproduces $^2\delta_{L=0}$, or the correlated $^2a_{\mathrm{nd}}$, in addition to being consistent with the low-energy NN observables. The correlation of the low-energy nd breakup result with $^2a_{\mathrm{nd}}$ was pointed out by Brayshaw.[6]

The Nd elastic cross-sections—both differential and total—have larger spin-quartet than spin-doublet contributions. According to universality, all spin quartet phase-shifts and the $L \neq 0$ spin-doublet phase shifts should be approximately model independent. Therefore, we expect that the differential and total cross sections will be weakly dependent on the 3N potential model and insensitive to the off-shell behavior of the NN potential. This is why almost any reasonable 3N interaction model fits the differential cross section data equally well.

The separable model calculation of Koike and Taneguchi[38] reproduces the experimental data for the nd elastic differential cross section at different energies satisfactorily, except for a small discrepancy at backward angles. This discrepancy is in agreement with the inability of most of the 3N potential models to reproduce the quartet part of the differential cross section at these angles, and reflects a breakdown of universality. Recent calculations with the Paris NN potential reduces this disagreement, possibly because of a better representation of the NN interaction for $L \geq 1$.[13,39]

The final-state interaction peak in the proton spectrum in nd breakup reaction deserves special attention because this effect occurs predominantly in the spin-doublet state, since the two neutrons will be in a spin-singlet state at low energies. Consequently, the final-state interaction peak for the proton, unlike the cross section, is expected to be not only sensitive to the NN interaction models employed, but also to ${}^2a_{\mathrm{nd}}$. The first sensitivity has been exploited to extract the nn-singlet scattering length, viz., $a_{\mathrm{nn}} = -18.31 \pm 0.22$ fm.[40] Since the peak in the proton spectrum is also expected to be sensitive to ${}^2a_{\mathrm{nd}}$, the experimental value of ${}^2a_{\mathrm{nd}}$ should be incorporated in the dynamical 3N model before extracting the value of a_{nn}. Whether the extracted value of a_{nn} suffers significant modification because of this effect is yet to be established.

7.3c.2. *Non-Central Trinucleon Observables*

7.3c.2A. Triton Asymptotic D-State to S-State Ratio. We have pointed out in Section 7.3b.2 that η^{d} provides a precise "measure" of the tensor force in the NN system, and it can be satisfactorily accounted for by including a tensor NN interaction alone. As η^{t} is also expected to be sensitive to the NN tensor force, any attempt to obtain its proper value in a 3N dynamical model should have the correct magnitudes of η^{d} and $B({}^3\mathrm{H})$ built in.

Ericson and Rosa-Clot[41] have found that, in a simple analytical model for a weak deuteron D-state, where a perturbative treatment is sufficient, η^{d} is essentially determined by the πNN coupling constant provided $B({}^2\mathrm{H})$ is held fixed. Because of the expected perturbative nature of the D-state of ${}^2\mathrm{H}$ and ${}^3\mathrm{H}$, all the few-nucleon observables that are predominantly determined by the NN tensor force are related to η^{d} in a definite way. They have also established that the deuteron quadrupole moment Q^{d} is given by

$$\frac{Q^{\mathrm{d}}}{\eta^{\mathrm{d}}} \sim (C_{\mathrm{s}}^{\mathrm{d}})^2 G, \tag{7.3.1}$$

where $C_{\mathrm{s}}^{\mathrm{d}}$ is the deuteron S-state ANP and the function G is determined mainly by $B({}^2\mathrm{H})$. In the context of a more recent analytic model of the ${}^3\mathrm{H}$ D-state, η^{t} is given by

$$\frac{\eta^{\mathrm{t}}}{\eta^{\mathrm{d}}} \sim (C_{\mathrm{s}}^{\mathrm{d}})^2 F, \tag{7.3.2}$$

where F is primarily a function of $B({}^2\mathrm{H})$ and $B({}^3\mathrm{H})$.[15] The manifest similarity of Eqs. (7.3.1–7.3.2) is a consequence of the presumed perturbative nature of the relevant D-states.

Assuming that $C_{\mathrm{s}}^{\mathrm{d}}$ is known, Eq. (7.3.2) allows us to predict: (a) a linear $\eta^{\mathrm{t}}-\eta^{\mathrm{d}}$ correlation at a fixed $B({}^3\mathrm{H})$, and (b) a $\eta^{\mathrm{t}}/\eta^{\mathrm{d}}-B({}^3\mathrm{H})$ correlation according to Eq. (7.3.2). In order to verify the linear $\eta^{\mathrm{t}}-\eta^{\mathrm{d}}$ correlation at a fixed $B({}^3\mathrm{H})$, Adhikari, Frederico, and Hussein (AFH) performed a separable-potential-model 3N calculation,[15] where they changed the parameters in a controlled

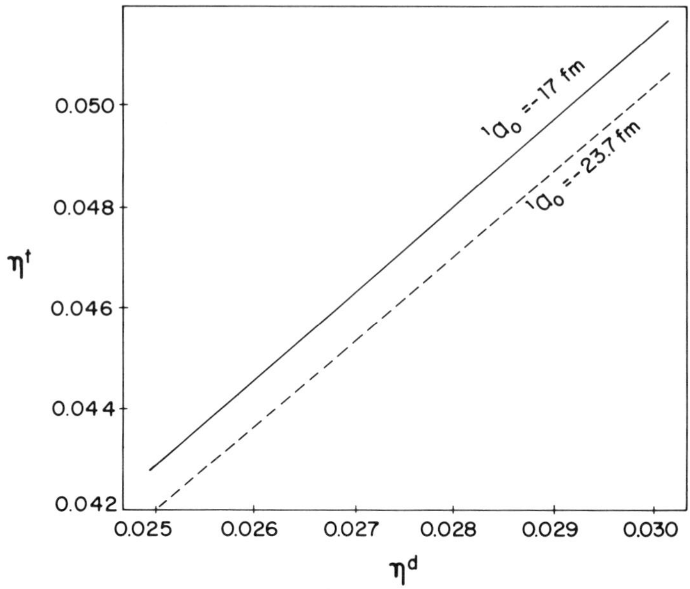

FIG. 7.6. The $\eta^{\mathrm{d}}-\eta^{\mathrm{t}}$ correlation at a fixed E_{t} (=8.48 MeV); separable-potential model calculation with ${}^{1}a_{0} = -17$ and -23.7 fm. (from Frederico (1988a)).

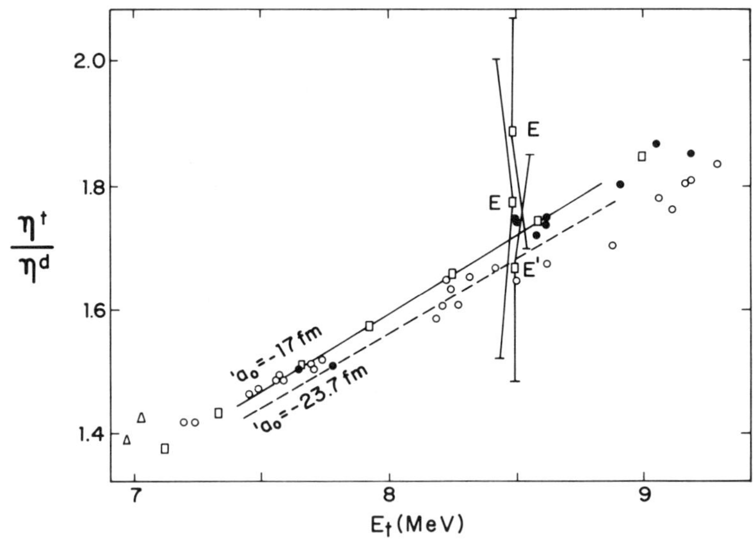

FIG. 7.7. The $\eta^{\mathrm{t}}/\eta^{\mathrm{d}}-E_{\mathrm{t}}$ correlation. Symbols (open circles, triangles, squares, and solid circles) are the results of various realistic calculations. Straight lines denote separable-potential model calculations. The points E and E' are experimental results (from Frederico (1988a)).

way so that, while η^{d} varied, $B(^{3}\mathrm{H})$ was held fixed at $B(^{3}\mathrm{H})_{\mathrm{exp}}$. This correlation was first suggested by Ericson and Rosa-Clot who, however, did not keep $B(^{3}\mathrm{H})$ fixed. AFH performed their calculations with two choices of the spin-singlet NN scattering lengths. As shown in Fig. 7.6, in each case they find a straight line passing through the origin, $\eta^{d} = \eta^{t} = 0$.

Next, in order to verify the η^{t}/η^{d}–$B(^{3}\mathrm{H})$ correlation implicit in Eq. (7.3.2), we plot in Fig. 7.7 the ratio η^{t}/η^{d} versus the triton binding energy $E_{t} \equiv B(^{3}\mathrm{H})$ for a wide variety of calculations, including both separable potential and realistic 3N models; all points fall on the same universal curve. This correlation is particularly important because of the large uncertainty in the experimental value of η^{d}, which has been claimed to be of the order of 10% by Londergan *et al.*[42] According to Eq. (7.3.2) and the universality argument, the ratio η^{t}/η^{d} is determined essentially by $B(^{3}\mathrm{H})$, C_{s}^{d}, and other low-energy NN on-shell properties, and is independent of further details of the short-range portion of the NN and 3N potentials. Most of the realistic phenomenological NN potentials have η^{d} ranging from 0.0261 to 0.0265, a very small variation. When η^{d} can be considered to be fixed in Eq. (7.3.2), one has a linear correlation between η^{t} and $B(^{3}\mathrm{H})$, as has been emphasized by Ishikawa and Sasakawa.[27] However, this correlation is a consequence of universality rather than the one-pion-exchange tail of the NN potential. AFH have demonstrated that if η^{d} is allowed to vary over a wide range (0.025 to 0.029), the width of the η^{t}–$B(^{3}\mathrm{H})$ plot increases substantially.

The universal η^{t}/η^{d} versus $B(^{3}\mathrm{H})$ plot has been utilized by AFH to predict a model-independent estimate of $\eta^{t}/\eta^{d} = (1.68 \pm 0.04)$ consistent with $B(^{3}\mathrm{H})_{\mathrm{exp}}$. Once an accurate experimental value of η^{d} becomes available, the "correct" theoretical η^{t} can be predicted from this universal estimate of η^{t}/η^{d} without the need for a new dynamical 3N calculation.

The preceding considerations are applicable to the $^{3}\mathrm{He}$ nucleus. The Coulomb effect in $^{3}\mathrm{He}$ modifies the result for $\eta(^{3}\mathrm{He})$ only slightly. The general arguments about universality and correlations still hold true.[23]

7.3c.2B. Nucleon-Deuteron Polarization Observables (P.O.'s). It has become apparent that the unpolarized elastic cross section and breakup data for Nd scattering do not provide information about the N-C part of the NN interaction. Accurate data on P.O.'s for NN and Nd systems are needed to investigate both the tensor part and the $L > 0$ states of the NN interaction. Accurate data for NN scattering exist only for the pp system, for which the P.O.'s are very small. The situation is different in the pd system, where the P.O.'s are larger by an order of magnitude and also high-precision data is available. This could provide more information on the NN tensor force and

P-wave interactions. Besides the first-order P.O.'s, a number of second-order Nd scattering P.O.'s can be measured with acceptable precision.

Let us briefly discuss some of the different P.O.'s.[43,44] Among the first-order P.O.'s, there are the proton and deuteron vector analyzing powers A_y and iT_{11}, and the deuteron tensor analyzing powers T_{20}, T_{21}, and T_{22}. The second-order P.O.'s include the proton-to-proton polarization-transfer parameters (PTP's) in ^{2}H($\mathbf{p},\mathbf{p}$)^{2}H and proton-to-deuteron PTP's in ^{2}H($\mathbf{p},\mathbf{d}$)^{1}H; both sets are denoted as $K_y^{y'}$, $K_x^{x'}$, and $K_z^{x'}$. There are also more complicated proton-to-deuteron vector-to-tensor PTP's in ^{2}H($\mathbf{p},\mathbf{d}$)^{1}H, label $K_x^{y'z'}$, $K_x^{x'y'}$, $K_z^{x'y'}$, $K_z^{x'y'}$, $K_y^{x'y'}$, $K_y^{x'z'}$, $K_y^{x'x'}$, $K_y^{x'y'}$, etc. The spin-correlation functions are second-order P.O.'s.[45]

We concentrate mainly on the experimental data at an incident proton energy (E_{p}) of 10 MEV. This is because accurate data on all the above observables exist at this energy and there are several numerical calculations at this energy. Though the Coulomb potential can be incorporated in such calculations,[46] we usually neglect it.

Polarization Transfer Parameters. The study of low-energy correlations involving P.O.'s began only recently. The P.O.'s that are expected to be correlated with $^2a_{\text{nd}}$ are the PTP's $K_y^{y'}$, $K_x^{x'}$, and $K_z^{x'}$, since they are determined principally by S-wave NN potentials; the $L \geq 1$ NN interactions contribute to the corrections.[16,43]

The proton-to-proton PTP's have been studied by Doleschall[43] and the Graz group[47] (D-GG). Both groups found that these PTP′s are dominated by the S-wave NN potentials. They also concluded that these parameters are very sensitive to the off-shell behavior of the NN interaction. In particular, they demonstrated that a 3N model with a separable Yamaguchi-type NN potential cannot reproduce these PTP's, whereas the more realistic 3N model of Doleschall, or ones based on separable representations of the Paris or the Bonn NN potentials, do so. The presumably incorrect off-shell behavior of the Yamaguchi potential has been thought to be responsible for this.

In a more recent study, a Brazilian group[16] found a strong correlation between $^2a_{\text{nd}}$ and the proton-to-proton PTP′s. They showed in a simple Nd model with Yamaguchi NN potentials that, if $(^2a_{\text{nd}})_{\text{exp}}$ is reproduced, this model yields values for these PTP's comparable to those obtained in 3N calculations using the Paris or Bonn NN potentials. In view of this, it seems unlikely that useful information about the NN interaction can be obtained from these parameters that is not already implicit in $^2a_{\text{nd}}$. Hence, one can conclude that it is not the off-shell behavior of the Paris or Bonn NN potential that is responsible for a good reproduction of these PTP's.

Although the $^1S_0$ and $^3S_1$ NN potentials are expected to dominate these types of PTP's, the tensor NN potential and the higher partial waves of the NN interaction are needed for a more accurate representation. For a central NN potential, these PTP's are related by[16,43]

$$K_z^{x'} = -K_y^{y'} \sin\theta_{\text{lab}},$$

$$K_x^{x'} = K_y^{y'} \cos\theta_{\text{lab}},$$

where θ_{lab} is the nucleon scattering angle in the laboratory system. The combinations

$$\Delta K = -K_y^{y'} \sin\theta_{\text{lab}} - K_z^{x'},$$

$$\Delta K' = K_y^{y'} \cos\theta_{\text{lab}} - K_x^{x'}$$

provide good measures of the effect of the noncentrality of the NN interaction. The experimental ΔK was reconstructed by the Brazilian group and compared with theoretical calculations. They found that ΔK is not very sensitive to $^2a_{\text{nd}}$, but $K_y^{y'}$ and $K_z^{x'}$ are.

The parameter ΔK is sensitive to the tensor force and the P-waves of the NN interaction, but it was poorly reproduced by the Yamaguchi model used by the Brazilians. Later, ΔK was inferred from the Paris-potential results provided by Plessas to Frederico; this potential did not reproduce ΔK any better than the Yamaguchi did. In Figs. 7.8 and 7.9, we present results for $K_y^{y'}$, $K_z^{x'}$, and ΔK, respectively, obtained with the Yamaguchi NN tensor potential and P-wave NN potentials of Doleschall. The agreement between the experimental $K_z^{x'}$ and the calculation with $^2a_{\text{nd}} = 0.61$ fm is quite satisfactory and is almost indistinguishable from the calculations of D-GG. The $^2a_{\text{nd}}$ calculated by D-GG ranged from 0.6 to 1.0 fm, whereas their earlier version of the reference Yamaguchi model yields a value of about -0.5 fm.[16] The agreement between $(^2a_{\text{nd}})_{\text{exp}}$ and the prediction by D-GG, and the disagreement with that of the earlier Yamaguchi model, are consistent with the inability of the earlier Yamaguchi model to reproduce the experimental PTP's.

The present conclusions about universality seem to be extendable to the case of $^2\text{H}(\mathbf{p},\mathbf{d})^1\text{H}$ vector-to-vector PTP's, which are also not expected to yield much information about the off-shell behavior of NN interactions as these parameters are also dominated by a $^1S_0$, $^3S_1$ NN potential model. Sperisen *et al.*, Koike *et al.*, and Witala *et al.* have reproduced the behavior of these PTP's in realistic models.[43,47,48] Our conclusions about these PTP's are expected to hold for other second-order P.O.'s such as the spin-correlation functions, which are well-accounted for in a $^1S_0$, $^3S_1$ NN potential model.[45]

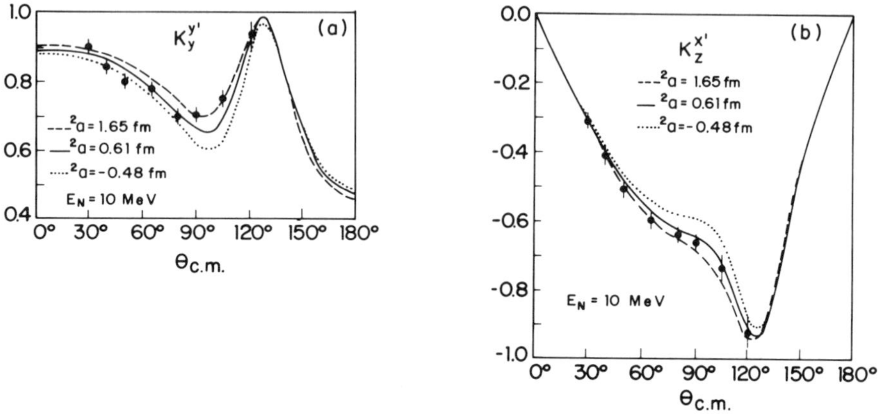

FIG. 7.8. The $K_y^{y'}-{}^2a_{nd}$ and $K_z^{x'}-{}^2a_{nd}$ correlations at $E_p = 10$ MeV in a separable-potential model (from Frederico (1988b)).

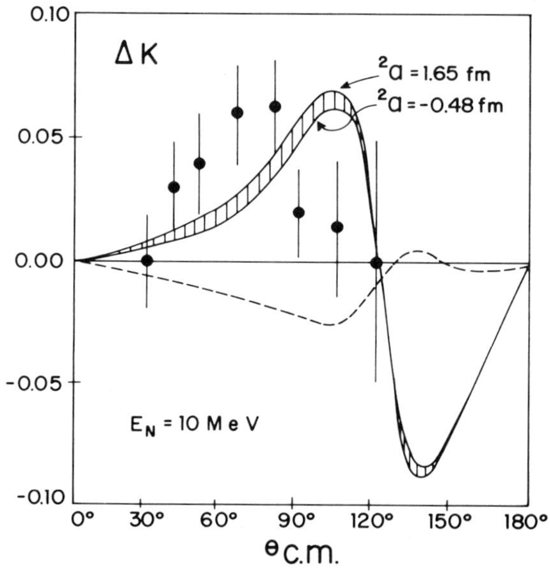

FIG. 7.9. Plot of ΔK at $E_p = 10$ MeV (from Frederico (1988b)).

FIRST ORDER POLARIZATION OBSERVABLES. A fundamental difference between the first-order and second-order P.O.'s is that the first type, viz., the deuteron and proton analyzing powers, are extremely sensitive[5] to the tensor and P-wave parts of the NN interaction. Indeed, a purely S-wave NN potential model yields zero values for these observables, but they are expected to remain unmodified provided that at least the low-energy NN observables, which should now include η^d, the $^3S-^3D$ mixing parameter ε_t, and the $L \geq 1$ NN phase-shifts, in addition to $^2a_{nd}$, are held fixed. Over and above this, it might be necessary to keep some other P-wave nd observables fixed, e.g., the nd P-wave scattering lengths. Since the sensitivity of the nd central observables to NN off-shell effects is contained entirely in $^2a_{nd}$, it is plausible to assume that the sensitivity of the analyzing powers on the off-shell behavior of the P-wave NN potentials is contained in the P-wave Nd scattering lengths. Even though universality requires that a correlation should exist between the first-order P.O.'s of the Nd system and the low-energy few-nucleon observables, no systematic study has been done to establish this claim.

The deuteron and nucleon analyzing powers have been calculated by several groups. Even the most sophisticated of these calculations by the Bochum group[13,48] using the Paris and Bonn potentials is unable to reproduce these observables exactly, especially the neutron vector-analyzing power A_y. The reason for the discrepancy between the theoretical prediction and the experimental value of A_y is unclear at present. The Bonn NN potential yields a value for $^2a_{nd}$ closer to experiment than the Paris NN potential. However, the Paris potential fits A_y better, possibly due to the more realistic P-wave part of this potential. Unless the P-wave NN observables are held fixed, it will be difficult to notice any correlation in the first-order P.O.'s.

The calculations by the Brazilian[5] and the Graz groups[49] can be used to draw some preliminary conclusions about the expected correlations in the first-order P.O.'s. The investigations of each group are complementary in this respect. The Brazilian group fixed the NN P-waves, and then studied the dependence of the first-order P.O.'s on (a) the off-shell behavior of the NN S-wave potential contained in $^2a_{nd}$, and (b) η^d. They found that both A_y and iT_{11} are sensitive to variations of η^d but not $^2a_{nd}$, whereas the deuteron tensor-analyzing powers are less (more) responsive to changes in η^d ($^2a_{nd}$).

Therefore, any consistent attempt to extract new information about the NN interaction, especially about its off-shell behavior, from a comparison of the experimental and calculated values of these observables should have the experimental values of $^2a_{nd}$ and η^d built in as constraints on the theoretical models. This will eliminate part of the discrepancy between the theoretical and experimental results that arises simply from the incorrect values of $^2a_{nd}$ and η^d implicit in the models. The remaining discrepancy between the experimental and theoretical results may be attributed to wrong "on-shell" and off-shell" characteristics, independently of $^2a_{nd}$ and η^d, of the NN potential models used in the theoretical calculations. We have seen that the same is true for the second-order P.O.'s.[5]

The Graz group, on the other hand, fixed the S-wave and tensor NN potentials, but varied the off-shell behavior of the P-wave NN potentials. They found that all of the first-order P.O.'s are sensitive to such variations.[49]

7.3d. Effect of the Coulomb Potential on Nd Scattering

Almost all of the calculations for Nd scattering pertain only to the nd system since they completely neglect the pp Coulomb potential. On the other hand, most of the accurate experimental data refer to the pd system. It is important to know when the neglect of the Coulomb potential is justified. (The polarization of the deuteron by the Coulomb potential gives rise to a potential with a r^{-4} behavior at large distances, but it can be shown that its effect can be neglected for all practical purposes.[14]) The work of the Graz group shows that, at an incident proton energy of 2.5 MeV, the effect of the Coulomb potential cannot be neglected, especially for calculating deuteron tensor polarizations.[50] The effect on the differential cross sections and the deuteron and nucleon vector polarizations is much smaller. At a proton incident energy of 10 MeV, Coulomb effects are expected to be less important, but whether they can be totally ignored remains to be established.

7.3e Breakdown of Universality

There are 3N observables that do not follow the definite trend expected from universality and appear to be sensitive to the details of the 3N models. This implies a breakdown of universality that arises from many effects of unknown magnitude, e.g., nucleonic isobars, relativistic dynamics, charge-symmetry breaking, and mesonic and quark degrees of freedom.[51] For example, the calculated 3N magnetic moment and Coulomb energy are in disagreement with experiment; a study of these quantities may yield information about some unknown effects, such as charge-symmetry breaking and relativistic corrections. Among these observables, the static Coulomb-energy contribution, as found in dynamical 3N calculations, has the behavior associated with universality, despite the disagreement with experiment of the value expected from the theoretical correlation. It appears that the 3N form-factors receive large contributions from the quark and mesonic degrees of freedom and relativistic effects, particularly at large momentum transfers.[51]

Correlations among these observables are unlikely because they are strongly dominated by universality-breaking effects. For example, Friar *et al.*[52] plot the position of the first diffraction minimum of ^{3}He and ^{3}H charge form-factors versus 3N binding energy. No correlation is observed in the case of the charge form-factor, as expected. However, the value of the form-factor at zero-momentum transfer is relatively insensitive to the triton binding energy, and is essentially the same for most models. This is substantiated by the calculation of the magnetic moment of the trinucleon system, which is related to the value of magnetic form-factor at zero momentum transfer.[53] The theoretical model calculations for the isoscalar and isovector magnetic moments fall in a narrow band with the experimental value lying somewhat outside this range. For example, the calculated values of the isoscalar magnetic moment μ_s lie between 0.40 and 0.41, while $(\mu_s)_{exp} = 0.426$.[53] The theoretical predictions for the isovector magnetic moment μ_v lie between -2.44 and -2.50, but $(\mu_v)_{exp}(= -2.5532)$ does not lie in this range.[53] A correct treatment of relativistic and other universality-breaking effects may reconcile these discrepancies.

7.3f. Future Perspectives

7.3f.1. Experimental

After two decades of extensive calculations, the 3N observables can be broadly divided into three categories: (a) those that are compatible with the correlations anticipated theoretically, (b) those that are not, and (c) those which have such large experimental errors that compatibility cannot be decided. Observables of each type are (a) ${}^2a_{nd}$, the r.m.s radii, and the S-wave ANP of ^{3}H and ^{3}He; (b) the Coulomb energy, magnetic moment, charge and magnetic form factors of ^{3}H and ^{3}He; (c) η^t and ${}^2a_{pd}$.

Type (a) observables, which can adequately be explained in nonrelativistic nucleons-only potential models, cannot provide much information about the NN and 3N interactions. Those of type (c) should be measured again more accurately to determine whether they fall into type (a) or (b). For example, new measurements of η^t and η^d are needed in order to see if the ratio η^t/η^d agrees with its universality-based estimate.[15] Also, a more accurate experimental value for ${}^2a_{pd}$ is needed.

7.3f.2. Theoretical

There is a need for more theoretical work on type (b) observables. The careful study of these observables may yield new physics. For example, a better understanding of the 3N Coulomb-energy difference may provide information about charge-symmetry breaking in nuclear interactions, or about the distortion of the pp Coulomb interaction at short distances because of the finite size of the proton. This is a difficult problem to handle technically as we are concerned with a small discrepancy (~ 0.1 MeV) in an observable, namely the Coulomb energy.

Even though high-precision data exist for the magnetic moment, this type (b) observable is much more difficult to study theoretically because of the exchange current and relativistic corrections that dominate it.

More effort should be devoted to the exact inclusion of the Coulomb potential in pd scattering, especially above the breakup threshold. The work of the Graz group indicates that the Coulomb force produces a nonnegligible effect on deuteron tensor polarizations below the breakup threshold. It yet remains to be seen how much of the discrepancy of the deuteron tensor polarizations at a proton incident energy of 10 MeV can be accounted for by a proper treatment of Coulomb potential alone. The first-order P.O.'s are extremely sensitive to both the on- and off-shell behavior of the P-wave NN amplitudes, and therefore their study is expected to provide information about the NN interaction.

We again emphasize that our qualitative discussion of the 3N problem should by no means be considered to be a criticism or a substitution for doing realistic Faddeev calculations. The Faddeev calculations are indispensable for understanding the 3N system. Our present discussion should only be considered to be complementary in the understanding and interpretation of the results obtained in realistic 3N calculations.

Evidently, at the present time, theoreticians should be very cautious about drawing conclusions from a 3N calculation because there are so many competing corrections that are either neglected or only treated approximately. It is now apparent that before making a theoretical estimate of any 3N observable, a given nonrelativistic model should reproduce $({}^2a_{\text{nd}})_{\text{exp}}$ and $(\eta^{\text{d}})_{\text{exp}}$. The inability of a nonrelativistic nucleons-only potential model to explain all of the experimental data may not necessarily indicate an inadequacy of the potential model, but might be due to an incomplete treatment of various corrections. The low-energy 3N correlations, which are a direct manifestation of the universal behavior of the 3N system, make it simple to classify results of theoretical calculations, while at the same time make the extraction of physically interesting information that much harder.

7.4. Photodisintegration of ^{3}He and ^{3}H

Another physical process involving the trinucleon system that has been studied using dynamical three-body equations is the photon-induced low-energy breakup of ^{3}He (^{3}H) expressed as

$$\gamma + {}^3\text{He}\,({}^3\text{H}) \rightarrow \text{p}\,(\text{n}) + \text{d},$$
$$\rightarrow \text{p}\,(\text{n}) + \text{n} + \text{p}.$$

Comprehensive reviews of this topic have appeared recently.[1]

Barbour and Phillips (BP) were the first to apply three-body dynamical equations to this problem.[2] They assumed an electric dipole (E1) disintegration mechanism for the fundamental amplitude $\gamma + {}^3\text{He}\,({}^3\text{H}) \rightarrow \text{p}\,(\text{n}) + \text{d}$; the continuum final states were generated by solving the Faddeev equations for the 3N system employing separable S-wave interactions.

The approximate wave function Ψ_B, employed by BP in their calculation, is similar to that obtained from separable S-wave NN potentials with two parameters in the spectator function that are fixed by fitting the binding energy and the radius of the 3N system to their experimental values. Consequently, the experimental data for both two-body and three-body photodisintegration cross sections for the trinucleon system were well reproduced. Later on, the NN tensor force was also included; this improved the agreement with the 90° two-body cross section.[3] The formalism also was extended to include the electric quadrupole (E2) interaction.[4]

Gibson and Lehman[5,6] (GL) improved upon the calculations derived from the BP work[2–4] by generating the trinucleon wave function exactly using a separable NN tensor potential, instead of taking a simple model fitted to the trinucleon binding energy and the r.m.s radius. The dynamically more complete treatment of GL not only showed some improvement in fitting the two- and three-body photodisintegration cross sections, but also reproduced the $^3\text{He}(\gamma, \text{n})2\text{p}$ and $^3\text{He}(\gamma, \text{p})\text{np}$ energy spectra. The 90° two-body cross sections determined by GL fit the data better than those of BP *et al.*

Once again, as in the Nd scattering problem, it is obvious that some reasonable NN potential, along with an exact treatment of the three-particle dynamics, can account for the two- and three-body photodisintegration of the 3N system. However, it is not at all obvious that treating the three-particle dynamics exactly is essential for explaining many of these results, in view of our discussion of universality in Section 7.3. Two-body models for the 3N bound states and final-state amplitudes might be enough to describe the two-body photodisintegration of ^{3}H and ^{3}He.

An effective two-body Nd wavefunction for the 3N bound state can be constructed with two partial-wave components ($L = 0$ and 2) to accomodate a 3N D-state resulting from the NN tensor force. This D-state is essential in reproducing the tensor P.O. T_{20} in the capture of tensor-polarized deuterons[7–10]

$$\mathbf{d} + \text{p} \rightarrow {}^3\text{He} + \gamma.$$

This is the time-reversal of the photodisintegration of ^{3}He. The disintegration (or capture) matrix element can then be constructed using the effective Nd wavefunction. The Nd initial- or final-state motion is treated either in plane-wave approximation or as a scattering state generated from an Nd optical potential; this approximation is not usually compatible with the effective Nd bound state, in that they are not generated from the same optical potential.

Arriaga and Santos[7] applied the preceding approach to calculate the tensor-analyzing powers in the ^{1}H($\mathbf{d}$, γ)^{3}He reaction using the Sasakawa wave function for the 3N bound state and neglecting the initial-state interactions. They obtained good agreement with experiment, and found that their result is sensitive to the ^{3}He D- to S-state asymptotic normalization ratio.

In other applications of the two-body approach to the radiative capture of a proton by a deuteron, King *et al.*[8] satisfactorily explained the ^{2}H(p, γ)^{3}He angular distribution, while Vetterli *et al.*[9] successfully accounted for the values of T_{20} extracted from ^{1}H($\mathbf{d}$, ^{3}He)γ measurements at an incident deuteron energy of 19.8 MeV. The results for T_{20} show very little sensitivity to either a correct treatment of the initial-state interaction in the rescattering amplitude or to the deuteron D-state percentage. The underlying reasons for these insensitivities are not entirely clear.

The success of the simple effective two-body model in explaining the photodisintegration of ^{3}H and ^{3}He and the time-reversed radiative-capture processes does not appear to be too surprising, as we have seen in Section 7.3 that two-body models of the trinucleon system are quite plausible. Nevertheless, it is reassuring that a fully consistent treatment of the three-body dynamics also explains the experimental results.

7.4a. *Future Perspectives*

More studies—both theoretical and experimental—need to be performed in order to draw definite conclusions about many interesting and relevant questions: (a) What are the minimum ingredients required of a theoretical model to explain the photodisintegration results? (b) What is the role of interactions other than E1 in this physical process? (c) How sensitive are the observables of the photodisintegration and radiative-capture processes to the underlying NN potentials, to the 3N binding energy, and to η^t? Answers to these questions will certainly enhance our understanding of the problem. To this end, future experimental efforts should concentrate not only on cross sections and the breakup spectrum, but also on the deuteron tensor-analyzing powers in the radiative capture of polarized deuterons on protons. Theoretical studies should deal with 3N dynamical calculations involving the most sophisticated meson-theoretic NN interactions as well as simple 3N models. By comparing the results of these two types of calculations, one might be able to determine the minimum ingredients required of a theoretical model.

7.5. Deuteron–Alpha Collisions

Apart from the trinucleon problem, many other effective three-body problems in nuclear physics have been solved using the Faddeev equations. The deuteron–alpha ($d\alpha$) system has been of particular interest. The alpha particle is strongly bound while the six-nucleon system is loosely bound. At low energies, this justifies the approximation of regarding the alpha particle as a fundamental particle with no structure. Hence the three-body $NN\alpha$ model of the $d\alpha$ system is expected to provide a good first-order representation of the actual state of affairs, at least for energies below the breakup threshold of the alpha particle. There has been a great deal of experimental activity concerning $d\alpha$ collision processes including deuteron breakup.

The $NN\alpha$ model of $d\alpha$ scattering was used by Shanley,[1] with simple (Yamaguchi-type) separable potentials for both the NN and the $N\alpha$ systems for deuteron energies up to 30 MeV. His results are in good agreement with experiment. Specifically, he calculated the $d\alpha$ elastic-scattering phase shifts for the S_1, $P_{2,1,0}$, $D_{3,2,1}$, $F_{4,3,2}$, $G_{5,4,3}$, and $H_{6,5,4}$ states, where the subscripts refer to the J values. He also calculated the theoretical absorption parameters for these partial waves, the differential cross sections, and the deuteron tensor polarizations.

The approximate calculation of Shanley was improved upon by Charnomordic *et al.*[2] In the case of the $N\alpha$ system, Shanley used rank-one separable potentials for $S_{1/2}$, $P_{1/2}$, and $P_{3/2}$ partial waves with Yamaguchi-type form-factors. Charnomordic *et al.* also used rank-one separable potentials for these partial waves, but with more realistic form-factors. The parameters of these potentials were determined by a fit to the experimental phase shifts for the $N\alpha$ system. For the $^1S_0$ partial wave of the NN system, they constructed a new two-term potential with an attractive and a repulsive form factor. For the $^3S_1-^3D_1$ partial wave, they used three versions of the separable potentials of Doleschall, Serduke, and Mongan.[2] They then calculated the differential cross sections for elastic $d\alpha$ scattering and obtained good agreement with experiment up to a deuteron energy of 17 MeV, but this fit became poorer at higher energies. Their results on the deuteron vector and tensor-analyzing powers are also in reasonable agreement with experiment.

Koike[3] has studied the reaction $^2H(\alpha, \alpha p)n$ using the contour-rotation technique for solving the Faddeev equation for breakup amplitudes. He employed rank-one separable potentials with Yamaguchi-type form factors for the $S_{1/2}$, $P_{1/2}$, and $P_{3/2}$ waves of the $N\alpha$ subsystem, and also for the NN S-waves. Considering the approximate nature of the treatment with simple separable potentials both for the NN and $N\alpha$ systems, Koike's calculations are in accord with experiment and a reasonable representation of the final-state interaction peak was obtained. This assures us that the three-body dynamical treatment of this reaction using the Faddeev equation is on the right track.

Koike[4] also applied his model to calculate the deuteron breakup reaction P.O.'s. He calculated the deuteron vector-analyzing power iT_{11} and the tensor-analyzing powers T_{20}, T_{21}, and T_{22} in the $^4He(\mathbf{d}, p)n^4He$ reaction at proton energies of about ~ 10–15 MeV and found agreement with experiment.

In Fig. 7.10, we compare Koike's results with experiment for the differential breakup cross section for the reaction $^4He(\mathbf{d}, p)n^4He$ at 11 MeV. In Fig. 7.11, we make a similar comparison for the vector-analyzing power iT_{11} for the

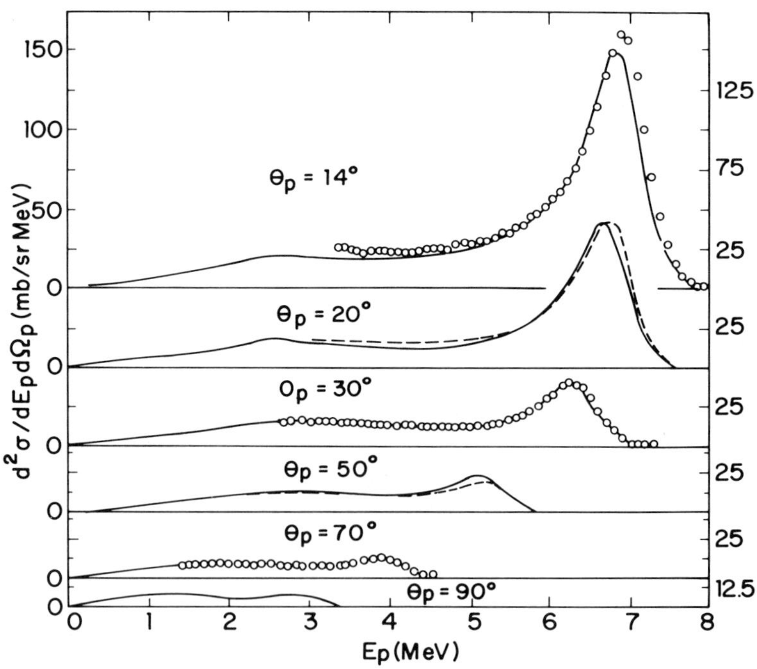

FIG. 7.10. The differential cross section of the $^4He(d, p)n^4He$ reaction at 11 MeV (from Koike (1980)).

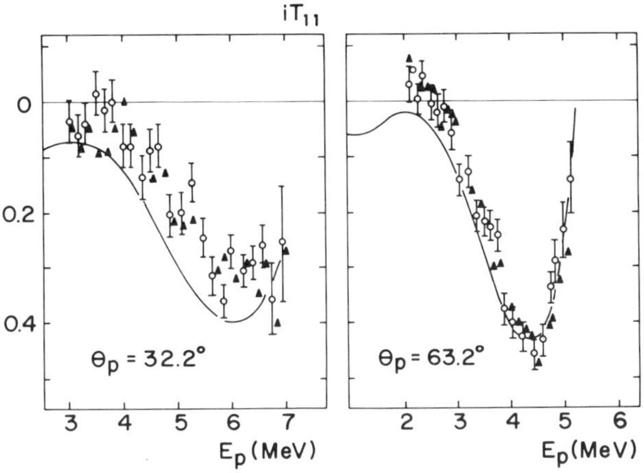

FIG. 7.11. The vector-analyzing power iT_{11} of the $^4He(\mathbf{d}, p)n^4He$ reaction at 11 MeV (from Koike (1980)).

same reaction initiated by polarized deuterons at 11 MeV. In both cases, the agreement between theory and experiment is impressive.

It should be remembered that none of these calculations included the Coulomb potential. At these energies, Coulomb effects are not expected to significantly modify the good fit with experiment.

The P.O.'s are sensitive to the deuteron D-state, or to η^d. Therefore, it should be interesting to study the effects that variations of η^d have on the P.O.'s.

We will not discuss the many αNN bound-state calculations. However, as in the trinucleon system, one can conjecture that the bound-state[5] and scattering results for the dα system are correlated and are not strongly dependent on all of the details of the NN and Nα interactions. The success of these approximate calculations seem to indicate that the major features of the Nα bound-state and scattering observables could be determined, to a good approximation, by the relevant NN and Nα on-shell observables and the ^6Li binding energy within the context of a theoretical model. Further investigations of these conjectures seems called for.

7.6. Three-Body Treatment of the NN−πd−πNN System

An important three-particle system that has been studied extensively at pion energies up to about 300 MeV using dynamical three-body models is the coupled πd−NN−πNN system. There is a basic difference between this and the 3N system. The low-energy 3N system can be described in terms of nonrelativistic scattering integral equations in which the c.m. motion has been separated out. The intermediate-energy, coupled πd–NN–πNN system has to be treated relativistically because the pion mass is so small. It is not known how to separate the c.m. and the relative motion even for only two interacting relativistic particles within conventional dynamical formulations. Moreover, for the coupled system, the number of particles is not conserved because pions can be spontaneously created or annihilated. Pions can appear on shell above the creation threshold. Virtual pions play a role in the dynamics even below the threshold for creation. Because of these, and a number of other problems, practical sets of integral equations for the coupled πd–NN–πNN system have been found only after considerable approximation.

The relativistic dynamics of this system are incorporated by modifying the nonrelativistic Green function slightly, so that relativistic unitarity is maintained both at the two-particle and "multiparticle" levels. The meaning of multiparticle is related to the maximum number of virtual pions one allows in the theory. In the simplest and the most commonly explored case, only one virtual pion is allowed; then, one has a three-body theory for the coupled NN–πd–πNN system. It is important in truncations of systems involving the creation and the destruction of pions that double counting be avoided.

The early attempts at constructing a relativistic three-body collision theory rely on the idea of isobar dominance.[1] Only the simplest πN dynamics are used, namely, a N $\rightleftarrows$ N + π pseudovector coupling, where the particle N is both the elementary particle and the quasi-particle (bound state) of N and π. The πN interaction is assumed to be dominated by an isobar that is represented by a separable interaction, as is the NN interaction.

In the simplest case of two-particle scattering, if one imposes relativistic unitarity, one has the following LS-type relativistic scattering equation due to Blankenbecler and Sugar:[1]

$$T(\mathbf{p}|\mathbf{q}) = V(\mathbf{p}|\mathbf{q}) + \int \frac{d^3\mathbf{k}}{2\omega_1\omega_2} V(\mathbf{p}|\mathbf{k}) \frac{\omega_1 + \omega_2}{w - (\omega_1 + \omega_2)^2} T(\mathbf{k}|\mathbf{q}), \tag{7.6.1}$$

where

$$\omega_1 = (\mathbf{k}^2 + m_1^2)^{1/2}, \qquad \omega_2 = (\mathbf{k}^2 + m_2^2)^{1/2}, \tag{7.6.2}$$

w is the square of the total energy as in Section 2.7, m_1 and m_2 are the particle masses, and $V(\mathbf{p}|\mathbf{q})$ can be thought of as the usual ladder potential taken from the Bethe–Salpeter equation but, in fact, any real symmetric $V(\mathbf{p}|\mathbf{q})$ will do. The three-particle equation for the coupled NN–πd system will have the form of the usual Amado-model equation with modified Green functions and interactions.

The relativistic analogues of the Faddeev equations, which are conjectured as three-body dynamical models, are similar in form to their nonrelativistic counterparts, but now depend on four-momentum variables. Blankenbecler and Sugar eliminate the time components of the relative momentum four-vectors in a covariant way by replacing the three-particle propagators by their δ-function parts and performing a dispersion integral in the total energy of the system. This leads to the relativistic propagator[1]

$$G_0(w) = \frac{-1}{\omega_i \omega_j \omega_k} \frac{\omega_i + \omega_j + \omega_k}{(\omega_i + \omega_j + \omega_k)^2 - w - i\varepsilon}. \tag{7.6.3a}$$

Aaron, Amado, and Young suggest the following alternative reduction[1]

$$G_0(w^{1/2}) = \frac{-1}{\omega_i \omega_j} \frac{1}{\omega_k^2 - [w^{1/2} - \omega_i - \omega_j]^2 - i\varepsilon}. \tag{7.6.3b}$$

These two propagators have the same pole positions and residues, so that the two reductions obey identical unitary relations.

Dynamical three-particle equations for the coupled πd–NN system were applied by Afnan and Thomas (AT)[2] to absorption (πd → pp) and elastic scattering (πd → πd), by introducing a bound state of one of the nucleons N′ and the pion; this state is required to be indistinguishable from the nucleon N′. They treated spin, isospin, and angular momentum exactly, but employed nonrelativistic kinematics since they only considered low energies. Their Amado-model type equations utilized the one-pion and one-nucleon-exchange Born (driving) terms for the NN ⇆ NN and NN ⇄ πd reactions, respectively. Then they constructed properly normalized and antisymmetrized amplitudes for the processes πd → πd and πd → pp. Obviously, similar sets of equations can be used for the time-reversed processes πd → πd and pp → πd.

The AT model for the coupled NN–πd system has since been improved upon by other workers. The main problem with the original AT model is that it treated one of the nucleons (N′) as a πN bound state, which implies that only one of the nucleons can emit a pion, so that the resultant model excludes a whole class of diagrams. Also, in intermediate states, the Pauli exclusion principle is violated because the two nucleons N and N′ are considered to be distinguishable. Recent formalisms are able to treat the two nucleons on the same footing and hence remove the proceeding undercounting problem, while at the same time maintaining both two- and three-particle unitarity.

There are other shortcomings of the AT model: (a) A more complete field-theoretic description of the problem would include indefinite numbers of baryons, antibaryons, and pions in intermediate states. However, a three-body theory is achieved by truncating this space to a space of two nucleons and a pion. (b) Crossing symmetry is lost; the theory preserves two- and three-particle unitarity only in the *w*-channel. (c) Blankenbecler–Sugar type relativistic scattering integral equations employ an effective potential between two particles and hence treat the left-hand cut structure of, for example, the πN effective potential only in a very phenomenological way. (d) Finally, there is an inherent ambiguity in the Blankenbecler–Sugar approach, which puts the spectator particle on its mass shell. There are other ways of reducing a four-dimensional relativistic scattering equation to a three-dimensional one,[3] but there is no fundamental reason to prefer one method over another.

Threshold-πd elastic scattering was calculated by AT. In their study of S-wave pion production and the elastic scattering of nucleons, the NN interaction was supposed to exist only in the $^3S_1$–$^3D_1$, P_1, and $^3P_{0,1,2}$ states. The P-waves were generated with Mongan's case II potentials while, in the $^3S_1$–$^3D_1$ channel, unitary-pole approximations (UPA) to the Bryan–Scott and Reid soft-core potentials were utilized. They also fit the low-energy πN phase-shifts in the S_{11}, S_{31}, P_{11}, and P_{33} channels with simple separable potentials. This three-body dynamical model yields results in reasonable agreement with experiment for G- and P-wave NN elastic scattering at lower energies, and for the πd scattering lengths; the predicted pion production was also reasonable.

The AT model was refined further by Thomas,[4] who used relativistic pion kinematics in a three-body treatment of πd scattering. He fit the elastic πd scattering data reasonably well up to 180 MeV incident pion energy, but the agreement between theory and experiment becomes worse at 256 MeV.

Later, Rinat and Thomas[5] (RT) solved a set of relativistic three-particle equations satisfying two- and three-body relativistic unitarity. RT treated the spin and isospin of the πd system correctly and used different ways—due to Aaron *et al.* and Freedman *et al.*—for implementing unitarity.[1] They took a separable P_{33} πN potential corresponding to a Δ-particle with bare mass 1.347 GeV, which provides a good fit to the P_{33} phase shifts. In addition, they also took separable $^3S_1$–$^3D_1$ NN potentials with Yamaguchi-type form-factors and varying (0, 4, and 6.7%) D-state probabilities. At pion energies lower than about 200 MeV ($T_\pi \leq 180$ MeV), their calculation for πd differential cross section is in close agreement with the experimental data; this favors a D-state percentage of 6.7% over the other two choices. At higher energy ($T_\pi = 256$ MeV), the agreement is not good. Possible reasons for the discrepancy between calculation and data at higher energies are: (a) the poor representation of both NN and πN interactions and (b) incorrect treatments of relativistic effects and pion-absorption in πd elastic scattering. It is also possible that the entire phenomenological dynamical framework becomes inadequate.

Although the different ways of including relativistic effects yield slightly different results, there are no means of prefering one method over another.

The calculations we have reported upon so far consider an isolated set of integral equations for the πd system. Other, later, calculations have been carried out based on sets of coupled integral equations for the reactions NN $\rightarrow$ πd, πd $\rightarrow$ πd, and NN $\rightarrow$ NN that have been proposed independently by Avishai and Mizutani and by Afnan and Blankleider; there were a number of earlier attempts is deriving such equations.[6–9] In these models, both the πN and the NN potentials are usually taken to be separable, so the relativistic dynamics of both the pion and the nucleons can be easily handled. A complete numerical analysis of these equations has been performed by Blankleider and Afnan[8] (BA) for all three relevant processes; eseentially the same set of equations has been used for elastic πd scattering by Fayard *et al.*[10] and by Rinat and Starkand[11] (RS).

BA treated the pion relativistically, but the nucleons were considered to be nonrelativistic. They employed the separable NN and πN interactions used by AT. In Fig. 7.12, the total cross section calculated by BA for $\pi^+ d \rightarrow pp$ is compared with the experimental data at various energies; the agreement is good. In Fig. 7.13, the corresponding differential cross sections at different proton laboratory energies are depicted. Above the πN resonance energy, the theoretical cross sections are systematically larger than the experimental results. According to RT, this discrepancy at higher energies seems to arise from the nonrelativistic modeling of the nucleons.[5]

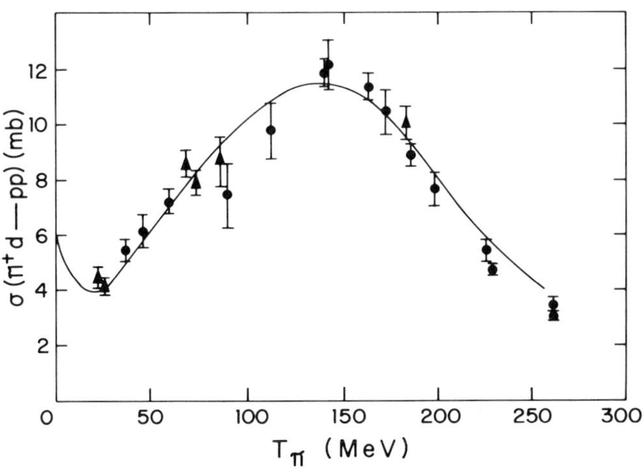

FIG. 7.12. Total cross section for $\pi^+ d \rightarrow pp$ (from Blankleider (1981)).

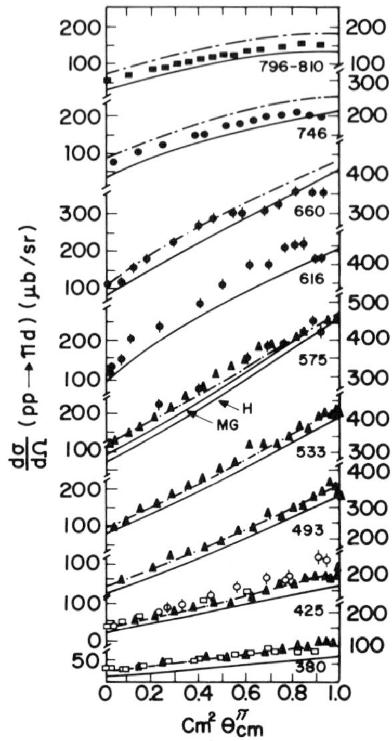

FIG. 7.13. Differential cross sections for pp → π^+d. The full lines correspond to the calculation of Rinat and Starkand multiplied by 1.5; the dash–dot lines correspond to the calculation of Blankleider and Afnan (from Rinat (1983)).

In addition, BA calculated differential cross sections for elastic πd scattering. Their results are compared with the experimental data at various energies in Fig. 7.14. The agreement is obviously better in the more complete calculation of BA than in earlier works by RT[5], especially at higher energies, $T_\pi = 232$ MeV and 256 MeV. This indicates that a full dynamical treatment of pion absorption, among other effects, is necessary at higher energies. BA also studied the sensitivity of the π^+d → pp and π^+d → π^+d differential scattering cross sections to changes of the πN interaction in the P_{11} channel and the NN D-state probability; no definite conclusion could be reached in favor of any πN interaction. The total π^+d cross section calculated by BA does not agree well with experiment, especially at higher energies.

BA also calculated the deuteron tensor and vector polarizations in πd scattering. The vector polarizations agree well with experiment at $T_\pi = 140$ MeV, where the πd differential cross section fits the data well, but not at

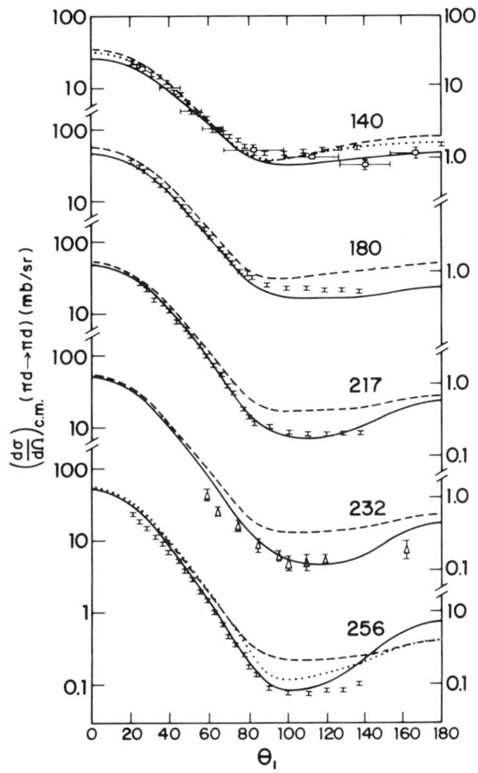

FIG. 7.14. Differential cross sections for πd elastic scattering. (from Blankleider (1981)).

$T_\pi = 256$ MeV, where the theoretical cross sections lead to poorer fits. They also calculated the NN phase shifts in the $^1S_0$, $^1D_2$, $^1G_4$, $^3P_1$, $^3F_3$, $^3F_4$, $^3P_2$, $^3F_2$ channels at laboratory proton energies up to 800 MeV; the agreement with phase-shift analyses of the data is at best fair under 400 MeV, and bad at higher energies.

Fayard *et al.*,[10] performed a fully relativistic calculation of πd scattering from threshold to $T_\pi = 260$ MeV based on a fully coupled set of integral equations employing the Blankenbecler–Sugar prescription with πN and NN interactions different from those of BA. The resulting πd differential cross sections are in reasonable agreement with the data. The pion-absorption results are also reasonable and provide credibility to the theory and the assumptions underying it.

Rinat and Starkand[11] (RS) essentially used the same theory, but employed a πN t-matrix different than BA's. The elastic differential πd cross sections obtained by RS are very similar to the BA calculation. The total-πd cross section of RS is in excellent agreement with experiment (Fig. 7.15), in contrast to BA's results. They also computed the pp $\to \pi$d differential cross section for the process and, as is apparent from Fig. 7.13, they are very different from those from the BA model. (Note that the results of RS have been multiplied by a factor of 1.5 before plotting.) The reason for this large discrepancy has not been investigated.

Calculations of the analyzing power A_{y0} in pp $\to \pi$d scattering have been carried out by RS and by Niskanen.[12] The input into Niskanen's calculation differs from RS's, so it is not easy to compare the two works. Both RS and Niskanen have determined other P.O.'s such as A_{xx}, A_{zz}, and A_{zx}. Niskanen's calculations seem to be superior to RS's for A_{y0}, but for A_{zx} the reverse is true.

There have been other approaches to the treatment of the πd system. For example, the calculations of P.O.'s in πd scattering by Garcilazo[13] have produced impressive agreement with experiment.

Except for Garcilazo's work, all other calculations of the πd P.O.'s treat spin nonrelativistically. Garcilazo not only treats the momentum variables relativistically, but the spin variables as well. He performs a partial-wave

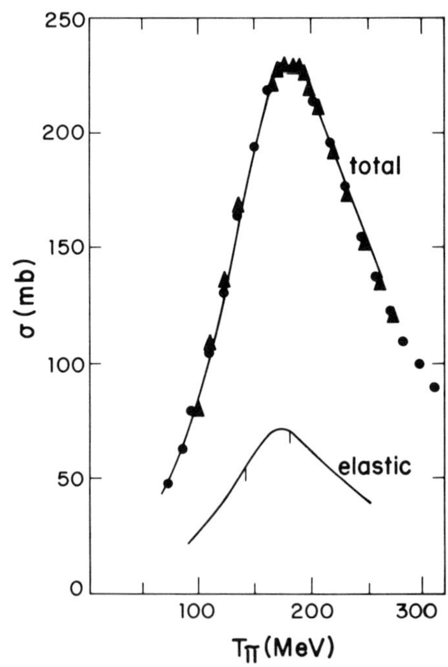

FIG. 7.15. Total and elastic πd cross sections (from Rinat (1983)).

decomposition of the dynamical equations using Wick's three-body helicity formalism that accounts for the Lorentz transformation of the spin between the two-body c.m. and the three-body c.m. frames. He claims that the relativistic treatment of spin is essential as one goes to higher energies.

Garcilazo does not consider coupling to other channels. His fully relativistic dynamical equations are solved with the πN interaction represented by the six S- and P-wave channels, and the NN interaction by the two S-wave channels. His results fit the experimental vector-analyzing power iT_{11} and tensor polarization t_{20} data quite well (Figs. 7.16, 7.17). In contrast to other calculations, the contribution of pion-absorption on these P.O.'s was found to be small. Apart from these P.O.'s, the elastic and breakup πd differential and total cross sections are also fit well, even though the differential cross sections

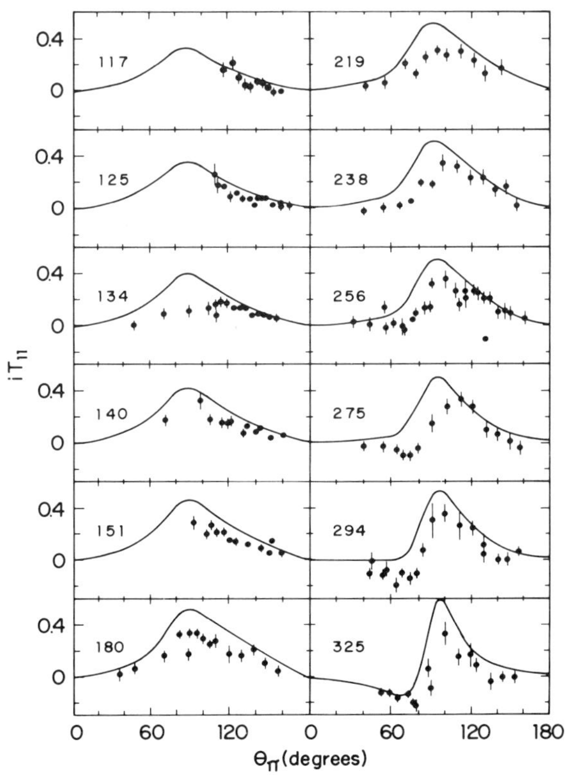

FIG. 7.16. The vector-analyzing power iT_{11} in the c.m. frame for T_π ranging from 117 MeV to 325 MeV (from Garcilazo (1984)).

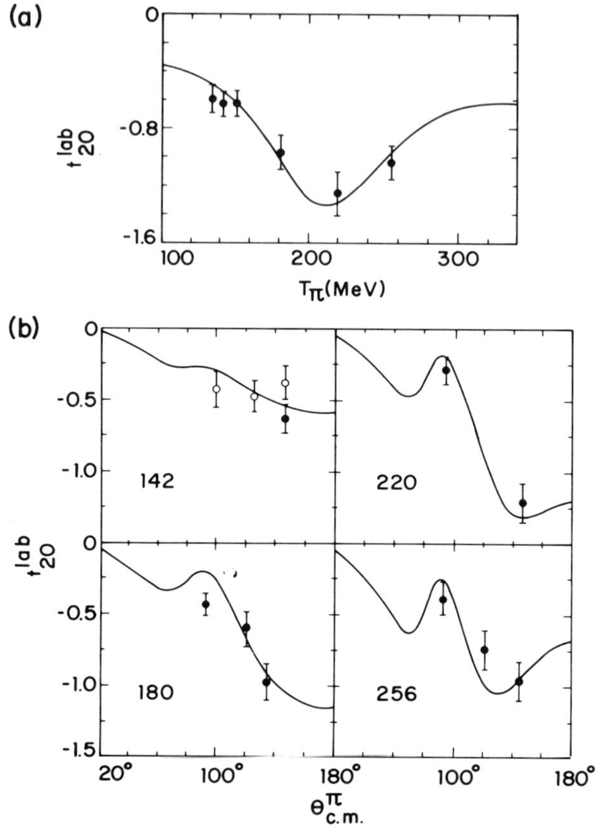

FIG. 7.17. The tensor polarization t_{20} in the laboratory system (a) for a fixed deuteron recoil angle of 18° as a function of T_π; and (b) at four values of T_π in MeV as a function of c.m. scattering angle of the pion (from Garcilazo (1984)).

are typical of one-channel pion–deuteron calculations in overestimating at backward angles at higher energies.

There have been other treatments of the πd problem. For detailed discussions of these approaches, we refer the reader to the original literature.[14,15]

Clearly Faddeev-type three-particle equations have been very successful in explaining the experimental data on the differential and total cross sections for elastic $\pi d \to \pi d$ scattering, reactions of the type $\pi d \rightleftarrows NN$, and pion-production, as well as for the different P.O.'s associated with these processes. The theoretical models of various groups differ in the treatment of the πNN dynamics and πN and NN input. Consequently, the greatly differing theoretical results obtained by these groups cannot easily be compared. The most important difference among these models is in the treatment of pion absorption, which is incorporated exactly through a coupled set of πd–πNN–NN equations by BA,[5] by RS[11] and by Fayard *et al.*,[10] but only approximately by Garcilazo.[13] In addition, for some of the calculations, relativistic kinematics are used only for the pion, while in others both the nucleons and the pions are treated relativistically. Consequently, it is not easy to discern the basic ingredients required of a theoretical model in order to describe the coupled NN–πd–πNN system. Some interesting questions in this regard are whether this system is sensitive to particular ways of including relativistic kinematics at intermediate energies, e.g., (a) whether treating only the pion as relativistic suffices at $T_\pi = 256$ MeV, or whether a full relativistic treatment of the nucleons is called for; (b) whether the numerical results distinguish among different ways of treating relativistic kinematics for three particles, for example, using the prescription of Blankenbecler and Sugar, or of Aaron, Amado, and Young; (c) whether the NN–πd–πNN system is sensitive to the form of the NN and πN interaction models used as input. We do not believe that these questions can be answered easily, but more controlled theoretical-model studies might shed light on them.

In the recent past, there has been a great deal of theoretical and experimental activity on the dibaryon-resonance problem.[16] The experimental cross section for the coupled πNN–NN system exhibits resonance-like behavior at a c.m. energy of about 200 MeV for the NN system in different partial waves such as $^3F_3$, $^1D_2$, and $^1G_4$. Similar structures have been observed in the πd system for T_π around 200–300 MeV. These structures of the coupled πd–NN system are known as dibaryon resonances. One reason that these studies on dibaryon resonances were considered important is that, though not excluded by the quark model, they are not naturally expected to result from the weak residual NN interaction of the quark model in the higher partial waves. The real question is whether the enhancements in the cross sections can be traced back to a pole in the partial-wave t-matrix on some unphysical sheet of the complex-energy plane. The experimental situation is confusing in regard to both the energy of such "resonances" as well as their existence. The theoretical calculations have not resolved the discrepancy. As these resonances are related to the NN system, they are also expected to occur in deuteron photodisintegration. The theoretical and experimental study of this reaction at appropriate energies has not clarified matters. Apparently, the experimental data, at the present level of sophistication of either the theory or the data analysis, is not conclusive as to the existence of dibaryon resonances. Many reviews have appeared on this topic.[17]

Before ending this section, we report on a recent formulation by Afnan and Araki[18] that not only extends the Blankleider–Afnan scheme for treating the coupled NN–πd–πNN system, but also allows a single photon in the initial or final states. In other words, they propose a dynamical set of equations for the coupled NN–πd–γd–πNN–γNN system. Their unified theory of photopion reactions includes photoabsorption and bremsstrahlung, obeys two-body unitarity in the BB sector, and three-body unitarity in the πBB and γBB sectors, where $B = N$ or $\Delta(1232)$. This work by Afnan and Araki should motivate new experiments and numerical calculations in the future.

In conclusion, the coupled πd–NN–πNN system is a very important one, and merits considerably more theoretical and experimental study.

7.7. Applications to Atomic/Molecular Collision Processes

The Faddeev equation has had widespread applications in nuclear physics at low and intermediate energies, but its use for atomic/molecular collision processes has been very limited. The main problem in electron–atom collision processes, or in any system involving charged particles, is that, because of the long-range potentials appearing in atomic/molecular problems, these equations either are not of the Fredholm type or converge very slowly in numerical applications, and hence do not provide an efficient solution algorithm. In the molecular atom–diatom system, the interatomic potential is weak and does not fall off sufficiently fast, which gives rise to several excited states of the diatomic system. In practice, this requires the handling of many partial waves, and the convergence is usually very slow. The electron–diatom system is also beset with similar difficulties. Nevertheless, we would like to summarize two efforts for solving the Faddeev equations in atomic/molecular collision processes.

7.7a. Atom–Diatom Systems

Usually atom–diatom collisions are treated by chemists using one-dimensional models, or other approximate techniques. For an account of such methods, see Section 9.4. Haftel and Lim[1,2] applied the Faddeev theory to a model of three helium-like atoms with one bound state for the diatom system. Their solution of the three-dimensional Faddeev equation in this case agreed with those obtained with the one-dimensional models, which suggests that these collision processes can be well described by one-dimensional models. They calculated the cross sections for both collision-induced dissociation of the diatom, and for the time-reversed process, recombination.

Haftel and Lim[2] also solved the realistic atom–diatom collision process with van der Waals type interaction. They used the Ernst–Shakin–Thaler expansion technique with (a) spin-dependent Morse potentials and (b) spin-independent Hulthen potentials. They found that the higher the initial state of the diatom, the higher the dissociation probability. Increasing the incident energy did not change this property, consistent with experimental findings, but in contrast with the result of the collinear model studies. This casts doubt on the collinear model studies at higher energies.

7.7b. Three-Body Recombination in Spin-Polarized Atomic Hydrogen

All the preceding applications of the Faddeev equation have been for three-dimensional systems, and in the following we consider an interesting application of the Faddeev equation to the solution of two-dimensional problems involving spin-polarized atomic hydrogen (H↓). The physical description of spin-polarized atomic hydrogen is usually made in two space dimensions, and physical processes involving H↓ are well described by the Schrödinger equation in two-dimensional space.

de Goey *et al.*[3] have studied the magnetic-field dependence of the 3H recombination rate in spin-polarized atomic hydrogen by solving the Faddeev equations in two dimensions. Though their calculation is approximate, the success and viability of an exact solution of Faddeev equations in two space dimensions deserves a special mention, and should lead to more comprehensive calculations in the future.

There are recent experimental activities for achieving Bose–Einstein condensation in H↓. For these attempts to be successful, it is important to understand decay mechanisms in H↓ samples. The use of two-body surface relaxations only, in explaining low-temperature decay in "precompression" experiments, leads to a large discrepancy between theory and experiment. Hess *et al.* have suggested that these discrepancies might be eliminated with the use of three-body processes.[4]

Explicitly, de Goey *et al.* study the recombination of two hydrogen atoms as a three-body collision in a strong external magnetic field:

$$H + H + H \rightarrow (HH) + H.$$

By performing approximate calculations, they have demonstrated the necessity of a full three-body dynamical calculation for this problem. In their calculation, they included the strong central singlet and triplet interactions exactly. The weak dipole interaction has, however, been treated only to first order.

After making a partial-wave projection in two dimensions, the resultant Faddeev equation has the form of an integral equation in two scalar momentum variables, which is then solved by summing the Neumann series by the technique of Padé approximants.

de Goey *et al.* calculate the three-body recombination rate essentially exactly as a function of the applied magnetic field. This work is approximate in nature. A more complete calculation should answer the question as to whether the three-body recombination rate is sufficiently suppressed at magnetic fields larger than about 25 T to allow Bose–Einstein condensation in H↓ to be achieved in a compression experiment.

References and Notes (7)

Section 7.1

1. Kloet (1973a, b), Takemiya (1987), Witala (1987, 1988).
2. Tang (1987), Austern (1987).
3. Faddeev (1960, 1965), Yakubovskii (1967).
4. Alt (1967, 1970), Grassberger (1967), Fonseca (1976a, b).
5. Weinberg (1964), Baer (1971–1973), Kouri (1974a, b; 1975a, b).

Section 7.2

1. The method was first proposed by Hetherington (1965), used in the trinucleon problem by Aaron (1964–1966) and refined by Cahill (1971).
2. Brayshaw (1968a, b).
3. Sohre (1971), Doleschall (1973), Larson (1974).
4. de Boor (1978).
5. Brannigan (1983), Alaylioglu (1984, 1986), Christie (1986), Osborn (1979), Pretorius (1987).
6. Glöckle (1982b), Witala (1987, 1988).
7. Kloet (1973a, b).
8. Haidenbauer (1983, 1986a), Koike (1985, 1987a, b), Plessas (1986a, b; 1987).
9. Tomio (1981).

Section 7.3

1. Mitra (1962, 1966, 1969), Bhakar (1965), Schrenk (1967), Amado (1963, 1969), Aaron (1965, 1966), Phillips (1965, 1968), Sitenko (1963, 1966), Kharchenko (1968), Cahill (1971), Duck (1968).
2. Reid (1968), Ueda (1968), Lacombe (1980), de Tourreil (1973, 1975), Wiringa (1984), Malfliet (1968, 1970), Machleidt (1987).
3. Coon (1979, 1981), Coelho (1983), McKellar (1984), Robilotta (1984, 1985).
4. Efimov (1981); see also Phillips (1965, 1968, 1977) and Levin (1990).
5. Frederico (1988b). The sensitivity of Nd analyzing powers to P-wave NN interactions was first noted in Pieper (1972a, b) and Doleschall (1973).
6. Brayshaw (1974), Haftel (1974).
7. Amado (1969), Kim (1974), Mitra (1969).
8. Sasakawa (1987), Friar (1987a).
9. Brandenburg (1988a, b).
10. Koike (1987a, b) Plessas (1986a, b; 1987).
11. Witala (1987, 1988, 1989a, b).
12. Friar (1983, 1984a), Chen (1986a, 1989).
13. Witala (1989b).
14. Tomio (1987), Adhikari (1988b), Petrov (1988). For a microscopic approach to nd effective interactions, see Delfino (1988) and Tomio (1990).
15. Frederico (1988a), Adhikari (1990a).
16. Frederico (1988b).
17. Efimov (1981, 1988); the review by Levin (1990) emphasizes the model independence of the 3N system.
18. Efimov (1970a, b).
19. Adhikari (1983a, 1984b); see also Whiting (1976).
20. Berthold (1986, 1988b).
21. Girard (1979a, b).
22. Friar (1984b).
23. Friar (1988a).
24. Rarita (1941), Gerjuoy (1942).
25. Amado (1978, 1979), Friar (1979).
26. Sperisen (1981, 1984), Zankel (1983a, b; 1984), Loiseau (1985), Koike (1985), Witala (1989a, b).
27. Sasakawa (1986), Ishikawa (1986b; 1987a, b).
28. Afnan (1977a), Sasakawa (1986), Ishikawa (1986a, b), Brandenburg (1975), Muslim (1983), Glöckle (1982a), Laverne (1973), Hajduk (1981), Payne (1980), Chen (1985).
29. Chen (1986b).
30. Hajduk (1981), Koike (1987b).
31. Dilg (1971).
32. Efimov (1985, 1988), Frederico (1988c).
33. Sasakawa (1986), Girard (1979a, b), Friar (1988a), Petrov (1988).

34. Brandenburg (1988b).
35. Friar (1987b).
36. Arvieux (1974), Bruinsma (1974).
37. Adhikari (1984b), Frederico (1987), Shapoval (1987).
38. Koike (1986).
39. Plessas (1986a, b).
40. Shirato (1973).
41. Ericson (1983–1985).
42. Londergan (1987), Weller (1988).
43. Sperisen (1981, 1984), Doleschall (1973, 1982).
44. Pieper (1972b), Sloan (1972d), Benayoun (1976a).
45. Chauvin (1975a, b).
46. Berthold (1986, 1988a, b).
47. Zankel (1983a, b; 1984), Loiseau (1985), Koike (1985, 1987a, b).
48. Witala (1987).
49. Haidenbauer (1986b).
50. Berthold (1988a).
51. Burov (1987), Sauer (1987).
52. Friar (1986).
53. Friar (1988b).

Section 7.4

1. Lehman (1987), Gibson (1981), and references cited therein.
2. Barbour (1967, 1970).
3. Hendry (1973).
4. Barbour (1972).
5. Gibson (1975, 1976).
6. Gibson (1977).
7. Arriaga (1984).
8. King (1983).
9. Vetterli (1985).
10. Jourdan (1985, 1986).

Section 7.5

1. Shanley (1969).
2. Charnomordic (1977), and references cited therein.
3. Koike (1978a, b).
4. Koike (1980).

5. For an account of the bound-state observables in the NNα system, see Lehman (1982) and references cited therein.

Section 7.6

1. Blankenbecler (1966), Freedman (1966), Aaron (1968).
2. Afnan (1974).
3. See, for example, Section 2.7.
4. Thomas (1976).
5. Rinat (1977a, b).
6. Avishai (1979, 1981).
7. Stelbovics (1978), Stingl (1978), Mizutani (1977).
8. Blankleider (1981), Afnan (1980, 1985a, b).
9. Comprehensive reviews are given in Afnan (1984) and Blankleider (1987).
10. Fayard (1980), Mizutani (1981).
11. Rinat (1983).
12. Niskanen (1978).
13. Garcilazo (1980, 1984, 1987, 1988a, b).
14. Betz (1981), Lee (1983, 1984, 1987).
15. Kloet (1977, 1980, 1981), Kopaleishvili (1978).
16. Kloet (1980), Jauch (1984), Araki (1982), Afnan (1985a, b).
17. Ueda (1987), Afnan (1984), Locher (1984).
18. Afnan (1988), Araki (1988).

Section 7.7

1. Lim (1984).
2. Lim (1981), Haftel (1982, 1986).
3. de Goey (1986), and references cited therein.
4. Hess (1984), Kagan (1981), de Goey (1984).

Chapter 8

Solution Methods and Techniques: Four-Particle Scattering

8.1. Introduction

Four-particle C-K scattering integral equations are intrinsically much more complicated than for three particles because of the increased dimensionality and number of channels. There are many different kinds of these C-K equations, but only the Yakubovskii-type (Y-type) equations have been employed in numerical applications.[1,2] As in the three-particle case, these equations, after partial-wave analysis, reduce to a set of one-variable integral equations if separable approximations are made for subsystem transition operators. In the case of the three-particle problem, the one-variable reduction of the AGS version[2] of the Y-type equations is quite straightforward, and has been performed in Ch. 5. This reduction is quite involved in the four-particle case, and we show in detail how this is done in this chapter.[3,4]

For $N = 4$ the two-cluster partitions are either of the $(2 + 2)$ or the $(3 + 1)$ variety. A $(m + n)$-type partition means that asymptotically one has two bound clusters of m and n particles, respectively. One of the complications of four-particle Y-type equations (5.5.45) is that they couple all seven two-cluster channels. However, a closed set of integral equations involving only the $(3 + 1)$ partitions in the intermediate state can be written for the $(3 + 1) \rightarrow (3 + 1)$ reactions.[3] These equations are then solved by making separable approximations to the three-particle sub-amplitudes. The $(3 + 1) \to (2 + 2)$ and $(2 + 2) \to (2 + 2)$ transition amplitudes are then calculated by quadrature rules.

This solution strategy heavily relies on techniques used in the three-particle problem, such as the methods of separable expansions, contour rotation, and matrix inversion. The final integral equations can be written in the LS form as in the $N = 3$ case, but there are some new technical difficulties arising from the energy-dependent effective potentials that appear at both the three- and four-particle levels. Furthermore, special considerations are needed for applying the methods of separable expansions for three-particle amplitudes as well as the contour-rotation technique.

Finally, we present the results of calculations for the four-nucleon and the four-atom systems. In both cases, the results are impressive, particularly considering the approximations that have been made, and they represent successes of the applications of C-K integral equations to the four-particle system.

8.2. Integral Equations for the (3 + 1) → (3 + 1) Reactions

For the four-particle system, the Eqs. (5.5.45) constitute a set of coupled equations for scattering initiated by two-cluster states of the $(3 + 1)$-type. There is coupling to the $(2 + 2)$ channels, so that the transition amplitudes for both the $(3 + 1) \to (3 + 1)$ and the $(3 + 1) \to (2 + 2)$ processes are involved. This set of equations can be decoupled to a set of integral equations involving only transition amplitudes of the $(3 + 1) \to (3 + 1)$ type, supplemented by quadrature rules for obtaining the amplitudes corresponding to the other physical processes.

In order to perform this decoupling, we first need to distinguish carefully between the two types of two-cluster partitions. The $(3 + 1)$ partitions will be denoted by Greek letters $\sigma, \rho, \gamma, \ldots$, as in Ch. 5. However, now the $(2 + 2)$ partitions will be denoted by Roman letters $a, b, \ldots$. Then for scattering initiated by a $(3 + 1)$ state, the Eqs. (5.5.45) become[1]

$$\begin{aligned} U_{ij}^{\sigma\rho} = {} & \bar{\delta}^{\sigma\rho}(G_0 t_i G_0)^{-1} \delta_{ij} \Delta_{\sigma i} \Delta_{\rho j} \\ & + \sum_{\gamma, k} \bar{\delta}^{\sigma\gamma} \Delta_{\gamma i} \Delta_{\sigma i} U_{ik}^{\gamma} G_0 t_k G_0 \Delta_{\gamma k} \Delta_{\rho j} U_{kj}^{\gamma\rho} \\ & + \sum_{a, k} \Delta_{ai} \Delta_{\sigma i} U_{ik}^{a} G_0 t_k G_0 \Delta_{ak} \Delta_{\rho j} U_{kj}^{a\rho}, \end{aligned} \tag{8.2.1a}$$

$$\begin{aligned} U_{ij}^{a\rho} = {} & (G_0 t_i G_0)^{-1} \delta_{ij} \Delta_{ai} \Delta_{\rho j} \\ & + \sum_{\gamma, k} \Delta_{\gamma i} \Delta_{ai} U_{ik}^{\gamma} G_0 t_k G_0 \Delta_{\gamma k} \Delta_{\rho j} U_{kj}^{\gamma\rho}. \end{aligned} \tag{8.2.1b}$$

As a (2 + 2)-type cluster a is always distinct from a (3 + 1)-type cluster, the $\bar{\delta}$-symbols in the Eqs. (5.5.45) are all unity. Also, in obtaining Eq. (8.2.1b), we note that the summation over the intermediate (2 + 2)-type states b vanishes because $\Delta_{ai}\Delta_{bi}\bar{\delta}_{ab} = 0$.

Equation (8.2.1a) couples (3 + 1) partitions to (3 + 1) and (2 + 2) partitions, while Eq. (8.2.1b) is a quadrature rule for the (3 + 1) → (2 + 2) amplitudes. If we substitute Eq. (8.2.1b) in Eq. (8.2.1a), the (2 + 2) partitions are eliminated, and we obtain the following set of coupled equations for the (3 + 1) → (3 + 1) four-particle amplitudes:[1,2]

$$\begin{aligned} U_{ij}^{\sigma\rho} &= \bar{\delta}^{\sigma\rho}(G_0 t_i G_0)^{-1}\delta_{ij}\Delta_{\sigma i}\Delta_{\rho j} \\ &+ \sum_{\gamma,k} \bar{\delta}^{\sigma\gamma}\Delta_{\gamma i}\Delta_{\sigma i} U_{ik}^{\gamma} G_0 t_k G_0 \Delta_{\gamma k}\Delta_{\rho j} U_{kj}^{\gamma\rho} \\ &+ \sum_{a} U_{ij}^{a}\Delta_{\sigma i}\Delta_{ai}\Delta_{aj}\Delta_{\rho j} \\ &+ \sum_{a,k}\sum_{\gamma,m} \Delta_{ai}\Delta_{\sigma i} U_{ik}^{a} G_0 t_k G_0 \Delta_{ak}\Delta_{\gamma k} U_{km}^{\gamma} G_0 t_m G_0 \Delta_{\gamma m}\Delta_{\rho j} U_{mj}^{\gamma\rho}. \end{aligned} \tag{8.2.2}$$

In order to rewrite Eq. (8.2.2) in the form of a LS equation, we define

$$V_{ij}^{\sigma\rho} \equiv \bar{\delta}^{\sigma\rho}(G_0 t_i G_0)^{-1}\delta_{ij}\Delta_{\sigma i}\Delta_{\rho j} + \sum_{a} U_{ij}^{a}\Delta_{ai}\Delta_{aj}\Delta_{\rho j}\Delta_{\sigma i}. \tag{8.2.3}$$

If we use Eq. (8.2.3) in Eq. (8.2.2), we have[2]

$$U_{ij}^{\sigma\rho} = V_{ij}^{\sigma\rho} + \sum_{\gamma,k\ m} V_{ij}^{\sigma\gamma}[G_0 t_k G_0 \Delta_{\gamma k} U_{km}^{\gamma} G_0 t_m G_0 \Delta_{\gamma m}] U_{mj}^{\gamma\rho}. \tag{8.2.4}$$

If we are considering the expression in the square bracket of Eq. (8.2.4) as an effective Green function, Eq. (8.2.4) is a LS-type equation involving only (3 + 1) → (3 + 1) amplitudes. The effects of the (2 + 2) sub-systems appear only in the effective potential $V_{ij}^{\rho\tau}$; we shall see how they are treated exactly by convolution. In Eq. (8.2.3), the partition a is uniquely fixed given the pair indices i, j; consequently, the summation over a is redundant, and will be dropped in the following. Equation (8.2.4) can be used as the basis for a calculation of (3 + 1) → (3 + 1) collisions.

Equation (8.2.4) is still a set of integral equations in three-vector-momentum variables. It is possible, however, to write a set of integral equations in one-vector-momentum variable if we assume that the two- *and* three-body subsystem amplitudes have separable forms. The transition amplitude $T^{\sigma\rho}$ for a reaction starting in the two-cluster channel ρ and ending in the two-cluster channel σ is defined by

$$T^{\sigma\rho} = \sum_{i,j} Z_{ij}^{\sigma\rho} \equiv \sum_{i,j} \langle \chi_i^{\sigma} | U_{ij}^{\sigma\rho} | \chi_j^{\rho} \rangle, \tag{8.2.5}$$

where χ_i^σ's are the bound-state wave-function components in channel σ, and satisfy the homogeneous Faddeev equations

$$|\chi_i^\sigma\rangle = \sum_k \bar{\delta}_{ki} G_0 t_i |\chi_k^\sigma\rangle. \tag{8.2.6}$$

The full wave function is defined by $|\chi^\sigma\rangle = \sum_i |\chi_i^\sigma\rangle$, where i runs over all pairs contained in partition σ.

In order to reduce the Eqs. (8.2.4) to one-variable integral equations, we assume that the two-body interaction for the pair i has the separable S-wave form:

$$v_i = \lambda_i |f_i\rangle\langle f_i|, \tag{8.2.7a}$$

so that the two-body t-matrix is given by

$$t_i = |f_i\rangle\tau_i\langle f_i|, \tag{8.2.7b}$$

with

$$\tau_i^{-1}(E) = \lambda_i^{-1} - \langle f_i|G_0(E)|f_i\rangle. \tag{8.2.7c}$$

If the pair i has a bound-state with binding energy B, the constant λ_i is given by

$$\lambda_i^{-1} = \langle f_i|G_0(-B)|f_i\rangle. \tag{8.2.7d}$$

With this two-body potential, the three-body AGS equations become the following one-vector-variable integral equation (*cf.* Eqs. (5.3.76–5.3.78)):

$$X_{ij}^\sigma = \bar{\delta}_{ij} B_{ij}^\sigma + \sum_k \bar{\delta}_{ik} B_{ik}^\sigma \tau_k X_{kj}^\sigma, \tag{8.2.8}$$

where

$$X_{ij}^\sigma = \langle f_i|G_0 U_{ij}^\sigma G_0|f_j\rangle, \tag{8.2.9a}$$

and

$$B_{ij}^\sigma = \langle f_i|G_0|f_j\rangle. \tag{8.2.9b}$$

Then, the homogeneous Faddeev equations (8.2.6) become

$$|\chi_i^\sigma\rangle = \sum_k \bar{\delta}_{ki} G_0|f_i\rangle\tau_i\langle f_i|\chi_k^\sigma\rangle, \tag{8.2.10a}$$

where the factorization with respect to the i-pair relative momentum is explicit. If we define the form-factor components $|g_i^\sigma\rangle$ by

$$|\chi_i^\sigma\rangle \equiv G_0|f_i\rangle\tau_i|g_i^\sigma\rangle, \tag{8.2.10b}$$

we have

$$|g_i^\sigma\rangle = \sum_k \bar{\delta}_{ki} B_{ik}^\sigma \tau_k |g_k^\sigma\rangle. \tag{8.2.10c}$$

Equation (8.2.10c) is the homogeneous version of the AGS equations (8.2.8).

In addition to the separable two-particle potentials given by Eqs. (8.2.7a–d), we assume that the three-particle operators X^{σ}_{ij} of Eq. (8.2.9a) also have the S-wave separable form

$$X^{\sigma}_{ij} = |g^{\sigma}_{i}\rangle D^{\sigma}\langle g^{\sigma}_{j}|. \tag{8.2.11}$$

Now we are prepared to derive one-vector-variable integral equations for the operators $Z^{\sigma\rho}_{ij}$ appearing in Eq. (8.2.5). In order to achieve this, we define

$$Y^{\sigma\rho}_{ij} \equiv \langle \chi^{\sigma}_{i}|V^{\sigma\rho}_{ij}|\chi^{\rho}_{j}\rangle. \tag{8.2.12}$$

If we use Eqs. (8.2.3), (8.2.7b), and (8.2.10b), Eq. (8.2.12) can be rewritten as

$$\begin{aligned} Y^{\sigma\rho}_{ij} = \langle g^{\sigma}_{i}|\tau_i\langle f_i|G_0\,\Delta_{\sigma i}\,\Delta_{\rho j}\,[\bar{\delta}^{\sigma\rho}\,(G_0 t_i G_0)^{-1}\,\delta_{ij} \\ + U^{a}_{ij}\,\Delta_{ai}\,\Delta_{aj}]\,G_0|f_j\rangle\tau_j|g^{\rho}_{j}\rangle. \end{aligned} \tag{8.2.13}$$

If we recall that, for the separable potential (8.2.7a–d),

$$\langle f_i|t_i^{-1}|f_i\rangle = \tau_i^{-1}, \tag{8.2.14}$$

the first term in the square bracket of Eq. (8.2.13) reduces to

$$Y^{\sigma\rho(1)}_{ij} \equiv \bar{\delta}^{\sigma\rho}\,\delta_{ij}\,\Delta_{\sigma i}\,\Delta_{\rho j}\,\langle g^{\sigma}_{i}|\tau_i|g^{\rho}_{i}\rangle. \tag{8.2.15}$$

The second term in the square bracket of Eq. (8.2.13) can be simplified with the explicit use of the formal solution (1.7.26) in the case of the AGS operator U^{a}_{ij}; then this term can be put in the form

$$\Delta_{ai}\,\Delta_{\sigma i}\,\Delta_{aj}\,\Delta_{\rho j}\,\langle g^{\sigma}_{i}|\tau_i\langle f_i|G_0[G_i^{-1}G_a G_j^{-1} - \delta_{ij}\,G_j^{-1}]G_0|f_j\rangle\tau_j|g^{\rho}_{j}\rangle. \tag{8.2.16}$$

In Eq. (8.2.16) and in the following, we assume, as in the original derivations,[1,2] that the right- and the left-handed inverses of all Green functions yield equivalent results, and so we do not distinguish between these two types of inverses.

If we use the identity

$$G_0 G_j^{-1} G_0 = G_0 - G_0\,V_j G_0,$$

the last term in expression (8.2.16) reduces to

$$-\Delta_{ai}\,\Delta_{\sigma i}\,\Delta_{aj}\,\Delta_{\rho j}\,\delta_{ij}\,\langle g^{\sigma}_{i}|\tau_i\langle f_i|(G_0 - G_0 v_j G_0)|f_j\rangle\tau_j|g^{\rho}_{j}\rangle,$$

which with the use of Eqs. (8.2.7a) and (8.2.14) becomes

$$Y^{\sigma\rho(2)}_{ij} \equiv -\delta_{ij}\,\Delta_{ai}\,\Delta_{\sigma i}\,\Delta_{aj}\,\Delta_{\rho j}\,\langle g^{\sigma}_{i}|(\tau_i - \lambda_i)|g^{\rho}_{j}\rangle. \tag{8.2.17}$$

The first term in expression (8.2.16) can be simplified using the convolution for the Green function G_a in terms of the Green functions for the pairs

G_i and G_j:

$$G_a(z) = \frac{1}{2\pi i} \oint dz'\, G_i(z')G_j(z - z'), \tag{8.2.18}$$

where the contour integration is performed in an anticlockwise sense around the singularities of the Green functions.[3]

For the separable potential of Eqs. (8.2.7a–d), we have the identity

$$G_0|f_j\rangle\tau_j = G_j|f_j\rangle\lambda_j, \tag{8.2.19}$$

which in conjunction with Eq. (8.2.18) can be used to rewrite the first term of expression (8.2.16) as

$$Y_{ij}^{\sigma\rho(3)} \equiv \Delta_{ai}\,\Delta_{\sigma i}\,\Delta_{aj}\,\Delta_{\rho j}\frac{1}{2\pi i}\oint dz'\,\langle g_i^\sigma|\tau_i(z')$$
$$\times \langle f_i|G_0(z')G_0(z - z')|f_j\rangle\tau_j(z - z')|g_j^\rho\rangle. \tag{8.2.20}$$

If we combine Eqs. (8.2.12), (8.2.15), (8.2.17), and (8.2.20), we see that

$$Y_{ij}^{\sigma\rho} = Y_{ij}^{\sigma\rho(1)} + Y_{ij}^{\sigma\rho(2)} + Y_{ij}^{\sigma\rho(3)}. \tag{8.2.21}$$

Now if we insert Eq. (8.2.7b), (8.2.9a), and (8.2.11) in Eq. (8.2.4), and multiply the resultant equation from the left by $\langle\chi_i^\sigma|$ and from the right by $|\chi_j^\rho\rangle$, as given by Eq. (8.2.10b), we arrive at

$$Z_{ij}^{\sigma\rho} = Y_{ij}^{\sigma\rho} + \sum_{\gamma,k,m} Y_{ik}^{\sigma\gamma}\,D^\gamma\,Z_{mj}^{\gamma\rho}. \tag{8.2.22}$$

This is an integral equation in one vector-variable, which after partial-wave projection involves only a single scalar-variable.

It is important to emphasize that Eq. (8.2.22) only involves two-cluster states of the (3 + 1)-type. The two-cluster intermediate states of the (2 + 2)-type have been eliminated by iteration in Eq. (8.2.4), and can be treated exactly, without resorting to separable expansions for the amplitudes U_{ij}^a, via Eqs. (8.2.20–8.2.21). This reduction procedure was first suggested in the context of the AGS equations by Haberzettl and Sandhas,[1] equivalent equations were derived earlier in a field-theoretic model by Fonseca and Shanley.[2] Instead, if we had worked with the original Yakubovskii equations (5.5.45), separable approximations for the two-cluster amplitude U_{ij}^a of the (2 + 2)-type are needed to obtain single-variable integral equations.

In Eqs. (8.2.7a–d) and (8.2.11), we have assumed a one-term separable approximation for the two- and three-particle t-matrices. This assumption can, of course, be relaxed. If multiterm, finite-rank separable approximations are employed, the final set of equations will then involve a correspondingly larger number of coupled equations.

Once the (3 + 1) → (3 + 1) problem is solved, the (3 + 1) → (2 + 2) and (2 + 2) → (2 + 2) amplitudes are obtained by quadrature. First, we consider the (3 + 1) → (2 + 2) reaction. If we multiply Eq. (8.2.1b) from the right by $|\chi_j^\rho\rangle$ and from the left by $\langle\chi_i^a|$, we find that

$$Z_{ij}^{a\rho} = \langle\chi_i^a|\,\delta_{ij}\,\Delta_{ai}\,\Delta_{\rho j}(G_0 t_i G_0)^{-1}|\chi_j^\rho\rangle + \sum_{\gamma,k}\langle\chi_i^a|\,\Delta_{\gamma i}\,\Delta_{ai}\,U_{ik}^{\gamma}\,\Delta_{\gamma k}\,G_0 t_k G_0\,U_{kj}^{\gamma\rho}|\chi_j^\rho\rangle, \tag{8.2.23}$$

where Eq. (8.2.5) holds with σ replaced by a. For the noninteracting (2 + 2) cluster, Eqs. (8.2.8–8.2.10) hold with the (3 + 1) partition label σ replaced by the (2 + 2) partition label a. Now, $|\chi^a\rangle \equiv \sum_i|\chi_i^a\rangle$ is the product of the two bound states internal to partition a; also, the components χ_i^a satisfy Eqs. (8.2.10a–c) with the partition label σ changed to a. The use of Eqs. (8.2.7a–d), (8.2.10a–c), and (8.2.11) in Eq. (8.2.23) now leads to

$$Z_{ij}^{a\rho} = Y_{ij}^{a\rho} + \sum_{\gamma,k} Y_{ii}^{a\gamma}\,D^{\gamma}\,Z_{kj}^{\gamma\rho}, \tag{8.2.24}$$

with

$$Y_{ij}^{a\rho} = \delta_{ij}\,\Delta_{ai}\,\Delta_{\rho j}\langle g_i^a|\tau_i|g_j^\rho\rangle. \tag{8.2.25}$$

Equation (8.2.24) is an expression for the transition amplitude for reactions of the type (3 + 1) → (2 + 2) in terms of transition amplitudes for the (3 + 1) → (3 + 1) reaction.

Finally, we consider the (2 + 2) → (2 + 2) reaction for which the relevant Yakubovskii equations are

$$U_{ij}^{ab} = \sum_{\gamma,k}\Delta_{\gamma i}\,\Delta_{ai}\,U_{ik}^{\gamma}\,G_0 t_k G_0\,\Delta_{\gamma k}\,U_{kj}^{\gamma b}. \tag{8.2.26}$$

Because of the restriction imposed by anti-Kronecker delta functions in the Eqs. (5.5.45), the homogeneous term and the summation over (2 + 2) partitions in the intermediate states are absent. If we multiply Eq. (8.2.26) from the left by $\langle\chi_i^a|$ and from the right by $|\chi_j^b\rangle$, and employ the usual separable approximations for the two- and three-body t-matrices given by Eqs. (8.2.7a–d) and (8.2.11), we see that

$$Z_{ij}^{ab} = \sum_{\gamma,k} Y_{ii}^{a\gamma}\,D^{\gamma}\,Z_{kj}^{\gamma b}. \tag{8.2.27}$$

This expresses the transition amplitudes for reactions of the type (2 + 2) → (2 + 2) in terms of those for the (2 + 2) → (3 + 1) reactions, which are related to time-reversed amplitudes for the (3 + 1) → (2 + 2) amplitudes. In both Eqs. (8.2.24) and (8.2.27), the intermediate-state integrations are over one vector-variable, which reduces to an integration in one scalar-variable after partial-wave decomposition.

Hence the calculational procedure is to solve Eq. (8.2.22) for $(3+1)\rightarrow(3+1)$ reactions, and then evaluate the observables of $(3+1)\rightleftarrows(2+2)$ and $(2+2)\rightarrow(2+2)$ reactions via Eqs. (8.2.24–8.2.27).[4]

8.3. Explicit Momentum-Space Scattering Equations

Further simplification of Eq. (8.2.22) results if we consider four identical particles. First, we indicate this simplification for four-identical bosons. Recalling that the physical transition amplitude $T^{\sigma\rho}$ is obtained by summing over the pair indices i, j of $Z_{ij}^{\sigma\rho}$ (see Eq. (8.2.5)), we rewrite Eq. (8.2.22) as[1]

$$T^{\sigma\rho} = W^{\sigma\rho} + \sum_{\gamma} W^{\sigma\rho} D^{\gamma} T^{\gamma\rho}, \tag{8.3.1}$$

with

$$W^{\sigma\rho} = \sum_{i,j} Y_{ij}^{\sigma\rho}. \tag{8.3.2}$$

The symmetrized amplitudes are defined by

$$T^{\mathrm{S}} = \sum_{\sigma} T^{\sigma\rho}, \tag{8.3.3a}$$

$$W^{\mathrm{S}} = \sum_{\sigma} W^{\sigma\rho}, \tag{8.3.3b}$$

where the sum over σ runs over all $(3+1)$-type two-cluster channels. By taking an appropriate linear combination of the Eqs. (8.3.1), it is possible to write down the integral equation for the symmetrized amplitude $Z(E)$ for the $(3+1)\rightarrow(3+1)$ reaction, and we quote the result here:[1]

$$\begin{aligned}\langle \mathbf{P}'|Z(E)|\mathbf{P}\rangle = \langle \mathbf{P}'|V^{\mathrm{ef}}(E)|\mathbf{P}\rangle + \int \frac{d^3P''}{(2\pi)^3}\langle \mathbf{P}'|V^{\mathrm{ef}}(E)|\mathbf{P}''\rangle \\ \times D(E - \tfrac{4}{3}\mathbf{P}''^2)\langle \mathbf{P}''|Z(E)|\mathbf{P}\rangle,\end{aligned} \tag{8.3.4}$$

where the effective potential V^{ef} is obtained by convolution and is given by

$$\begin{aligned}\langle \mathbf{P}'|V^{\mathrm{ef}}(E)|\mathbf{P}\rangle = \langle \mathbf{P}'|V(E)|\mathbf{P}\rangle + \sum_{\mu=1}^{2} 2\int \frac{d^3Q}{(2\pi)^3} \\ \times \langle \mathbf{P}'|\bar{B}^{12}(E)|\pm\mathbf{Q}\rangle G_{\mu}(E; \mathbf{P}', \mathbf{Q}, \mathbf{P})\langle \mathbf{Q}|\bar{B}^{21}(E)|\mathbf{P}\rangle.\end{aligned} \tag{8.3.5}$$

The + (−)sign is associated with $\mu = 1(2)$. In Eq. (8.3.5),

$$\langle \mathbf{P}'|V(E)|\mathbf{P}\rangle = g(E - \tfrac{4}{3}P'^2; \mathbf{P} + \tfrac{1}{3}\mathbf{P}') \times \tau(E - P'^2 - P^2 - (\mathbf{P} + \mathbf{P}')^2/2) g(E - \tfrac{4}{3}P^2; \mathbf{P}' + \tfrac{1}{3}\mathbf{P}), \tag{8.3.6a}$$

and

$$\langle \mathbf{Q}|\bar{B}^{21}(E)|\mathbf{P}\rangle = \frac{f(|\mathbf{P} + \frac{1}{2}\mathbf{Q}|) g(E - \frac{4}{3}\mathbf{P}^2; \mathbf{Q} + \frac{2}{3}\mathbf{P})}{E + \varepsilon_2 - \frac{1}{2}Q^2 - (\mathbf{Q} + \mathbf{P}')^2 - P'^2}, \tag{8.3.6b}$$

where in Eqs. (8.3.5–8.3.6) the superscripts 1(2) over $\bar{B}$ denote a (3 + 1)-((2 + 2)-) type two-cluster channel and consequently, $\bar{B}^{12}$ is the transposed $\bar{B}^{21}$. If we assume that there is a two-body bound state with energy ε_2, the generalized Green functions G_1 and G_2 are given by

$$G_1(E; \mathbf{P}', \mathbf{Q}, \mathbf{P}) = \tau(Y - \varepsilon_2) - \frac{(Y - U)(Y - U')}{\pi} \int_0^\infty dx \frac{\operatorname{Im}[\tau(x)]\, \tau(Y - 2\varepsilon_2 - x)}{(Y - U - \varepsilon_2 - x)(Y - U' - \varepsilon_2 - x)},$$

$$G_2(E; \mathbf{P}', \mathbf{Q}, \mathbf{P}) = \frac{(Y - U'')}{-U''} \tau(Y - \varepsilon_2) + \frac{(Y - U)(Y - U'')}{Y - U - U''} \tau(U'' - \varepsilon_2) \tau(Y - U'' - \varepsilon_2) - \frac{(Y - U)(Y - U'')}{\pi} \int_0^\infty dx \frac{\operatorname{Im}[\tau(x)]\, \tau(Y - 2\varepsilon_2 - x)}{(x + \varepsilon_2 - U'')(Y - U - \varepsilon - x)},$$

where

$$Y = E + 2\varepsilon_2 - \mathbf{Q}^2,$$

$$Y - U = E + \varepsilon_2 - \tfrac{1}{2}\mathbf{Q}^2 - (\mathbf{P} + \mathbf{Q})^2 - \mathbf{P}^2,$$

$$Y - U' = E + \varepsilon_2 - \tfrac{1}{2}\mathbf{Q}^2 - (\mathbf{P}' + \mathbf{Q})^2 - \mathbf{P}'^2,$$

$$Y - U'' = E + \varepsilon_2 - \tfrac{1}{2}\mathbf{Q}^2 - \mathbf{P}'^2 - (\mathbf{P}' - \mathbf{Q})^2.$$

Equation (8.3.4) is a one-variable LS-type equation with the Born term given by Eq. (8.3.5). This equation is usually solved by the technique of contour rotation.

In the usual numerical applications of the four-particle Yakubovskii equations, one employs a one-term separable representation for the two-particle t-matrix, and a finite-rank representation for the three-particle

t-matrix generated by this two-particle interaction. Then, the four-particle scattering equation is solved using this input. Since the effective potential appearing in the three-particle scattering equation is energy-dependent, the special techniques of Section 3.8c, such as the GUPE, EDPE and some other methods,[2] are used for making separable expansion for the three-particle t-matrix.

The same simplification of Eq. (8.3.4) can be achieved in the four-particle bound-state problem that corresponds to the homogeneous version of this equation. Usually, the bound-state problem is solved first using a two-particle separable potential, where the three-particle subamplitude is obtained by the method of separable expansion. The convergence of this expansion scheme is tested by increasing the number of separable terms. Then, the scattering integral equation is solved in the same model, by the technique of contour rotation. The validity of the contour-rotation technique has not been established above the four-particle breakup threshold; consequently, all the calculations using this method are done for lower energies.[3] One can use direct integration along the real axis in this case.[4]

In the case of the four-nucleon problem, we have four identical fermions rather than bosons. The modifications arising from this are irrelevant to the reduction to one-variable integral equations.[1]

8.4. Numerical Calculations

Because of the complexity of the four-particle scattering equations, there has been much less work on their numerical solution than for $N = 3$. Extant calculations have been carried out using only separable rather than realistic two-particle potentials. Most of this work has been concerned with the four-nucleon (4N) system at low energies, but there have also been some applications to the four-atom system. In this section, we review the results obtained in these calculations at low energies.

8.4a. The Four-Nucleon System

As in the case of the 3N system (*cf.* Section 7.3), the 4N observables can also be classified into two categories—central and noncentral observables. Until recently, all calculations have been confined to the 4N central observables, and have neglected the Coulomb potential. The correct inclusion of the Coulomb potential and the NN tensor potential in numerical 4N calculations requires a major increase in the complexity of the solution algorithm.[1] There are also theoretical difficulties in the correct treatment of the Coulomb potential. In addition to this, it is a nontrivial problem to make a separable expansion of the 3N t-matrix in the presence of the NN tensor potential. This is essential for the reduction of the dimensionality of the four-particle scattering equations. The NN tensor potential, however, has been included approximately in some 4N calculations via a truncated t-matrix approximation.[2]

We consider both bound-state and scattering results for the 4N system because, as in the 3N problem of Section 7.3, they seem to be correlated.[2] Thus, the bound-state calculations are needed for a complete understanding.

The 4N system may appear in the following four spin (S) isospin (I) states: SI = 00, 01, 10, and 11. Because of the Pauli exclusion principle, only the S-wave spin–isospin-singlet state (SI = 00) is strongly attractive in nature, and the other three spin–isospin states are repulsive as in the spin-quartet 3N system. Consequently, the 4N system has only one bound-state with SI = 00, and this state is very sensitive to the off-shell behavior of the NN interaction. Analogous to the 3N system, attractive, separable NN potentials overbind ^{4}He, whereas realistic NN potentials underbind it.

The lowest scattering threshold in the SI = 00 channel is the nucleon trinucleon (NT) one, which allows two possibilities—(n–^{3}He) and (p–^{3}H). With the neglect of the Coulomb potential, as is done in all calculations, these two thresholds become degenerate. The experimentally measured coherent and incoherent scattering lengths, A_c and A_i, respectively, and the threshold cross section σ for these systems are related to the scattering lengths a_{SI} by

$$A_c(\text{n}-{}^3\text{H}) = \frac{1}{4}(a_{01} + 3a_{11}),$$

$$A_i(\text{n}-{}^3\text{H}) = \frac{\sqrt{3}}{4}(a_{11} - a_{01}),$$

$$\sigma(\text{n}-{}^3\text{H}) = \pi(a_{01}^2 + 3a_{11}^2),$$

$$A_c(\text{n}-{}^3\text{He}) = \frac{1}{8}[a_{00} + a_{01} + 3(a_{10} + a_{11})],$$

$$A_i(\text{n}-{}^3\text{He}) = \frac{\sqrt{3}}{8}(a_{10} + a_{11} - a_{00} - a_{01}),$$

$$\sigma(\text{n}-{}^3\text{He}) = \frac{\pi}{4}[(a_{00} + a_{01})^2 + 3(a_{10} + a_{11})^2].$$

The Coulomb potential, as in the 3N system, may quantitatively modify the NT scattering lengths. Hence, the following discussion about these scattering lengths, especially a_{00}, is only qualitative in simple theoretical models, which may turn out to be unrealistic.

Most of the calculations of the scattering lengths a_{SI} and the binding energy $B(^4\text{He})$ employ only the $(3+1)$ subamplitude for $L = 0$. The contribution of $L = 1$ $(3+1)$ subamplitudes to these observables is very small.[3]

It has been found that the S-wave scattering lengths a_{01}, a_{10}, and a_{11} are insensitive to the choice of NN potentials, whereas in the attractive SI = 00 channel, the S-wave scattering length a_{00} is very sensitive. Moreover, as in the trinucleon problem, a_{00} seems to be correlated with $B(^4\text{He})$, which was first observed by Kharchenko and Levashev.[4] However, because of the large width of this correlation, the experimental a_{00} could not be predicted from it. The other three scattering lengths are basically model independent, so approximate theoretical estimates can be obtained. Threshold scattering experiments have not yet determined the accurate values of these scattering lengths,[1]

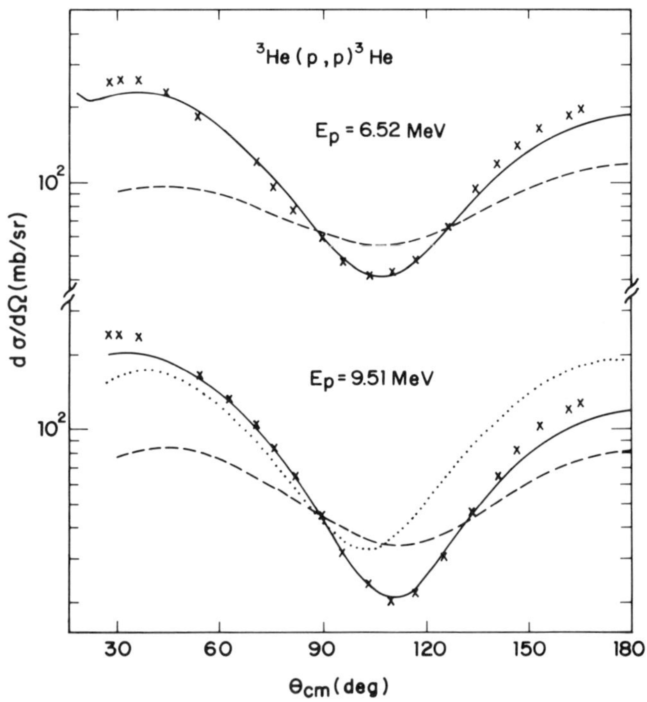

FIG. 8.1. Angular distribution for ^{3}He(p, p)^{3}He at different proton laboratory energies. The solid line includes both $L = 0$ and $L = 1$ $(3+1)$ subamplitudes, while the dashed line includes the $L = 0$ subamplitude only. The dotted line corresponds to including the $L = 1$ subamplitude in first-order perturbation theory. The crosses denote experimental points (from Fonseca (1986c)).

particularly a_{00}. Scattering-length calculations were also performed by Tjon and Fonseca.[1,2]

The physical ^{4}He system has an excited resonance just above the lowest scattering threshold (p + ^{3}H). With the neglect of Coulomb effects, the resonant state becomes an excited bound state in most of the 4N model calculations that employ NN separable potentials. Because this excited state is very near the scattering threshold, these calculations predict a large value of a_{00} with large NN-potential-dependent variations. In certain 4N models, the excited bound state is absent, and this leads to a large a_{00} of the opposite, negative, sign.

Most of the 4N scattering calculations are confined to the NT system below the dd threshold. It has been established by Fonseca that a good fit to the experimental scattering cross section data for this system can be achieved,[3] provided that the P-wave components of the (3 + 1)-subsystem amplitudes are included. The inclusion of these P waves modifies the P-wave NT phase-shifts, and hence the cross sections, substantially without changing the S-wave observables a_{00} and $B(^4\mathrm{He})$ significantly. The importance of the P-wave phase shifts in explaining NT cross section results was already apparent from a previous study of this problem by Tjon.[2]

Calculations of the cross sections for the dd $\rightarrow$ p + ^{3}H reaction have also been carried out.[1,3] In this case, the agreement with experiment at backward angles is not good; the neglect of the D-wave components of the tensor force has been held responsible for this. The inclusion of the P-wave components of the (2 + 2)-subsystem amplitudes is also expected to improve the agreement with data. Calculations should be done in the future to test these conjectures.

We present in Figs. 8.1 and 8.2 the angular distributions of the ^{3}He(p, p)^{3}He and ^{2}H(d, p)^{3}H reactions taken from the calculation of Fonseca[3] at different incident projectile energies in the laboratory system. These figures illustrate the dramatic effect of the P waves of the (3 + 1) subamplitudes.

At energies above the four-body breakup threshold, the solution of the relevant one-variable integral equations is difficult because the contour rotation method has not been justified. A careful study of the type presented in Section 7.2 of the singularity structure of the kernel and the solution of the integral equation has not been performed to find an appropriate contour in this case. Such a study is expected to be substantially more difficult one than for three particles because of the much more complex singularity structure associated with the four-body problem. However, the real-axis integration technique of Section 7.2 is not beset with these difficulties. Also, the poor convergence of the EDPE- and GUPE-type separable expansions for the (3 + 1)-subsystem amplitudes at higher energies suggests that we follow

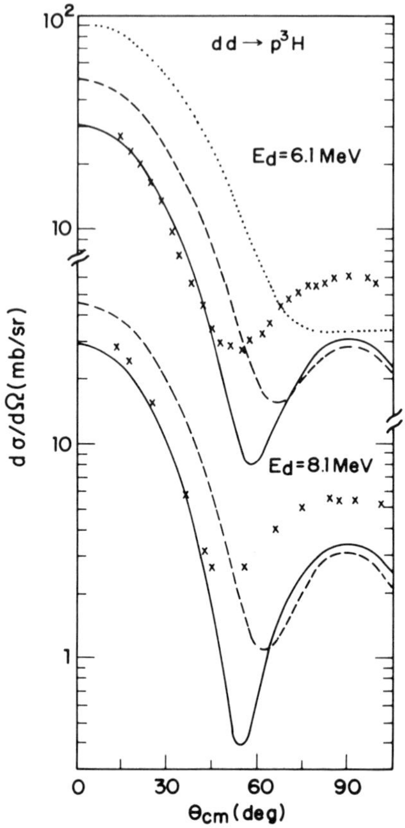

FIG. 8.2. Angular distribution for $^2H(d, p)^3H$ at different deuteron laboratory energies. The meaning of the various lines and crosses is the same as for Fig. 8.1 (from Fonseca (1986c)).

alternate strategies.[3] One strategy is to use the GEST expansion technique of Section 3.8c, which is expected to provide better convergence; the other is to abandon the technique of separable expansion for the (3 + 1)-subsystem amplitude, and treat this amplitude exactly by solving a two-variable integral equation.[1] The use of a K-matrix type of approximation by Sandhas and collaborators[5] at these energies results in a simpler set of integral equations to be treated numerically; this method has yielded reasonable qualitative agreement with the cross-sections data for the p + ^{3}He → p + ^{3}He, d + d → N + T, and the breakup ^{3}He(p, pd)p reactions. However, the agreement is much poorer, especially in the p + ^{3}He → p + ^{3}He case, at energies below the breakup threshold compared to the results obtained by Fonseca with a more accurate solution of dynamical equations that included the P-wave (3 + 1) subamplitudes.

It is obvious that the connected-kernel Yakubovskii integral equations have provided a major breakthrough in understanding the dynamics of the 4N system for the central observables using only the NN potential as input. Similar calculations should be carried out in the future for the noncentral observables, such as the D-state to S-state asymptotic normalization parameter for ^{4}He, and the different polarization observables of 4N scattering.[1]

It is worth mentioning that in addition to the a_{00}–$B(^4\text{He})$ correlation another correlation has been observed in 4N calculations. It was pointed out by Tjon[2] that in dynamical calculations the binding energies $B(^3\text{H})$ and $B(^4\text{He})$ are correlated. Different types of NN potentials generate slightly different correlations, so that if the results for all types of NN potentials are collected one encounters a band rather than a straight line.[1] Not much work has been done, in contrast to the 3N case, about how to exploit such correlations in order to obtain information about the NN potentials or experimental values of 4N observables. In the following, we make some suggestions about how to perform future calculations so that such information might be obtained.

The S-wave (SI = 00) 4N observables are very sensitive to underlying NN and 3N interactions. The nonrelativisitic nucleons-only potential models, which are considered equivalent so far as NN and 3N observables are concerned, may yield qualitatively different results for the spin–isospin-singlet 4N observables. Consequently, a study of these observables may yield new information about NN and 3N interactions. The study of the 4N spin–isospin-singlet system, with potential models that yield identical NN and 3N results, is expected to yield low-energy correlations as in the 3N system. The study of correlations is expected to be effective in deciding which potential model is superior in deciding the experimental values of certain 4N observables and whether the 4N interaction is indispensable in the 4N system.

Low-energy correlations have been studied before in the 4N system.[2,4] However, in these earlier studies, potential models, which produce identical results at the two-nucleon level only, have been employed, and no conclusions about the effectiveness of a potential model could be drawn because of poor fits to the experimental values of the trinucleon observables.

The correlation between $B(^3\text{H})$ and $B(^4\text{He})$ suggests that for a correct reproduction of $B(^4\text{He})$, a model with an experimental value of $B(^3\text{H})$ is to be preferred. If this correlation were entirely model independent, then it is unlikely that a study of the 4N system would yield new information about the NN and 3N interactions. However, the strongly bound ^{4}He system is expected to be more sensitive to the short-range part of the NN and 3N interactions than the weakly bound ^{3}H and ^{2}H systems. This means that the $B(^4\text{He})$–$B(^3\text{H})$ correlation is expected to be model dependent and consequently have a large width.

A different type of correlation may be more useful to investigate instead.[6] While keeping $B(^3\text{H})$ constant, the sensitivity of $B(^4\text{He})$ and other 4N observables to the potential models employed would be interesting to observe. In particular, the $a_{00}-B(^4\text{He})$ correlation, while keeping $B(^3\text{H})$ and other low-energy NN observables fixed, is expected to be of narrow width; this could then be used to predict the theoretical estimate of the yet unknown observable a_{00}. Hence, future studies should be aimed at exploiting correlations among different low-energy 4N observables, keeping the low-energy NN and 3N observables fixed.

For a complete understanding of the 4N system, the pp Coulomb potential has to be included. This is not an easy task in the usual framework of collision integral equations, and might be an interesting future project.

8.4b. The Molecular System

A realistic treatment of the scattering process involving the four-atom system is plagued by the same difficulties of the three-atom system as discussed in Section 7.7. Fonseca and Lim[7] have performed a model study of the scattering process involving four spinless helium-like atoms interacting in the S-wave state through a one-term separable representation of the Lennard–Jones potential. They used the one-variable Yakubovskii integral equation and treated the (2 + 2)-subsystem amplitude by convolution. Only the S-wave (3 + 1)-subsystem amplitude was included by means of EDPE. They calculated the phase shifts and cross sections for elastic and rearrangement scattering initiated by a (2 + 2) state, and were able to identity certain interesting qualitative features from this model study. Their model had a single diatom 0^+ bound state and two triatom 0^+ bound states. The $L = 0$, $(2 + 2) \rightarrow (2 + 2)$ phase-shift shows a cusp at the $(1 + 3^*)$ threshold, and the $L = 2$ phase-shift, a resonance-like behavior, where the asterisk represents the excited state. Both of the preceding behaviors are associated with the strong coupling between the (2 + 2) and $(1 + 3^*)$ channels. The total cross sections obey the condition $\sigma_{22\rightarrow 13^*} \rightarrow \sigma_{22\rightarrow 33}$, which indicates that vibrational excitation is effective for reactions at low energies.

References and Notes (8)

Section 8.1

1. Yakubovskii (1967).
2. The AGS-type equations for the three-particle system are to be found in Alt (1967). The four-particle dynamical equations in Grassberger (1967) are not of the C-K type except in approximation. The exact C-K AGS integral equations for any N are established in Alt (1972); their equivalence to the Yakubovskii (1967) equations is shown in Karlsson (1974a, b).
3. Haberzettl (1981a, b).
4. Fonseca (1976a, b).

Section 8.2

1. Haberzettl (1981a).
2. Fonseca (1986a, b).
3. Bianchi (1964).
4. Alt (1970).

Section 8.3

1. Fonseca (1986a, b).
2. Sofianos (1979), Casel (1982), Fonseca (1983).
3. Fonseca (1984), Fiedeldey (1987).
4. Tjon (1975, 1976, 1978)

Section 8.4

1. For recent reviews see Fonseca (1984, 1986a, b) and Fiedeldey (1987). The inclusion of NN tensor potential in 4N bound-state and scattering calculations is discussed in Fonseca (1989a, b).
2. Tjon (1975, 1976, 1978).
3. Fonseca (1986c).
4. Kharchenko (1976, 1980).
5. Alt (1970), Sofianos (1985), Mdlalose (1986, 1988), Fiedeldey (1987).
6. Barlette (1991).
7. Fonseca (1985).

Chapter 9

Solution Methods and Techniques: Multiparticle Scattering

9.1. Introduction

For collisions involving more than four constituents, any solution strategy involves nontrivial approximations because of the complexity accompanying more than twelve active degrees of freedom. This complexity leads to dynamical strategies that are very different than those discussed in the preceding chapters.

In order to appreciate these remarks, let us review what we have learned about solving collision problems thus far. The two- and three-body collision problems, as defined within a nonrelativistic Schrödinger statement of the dynamics, can be calculated with an accuracy that exceeds that of the underlying physical model. Any disagreements with experiment necessarily reflect on the adequacy of the model and its phenomenological input. This is not the present state of affairs for collisions involving more particles except when the model is understood to involve the truncation to an effective two- or three-body dynamics.

The collision problem for $N = 4$ indicates the extent to which the strategy used to solve the $N = 3$ problem has been carried up to now. At present, the $N = 4$ calculations done using connected-kernel (C-K) equations with $N = 2$

and $N = 3$ input as off-shell amplitudes have been quite crude, as we have seen in Ch. 8, compared to extant $N = 3$ work.

This solution strategy, where one calculates a hierarchy of collision subamplitudes for insertion into a full complement of dynamical equations, is the earmark of the point of view that accompanies the practical use of C-K equations. It is important to distinguish this hierarchical approach from one that is often used in field theory,[1] many-particle theory,[2] and statistical mechanics.[3] In these cases, a low-order quantity, for instance, a single-particle Green function G_1, is defined in terms of a set of dynamical equations in which the driving term in the equation for G_n is the next higher-order Green function G_{n+1}. For systems with a large number of particles, this goes on indefinitely for all practical purposes. The hierachical approximation in this context is to set $G_n = 0$ for all $n \geq n_{\max}$, where $n_{\max}$ defines the degree of approximation. The G_n for $1 < n \leq n_{\max}$ may be approximated further, as in the Hartree–Fock and self-consistent Hartree–Fock methods.[4]

This is all very different in style, motivation, and execution than what may be called the conventional techniques for calculating collision amplitudes involving many particles. We include under the rubric "conventional" fairly well-defined dynamical procedures such as the coupled reaction channel (CRC) technique, the resonating-group method (RGM), some of the approaches to optical potential (OP), including the impulse (IA) and multiple-scattering (MS) approximations, as well as purely phenomenological procedures, such as R-matrix theory[5] and the shell-model approach to nuclear reactions,[6] that have relatively unspecific dynamical foundations.

The simplest characterization of conventional approaches is that they involve truncations of the complete Hilbert space in some fashion rather than the approximation of identifiable subamplitudes. One of the results of recent progress in collision theory has been to establish, in a few instances, links between this approach and the C-K hierarchical approach. Perhaps more significantly, there have appeared a number of formulations where the Hilbert-space truncation idea can be systematically improved, a feature that is notably absent from conventional treatments.

The conventional approaches to the approximation of multiparticle collisions, almost without exception, are incapable of systematic improvement within their own terms simply because they do not begin with a large enough set of equations. This has the effect of leaving asymptotic conditions in some portions of phase space entirely indeterminate. If those portions of phase space have little effect on the regions of interest, the approximation may be successful. However, one should have the opportunity, in principle, to consider the magnitude of higher-order corrections, and to be able to calculate them in a definite manner.[7]

Two very important questions are: How good are the "standard" calculational algorithms, and what should we aim for in attempting to improve upon them? A hint of where the answers may be found is contained in the following, still relevant, observation:[8]

> "... When the standard methods are applied to real reactions they can give acceptable results only because of the flexibility introduced by the use of effective interactions. These difficulties should be borne in mind when evaluating the apparent success of nuclear reaction theories."

Evidently, to improve on this situation what is needed is a facility with the execution of conventional Hilbert-space truncations and an understanding of their significance in the wider context of a dynamically complete set of collision integral equations. Neither such facility or understanding has yet been fully achieved, but a major objective of this chapter is to describe to what extent these objectives have been reached and what the future prospects are for fully realizing them, especially concerning the definition and calculation of "effective interactions."

Consequently, our selection of topics and their developments are far from inclusive. Generally, we have selected topics for which some insight into the preceding questions has been attained at the present time.

Much of the analysis that was used to formulate the subject we review was originally directed towards nuclear problems, where Coulombic forces play a secondary role. In fact, it is not fully understood how to incorporate the Coulomb force into the C-K collision integral equations approach.[9] Thus, the application of this method to atomic problems is hardly as secure as it is when there are only short-range interactions. Typically, it can be applied only to few-body models where the effective particles are uncharged.

The Coulomb force is responsible for another effect that distinguishes atomic from nuclear problems, namely the large numbers of atomic excited states compared to the nuclear case. This discourages the use of Faddeev methods, for example, because the method of separable expansions of the two-particle subamplitudes is not effective.[10] A far more practical alternative up to now has been to employ methods that are designed to exploit the dynamical situation where there is such a proliferation of bound states. We review two major techniques that do this, namely projection-operator methods for elastic scattering,[11] and various coupled-channel approximation schemes.[12] In both cases, it is possible to establish a relationship to particular sets of C-K equations.

There are many unanswered questions about all of the topics that are discussed in this chapter. This should bode well for the vitality of future research efforts on this important problem of enhancing the practical effectiveness of general collision theories.

9.2. The Method of the Optical Potential

The OP is an effective interaction (EI) for describing two-cluster elastic scattering. Usually, one of the clusters is a composite of many particles, while the other contains only a single particle, such as in nucleon–nucleus, pion–nucleus, and electron–atom scattering. Throughout the entire history[1] of its progressive development, the OP has been an extraordinarily useful phenomenological device in search of an adequate theoretical description. Part of the difficulty in attaining such a description has had to do with the question of whether the OP was simple enough to represent the validity of the so-called optical model.[2] In this section, we consider the major theoretical vehicles that have been used to define and calculate OP's irrespective of concerns about whether an "optical model" is valid.

Apart from its role as an EI for two-cluster elastic scattering, there have been a number of different formulations of the OP using wave-function, transition-matrix, Fock-space, and path-integral techniques. Each of these has been accompanied by markedly different methods for incorporating the permutation symmetries that are always part of a problem involving composites. While each of these methods will generally lead to different OP's, within our usual nonrelativistic dynamical scheme it is possible to understand the differences among them.

On the other hand, the modeling that is facilitated by the use of a Dirac equation, along with a relativistic description of the nucleus, for describing intermediate-energy (<1 GeV) nucleon–nucleus scattering[3] is the most recent in a long series of OP calculations that have been suggested using incomplete dynamical formalisms.[4] The calculational successes achieved with what is called the *Dirac approach* to the OP seem to indicate the need for a theoretical framework in nuclear physics incorporating more realistic physics than is included in the usual nonrelativistic collision theory. Similar remarks apply to the large body of pion–nucleus OP calculations.[5] In spite of this, most of our discussion is confined to conventional nonrelativistic dynamics, primarily because this presents a well-defined dynamical scheme and not purely a phenomenology. In addition:

(i) The nonrelativistic formalism is physically acceptable for atomic processes, where the OP approach has been frequently applied.

(ii) The effects that are characteristic of the Dirac phenomenology are corrections to a roughly accurate nonrelativistic picture.
(iii) The nonrelativistic formalism has provided a useful vehicle for modeling both nucleon–nucleus and pion–nucleus scattering even when certain aspects of the dynamics have not been entirely understood.

In principle, for any elastic two-cluster process there is *always* a two-body EI, $U(\mathbf{p}, \mathbf{p}'; E)$, that generates the (off-shell) elastic transition matrix $T(\mathbf{p}, \mathbf{p}'; E)$ by means of the integral equation

$$T(\mathbf{p}, \mathbf{p}'; E) = U(\mathbf{p}, \mathbf{p}'; E) + \int d^3q \frac{U(\mathbf{p}, \mathbf{q}; E)T(\mathbf{q}, \mathbf{p}'; E)}{E - E(\mathbf{q}) + i0}. \qquad (9.2.1)$$

Here, $\mathbf{p}$, $\mathbf{p}'$, and $\mathbf{q}$ denote the relative momenta of the two clusters, $E(\mathbf{q})$ is the total energy corresponding to relative momentum $\mathbf{q}$, and E is arbitrary, but is usually identified with $E(\mathbf{p}')$. We have suppressed possible dependences upon spin and internal symmetries. We have also restricted ourselves here to the nonrelativistic situation, but contemporary relativistic OP formalisms yield equations that are not significantly different from Eq. (9.2.1) (*cf.* Section 9.2c). Generally, it is assumed that all possible permutation symmetries have been absorbed into the OP, so that $T(\mathbf{p}, \mathbf{p}'; E)$ is a correctly symmetrized amplitude.

Once U is given, T is uniquely defined, and vice versa, both in principle and in practice since we can regard the question of finding the solution of Eq. (9.2.1) for either alternative as a laborious, but straightforward, two-body problem.[6] The real difficulty is to find $U(\mathbf{p}, \mathbf{p}'; E)$, and there is no unique or even generally accepted method for doing this. If we suppose a given dynamical structure with fixed microscopic interactions, then this freedom in the choice of U can only arise from the freedom of the off-shell extension of the many-body, two-cluster transition matrix. Only recently has it been realized explicitly that one could exploit this arbitrariness in off-shell behavior to obtain an OP with specified properties.[7] We remark in this regard that there is a class of off-shell extensions of T for which $U(\mathbf{p}, \mathbf{p}'; E)$ is independent of the parametric energy; we will not be concerned with such OP's.[8]

A strategy for determining $U(\mathbf{p}, \mathbf{p}'; E)$ depends very much on what additional constraints are imposed on it. Before we discuss these possibilities, we emphasize that we are concerned solely with what are usually referred to as *microscopic*[9] (or *generalized*[10]) OP's as contrasted with *phenomenological* OP's. The simple structure that often appears in the latter type does not necessarily suggest a close relationship to any reasonably defined microscopic OP.[11] The test of a microscopic OP is in reproducing the elastic scattering cross sections and other observables; from this standpoint, phenomenological

OP's represent information of uncertain relevence. The significance of the energy averaging often used to relate the microscopic and phenomenological OP's is, therefore, not clear.

Traditionally, an OP is distinguished from an arbitrary EI that is inserted into Eq. (9.2.1) by its singularity and reality structure as related to elastic unitarity. Below any inelastic thresholds, because the composite structure of the clusters is inaccessible, the OP, according to conventional criteria,[12] should correspond to a Hermitian operator with no discontinuity in total energy across the elastic unitarity cut. Only the anti-Hermitian part of that operator is supposed to reflect a loss of flux from the elastic channel due to inelastic processes. Therefore, for arbitrary total energy, any discontinuity of the OP should arise solely from the inelastic branches of that cut. Thus, if we express $U(\mathbf{p}, \mathbf{p}'; z)$ for complex z in terms of an operator $U_{\text{opt}}(z)$,

$$^{c}\langle \mathbf{p} | U_{\text{opt}}(z) | \mathbf{p}' \rangle^{c} = U(\mathbf{p}, \mathbf{p}'; z), \tag{9.2.2}$$

where $|\mathbf{p}\rangle^{c}$, for example, is some sort of cluster state, then we should have

$$U_{\text{opt}}(z^*)^\dagger = U_{\text{opt}}(z) \tag{9.2.3a}$$

and

$$\{\text{disc}\, U_{\text{opt}}(z)\}_{\text{elastic cut}} = 0, \tag{9.2.3b}$$

for all $\text{Re}\, z \equiv E \geq E_t$, where E_t is the elastic threshold energy. Property (9.2.3a) is called *Hermitian analyticity*, and OP's that satisfy both of Eqs. (9.2.3a, b) are referred to as *quasi-Hermitian* for the sake of brevity.

Equations (9.2.3a, b) together imply that, for $E < E_{\text{in}}$, where E_{in} is the lowest inelastic threshold, the amplitude $T(\mathbf{p}, \mathbf{p}'; E)$ defined by Eq. (9.2.1) satisfies unitarity exactly. For $E > E_{\text{in}}$, this amplitude will be consistent with unitarity ($\equiv$ quasi-unitary) if $-\,\text{Im}\, U_{\text{op}} \equiv (1/2i)[U_{\text{op}}^\dagger - U_{\text{op}}]$ is a positive semi-definite operator corresponding to a positive contribution to the total cross section from the inelastic channels. For energies very much above E_{in}, a number of formalisms lead to OP's in low-order approximations that contain elastic flux by violating Eqs. (9.2.3a, b), and yet provide reasonably successful representations of the scattering.[13] The violations of Eqs. (9.2.3a, b) in these cases may be insignificant. Generally, the properties (9.2.3a, b) clearly isolate the singularity structure associated with the outgoing wave in the elastic channel in the kernel of Eq. (9.2.1) exactly as in the two-body potential scattering problem; to do otherwise would seem to require special dynamical motivation.

Although an OP is not a fundamental collision observable, but only an auxiliary amplitude, it is often supposed to possess something of a universal character allowing it to be calculated in any number of ways. It is not possible, at present, to do this within the Dirac phenomonology because the underlying physics is not fully understood, but it should be possible to do so to a certain extent in the nonrelativistic formalism. One way to achieve this is to define the OP in a formalism-independent way. We will say that an operator is *formalism-independent* if it is specified solely in terms of the relevant two-cluster channel Hamiltonian and interaction, H_α and V^α, respectively. While this still leaves open considerable off-shell freedom, such operators have the same form whether there are multiparticle forces or not, and are independent of the particular multiparticle collision theories that are used to calculate them. The satisfaction of both of Eqs. (9.2.3a, b) and formalism independence seem closely related.

9.2a. *Operator Formalisms*

The Feshbach[10] OP is the original formalism-independent realization of Eqs. (9.2.1) and (9.2.3a, b) for the scattering of two bound clusters of distinguishable particles labeled by a partition index α. The transition operator

$$T_{\alpha\alpha}(z) = V^\alpha + V^\alpha G(z) V^\alpha \tag{9.2.4}$$

satisfies the two-body integral equations

$$T_{\alpha\alpha}(z) = U_{\mathrm{FO}}(\alpha, z) + U_{\mathrm{FO}}(\alpha, z) P_\alpha G_\alpha(z) T_{\alpha\alpha}(z) \tag{9.2.5a}$$

$$= U_{\mathrm{FO}}(\alpha, z) + T_{\alpha\alpha}(z) P_\alpha G_\alpha(z) U_{\mathrm{FO}}(\alpha, z), \tag{9.2.5b}$$

where

$$U_{\mathrm{FO}}(\alpha, z) = V^\alpha + V^\alpha Q_\alpha [z - Q_\alpha H Q_\alpha]^{-1} Q_\alpha V^\alpha \tag{9.2.6}$$

is the Feshbach[10] OP. The notation follows that of Chs. 1, 4, and 5. In particular, P_α is the projection on the space of cluster states $|\phi_\alpha\rangle$ that are eigenstates of the channel Hamiltonian H_α, and $Q_\alpha = I - P_\alpha$. Equations (9.2.5a, b) were obtained in Section 1.8 from the LS equation for $T_{\alpha\alpha}(z)$ as an example of a generalized Heitler-equation subtraction technique. They can be obtained directly along with the formal operator expression (9.2.6) for $U_{\mathrm{FO}}(\alpha, z)$ by means of the identity

$$\{I + Q_\alpha [Q_\alpha G(z)^{-1} Q_\alpha]^{-1} Q_\alpha V^\alpha\} P_\alpha = \{I - Q_\alpha [Q_\alpha G(z)^{-1} Q_\alpha]^{-1} Q_\alpha G(z)^{-1}\}, \tag{9.2.7}$$

where the order of the inverses is immaterial.[14] In the scattering limit, $z \to E + i0$. Clearly, $U_{\mathrm{FO}}(\alpha, z)$ has poles, called *Feshbach resonances*, when z is the spectrum of $Q_\alpha H Q_\alpha$.[15]

The preceding OP formalism is easily generalized to the case in which each cluster consists of a distinct species of identical particles. This is because the (anti-) symmetrization operator, R_α in this instance (*cf.* Section 1.11), commutes with H_α and H. Then V^α, P_α, and Q_α are simply replaced by $V^\alpha R_\alpha$, $P_\alpha R_\alpha$, and $Q_\alpha R_\alpha$, respectively, in Eqs. (9.2.5–9.2.7).

The permutation symmetry can be used to simplify the LS equation for $U_{\text{FO}}(\alpha, z)$ on the R_α-projected space:

$$U_{\text{FO}}(\alpha, z) = V^\alpha + V^\alpha Q_\alpha R_\alpha G_\alpha(z) U_{\text{FO}}(\alpha, z). \tag{9.2.8}$$

Here, we suppose that Eq. (9.2.8) is defined on the R_α-projected space, so when there are only pair interactions, V^α can be replaced by $A_1 A_2 V_i$, where $V_i, i \subset \alpha$, is any one of the pair interactions linking the two clusters, and A_1, A_2 denote the numbers of particles within the two clusters. This is because $R_\alpha V_i R_\alpha = R_\alpha V_j R_\alpha$ for all $i, j \subset \alpha$. If we multiply Eq. (9.2.8) on the right by $[I + \hat{\tau}_i(z) G_\alpha(z) Q_\alpha R_\alpha]$, where

$$\hat{\tau}_i(z) = V_i + V_i G_\alpha(z) R_\alpha Q_\alpha \hat{\tau}_i(z) \tag{9.2.9a}$$

$$= V_i + \hat{\tau}_i(z) G_\alpha(z) R_\alpha Q_\alpha V_i, \tag{9.2.9b}$$

we obtain

$$U_{\text{FO}}(\alpha, z) = A_1 A_2 \hat{\tau}_i(z) + (A_1 A_2 - 1)\hat{\tau}_i(z) Q_\alpha G_\alpha(z) R_\alpha U_{\text{FO}}(\alpha, z). \tag{9.2.10}$$

If we define

$$U'_{\text{FO}}(\alpha, z) \equiv \frac{(A_1 A_2 - 1)}{A_1 A_2} U_{\text{FO}}(\alpha, z), \tag{9.2.11}$$

then

$$U'_{\text{FO}}(\alpha, z) = \hat{u}(\alpha, z) + \hat{u}(\alpha, z) Q_\alpha G_\alpha(z) R_\alpha U'_{\text{FO}}(\alpha, z), \tag{9.2.12}$$

where

$$\hat{u}(\alpha, z) \equiv (A_1 A_2 - 1)\hat{\tau}_i(z), \qquad i \subset \alpha. \tag{9.2.13}$$

The same trick that led to Eqs. (9.2.10–9.2.13) can obviously be applied to the LS equation for $T_{\alpha\alpha}(z)$ defined on the R_α-projected space. This leads to the result

$$T_{\alpha\alpha}(z) = A_1 A_2 \tau_i(z) + (A_1 A_2 - 1)\tau_i(z) G_\alpha(z) R_\alpha T_{\alpha\alpha}(z), \tag{9.2.14}$$

where

$$\tau_i(z) = V_i + V_i G_\alpha(z) R_\alpha \tau_i(z) \tag{9.2.15a}$$

$$= V_i + \tau_i(z) G_\alpha(z) R_\alpha V_i. \tag{9.2.15b}$$

Thus,

$$T_{\alpha\alpha}(z) = \frac{A_1 A_2}{(A_1 A_2 - 1)} T'_{\alpha\alpha}(z), \tag{9.2.16}$$

with

$$T'_{\alpha\alpha}(z) = u(\alpha, z) + u(\alpha, z) G_\alpha(z) R_\alpha T'_{\alpha\alpha}(z), \tag{9.2.17}$$

where

$$u(\alpha, z) = (A_1 A_2 - 1)\tau_i(z), \qquad i \subset \alpha. \tag{9.2.18}$$

The Feshbach subtraction technique can be applied to Eq. (9.2.17), yielding

$$T'_{\alpha\alpha}(z) = U_{\mathrm{KMT}}(z) + U_{\mathrm{KMT}}(z) G_\alpha(z) R_\alpha P_\alpha T'_{\alpha\alpha}(z), \tag{9.2.19}$$

with

$$U_{\mathrm{KMT}}(z) = u(\alpha, z) + u(\alpha, z) R_\alpha Q_\alpha [z - Q_\alpha H_{\mathrm{KMT}}(\alpha, z) Q_\alpha]^{-1} R_\alpha Q_\alpha u(\alpha, z), \tag{9.2.20}$$

which is the OP, relative to T' but *not* to T, first introduced by Kerman, McManus, and Thaler (KMT)[13] for the case of the elastic scattering of a single distinguishable particle from a nucleus. Here,

$$H_{\mathrm{KMT}}(\alpha, z) \equiv H_\alpha + u(\alpha, z). \tag{9.2.21}$$

The EI given by (9.2.20) does not satisfy property (9.2.3b) because $u(\alpha, z)$ does not. This complicates the usual physical interpretation of Eq. (9.2.19) as representing propagation in the "potential" $U_{\mathrm{KMT}}(z)$ with the clusters kept in their ground states through successive iterations. When approximations are made, $U_{\mathrm{KMT}}(z)$ will generate physical scattering amplitudes with different unitarity characteristics than those generated by $U_{\mathrm{FO}}(z)$.

Since all matrix elements of $R_\alpha \tau_i R_\alpha$ are independent of i, successive iterations of (9.2.17) do not have the structure of a multiple-scattering (MS) theory except as to repetition of a low-order interaction. One must be careful about defining higher-order correlations on the basis of Eqs. (9.2.17–9.2.20) without going out of the R_α-projected space as in Watson-type theories.[16] Various cumulant-type reorderings of MS expansions[17] as well as methods for incorporating projectile-target permutation symmetries in this way[18] are designed to systematically deal with these problems.

None of the preceding formalisms provides a complete theory for calculating an OP. Therefore, we next consider dynamical theories that are capable of doing this.[19] We consider the C-K method of Ch. 5 for which the intercluster interactions are replaced by potential matrices. We suppose for simplicity that we are dealing with distinguishable particles.

We learned in Ch. 5 that a general class of C-K equations for the post transition operators can be written in the form (*cf*. Eq. (5.2.27a))

$$\underline{T}^{(+)} = \mathscr{V}\underline{G}\mathscr{S}\overleftarrow{\underline{G}}^{-1} + \mathscr{V}\underline{G}\underline{T}^{(+)}. \tag{9.2.22}$$

The notation is completely defined in Ch. 5; we have suppressed the z-dependence in Eq. (9.2.22) and in what follows. Let us introduce the projection $\underline{P}(\alpha)$ onto the distinguished two-cluster channel α:

$$[\underline{P}(\alpha)]_{ab} \equiv P_\alpha \delta_{a\alpha} \delta_{ab}. \tag{9.2.23}$$

Then, by our standard OP subtraction technique:

$$\underline{T}^{(+)} = \underline{\Gamma}(\alpha)\underline{G}\mathscr{S}\overleftarrow{\underline{G}}^{-1} + \underline{\Gamma}(\alpha)\underline{G}\underline{P}(\alpha)\underline{T}^{(+)}, \tag{9.2.24}$$

where

$$\underline{\Gamma}(\alpha) = \mathscr{V} + \mathscr{V}\underline{G}\underline{Q}(\alpha)\underline{\Gamma}(\alpha) \tag{9.2.25a}$$

$$= \mathscr{V} + \underline{\Gamma}(\alpha)\underline{G}\underline{Q}(\alpha)\mathscr{V}, \tag{9.2.25b}$$

and $\underline{Q}(\alpha) = I - \underline{P}(\alpha)$, with I the identity on the Hilbert and partition spaces. Using the technique of Section 1.9, we find that $\underline{\Gamma}(\alpha)$ is continuous across the elastic unitarity cut, although it is not Hermitian-analytic.

At this point, one has two options for obtaining an OP potential from Eq. (9.2.24). One is to recall that the component $T^{(+)}_{\alpha\alpha}$ satisfies Eqs. (9.2.5a, b) in terms of the Feshbach OP. It then follows from Eq. (9.2.24) that

$$U_{\mathrm{FO}}(\alpha) = [\underline{\Gamma}(\alpha)\underline{G}\mathscr{S}\overleftarrow{\underline{G}}^{-1}]_{\alpha\alpha} - [\underline{\Gamma}(\alpha)\underline{G}(\mathscr{S} - \underline{I})\underline{P}(\alpha)]_{\alpha\alpha} U_{\mathrm{FO}}(\alpha). \tag{9.2.26}$$

Since only the matrix elements of $U_{\mathrm{FO}}(\alpha)$ between the cluster states $|\phi_\alpha\rangle$ are needed in Eq. (9.2.5), we see that Eq. (9.2.26) is an off-shell two-body dynamical equation for $U_{\mathrm{FO}}(\alpha)$. Therefore, given the input in Eq. (9.2.26), the calculation of $U_{\mathrm{FO}}(\alpha)$ is straightforward.

The remaining question is whether Eqs. (9.2.25a, b) are of the C-K type. Because of the $\underline{Q}$ operator, the kernels of Eqs. (9.2.25a, b) do not necessarily become connected after a finite number of iterations even if some power of the kernel of Eq. (9.2.22) does have this property. A sufficient condition is that $\mathscr{V}^{ab}$ involve only operators with connectivities external to α.[20] Henceforth, we assume that this condition, which is satisfied by the arrays we use, is realized.

With this understood, Eq. (9.2.26), together with Eqs. (9.2.25a, b), constitutes a definitive, complete dynamical strategy for the calculation of the Feshbach OP, which is something all of the conventional approaches to the OP fail to provide.

There is an alternative way to deal with Eq. (9.2.24). Since we are only interested in the elements $T^{(+)}_{\alpha\alpha}$, we need consider only

$$T^{(+)}_{\alpha\alpha} = [\underline{\Gamma}(\alpha)\underline{G}\mathscr{S}\overleftarrow{\underline{G}}^{-1}]_{\alpha\alpha} + \Gamma_{\alpha\alpha}(\alpha)G_\alpha P_\alpha T^{(+)}_{\alpha\alpha}. \tag{9.2.27}$$

Furthermore, since we only require the half-shell matrix elements, $\langle\phi_\alpha(\mathbf{p})|T^{(+)}_{\alpha\alpha}|\phi_\alpha(\mathbf{p}')\rangle$, where $\mathbf{p}'$ corresponds to an on-shell state, we can use the Lippmann identity to simplify Eq. (9.2.27). One obtains

$$T^{(+)}_{\alpha\alpha}|\phi_\alpha(\mathbf{p}')\rangle = \Gamma_{\alpha\alpha}(\alpha)|\phi_\alpha(\mathbf{p}')\rangle + \Gamma_{\alpha\alpha}(\alpha)G_\alpha P_\alpha T^{(+)}_{\alpha\alpha}|\phi_\alpha(\mathbf{p}')\rangle. \tag{9.2.28}$$

This is equivalent to using, instead of Eq. (9.2.27), the equation

$$\tilde{T}^{(+)}_{\alpha\alpha} = \Gamma_{\alpha\alpha}(\alpha) + \Gamma_{\alpha\alpha}(\alpha)G_\alpha P_\alpha \tilde{T}^{(+)}_{\alpha\alpha} \tag{9.2.29}$$

for the operator $\underline{\tilde{T}}^{(+)}$, which is half-on-shell equivalent to $\underline{T}^{(+)}$, at least with respect to two-cluster states $[\underline{T}^{(+)} = \underline{\tilde{T}}^{(+)}\underline{G}\mathscr{S}\overleftarrow{\underline{G}}^{-1}]$. Evidently, $\Gamma_{\alpha\alpha}(\alpha)$ is an OP for elastic scattering although it is not formalism-independent, in contrast to U_{FO}; it is also not Hermitian-analytic.

In any except the simplest physical applications, Eqs. (9.2.25a, b) will be solved approximately to determine the components of $\underline{\Gamma}(\alpha)$ needed in Eq. (9.2.26) or in Eq. (9.2.29). The resultant approximations to $U_{\text{FO}}(\alpha)$ and $\Gamma_{\alpha\alpha}(\alpha)$ correspond to different elastic transition matrices. It is unclear which approach will yield superior results. Also, it should be kept in mind that, for arbitrary approximations to $\underline{\Gamma}(\alpha)$, one cannot be assured that the solution of Eq. (9.2.26) will yield a Hermitian-analytic OP, or even one that is formalism-independent. Nevertheless, scenarios based on Eq. (9.2.26) have been proposed[21] that yield an approximate U_{FO} with the requisite properties.

It is also possible to develop these OP formalisms from a wave function point of view.[22] This is done most directly by starting with the prior operators $\underline{T}^{(-)}$, and instead of $\underline{\Gamma}(\alpha)$ the quantity, $\underline{\Gamma}^{\text{t}}(\alpha)$ given by the Eqs. (9.2.25a, b) except with $\mathscr{V}$ replaced by $\mathscr{V}^{\text{t}}$. (Actually, $\underline{\Gamma}^{\text{t}}(\alpha, z^*) = \underline{\Gamma}(\alpha, z)^\dagger$, where we have inserted the complex parametric energy in order to manifest the non-Hermitian analyticity of both Γ-operators.) Then, in terms of the component wave functions $|\chi^{(+)}_b(\alpha)\rangle$ of Ch. 5 (*cf.* Eq. (5.2.37)), one finds that, half-on-shell,

$$P_\alpha \tilde{T}^{(-)}_{\alpha\alpha}|\phi_\alpha\rangle = P_\alpha \Gamma^{t}_{\alpha\alpha}(\alpha)P_\alpha|\chi^{(+)}_\alpha(\alpha)\rangle, \tag{9.2.30}$$

so

$$[\vec{G}_\alpha^{-1} - P_\alpha \Gamma^{t}_{\alpha\alpha}(\alpha)]P_\alpha|\chi^{(+)}_\alpha(\alpha)\rangle = 0, \tag{9.2.31}$$

where $\underline{T}^{(-)} = \vec{\underline{G}}^{-1}\mathscr{S}\underline{G}\underline{\tilde{T}}^{(-)}$. This should be contrasted with the wave equation

$$[\vec{G}_\alpha^{-1} - P_\alpha U_{\text{FO}}(\alpha)]P_\alpha|\psi^{(+)}_\alpha\rangle = 0. \tag{9.2.32}$$

Generally speaking, any wave-function approach obscures the crucial role of the off-shell extension in the definition of the OP.

The generalization of the Feshbach OP to include permutation symmetries was first discovered using a C-K, transition-operator collision theory.[7] This was the first result obtained using multiparticle scattering theory that was anticipated, but not found, using conventional LS- or multiple-scattering-type operator formalisms. Actually, the crucial idea[7] involves an appropriate choice of off-shell extension for the multichannel rearrangement operators that are physically equivalent by permutation symmetry. Once this is recognized, the relevant "conventional" operator formalism can be constructed,[23] yielding fully-symmetrized generalizations of Eqs. (9.2.4–9.2.6).

We next review the essential ideas for constructing a permutation-symmetric OP. For the sake of simplicity, let us suppose that all N particles in each of the two clusters are identical fermions.[24]

In order to provide a means for comparing our treatment with the extant literature in atomic and nuclear physics on these problems, we outline Feshbach's treatment of identity in elastic two-fragment scattering.[25] An essential role is played by projectors $P(\Lambda)$ and $Q(\Lambda) = \Lambda - P(\Lambda)$ on the space $\mathscr{H}(\Lambda)$ of fully antisymmetrized states. Here, Λ is the identity on $\mathscr{H}(\Lambda)$ and the projection of the full N-particle Hilbert space $\mathscr{H}$ onto $\mathscr{H}(\Lambda)$.

The operator $P(\Lambda)$, which satisfies[26]

$$P(\Lambda)^2 = P(\Lambda), \tag{9.2.33}$$

on $\mathscr{H}(\Lambda)$, was originally[25] stipulated to be the projection onto those fully antisymmetric states that correspond to the two noninteracting fragments in an asymptotic limit. This stipulation is not restrictive enough to define $P(\Lambda)$ uniquely.[25,27]

Feshbach's strategy is to look for a scattering solution of the Schrödinger equation,

$$(E - H)|\psi^{(+)}(\Lambda)\rangle = 0, \tag{9.2.34}$$

satisfied by the antisymmetrized wave function $|\psi^{(+)}(\Lambda)\rangle \equiv \Lambda|\psi^{(+)}\rangle$. It is, of course, assumed that $[H, \Lambda] = 0$. Since only the elastic two-fragment transition amplitudes are sought, it suffices to work with the following projected version of Eq. (9.2.34):

$$[E - P(\Lambda)H_{\text{eff}}(\Lambda)P(\Lambda)]P(\Lambda)|\psi^{(+)}(\Lambda)\rangle = 0, \tag{9.2.35}$$

where

$$H_{\text{eff}}(\Lambda) = H + HQ(\Lambda)[E - Q(\Lambda)HQ(\Lambda) + i0]^{-1}Q(\Lambda)H. \tag{9.2.36}$$

The elastic transition amplitude is identified using the LS equation

$$P(\Lambda)|\psi^{(+)}(\Lambda)\rangle = |\chi_p^{(+)}(\Lambda)\rangle + G_p^{(+)}(\Lambda)P(\Lambda)[H_{\text{eff}}(\Lambda) - H]P(\Lambda)|\psi^{(+)}(\Lambda)\rangle, \tag{9.2.37}$$

and the two-potential formalism (*cf.* Section 1.8). Here, $G_p^{(+)}(\Lambda) \equiv [E - P(\Lambda)HP(\Lambda) + i0]^{-1}$ and

$$[E - P(\Lambda)HP(\Lambda)]|\chi_p^{(+)}(\Lambda)\rangle = 0 \tag{9.2.38}$$

with outgoing-wave boundary conditions, where again we have suppressed the dependence on the total energy E.

The preceding symmetric treatment of elastic scattering differs from the distinguishable-projectile case in that it does not define an operator realization of the OP. The Feshbach identification[28] of $P(\Lambda)[H_{\text{eff}}(\Lambda) - H]P(\Lambda)$ as an OP is inconsistent, even though it functions as a partial EI in Eq. (9.2.37), because $P(\Lambda)HP(\Lambda)$ includes inter-fragment interactions. Effective interactions much like the OP do appear in the equation for the relative wave function $u(\mathbf{R})$ of the two clusters with a separation $\mathbf{R}$ between their c.m.'s, when $|\psi^{(+)}(\Lambda)\rangle$ is supposed to have the form of an antisymmetrized product of $|u(\mathbf{R})\rangle$ and the cluster state $|\phi_\alpha\rangle$, but this does not provide us with the general expression we would like.[29]

Let us call this desired expression $\mathscr{V}_{\text{opt}}(\alpha, z)$, where α refers to any one of the two-cluster channels that are physically equivalent by permutation symmetry. We expect $\mathscr{V}_{\text{opt}}(\alpha, z)$ to reduce to U_{FO} in the case when each cluster contains a different species of particle, to satisfy Eqs. (9.2.3a, b), and to have the same Feshbach-resonance content as $H_{\text{eff}}(\Lambda)$.

One can deduce $\mathscr{V}_{\text{opt}}(\alpha, z)$ by using a wave-function starting point.[30] However, it is difficult to see how one could arrive at this result without knowing the outcome beforehand.

The essential physical ideas are much more transparent in a transition-matrix framework. The important point is that identity among the constituent particles in a collision necessarily involves the consideration of rearrangement scattering into all physically indistinguishable channels. Thus, identity is naturally encompassed in a general treatment of rearrangement collisions.

To this end, we introduce a matrix of EI operators $\mathscr{U}_{ba}$ by means of the following integral equations for the transition operators T_{ba} of arbitrary off-shell extension:

$$T_{ba} = \mathscr{U}_{ba} + \sum_{\gamma \in S} \mathscr{U}_{b\gamma} P_\gamma G_\gamma T_{\gamma a} \tag{9.2.39a}$$

$$= \mathscr{U}_{ba} + \sum_{\gamma \in S} T_{b\gamma} P_\gamma G_\gamma \mathscr{U}_{\gamma a}. \tag{9.2.39b}$$

Here we have singled out a subset S of two-cluster channels that we assume is not overcomplete.[31] Also, P_γ denotes our usual projector on the two-cluster states; we have suppressed both the parametric energy z and the dependence of $\mathscr{U}$ on S. If S contains only a single partition α, then $\mathscr{U}_{\alpha\alpha} = U_{FO}(\alpha, z)$. When $a, b \in S$, Eqs. (9.2.39a, b) are closed within S. In the identical-particle case, where we are seeking an OP, we take $S = \bar{\alpha}$, the equivalence class of channels related to α by permutation, and $a, b \in S$. However, for the moment, we let a, b, and S be arbitrary in order to indicate how some of the essential ideas can be utilized for general rearrangement collision theory.

Given T_{ba}, the EI operators $\mathscr{U}_{ba}$ are defined by Eqs. (9.2.39a, b). As a rule, approximation scenarios for rearrangement collisions involve the explicit or implicit introduction of operators such as the $\mathscr{U}_{ba}$ with some characteristic property related to the physical situation of interest. The new idea we now propose is that we attempt to exploit the off-shell freedom in the choice of T_{ba} to define EI operators that possess that characteristic property. In general, it may not be possible to do this in a practical way, but for the OP that property consists of the conditions (9.2.3a, b) for quasi-Hermiticity, and then the corresponding T_{ba} can be found.

The relevant attributes of T_{ba} relate to its off-shell unitarity relations. We know from Section 1.9 that these relations have the general form

$$T_{ba} - T^{\dagger}_{ab} = \sum_{\gamma \in S} T_{b\gamma}\rho_\gamma(s)T^{\dagger}_{a\gamma} + \sum_{c \notin S} T_{bc}\bar{\rho}_c(s)T^{\dagger}_{ac} + R_{ba}, \tag{9.2.40}$$

where $\rho_\gamma(s)$ and $\bar{\rho}_c(s)$ denote the phase-space factors relevant to channels contained within and outside the subset S, respectively. Whatever the off-shell extension, the on-shell matrix elements of the remainder R_{ba} must vanish in order to recover the physical unitarity condition from Eq. (9.2.40).

Equation (9.2.40) can be converted into a unitarity relation for the EI operators with the aid of the integral equations (9.2.39a, b):

$$\mathscr{U}_{ba} - \mathscr{U}^{\dagger}_{ab} = \sum_{c \notin S} \mathscr{U}_{bc}\bar{\rho}_c(s)\mathscr{U}^{\dagger}_{ac} + R\text{-term}, \tag{9.2.41}$$

where the last term is a complicated product involving the remainder terms in Eq. (9.2.40). If $R = 0$, then we can be sure that no channels $\gamma \in S$ appear on the right-hand side of Eq. (9.2.41). Finally, and in addition, it follows from Eqs. (9.2.39a, b) that if T_{ba} is Hermitian-analytic, so is $\mathscr{U}_{ba}$.

It is easy to verify that $R_{ba} \equiv 0$ if T_{ba} is identified with the AGS off-shell extension (*cf.* Section 1.7):[23]

$$U_{ba}(z) = \bar{\delta}_{ba} G_a(z)^{-1} + V^b + V^b G(z) V^a, \tag{9.2.42}$$

which is also Hermitian-analytic,

$$U_{ba}(z)^{\dagger} = U_{ab}(z^*). \tag{9.2.43}$$

When this off-shell extension is used, the R-term is absent in Eq. (9.2.41), the EI operators are Hermitian-analytic, and they have no discontinuities across any of the elastic unitarity cuts that are associated with the channels $\gamma \in S$. As with OP's satisfying Eqs. (9.2.3a, b), we call such EI operators quasi-Hermitian.

The preceding analysis beginning with Eqs. (9.2.39a, b) is general, and provides a means of constructing a two-cluster multichannel theory with quasi-Hermitian EI operators. We apply this theory to the special case where all of the channels $\gamma \in S$ are physically equivalent by virtue of permutation symmetry.

Thus, we consider Eqs. (9.2.39a, b) with the AGS off-shell extension and with all partition indices restricted to $S = \bar{\alpha}$:

$$U_{\beta\alpha} = \mathscr{U}_{\beta\alpha} + \sum_{\gamma \in \bar{\alpha}} \mathscr{U}_{\beta\gamma} P_{\gamma} G_{\gamma} U_{\gamma\alpha} \tag{9.2.44a}$$

$$= \mathscr{U}_{\beta\alpha} + \sum_{\gamma \in \bar{\alpha}} U_{\beta\gamma} P_{\gamma} G_{\gamma} \mathscr{U}_{\gamma\alpha}. \tag{9.2.44b}$$

with $\alpha, \beta \in \bar{\alpha}$. The identical-particle formalism of Section 1.11 can be applied directly to Eqs. (9.2.44a, b), which are regarded as equations for the EI operators $\mathscr{U}_{\beta\alpha}$. Since the AGS operators are manifestly label-transforming (*cf.* Section 1.11 and Section 5.4f), so are the $\mathscr{U}_{\beta\alpha}$.

The fully symmetrized AGS transition operator is

$$T_{\mathrm{AGS}}(\bar{\alpha}) \equiv \sum_{\alpha' \in \bar{\alpha}} \mathscr{A}^{\dagger}_{\alpha'\alpha} U_{\alpha'\alpha}, \tag{9.2.45}$$

where the sum is over the set $\bar{\alpha}$ of all partitions related by permutations to the canonical (fixed) partition α. Here,

$$\mathscr{A}^{\dagger}_{\alpha'\alpha} = R_{\alpha} \bar{U}(P_{\alpha'\alpha})^{\dagger} \tag{9.2.46}$$

is simply the product of the adjoint of the parity-weighted unitary operator that accomplishes the permutation $\alpha \to \alpha'$, and the symmetrizer R_{α} with respect to all permutations that map α into itself. If P_{α} is chosen so that $P_{\alpha} R_{\alpha} = P_{\alpha}$, then

$$T_{\mathrm{AGS}}(\bar{\alpha}) = \mathscr{V}_{\mathrm{opt}}(\bar{\alpha}) + \mathscr{V}_{\mathrm{opt}}(\bar{\alpha}) G_{\alpha} P_{\alpha} T_{\mathrm{AGS}}(\bar{\alpha}) \tag{9.2.47a}$$

$$= \mathscr{V}_{\mathrm{opt}}(\bar{\alpha}) + T_{\mathrm{AGS}}(\bar{\alpha}) G_{\alpha} P_{\alpha} \mathscr{V}_{\mathrm{opt}}(\bar{\alpha}). \tag{9.2.47b}$$

Here,

$$\mathscr{V}_{\mathrm{opt}}(\bar{\alpha}) \equiv \sum_{\alpha' \in \bar{\alpha}} \mathscr{A}^{\dagger}_{\alpha'\alpha} \mathscr{U}_{\alpha'\alpha} \tag{9.2.48}$$

is the symmetrized version of the Feshbach OP; it satisfies Eqs. (9.2.3a, b) by construction.

The preceding is entirely general, and leaves open the dynamical strategy one wishes to use to calculate $\mathscr{V}_{\text{opt}}(\bar{\alpha})$. There are a number of possible C-K formalisms for doing this. However, it is of considerable interest to inquire as to the nature of the fully symmetrized counterpart of the Feshbach closed-form solution (9.2.6) and the relationship of $\mathscr{V}_{\text{opt}}(\bar{\alpha})$ to H_{eff} defined by Eq. (9.2.36). We next outline how these results can be derived.[30]

Let

$$\mathscr{A}(\bar{\alpha}) \equiv \sum_{\alpha' \in \bar{\alpha}} \mathscr{A}^{\dagger}_{\alpha'\alpha}. \tag{9.2.49}$$

From Section 1.11, we have that

$$\mathscr{A}(\bar{\alpha}) = a_{\bar{\alpha}}\Lambda, \tag{9.2.50}$$

where Λ is the projection onto the symmetrized space $\mathscr{H}(\Lambda)$ and $a_{\bar{\alpha}}$ is a normalization constant. We need two more operators:

$$\bar{\mathscr{A}}(\bar{\alpha}) \equiv \mathscr{A}(\bar{\alpha}) - R_{\alpha} \tag{9.2.51}$$

and

$$\mathscr{N}_{\alpha} \equiv \bar{\mathscr{A}}(\bar{\alpha})P_{\alpha}. \tag{9.2.52}$$

It then follows that

$$(1 + \mathscr{N}_{\alpha})^{-1}\mathscr{A}(\bar{\alpha})P_{\alpha} = P_{\alpha}, \tag{9.2.53}$$

where we recall that $P_{\alpha}R_{\alpha} = R_{\alpha}P_{\alpha} = P_{\alpha}$.

The calculation of the inverse operator in Eq. (9.2.53) is equivalent to solving a nondynamical integral equation with the kernel $-\mathscr{N}_{\alpha}$, which is a problem that appears in a variety of treatments of permutation symmetry in two-cluster collisions, but usually for the bounded Hermitian operator $-P_{\alpha}\mathscr{N}_{\alpha}$ rather than $-\mathscr{N}_{\alpha}$. All that we require is the knowledge that $(1 + \mathscr{N}_{\alpha})^{-1}$ exists, and this can be demonstrated assuming only that the cluster states entering into P_{α} have the correct physical structure with respect to spatial translations and correlations.[32] The existence of this inverse operator is equivalent to the statement that the set of exchange equivalent partitions is not overcomplete.[31]

The essential results concerning a "closed form" expression for $\mathscr{V}_{\text{opt}}(\bar{\alpha})$ and its relationship to H_{eff} can succinctly be stated as a series of theorems:[30]

Theorem 1. *The Feshbach realization of the idempotent projector $P(\Lambda)$ is*

$$P(\Lambda) = \frac{1}{a_{\bar{\alpha}}}(1 + \mathscr{N}_{\alpha})P_{\alpha}(1 + \mathscr{N}_{\alpha})^{-1}P_{\alpha}(1 + \mathscr{N}^{\dagger}_{\alpha}). \tag{9.2.54}$$

Let $|\psi^{(+)}(\bar{\alpha})\rangle$ be a fully symmetrized scattering solution that corresponds to two fragments in the class $\bar{\alpha}$ in the infinite past. Then, corresponding to Eq. (9.2.32),

Theorem 2.

$$[\vec{G}_\alpha(z)^{-1} - \mathscr{V}_{\text{opt}}(\bar{\alpha})]P_\alpha|\psi^{(+)}(\bar{\alpha})\rangle = 0, \tag{9.2.55}$$

where

$$\mathscr{V}_{\text{opt}}(\bar{\alpha}) = P_\alpha\{V^\alpha_{\text{ex}} + V^\alpha_{\text{ex}}Q_\alpha[\vec{G}_\alpha(z)^{-1} - Q_\alpha V^\alpha_{\text{ex}}Q_\alpha]^{-1}Q_\alpha V^\alpha_{\text{ex}}\}P_\alpha, \tag{9.2.56}$$

and

$$V^\alpha_{\text{ex}} \equiv (1 + \mathscr{N}_\alpha)^{-1}[V^\alpha + \mathscr{N}_\alpha G_\alpha(z)^{-1}], \tag{9.2.57}$$

where $z = E + i0$. (The directionality of $G_\alpha(z)^{-1}$ in Eq. (9.2.57) is irrelevant.)

The preceding theorem reduces to the standard Feshbach form (9.2.6) when there is no permutation symmetry between the two clusters.

Finally, we have interpolative identity

Theorem 3.

$$P_\alpha[\vec{G}_\alpha(z)^{-1} - \mathscr{V}_{\text{opt}}(\bar{\alpha})]P_\alpha = w_{\bar{\alpha}}P(\Lambda)[z - H_{\text{eff}}(\Lambda)]P(\Lambda)(w_{\bar{\alpha}})^\dagger, \tag{9.2.58}$$

by means of the nondynamical similarity transformations, where

$$w_{\bar{\alpha}} = \sqrt{a_{\bar{\alpha}}}(1 + \mathscr{N}_\alpha)^{-1}, \tag{9.2.59}$$

and $z = E + i0$.

As a consequence of Theorems 1–3 we conclude that

Theorem 4. *The physical implications of Feshbach's projected Schrödinger equation (9.2.35) for elastic two-fragment scattering of identical particles are equivalent to the use of the OP $\mathscr{V}_{\text{opt}}(\bar{\alpha})$ in the two-body equations (9.2.47a, b) and (9.2.55).*

Corollary. *The (Feshbach) resonance structures of H_{eff} and $\mathscr{V}_{\text{opt}}(\bar{\alpha})$ are identical.*

Besides providing the identification of the unique OP that is implicit in Feshbach's effective Hamiltonian approach, Theorem 4 provides the clean resolution of the question of how one solves the two-body scattering problem presented by Eq. (9.2.35). The real difficulty is that $P(\Lambda)H_{\text{eff}}(\Lambda)P(\Lambda)$ cannot be separated into a noninteracting two-fragment piece plus an interacting part, except asymptotically. On the other hand, we see that with Eqs. (9.2.47a, b) and (9.2.55) this separation is unequivocal, and the canonical partition α labels the distinguished two-fragment channel. By the result summarized in Eq. (1.11.41), the physical matrix elements will be independent of this choice for α. This circumvents a major complicating feature of the fully symmetrized Feshbach formalism.[23,33]

The preceding results depend crucially upon the use of the AGS off-shell extension which, in turn, is distinguished from the more widely used post and prior off-shell extensions by its symmetry with respect to the entrance and exit channel labels. When the post or prior off-shell extensions are used, the resultant OP's do not satisfy Eqs. (9.2.3a, b), and thus contain elastic flux.[23] They also do not contain obvious Feshbach-resonance signatures—Hermitian-analyticity seems crucial to obtain this. These OP's are nevertheless formalism-independent and they seem to present some advantage for the introduction of multiple-scattering algorithms.[34]

The lesson to be learned from all of this is that off-shell extensions are quite important for the formulation of collision theory and, especially, for the statement of approximations. This last point has been long realized in the form of the famous post-prior ambiguity in the choice of low-order distorted-wave calculations.[34] However, we have shown that it is possible for one to turn the off-shell freedom to one's advantage. It is not yet known how to best exploit this in the general multichannel case, but some proposals have been put forth.[35] It should be mentioned that in the OP formalism of Ref. 22 still another off-shell extension is used, one that is related to the prior form half-on-shell. The fact that it is not *equal* to the prior extension is why OP's with no elastic cuts can be defined.

The methods summarized in Eqs. (9.2.22–9.2.32) can be extended using the results of Section 1.11 to the fully symmetrized case to obtain C-K dynamical equations for the determination of $\mathscr{V}_{\text{opt}}(\bar{\alpha})$ or other symmetrized two-body elastic EI operators that correspond to different off-shell extensions. All of the remarks that we made previously concerning conditions (9.2.3a, b) and formalism independence apply with only minor modifications to the fully symmetrized case. We refer to the original literature[19] for the details of the development of these symmetrized C-K OP formalisms. A few of the results of this work will be utilized when we take up certain systematic methods for approximating the OP.

Instead of dealing with the explicit symmetrization of the Feshbach OP, there is another approach to the symmetrized OP based on the C-K multiparticle equations. In the original application of this idea wave-function equations were used,[22] although it can be formulated entirely in terms of transition operator equations.

The Feshbach OP can be derived by projecting the LS or the Schrödinger equation onto the elastic-channel space. The same trick can be applied to a special class of C-K equations for wave-function components. If these components, when projected onto a certain channel space, carry the elastic scattering phase for that channel, then this projection of the C-K equations will also lead to a definition of an OP. As these C-K equations usually have an asymmetric structure in channel space, such an OP will not be Hermitian-analytic, but nevertheless will be free of elastic unitarity cuts, and hence real in this sense, below the lowest inelastic threshold. This OP was anticipated in Eqs. (9.2.29–9.2.31) for the unsymmetrized case, but of course the same result is obtained in the manner just suggested by beginning with the wave-function equations (5.2.37) and applying the Feshbach projection-operator formalism in the standard fashion.[22]

The reason this is an interesting thing to do is that the components $P_\alpha|\chi_\alpha^{(+)}(\alpha)\rangle$, by virtue of the Green function $G_\alpha^{(+)}$ in the kernel of the Eqs. (5.2.37), will have a coordinate-space function $\chi_\alpha(\mathbf{x})$ with an outgoing-wave part that is proportional to the exact elastic amplitude and so in this sense can be said to carry the elastic phase. So in view of Eq. (9.2.31), it is entirely consistent to identify

$$\tilde{\mathscr{V}}_{\text{opt}}(\alpha) \equiv P_\alpha \Gamma^t_{\alpha\alpha}(\alpha) P_\alpha \tag{9.2.60}$$

as the (projected) OP for elastic scattering. One has the advantage of working only with quantities defined in terms of C-K equations, although clearly definition (9.2.60) depends on the formalism one chooses to use. The same procedure can be carried out using the symmetrized form of the Eqs. (5.2.37), and the corresponding OP is then (*cf*. Eqs. (5.4.80–5.4.81))[22]

$$\tilde{\mathscr{V}}_{\text{opt}}(\bar{\alpha}) \equiv P_{\alpha_0} \Gamma^t_{\bar{\alpha}\bar{\alpha}}(\bar{\alpha}) P_{\alpha_0}, \tag{9.2.61}$$

where α_0 is a canonical partition of the equivalence class $\bar{\alpha}$.

We remark that the OP's (9.2.60–9.2.61) are one-body potentials, so the entire array of potential scattering expressions either in wave function or transition operator form can be written down directly.[22] Also Feshbach-like closed forms can be written down for the Γ-operators.[20,22]

9.2b. Second-Quantized Formalisms

In Section 1.11, a brief outline is given of the application of Fock-space, or second-quantized, techniques to nonrelativistic collision theory. In this section, we discuss a few aspects of their use in the definition and determination of the OP. The relativistic case is taken up in the next subsection.

The identification of an OP for the elastic scattering of a particle from a bound state of particles identical to it was an early success of many-particle theory.[1,36] This bound state is typically either a nucleus or an atom,[37] so we deal with fermions. The same ideas have been applied to pion–nucleus scattering where the pion, of course, is a boson but nonetheless the entire problem is treated using field-theoretic techniques.[38] Extensions of the idea to elastic scattering of, say, two nuclear fragments have been less frequent and accompanied by a much less transparent formalism.

Things are really only simple for the particle-bound-state case, where the motion of the target is ignored. For a many-particle target and low projectile energies, this is a reasonable approximation. With this understood, the OP is identified by relating the wave function for the elastic scattering of the projectile to the propagator[36,39–42]

$$G(x|x') \equiv -i < 0|T[\psi(\mathbf{x},t)\psi^{\dagger}(\mathbf{x}',t')]|0\rangle, \tag{9.2.62}$$

which is the amplitude for creating a particle at the space–time point $x' \equiv (\mathbf{x}', t')$ and destroying it at $x = (\mathbf{x}, t)$ with the medium ($\equiv$target) remaining in its ground state ($\equiv|0\rangle$). We ignore all spin-indices.

The Dyson equation satisfied by the propagator,

$$G(x|x') = G_0(x|x') + \int d^4y \int d^4z\, G_0(x|y)\Sigma(y|z)G(z|x'), \tag{9.2.63}$$

where $G_0(x|x')$ is equal to $G(x|x')$ with all the mutual interactions set equal to zero, can serve as the definition of the mass operator or self-energy, $\Sigma(x|x')$. Alternatively, Eq. (9.2.63) can be inferred by defining $\Sigma(x|x')$ as the sum of all single-particle Green-function graphs, but with no external lines, that are irreducible with respect to the cutting of an internal, single-particle line. Equation (9.2.63) then generates all of the reducible graphs that are needed in addition to $\Sigma(x|x')$ to obtain $G(x|x')$. One can also derive Eq. (9.2.63) by functional techniques that are not explicitly dependent on perturbative ideas.

Clearly, $\Sigma(x|x')$ represents a propagation in which the target begins and ends in its ground state, but is in excited states on its ensuing space–time trajectory. It is the object we would expect to identify with the nonlocal quasi-Hermitian OP in coordinate space, and this indeed turns out to be the case. In momentum space the elastic transition amplitude is essentially $G_0^{-1}GG_0^{-1} - G_0^{-1}$. Thus, Eq. (9.2.63) can be converted into a transition matrix LS equation with Σ as the effective potential.[40,42]

The details of these identifications, whether in coordinate or momentum space, are of no intrinsic interest for our development. We confine ourselves to some remarks that are pertinent to the question of viable dynamical strategies. First, we note that the definition of Σ involves no P, Q projection operators so that problems concerning the explicit construction of these quantities are bypassed. The expression of Σ in second-quantized form opens the door to a host of many-particle techniques that can be exploited for its approximate evaluation.[40–42] This can be done to arbitrary accuracy using the hierarchy of propagators strategy outlined in Section 9.1. In order to solve for the higher-order propagators systematically, one will be led to the same type of general problem of constructing C-K integral equations that we encounter in the fixed number of particles, Hilbert-space treatment. The formalism entailed in doing this has not yet been worked out in detail.

The restriction of no target recoil is a serious one in the second-quantized formalism. Things become quite complicated when recoil is included.[43] Indeed, some of the standard second-quantized approaches do not lead to conventional OP's when recoil is included.[44]

It is tempting to regard the nucleon–nucleus, for example, elastic scattering problem in Fock space as a problem that can be handled analogously to the establishment of Bethe–Salpeter or Dyson–Schwinger equations. Here, one introduces two-body irreducibility with respect to graphs with an internal nucleon *and* a ground-state target-nucleus line.[45] Then, with all kinematical variables and intermediate integrations suppressed, one obtains for the elastic transition matrix T_{EL} an OP-type integral equation

$$T_{\mathrm{EL}} = \Sigma_2 + \Sigma_2 g_2 T_{\mathrm{EL}}, \tag{9.2.64}$$

where Σ_2 is identified as a two-particle irreducible self-energy. Here g_2 is a simple two-body Green function of the form $(E_0 - \mathbf{p}_{\mathrm{rel}}^2 + i0)^{-1}$ after the c.m. motion has been separated out, where E_0 is the total c.m. energy with the nucleus in its ground state, and $\mathbf{p}_{\mathrm{rel}}$ is the relative momentum.

Again Σ_2 functions as a quasi-Hermitian OP that requires no P and Q operators for its definition. However, in contrast to the self-energy Σ appearing in Eq. (9.2.63), one requires a complete composite-particle collision theory in second-quantized form in order to define Σ_2 unequivocally. Graphs with internal bound-state lines do not appear in the usual Fock-space formulations of many-particle theory. A definitive formalism that would provide the justification for Eq. (9.2.64) and a means of calculating Σ_2 has yet to be worked out. One encounters these questions again in the relativistic case.

9.2c. The Dirac Optical Potential

Perhaps the most extraordinary development in the history of the OP was the success achieved in 1983 in explaining some puzzling data on intermediate-energy proton–nucleus elastic scattering.[3] This was accomplished using a simple model that seemed to incorporate negative-energy states of the projectile, viewed as a point Dirac particle, in a dynamically significant manner. In the words of one of the original groups of authors:[46]

> "The results of this and other investigations demonstrate the dramatic success of the Dirac impulse approximation optical potential above 500 MeV. Clearly a formal discussion of relativistic multiple-scattering theory is called for by this work."

Many have heeded this call; very few seem to have been chosen to reveal the entire truth.[47] In this subsection, we review those aspects of the Dirac approach to the OP that seem most relevant to our particular dynamical point of view. The so-called impulse approximation is discussed in the next section.

It has long been appreciated that a dynamical picture of nuclear phenomena based on the nonrelativistic Schrödinger equation involving nucleon–nucleon potentials is of limited physical validity. The extent of the limits that are involved has been more sharply delineated over the past decade by a number of important experimental and theoretical developments.[47] This has led to a new picture that has at least phenomenological validity. In this picture, the interaction of a nucleon (N), whether bound or unbound, with the field of a nucleus is intrinsically relativistic in the sense that transitions with virtual $N\bar{N}$-pair production appear to be important.

Effective potentials with strengths on the order of the nucleon mass, but of opposite sign and nearly compensating average effect, seem to provide a plausible rationale for a dynamics in which the lower, or "negative-energy" components of the nucleon are important. The motion of the nucleon of energy-momentum $(E, \mathbf{p})$ and mass m_N propagating in a "mean-field" potential U_{MF} is then described phenomenologically by a Dirac equation

$$[E\gamma^0 - \boldsymbol{\gamma} \cdot \mathbf{p} - m_N - U_{MF}]\psi(x) = 0, \tag{9.2.65}$$

where $(\gamma^0, \boldsymbol{\gamma})$ are the four-dimensional Dirac matrices. The third and fourth ("lower") components of the Dirac-spinor wave function $\psi(x)$, in a conventional representation for the γ-matrices, are associated with negative-energy states when the nucleon can be regarded as a point particle. Since the nucleon is composite and larger than the range of the interactions arising from the production of virtual $N\bar{N}$ pairs, this association is open to question.

At the present time, it is unclear whether or not the apparent occurrence of virtual antinucleons in intermediate-energy nuclear physics is simply an artifact arising from a too literal interpretation of Eq. (9.2.65). Interpretations that are consistent with the composite structure of nucleons have been proposed, but they appear as yet to be unsubstantiated.[48] Thus far, $N\bar{N}$ states have played a negligible role in constructing models of the NN interaction from low to intermediate energies, but there is some indication that the effects of virtual pairs may be important.[49]

The potential in Eq. (9.2.65) has an approximate form

$$U_{MF} = S + \gamma^0 V, \tag{9.2.66}$$

where the scalar (vector) potential $S(V)$ has no Dirac matrix structure and is strongly attractive (repulsive). The potentials S and V are thought to be associated with the exchange of a hypothetical scalar (σ) meson and the isoscalar vector (ω) meson, respectively. The potential (9.2.66) is replaced in Eq. (9.2.65) by a density-dependent self-energy operator for a nucleon propagating in a nuclear medium. In such a medium, the reduction of the effective mass of the nucleon, compared to its free value, changes the ratio of the lower-components to the upper-components of the nucleon wave function and leads, by a confluence of several effects, to a plausible explanation of nuclear saturation.[50]

Both in the bound and continuum states, there are strong indications that the lower Dirac components are responsible for obtaining hitherto unexplained effects at least in comparison with a conventional conception of the physics when the characteristic Dirac structure is subtracted.[51] It seems that there is little direct evidence, nor any dynamical imperative, however, for the idea that virtual $N\bar{N}$ pairs are associated with the appearance of these lower components. In spite of the considerable controversy about these questions of interpretation,[52] there appears to be a partial consensus that an indirect effect of the lower components is the suppression of spurious short-ranged effects arising from the particle part of the Dirac propagator.[53] Perhaps the same effect could be achieved by alternative means. The possible role of virtual $N\bar{N}$ pairs in higher-order effects, such as in three-nucleon forces, needs to be considered carefully in order to test the consistency of the idea in other contexts.[54]

Some attempts have been made to justify the Dirac approach to the OP on more fundamental grounds using a relativistic version of the Fock-space technique of Section 9.2b.[45] The basic idea is simple: One adapts a Bethe–Salpeter strategy for the collision of a nucleon with, say, a spin-zero nucleus. Irreducibility is defined with respect to the appearance of nucleon–nucleus lines in intermediate states, where the nucleus is regarded as being in its ground state. This procedure is covariant and includes the full effect ofnucleus recoil, an effect ignored in early phenomenological applications of Eq. (9.2.65). Then, one obtains a two-body integral equation for the covariant transition amplitude M^{ij} of the generic form

$$M^{ij} = V^{ij} + \sum_k V^{ik}G^k M^{kj}, \tag{9.2.67}$$

where $i, j, k = +, -$ refer to the upper $(+)$ and lower $(-)$ components. Here, M^{ij}, the irreducible amplitudes V^{ij}, and the propagators

$$G^k = (i/2\pi)[(\not{W} - \not{p}) - m_N]^{-1}(p^2 - M^2)^{-1}$$

retain a two-by-two Pauli spin structure. Also W and p are the total and intermediate nuclear four-momenta, respectively, and $\not{p} = \gamma^\mu p_\mu$, for example. The amplitudes V^{ij} are the relativistic generalization of the quasi-Hermitian OP considered earlier.

The reduction of the four-dimensional (integration over d^4p) integral equation (9.2.67) to three-dimensional form is achieved in a plausible fashion by supposing that the intermediate ground-state nucleus is kept on its mass shell. This preserves the elastic unitarity characteristics of M^{ij}, and leads to the exact relation

$$M^{ij} = \Gamma^{ij} + \sum_k \Gamma^{ik} g^k M^{kj}, \tag{9.2.68}$$

where g^k is the three-dimensional propagator and, as a consequence of the subtraction, the quantity Γ^{ij}, which satisfies

$$\Gamma^{ij} = V^{ij} + \sum_k V^{ik}(G^k - g^k)\Gamma^{kj}, \tag{9.2.69}$$

can be thought of as a nucleon self-energy. The actual covariant OP that is Hermitian at very low energies is identified[45] with $(\gamma^0\Gamma)^{ij}$. In applications, one takes $\Gamma^{ij} \simeq V^{ij}$, and then retains only some low-order approximation to V^{ij}.

It is commonly expected that the nonrelativistic limit of (9.2.69) will involve only the quantity[45]

$$M^{++} = U^{++} + U^{++} g^+ M^{++}, \tag{9.2.70}$$

where

$$U^{++} \equiv \Gamma^{++} + \Gamma^{+-} g^- [1 - \Gamma^{--} g^-]^{-1} \Gamma^{-+}. \tag{9.2.71}$$

Whether that limit will coincide with some special model for low-energy scattering such as the KMT phenomenology is a matter of conjecture.

The preceding provide what appears to be a correct framework for describing a relativistic nucleon–nucleus OP. The loose ends become apparent when one attempts to interpret the results of a calculation that is made by using some approximation to this very general formalism. Closely related to the fact that the microscopic relativistic field theory underlying this is not carefully specified, there are serious questions concerning the Bethe–Salpeter approach to collision problems involving bound states such as nuclei. For example, the irreducibility classifications with respect to intermediate bound states are likely to be ambiguous. The various Feynman graphs depicting intermediate propogation of nuclei are likely to have only heuristic value, and the associated use of nucleus–nucleon vertex functions may involve subtle double-counting problems.[55] Finally, there is no solution algorithm for calculating the irreducible amplitudes to arbitrary accuracy.

Thus, in the end, the preceding Bethe–Salpeter formalism serves as a framework for constructing models for covariant, quasi-Hermitian OP's for elastic nucleon–nucleus scattering. This is a very useful role and a considerable improvement over the patching up of the nonrelativistic phenomenologies by changing to relativistic kinematics and monitoring subsystem-frame transformations. Moreover, this is all these covariant formalisms were designed to do. Yet, it would be an important advantage to possess a dynamical starting point as definitive as the Faddeev–Yakubovskii equations are for the nonrelativistic problem, irrespective of their intractibility. Then, some of the lower-order approximations could be defined precisely, and questions of principle such as the role of the virtual $N\bar{N}$-pair contributions could be formulated more clearly.

9.3. Multiple-Scattering Algorithms

When a particle or a wave propagates by successive collisions from a number of scattering centers embedded within a relatively inactive surrounding medium, the process is called *multiple-scattering* (MS). A rich variety of phenomena can be interpreted as MS processes. In this section, we discuss a few topics of dynamical interest concerning the MS approach in collision theory. For the sake of simplicity, we suppose that the projectile is "elementary" and that it scatters elastically from the "medium," which is a target nucleus consisting of A identical, constituent nucleons.

Let us first review some approaches to the subject. Methods for dealing with MS quantum mechanically were introduced by Foldy[1] and Watson.[2] Since then an enormous literature has grown up about the subject, particularly in nuclear physics.[3] There are three major MS theories due to Watson,[2,4–6] Kerman, McManus, and Thaler (KMT),[7] and Glauber,[8] respectively. A number of investigations have been carried out which explore the differences and interelationships among these three formalisms.[9]

Of the three, the Glauber approach is distinctive in being phrased as a high-energy approximation. The physics underlying this seems to become operative as the incident projectile enters the diffractive regime at about 1 GeV incident nucleon energy. The relationship of the Glauber approximation to the method of the eikonal[9] gives it a generality that has allowed its successful application far outside the scope of the Watson–KMT theories. This strength, which lies in the nondynamical prescription that results from the Glauber approximation, is also its main drawback. That is, it is not always clear on how to extend this approximation outside the high-energy diffractive regime, and for this reason Glauber theory enters our discussion only incidently as an exemplar of an algorithm for summing the contributions of designated projectile–constituent collisions.

The Watson–KMT theories, on the other hand, directly confront the fact that, unlike the canonical MS picture, the target nucleons interact with the rest of the "medium" in a complicated way in response to a collision with the projectile.[10] Any MS theory that takes the dynamics of the individual "scattering centers" into account involves a convention about the representation of the individual two-particle collisions in terms of auxiliary transition operators. Since the forms of these operators that appear in the Watson–KMT theories involve the full N-particle dynamics, differences between them involve complicated multiparticle effects that are difficult to estimate accurately. Further variations arise when the OP summation is combined with the MS picture, so that scatterings that occur with the target in the ground state must also be identified. Neither MS picture survives the introduction of permutation symmetry between a nucleon projectile and the target nucleons except in the approximation where these effects are included only in quasi-free two-nucleon collisions.

Our objective in this section is to see what we have learned about multiparticle dynamics allows us to say about the MS algorithms.

9.3a. Standard Low-Order Approximations

What matters about conventional MS formalisms is how they are approximated. In this subsection, we compare the differences between the Watson and KMT MS algorithms along with the dynamical rationale underlying their eventual approximation.

The original work by KMT involved the exploitation of two closely related ideas for dealing with MS from a nuclear target. The first was to use the permutation symmetry among the target nucleons to simplify the basic equations in the manner indicated in Section 9.2. The second was to designate the operator $\tau_i(z)$ defined by Eqs. (9.2.15a, b) as the object representing individual projectile/target–nucleon collisions. The Watson formulation can easily accommodate the first idea, but differs from KMT in choosing $\hat{\tau}_i(z)$ (*cf.* Eqs. (9.2.9a, b)) instead of $\tau_i(z)$.

There seems to be little formal or calculational grounds for preferring $\tau_i(z)$.[11] Formally, the KMT organization of the MS sequence does not appear naturally in any other way of dealing with the elastic scattering problem using an OP. Besides the original Watson method, including a C-K version,[12] the alternatives include the Fock-space and the C-K methods of Section 9.2, and a field-theoretic Low-equation formalism.[13]

The practical difference in standard low-order approximations arises from the suggestion of replacing both $\tau_i(z)$ and $\hat{\tau}_i(z)$ by the *same* free-particle transition operator t^i_{free} as advocated by KMT[7] and Watson,[4–6] respectively. This is reflected in the scaling factor $A(A-1)^{-1}$ (*cf.* Eq. (9.2.16) for $A_1 = 1$, $A_2 = A$) that links the two formalisms after target–nucleon permutation symmetry is taken into account. From Eqs. (9.2.5–9.2.18), it follows that if we approximate

$$\tau_i(z) \simeq t^i_{\text{free}}(\hat{z}), \qquad \text{(I.A.)} \tag{9.3.1a}$$

$$\hat{\tau}_i(z) \simeq t^i_{\text{free}}(\hat{z}), \tag{9.3.1b}$$

where $\hat{z}$ is a parametric energy, in general shifted from z, then we obtain

$$T_{\alpha\alpha}(z) \simeq A t^i_{\text{free}}(\hat{z}) \tag{9.3.2}$$

in the single-scattering approximation from *either* the KMT or Watson algorithms. We refer to the replacements (9.3.1a, b) as the *impulse approximation* (I.A.). Historically,[14] the I.A. consisted of both the replacements (9.3.1a, b) and the single-scattering approximation (9.3.2), but the identifications (9.3.1a, b) are in accord with contemporary usage.[3]

In the I.A., the differences between Watson and KMT will show up only in the contributions of higher-order multiple scatterings. Most calculations[15] employ the KMT algorithm and do not separate out the effects of higher-order multiple scatterings for which the KMT scaling factor is operative. However, in the realm where we would expect Eq. (9.3.1a, b) to provide a good single-scattering approximation to U_{F0} or U_{KMT}, given by Eqs. (9.2.10) and (9.2.20), respectively, it is found that even for $A = 3$, these differences are smaller than the other uncertainties in the calculation.[15,16] It would be useful if in the future similar comparisons could be made between the two methods, especially for sensitive spin observables.

However one chooses to identify an elementary projectile-constituent collision, there remains the approximation of reducing the N-particle integral relations satisfied by $\hat{\tau}_i(z)$ and $\tau_i(z)$,

$$\hat{\tau}_i(z) = V_i + V_i G_\alpha(z) Q_\alpha \hat{\tau}_i(z), \tag{9.3.3}$$

$$\tau_i(z) = V_i + V_i G_\alpha(z) R_\alpha \tau_i(z), \tag{9.3.4}$$

respectively, to a managable form. This is achieved by replacing the restricted, multiparticle Green functions, $G_\alpha(z)Q_\alpha$ or $G_\alpha(z)R_\alpha$, by a free Green function defined on the whole space:

$$G^i_{\text{free}}(\hat{z}) \equiv (\hat{z} - H^i_0)^{-1}, \tag{9.3.5}$$

where H^i_0 is a sum of the projectile kinetic-energy operator and some combination, depending on i, of the kinetic-energy operators of the target particles.

The justification of the preceding replacement is no trivial endeavor. This is because it amounts to the replacement of the unsolvable $(A + 1)$-particle integral relations (9.3.3–9.3.4) by a completely defined two-body collision integral equation

$$t^i_{\text{free}}(\hat{z}) = V_i + V_i G^i_{\text{free}}(\hat{z}) t^i_{\text{free}}(\hat{z}). \tag{9.3.6}$$

This illustrates the incomplete dynamical strategies used in all conventional approaches to collision problems. Namely, a object such as $\hat{\tau}_i(z)$ or $\tau_i(z)$, which cannot be calculated using only Eq. (9.3.3) or Eq. (9.3.4), respectively, because of nonuniqueness and noncompactness problems, is rendered calculable by reducing these $(A + 1)$-body integral relations, by approximation, to well-defined two-body integral equations. One can contemplate more sophisticated approximations[17] whereby well-defined auxiliary three-body equations are obtained to calculate $\hat{\tau}_i(z)$ or $\tau_i(z)$, but the rationale is exactly the same.

There have been a host[18] of attempts to justify the approximations (9.3.1a, b), where $t^i_{\text{free}}(\hat{z})$ is defined by Eqs. (9.3.5–9.3.6), but where there is considerable variation in choices of $\hat{z}$ and H^i_0. These arguments have varying degrees of plausibility in trying to quantify the expectation that the target remains bound during the projectile–constituent collision, and that the projectile energy is much higher than the range of target energies that become excited as part of the response of the nucleus to the individual collision that $\hat{\tau}_i(z)$ or $\tau_i(z)$ represents. This is often realized as some form of closure approximation reflecting effective excited-state degeneracy. Recoil is often difficult to take into account correctly by these methods. None of these arguments usually use more than the simplest spectral features of $G_\alpha(z)$; the projections Q_α and R_α are, as a rule, disposed of without too much substantiation.

For a projectile that is distinguishable from the target nucleons, such as a pion, there is only one further approximation of major dynamical significance, namely to what degree one will truncate or sum the MS for the relevant OP. Even in lowest order, however, the evaluation of the matrix elements of the form ${}^{c}\langle \mathbf{p}|t^{i}_{\text{free}}(\hat{z})|\mathbf{p}'\rangle^{c}$, with respect to the cluster states $|\mathbf{p}\rangle^{c}$, is a difficult integration. In fact, this integral is so difficult, not least because of the lack of detailed nuclear wave-function information, that further approximations are made that in effect amount to an alteration of the definition of $t^{i}_{\text{free}}(\hat{z})$. (These "folding" integrals are discussed in Section 9.3b.) At this stage, the remainder of the calculation is nontrivial in scope, but thoroughly routine dynamically because it amounts to the solution of the two-body integral equation (9.2.1) with given input $U(\mathbf{p}, \mathbf{p}'; E)$.[15,19]

If the projectile is a nucleon, then the final calculational prescription usually used in practice is exactly the same as in the pion case, but the entire approximation scenario has to be replayed with the Pauli principle imposed in order to justify the use of antisymmetrized forms for $t^{i}_{\text{free}}(\hat{z})$, a prescription originally due to Takeda and Watson (TW).[20] Like most of the approximations we have discussed that are used in practical MS algorithms, the TW prescription is intuitively appealing at nucleon energies that are relatively high compared to the average binding energy per nucleon.[21]

The TW prescription is realized differently in the KMT and Watson algorithms. It seems to be generally assumed[7] that the TW prescription is directly applicable to the KMT operator $\tau_i(z)$, although the TW analysis did not identify this operator with an individual projectile–constituent collision.

All conventional MS approaches that impose the full Pauli principle in the case of nucleon–nucleus collisions do so in a direct fashion rather than as a rearrangement problem with Pauli-equivalent channels. We wish to emphasize that these direct antisymmetrization procedures are, of course, exact. The problem (*cf.* Section 1.11) is that they do not capitalize on the considerable simplifications that arise from permutation symmetry, nor are they well positioned to make approximations about properly antisymmetrized subsystem dynamics.

It is not entirely disadvantageous to use direct antisymmetrization as the analysis of Picklesimer and Thaler[22] reveals. In their treatment, a cumulant-type reordering of the MS series (the spectator expansion (S.E.)) that groups terms by the number of active target nucleons is employed. That number is correlated with the number of different explicit target–nucleon indices, i, that appear in the operator groupings. This provides a neat identification of what appear to be successively higher-order effects of the Pauli principle. All of the constituent operators in each of the terms of the S.E. involve the propagator

$G_\alpha R_\alpha$, which engages all of the target nucleons dynamically and equivalently. Thus, the S.E. provides an efficient organization of correlative effects among the target nucleons, but only *after* the TW prescription is imposed, *not* before. Therefore, the exactly antisymmetrized spectator expansion, at least in its original form, cannot shed any light on the magnitude of any of the dynamical effects of the Pauli principle that are ignored in the TW prescription.

9.3b. Folding Integrals

All MS algorithms involve the following lowest-order contributions to the OP:

$$u_i(\mathbf{P}_A, \mathbf{k}|\mathbf{P}'_A, \mathbf{k}') = \langle \phi_\alpha(\mathbf{P}_A, \mathbf{k})| t^i_{\text{free}}(\hat{z}) | \phi_\alpha(\mathbf{P}'_A, k')\rangle. \tag{9.3.7}$$

Here the cluster states $|\phi_\alpha(\mathbf{P}_A, \mathbf{k})\rangle$ are products of the projectile and target–nucleus wave functions, and depend upon the projectile ($\mathbf{k}$) and nuclear c.m. ($\mathbf{P}_A$) momenta. We confine ourselves throughout to nonrelativistic kinematics; the essential points that we are concerned with apply to the standard relativistic modelings of MS processes.[23]

Because of the invariance of $t^i_{\text{free}}(\hat{z})$ with respect to spatial translations that maintain the relative positions of the projectile and particle i, the matrix element (9.3.7) reduces to a three-dimensional (folding) integral over the momentum distribution of the struck target nucleon. In this subsection, we consider the structure of this integral and some of the means that have been proposed to evaluate it.

The simplest instance of Eq. (9.3.7) is, of course, for a deuterium target. This case will suffice to illustrate both the structure of the folding integral as well as the strategies that have been proposed for its approximate evaluation. At the present time, this case remains the only situation where the accuracy of these approximations can be fully tested. For the sake of definiteness, we suppose that the projectile is a nucleon.

For nucleon–deuteron scattering we identify $t^i_{\text{free}}(z)$ with the two-nucleon transition operator

$$t_\alpha(z) = V_\alpha + V_\alpha G_\alpha(z) V_\alpha, \tag{9.3.8}$$

where we revert to our usual three-body notation $\alpha = 1, 2, 3$ so that, e.g., V_1 is the potential acting between particles 2 and 3. In what follows, we can regard the nucleons as distinguishable without any loss of generality.

The translational invariance properties of $t_\alpha(z)$ follow from those of the potential, which can be expressed in terms of Jacobi coordinates as

$$\langle \mathbf{k}_1, \mathbf{k}_2, \mathbf{k}_3 | V_\alpha | \mathbf{k}'_1, \mathbf{k}'_2, \mathbf{k}'_3\rangle = \delta^{(3)}(\mathbf{P} - \mathbf{P}')\,\delta^{(3)}(\mathbf{q}_\alpha - \mathbf{q}'_\alpha)\langle \mathbf{p}_\alpha | V_\alpha | \mathbf{p}'_\alpha\rangle. \tag{9.3.9}$$

The plane-wave states $|\mathbf{k}_1, \mathbf{k}_2, \mathbf{k}_3\rangle$ are defined by the wave vectors of the three nucleons, in an arbitrary frame, that we denote as $\mathbf{k}_\lambda$, where $\lambda = 1, 2, 3$ labels the relevant particle. Thus, the total momentum $\mathbf{P}$ is

$$\mathbf{P} = \sum_{\lambda=1}^{3} \mathbf{k}_\lambda, \tag{9.3.10}$$

while the other Jacobi wave vectors are

$$\mathbf{q}_\alpha = \tfrac{2}{3}[\mathbf{k}_\alpha - \tfrac{1}{2}(\mathbf{k}_\beta + \mathbf{k}_\gamma)], \tag{9.3.11}$$

$$\mathbf{p}_\alpha = \tfrac{1}{2}(\mathbf{k}_\beta - \mathbf{k}_\gamma), \tag{9.3.12}$$

where it is understood that α, β, γ are in cyclic order. One finds that

$$\sum_{\lambda=1}^{3} \frac{\mathbf{k}_\lambda^2}{2m} = E_{c.m.} + \omega_\alpha + v_\alpha \tag{9.3.13}$$

for any α, where

$$E_{\text{cm}} \equiv \frac{1}{2(3m)}\mathbf{P}^2, \qquad \omega_\alpha \equiv \frac{3}{4m}\mathbf{q}_\alpha^2, \qquad v_\alpha \equiv \frac{\mathbf{p}_\alpha^2}{m}. \tag{9.3.14}$$

Next we see that

$$\begin{aligned}
&\langle \mathbf{P}, \mathbf{q}_\alpha, \mathbf{p}_\alpha | G_\alpha(z) | \mathbf{P}', \mathbf{q}'_\alpha, \mathbf{p}'_\alpha \rangle \\
&\quad = \delta^{(3)}(\mathbf{P} - \mathbf{P}')\,\delta^{(3)}(\mathbf{q}_\alpha - \mathbf{q}'_\alpha)\langle \mathbf{p}_\alpha | (z - E_{\text{cm}} - \omega_\alpha - \hat{v}_\alpha - V_\alpha)^{-1} | \mathbf{p}'_\alpha \rangle,
\end{aligned} \tag{9.3.15}$$

where $\hat{v}_\alpha$ is equal to v_α with $\mathbf{p}_\alpha$ replaced by its operator counterpart. Therefore, using Eq. (9.3.9) we see that

$$\langle \mathbf{k}_1, \mathbf{k}_2, \mathbf{k}_3 | t_\alpha(z) | \mathbf{k}'_1, \mathbf{k}'_2, \mathbf{k}'_3 \rangle = \delta^{(3)}(\mathbf{P} - \mathbf{P}')\,\delta^{(3)}(\mathbf{q}_\alpha - \mathbf{q}'_\alpha)\langle \mathbf{p}_\alpha | \hat{t}_\alpha(\hat{z}) | \mathbf{p}'_\alpha \rangle, \tag{9.3.16}$$

with $\hat{t}_\alpha(z)$ defined on the relative two-particle space as

$$\hat{t}_\alpha(\hat{z}) = \hat{V}_\alpha + \hat{V}_\alpha \hat{G}_\alpha(\hat{z}) \hat{V}_\alpha, \tag{9.3.17}$$

and

$$\hat{z} = z - E_{\text{c.m.}} - \omega_\alpha. \tag{9.3.18}$$

The restriction of V_α to the relative two-particle space is called $\hat{V}_\alpha$.

The cluster states can be labeled in terms of the total energy E, $\mathbf{P}$, and $\mathbf{q}_\alpha$, since

$$\mathbf{q}_\alpha = \mathbf{k}_\alpha - \tfrac{1}{3}\mathbf{P}. \tag{9.3.19}$$

So we have

$$H_\alpha|\phi_\alpha(E,\mathbf{P},\mathbf{q}_\alpha)\rangle = E|\phi_\alpha(E,\mathbf{P},\mathbf{q}_\alpha)\rangle, \tag{9.3.20a}$$

$$\mathrm{E} = \mathrm{E}_{\mathrm{c.m.}} + \varepsilon_\alpha + \omega_\alpha, \tag{9.3.20b}$$

$$|\phi_\alpha(E,\mathbf{P},\mathbf{q}_\alpha)\rangle = |\mathbf{P}\rangle|\mathbf{q}_\alpha\rangle|\hat{\phi}_\alpha\rangle, \tag{9.3.20c}$$

$$(\hat{v}_\alpha + \hat{V}_\alpha)|\hat{\phi}_\alpha\rangle = \varepsilon_\alpha|\hat{\phi}_\alpha\rangle, \tag{9.3.20d}$$

where ε_α is the deuteron ground-state energy. Evidently,

$$\langle\mathbf{P}'_\alpha,\mathbf{q}'_\alpha,\mathbf{p}'_\alpha|\phi_\alpha(E,\mathbf{P},\mathbf{q}_\alpha)\rangle = \delta^{(3)}(\mathbf{P}'-\mathbf{P})\,\delta^{(3)}(\mathbf{q}'_\alpha-\mathbf{q}_\alpha)\,\phi_\alpha(\mathbf{p}_\alpha), \tag{9.3.21a}$$

with

$$\phi_\alpha(\mathbf{p}_\alpha) \equiv \langle\mathbf{p}_\alpha|\hat{\phi}_\alpha\rangle \tag{9.3.21b}$$

denoting the deuteron wave function in momentum space.

We next return to the calculation of the matrix element (9.3.7), which in the present case amounts to $\langle\phi_\alpha(E,\mathbf{P},\mathbf{q}_\alpha)|t_\beta(\tilde{E})|\phi_\alpha(E',\mathbf{P}',\mathbf{q}'_\alpha)\rangle$, $\beta\neq\alpha$, and $\tilde{E}$ is a fixed total parametric energy. If we work in the total c.m. frame and omit the Dirac δ-function in the total momentum, this matrix element of $t_\beta(\tilde{E})$ reduces to

$$u_\beta(\mathbf{q}_\alpha,\mathbf{q}'_\alpha;\tilde{E}) \equiv \int d^3P_\alpha\,\phi^*_\alpha(\mathbf{P}_\alpha+\tfrac{1}{2}\boldsymbol{\Delta})\phi_\alpha(\mathbf{P}_\alpha) \tag{9.3.22}$$

$$\times\,\langle\tfrac{3}{4}\mathbf{q}_\alpha+\tfrac{1}{4}\boldsymbol{\Delta}+\tfrac{1}{2}\mathbf{P}_\alpha|\hat{t}_\beta(\hat{E})|\tfrac{3}{4}\mathbf{q}'_\alpha+\tfrac{1}{2}\mathbf{P}_\alpha\rangle,$$

where $\boldsymbol{\Delta}\equiv\mathbf{q}_\alpha-\mathbf{q}'_\alpha$, $\mathbf{P}_\alpha\equiv\mathbf{q}_\beta+\tfrac{1}{2}\mathbf{q}'_\alpha$, and

$$\hat{E} = \tilde{E} - \tfrac{3}{4}(\mathbf{P}_\alpha-\tfrac{1}{2}\mathbf{q}'_\alpha)^2. \tag{9.3.23}$$

We keep in mind that in general $|\mathbf{q}_\alpha|\neq|\mathbf{q}'_\alpha|$ when u_β is used as part of OP.

The three-dimensional integral (9.3.22) "folds" the triply off-shell two-particle transition amplitude with respect to the initial and final momentum amplitudes of the struck particle. Despite the fact that only two-nucleon functions enter into this integral, it has not been evaluated except at relatively low energies as part of exact three-nucleon calculations, or for on-shell scattering with special two-nucleon amplitudes.[24–26] At the projectile energies one would expect the single-scattering IA to be an adequate representation of the OP, the matrix elements of $\hat{t}_\beta$ that enter into the integral (9.3.22) are relatively complicated; nevertheless, exact calculation is certainly feasible at the present time.[27]

Historically, the complexity of the folding integrals, even for deuteron targets, along with incomplete information about the nuclear wave functions (for $A > 2$) and the off-shell transition amplitudes necessitated approximate evaluation. The dynamical basis for these approximations is only partially understood, but their accuracy is expected to be highest when the projectile energy is large compared to the average nucleon binding energy. The net effect of the approximations is to replace the matrix element of $\hat{t}_\beta(\hat{E})$ in Eq. (9.3.22) by a $\mathbf{P}_\alpha$-independent function $t_\beta(\mathbf{q}_\alpha, \mathbf{q}'_\alpha; \tilde{E})$. Then,

$$u_\beta(\mathbf{q}_\alpha, \mathbf{q}'_\alpha; \tilde{E}) \simeq t_\beta(\mathbf{q}_\alpha, \mathbf{q}'_\alpha; \tilde{E}) F(\mathbf{\Delta}), \tag{9.3.24}$$

where

$$F(\mathbf{\Delta}) \equiv \int d^3P_\alpha\, \phi^*_\alpha(\mathbf{P}_\alpha + \tfrac{1}{2}\mathbf{\Delta})\phi_\alpha(\mathbf{P}_\alpha) \tag{9.3.25}$$

is referred to as the target form-factor.

The phenomenological advantages of a prescription such as (9.3.24) are apparent if $t_\beta(\mathbf{q}_\alpha, \mathbf{q}'_\alpha; \tilde{E})$ can be identified with an on-shell NN amplitude. For then, the OP can be determined from phenomenological parametrizations of the NN data, and the results of electron-scattering and other experiments which determine $F(\mathbf{\Delta})$. This type of approximation was originally proposed by Chew,[14] and has been extensively used in the analysis of hadron–nucleus scattering in the energy range from about 50 MeV to 800 MeV. We point out that the form (9.3.24–9.3.25) is obtained with an arbitrary nuclear target.

The original forms[7,14] of the factorized approximation (9.3.24) used a particularly simple on-shell NN amplitude for $t_\beta(\mathbf{q}_\alpha, \mathbf{q}'_\alpha; \tilde{E})$ that is kinematically undefined for a large part of the scattering hemisphere (in the c.m. frame). Thus, it was not possible to execute the OP solution strategy directly in momentum space.[7] Rather, equivalent coordinate-space forms were inferred, and these were used to solve the two-body elastic projectile–nucleus scattering problem. These procedures are of little dynamical interest at present.

It was appreciated early on how to achieve a form like Eq. (9.3.24) that was kinematically consistent. The essential idea is to expand the matrix element of $\hat{t}_\beta$ occurring in the integral (9.3.22) in powers of $\mathbf{P}_\alpha$ about a point, $\langle \mathbf{P}_\alpha \rangle$, such that the integral of the first-order term in the Taylor series vanishes.[28] The rationalization for neglecting the contributions of the higher-order terms is based on the fact that $\phi(\mathbf{p})$ is peaked about $\mathbf{p} = 0$; thus, it is reasonable to expect that the maximum contribution to the integral (9.3.22) will arise from the midpoint between the peaks of $\phi_\alpha(\mathbf{P}_\alpha + \frac{1}{2}\mathbf{\Delta})$ and $\phi_\alpha(\mathbf{P}_\alpha)$, namely, $\mathbf{P}_\alpha = -\frac{1}{4}\Delta \equiv \langle \mathbf{P}_\alpha \rangle$.

A variety of names have been proposed[28–31] for this procedure, some of which convey a degree of optimism concerning its success; perhaps the most descriptive of these is the *linear approximation* (LA)[28,32] to the folding integral.

In its early applications,[28] the LA was used in conjunction with the identification of the IA to the OP with the full transition matrix, so $|\mathbf{q}_\alpha| = |\mathbf{q}'_\alpha|$. The initial and final t-matrix energies in Eq. (9.3.22) are then identical if $\mathbf{P}_\alpha = \langle \mathbf{P}_\alpha \rangle$. One then obtains a fully on-shell amplitude $t_\beta(\mathbf{q}_\alpha, \mathbf{q}'_\alpha; \tilde{E})$ in Eq. (9.3.24) provided the energy shift implicit in $\tilde{E}$ is chosen to guarantee this. Generally, however, one needs $u_\beta(\mathbf{q}_\alpha, \mathbf{q}'_\alpha; \tilde{E})$ off-shell to affect the OP program in momentum space as achieved in the pioneering calculations of Landau *et al.*[29] for light nuclei. Similar calculations have also been carried out for heavier nuclei and with relativistic kinematics.[23,33]

A number of calculations have been performed that indicate that even kinematically consistent versions of the factorized approximation (9.3.24) are not always quantitatively adequate.[24–27,34] This seems to be particularly true for large momentum transfers and for spin observables. In making assessments of the factorized (or any other) approximation to $u_\beta(\mathbf{q}_\alpha, \mathbf{q}'_\alpha; \tilde{E})$, the principal question, whose answer can be interpreted unequivocally, is how well the integral (9.3.22) is *mathematically* approximated once one has fixed the choice of $\hat{t}_\beta(\hat{E})$ including the implicit energy shift, etc. Comparisons of the factorized with the exact forms of $u_\beta(\mathbf{q}_\alpha, \mathbf{q}'_\alpha; \tilde{E})$ based on their consequences for the scattering observables can be misleading because each choice then represents different physics.[25]

It seems clear that there is considerable need for detailed evaluations of the folding integrals in which various options, such as choices of energy shifts in particular, are explored. Because of the availability of target-wave-function information, it would appear that the most conclusive results could be achieved for light nuclei with a nucleon projectile to minimize the uncertainty in off-shell extrapolation. Alternatively, given sufficiently accurate folding-integral evaluation techniques, one could finally begin to exploit nucleon–nucleus scattering as a probe of the behavior of off-shell NN scattering.

9.3c. *Other Approaches*

In this subsection we review attempts to describe MS phenomena from the viewpoint of C-K collision integral equations. Our objective is to indicate what can be learned in this way that is beyond the scope of the conventional approaches. For most of our discussion we consider the elastic scattering of a nucleon from a target nucleus with $A > 2$.

For $A = 2$, the Faddeev equations have been used to generate MS algorithms for both hadronic and triatomic collisions. We note that for $N = 3$ the Watson equations are of the C-K type but the KMT equations are not; neither formalism has been used much for three-hadron physics.[17] In a number of triatomic scattering problems arising in hot-atom chemistry, reactions of photolysis products, and nozzle-beam scattering, the collision energies are hyperthermal and so these processes can be depicted as a series of impulsive multiple scatterings describable in terms of the first few terms in the Faddeev iteration series.[35]

The conventional MS algorithms for $A > 2$ make essential use of the device of the OP to achieve a partial summation of the MS series. The OP is the central focus of our analysis as well. We will use the formalism-independent OP $\mathscr{V}_{\text{opt}}(\bar{\alpha})$ introduced in Section 9.2. This is the appropriate generalization of the Feshbach potential U_{FO} when the projectile identity is neglected, and thus it can be compared relatively easily to the various approximations of the conventional formalisms.

Other alternatives can be readily considered. For example, one could use any one of the formalism-independent OP's introduced by Goldflam and Kowalski,[36] such as the one corresponding to the post off-shell extension. This last OP formed the basis of the conventional operator treatment of Picklesimer and Thaler.[22] On the other hand, one could use the formalism-dependent OP's such as $\tilde{\mathscr{V}}_{\text{opt}}(\bar{\alpha})$ (*cf.* Eq. (9.2.61)) advocated by Adhikari, Kozak, and Levin,[37] which is not related to any of the customary off-shell extensions. The C-K dynamical equations underlying each of these different possibilities are the same, so that the basic analysis can be applied to each case with little additional effort and similar conclusions. The principal differences will relate to quasi-Hermiticity and formalism dependence.

The basic dynamical issue concerns the approximation of the OP in the MS picture. The OP satisfies a dynamical equation of the form (*cf.* (9.2.26)

$$\mathscr{V}_{\text{opt}}(\bar{\alpha}) = B(\bar{\alpha}) + C(\bar{\alpha})\mathscr{V}_{\text{opt}}(\bar{\alpha}). \tag{9.3.26}$$

As an initial step toward a consistent MS picture of the OP, a class of "nonresonant" approximations is defined by ignoring the fully connected parts of the driving and kernel terms of Eq. (9.3.26).[38]

One would expect that the nonresonant assumption would be valid for moderately high incident nucleon energies, so that the dominant elementary reaction mechanism is an impulsive direct interaction. The net effect of this assumption is an approximation to $\mathscr{V}_{\text{opt}}(\bar{\alpha})$ in terms of subsystem scattering operators:

$$\mathscr{V}_{\text{opt}}(\bar{\alpha}) \simeq \tilde{W}_{\text{MS}}^{\bar{\alpha}\bar{\alpha}} + \tilde{W}(\bar{\alpha}) + V_{\text{ex}}(\bar{\alpha}). \tag{9.3.27}$$

Here, $\tilde{W}_{\text{MS}}^{\bar{\alpha}\bar{\alpha}}$ is the quantity appearing in Eq. (5.4.75); it represents what corresponds to a MS series in a C-K theory. The contributions of the excited states of the target nucleus are contained in $\tilde{W}(\bar{\alpha})$, which includes only terms of connectivity in the two-cluster equivalence class $\bar{\alpha}$. Finally, $V_{ex}(\bar{\alpha})$ is the total contribution of the pickup or heavy-particle exchange terms; its matrix elements with respect to the cluster states $|\phi_{\alpha_0}(\mathbf{q})\rangle$, corresponding to a canonical partition $\alpha_0 \in \bar{\alpha}$ and a relative projectile momentum $\mathbf{q}$, are

$$\langle \phi_{\alpha_0}(\mathbf{q})| V_{ex}(\bar{\alpha}) |\phi_{\alpha_0}(\mathbf{q}')\rangle = -A(A-1)\langle \phi_{\alpha_0}(\mathbf{q})| P_{0j} V_{ij} |\phi_{\alpha_0}(\mathbf{q}')\rangle. \tag{9.3.28}$$

Here, V_{ij} is the pair potential between any two target particles i and j, and P_{0j} is the 0–j exchange operator, where the incident particle is denoted by 0. We have used the antisymmetry of the target wave-function part of $|\phi_{\alpha_0}(\mathbf{q}')\rangle$ in order to obtain the right-hand side of Eq. (9.3.28). The occurrence of $V_{ex}(\bar{\alpha})$ in Eq. (9.3.27) provides an unequivocal answer to the question of how such exchange terms should be handled in an OP formalism.

All of the conventional low-order MS algorithms involve a substantial neglect of the excited states of the target. Thus, we will ignore the effect of $\tilde{W}(\bar{\alpha})$ on the OP. It is possible to construct simple low-order approximations and by means of this calculate corrections to the nonresonant approximation. It is worth remarking that this can be done using solvable integral equations and without the introduction of unphysical projected Hamiltonians as in the Feshbach and related formalisms. All of the conventional formalisms provide no systematic means for dealing with the effects of excited states.

An important feature implicit in the terms $B(\bar{\alpha})$ and $C(\bar{\alpha})$ appearing in the exact dynamical equation (9.3.26), as well as the operators $\tilde{W}_{MS}^{\bar{\alpha}\bar{\alpha}}$ and $\tilde{W}(\bar{\alpha})$ that enter into the approximation (9.3.27), is that the subsystem dynamics takes place only in the properly antisymmetrized subspaces.

We expect to recover the counterparts of the conventional MS prescriptions for the OP within the term $\tilde{W}_{MS}^{\bar{\alpha}\bar{\alpha}}$. Only now we have *all* of the subsystem dynamics properly antisymmetrized, and we are not placed in a position of approximating improperly symmetrized dynamics.

In an impulsive situation for elastic scattering, the parameter of smallness is the average single-particle binding energy compared to the energy of the incident nucleon. A simple hierarchy of impulsive approximations is then suggested via the successive "ionization" of clusters of the target particles, where only these ionized nucleons interact with the projectile. The term "ionization" means that the nucleon is dynamically free of the residual target nucleons, but it does imply that the latter are bound together in all intermediate states. The projectile interaction with the ionized nucleons is made unambiguous if we require that it be described by a subamplitude that is fully connected among this subset of particles. For one ionized target nucleon, we are speaking of a two-particle connected amplitude. If we are considering two ionized target nucleons, the interactions with the projectile are represented by a fully connected three-to-three amplitude, and so on. The dynamics of the un-ionized nucleons are entirely arbitrary. As a consequence, we term this decomposition of $\widetilde{W}_{\mathrm{MS}}^{\bar{\alpha}\bar{\alpha}}$ *the inclusive connectivity expansion* (ICE).

The terms in $\widetilde{W}_{\mathrm{MS}}^{\bar{\alpha}\bar{\alpha}}$ can be systematically reordered in terms of operators that represent the two-, three-,..., particle-connected portions of the ICE. The two-particle-connected portions involve the operators

$$t_{\gamma}^{\alpha,\alpha_0}(z) = V_{\gamma}^{\alpha,\alpha_0} + V_{\gamma}^{\alpha} G_{\gamma}(z) V_{\gamma}^{\alpha_0}, \tag{9.3.29}$$

where, $\alpha, \alpha_0 \in \bar{\alpha}, \alpha_0$ is the canonical partition, and $\gamma = (0, j)\ (2, \ldots, j-1, j+1, \ldots A)$ is a two-cluster partition, where one cluster, $(0, j)$, refers to the (connected) interaction of the projectile (0) with some target particle (j), while the other cluster serves as a reference for all of the dynamical states (bound and continuum) of the residual nucleus. Now,

$$H_{\gamma} = K_0 - K_j - V_{0j} - H_{\gamma(j)}, \tag{9.3.30}$$

where $H_{\gamma(j)}$ is the Hamiltonian of the residual nucleus:

$$H_{\gamma(j)} = K_{\gamma(j)} + H_{\gamma(j)}^{\mathrm{I}}. \tag{9.3.31}$$

Here, $K_{\gamma(j)}$ refers to the kinetic-energy operator of the c.m. motion of the residual nucleus and $H_{\gamma(j)}^{\mathrm{I}}$ is the internal Hamiltonian.

When $t_{\gamma}^{\alpha,\alpha_0}(z)$ appears in $\widetilde{W}_{\mathrm{MS}}^{\bar{\alpha}\bar{\alpha}}$, it operates on the right on R_{α_0}. However, we observe that

$$R_{\alpha_0} = R_{\gamma(j)} R_{\alpha_0}, \tag{9.3.32}$$

where $R_{\gamma(j)}$ is the antisymmetrizer on the subspace of states of the residual nucleus (but not 0 and j). Since $R_{\gamma(j)}$ commutes with both $H_{\gamma(j)}$ and V_{0j}, we can absorb $R_{\gamma(0)}$ into $G_{\gamma}(z)$ in (9.3.29), which therefore will involve only physical residual-nucleus states.

A simple approximation to $t_{\gamma}^{\alpha,\alpha_0}(z)$ that is consistent with the impulsive nature of the MS scenario is to suppose that closure can be imposed with respect to the residual nucleus. This is equivalent to the condition

$$\varepsilon_n \simeq \varepsilon_0 \tag{9.3.33}$$

for all n, where n labels the energy eigenvalues ε_n of $H_{\gamma(j)}^{\mathrm{I}}$, and $n = 0$ denotes the ground state. Approximation (9.3.33) corresponds to the situation, where in comparison with the large initial projectile energies, there are only relatively small differences among the internal energies of those states that contribute significantly in the impulsive collision.

With the preceding closure approximation, we obtain as our lowest-order single-scattering approximation to the nucleon–nucleus OP:

$$\langle \phi_{\alpha_0}(\mathbf{q}) | [\mathscr{V}_{\mathrm{opt}}(\bar{\alpha})]_1 | \phi_{\alpha_0}(\mathbf{q}') \rangle \equiv A \langle \phi_{\alpha_0}(\mathbf{q}) | \tilde{t} | \phi_{\alpha_0}(\mathbf{q}') \rangle, \tag{9.3.34}$$

where

$$\tilde{t} \equiv (1 - P_{0j}) t_{\mathrm{D}}[\gamma(j)], \tag{9.3.35}$$

$$t_{\mathrm{D}}[\gamma(j)] \equiv V_{0j}\{I + [G_{\gamma(j)}(z)]_{\mathrm{D}} V_{0j}\}, \tag{9.3.36}$$

$$[G_{\gamma(j)}(z)]_{\mathrm{D}} \equiv (z - \varepsilon_0 - K_0 - K_j - V_{0j} - K_{\gamma(j)})^{-1}. \tag{9.3.37}$$

Clearly, $\tilde{t}$ is a properly antisymmetrized two-nucleon transition operator with three-body kinematics, and with a parametric energy shifted by the ground-state energy of the residual nucleus. We note that the recoil of both the struck nucleon and the residual nucleus are treated correctly.

The significant aspect of the preceding approximation scenario is that we recover not only the Takeda–Watson[20] prescription in a natural manner, but we do it with definitive *predictions* for the energy shift and the kinematical structure of the two-nucleon transition matrix. This reminds us of perhaps the most unwieldly feature of conventional MS formalisms, namely the difficulty of separating off in a clean and systematic fashion the struck particles from the target nucleus. The spectator-type expansions[22] attempt to do this but fall short because all of the auxiliary operators still involve the full target Hamiltonian rather than those of successively less complicated residual nuclei. It would appear that only in C-K theories can the rich collection of physical ideas implicit in MS theory be given a precise realization.

Further development of these ideas, including an unambiguous identification of the two-nucleon correlation terms and the effects of target excitations, is contained in Ref. 38.

9.4. Bound-State Approximations

9.4a. General Considerations

A number of widely-used approximations in multiparticle collision theory involve the truncation of the Hilbert space in terms of two-cluster states. The description of the multiparticle collision is then reduced to a multichannel two-body problem. There are basically only a few ways of achieving this simplification, and all of them are broadly classified under the title bound-state approximation (BSA). Different versions of the BSA range from the coupled-reaction-channel (CRC)[1] and the resonating-group method (RGM)[2] models in nuclear physics to the close-coupling (CC) scheme[3] of atomic physics. In all these approaches, the full scattering wave function is expanded in terms of an overcomplete set of two-cluster basis states and then inserted into the Schrödinger equation. This equation is projected onto the basis states, resulting in a set of coupled (integro-)differential equations for the expansion coefficients. This set is then solved, yielding an approximate solution of the multiparticle problem using only two-body dynamical equations.

The CRC, RGM, and CC models rest on the same idea, but their detailed realizations are quite different. In the CRC scheme, one expands the full wave function in terms of various two-body cluster states that correspond to possible rearrangements of the full N-particle system. When the RGM technique is applied in nuclear physics, one imposes, in addition, antisymmetrization among the identical nucleons in the two clusters. With the CC scheme of atomic physics, the expansion of the full wave function involves only the ground and the excited states of the target because there is no possibility of rearrangement. These three methods have been successfully used in different areas of physics as well as chemistry, but the reasons for these successes are not entirely clear. Also, such schemes do not provide us with either a systematic way of making improvements on the initial approximation or an estimate of the error in that approximation. BSA schemes are traditionally stated in terms of a wave-function description of scattering, but they can also be formulated in terms of transition-matrix integral equations, thus incorporating the necessary boundary conditions. The essential idea behind these different types of BSA was given by Wheeler[2] long ago, and they have been used extensively since.

The BSA cannot easily be classified under approximations made to C-K integral equations. However, the BSA has been successfully used recently in conjunction with C-K equations in describing scattering processes in atomic/molecular physics. We present an account of these applications. Before doing this, we develop the general idea behind the BSA and the advantages of different ways of implementing it.

9.4a.1. BSA in the Schrödinger Equation

All of the phenomenologically successful ways of implementing the BSA for two-cluster collisions can be classified under this category. This includes the CRC, CC, and the RGM methods. In all these approaches, a decomposition into two-cluster-labeled components,

$$|\psi\rangle = \sum_{\alpha} |\psi_\alpha\rangle, \tag{9.4.1}$$

is inserted into the Schrödinger equation

$$(E - H)|\psi\rangle = 0, \tag{9.4.2}$$

which then yields

$$(E - H_\alpha)|\psi_\alpha\rangle = -(E - H_\alpha)\sum_\beta \bar{\delta}_{\alpha\beta}|\psi_\beta\rangle + V^\alpha \sum_\beta |\psi_\beta\rangle. \tag{9.4.3}$$

The sums in Eqs. (9.4.1) and (9.4.3) extend over the ground and excited states in each two-cluster channel, and also over various two-cluster channels. The indices α, β label not only the two-cluster channels (as partitions) but also the excited states. In the CC approach[3] to electron–atom scattering, the components $|\psi_\alpha\rangle$ correspond to the excited states of the target atom; in the CRC description[1] of nuclear reactions, they correspond to different states, both ground and excited, of the possible two-cluster rearrangement channels. Finally, in the RGM method,[2] the two-cluster states of the CRC approach are antisymmetrized.

The crucial assumption is that the components $|\psi_\alpha\rangle$ have the form

$$|\psi_\alpha\rangle = P_\alpha|\hat{\psi}_\alpha\rangle, \tag{9.4.4}$$

where P_α is the projector onto the two-cluster state. When the ansatz (9.4.4) is substituted into the Eqs. (9.4.3), one obtains a set of coupled equations for the component states $P_\alpha|\hat{\psi}_\alpha\rangle$:

$$(E - P_\alpha H_\alpha P_\alpha)P_\alpha|\hat{\psi}_\alpha\rangle = \sum_\beta P_\alpha[V^\alpha - (E - H_\alpha)\bar{\delta}_{\alpha\beta}]P_\beta|\hat{\psi}_\beta\rangle. \tag{9.4.5}$$

There are two problems with the set (9.4.5). First, the expansion

$$|\psi\rangle = \sum_\alpha P_\alpha|\hat{\psi}_\alpha\rangle \tag{9.4.6}$$

that is implied by Eqs. (9.4.1) and (9.4.4) typically introduces a overcomplete and nonorthogonal set of basis states. Second, and closely related to this, the corrections to the approximation (9.4.6) are left undefined.

9.4a.2. BSA in Scattering Integral Equations for the Wave Function

The BSA can also be made in the context of scattering integral equations both of the LS type and of the C-K type as noted in Chs. 1, 4 and 5. We know that a single multiparticle LS equation has a nonunique solution, but the effect of the BSA is to truncate the full Hilbert space in such a way that the nonuniqueness problem is eliminated. Let us see how this is done.

The full wave function $|\psi_\alpha^{(+)}\rangle$ for describing scattering initiated in the two-cluster channel α satisfies

$$|\psi_\alpha^{(+)}\rangle = |\phi_\alpha\rangle + G_\alpha^{(+)}(E)V^\alpha|\psi_\alpha^{(+)}\rangle. \tag{9.4.7}$$

Now the spectral decomposition for the Green function $G_\alpha^{(+)}(E)$ is given by

$$G_\alpha^{(+)}(E) = \sum \frac{|\phi_\alpha\rangle\langle\phi_\alpha|}{E + B_\alpha - h_{0\alpha}} + \sum \frac{|\phi_\alpha^{(\pm)}\rangle\langle\phi_\alpha^{(\pm)}|}{E - E_\alpha - h_{0\alpha}}, \tag{9.4.8}$$

where the sums are over discrete indices and also represent integrations over continuous variables, as in Eq. (1.9.2). The states $|\phi_\alpha\rangle$ ($|\phi_\alpha^{(\pm)}\rangle$), as usual, represent the product of the bound-states (scattering-states) of each cluster, and $h_{0\alpha}$ is the kinetic energy representing the free relative motion between the c.m. of the clusters. The sum over the first term on the right-hand side of Eq. (9.4.8) includes the various ground and excited states of the clusters in channel α, and $B_\alpha(E_\alpha)$ is the sum of the binding (scattering) energies of each of the clusters. The BSA then consists in the neglect of the last term in Eq. (9.4.8) when $G_\alpha^{(+)}(E)$ is used in Eq. (9.4.7). Consequently, $|\psi_\alpha^{(+)}\rangle$ as defined by Eq. (9.4.7) must now have the form of the expansion (9.4.6), where the sum now includes all of the bound states only in one specific two-cluster channel α. In this special case, Eq. (9.4.7) reduces to a set of coupled two-body integral equations, which is entirely equivalent to the Eqs. (9.4.5). The only difference is that the Eqs. (9.4.5) represent a set of coupled differential equations and additional boundary conditions are needed for its solution, whereas the set of coupled integral equations that we obtain from Eq. (9.4.7) involves only two-cluster Green functions and has a unique solution. The expansion states included in Eqs. (9.4.7–9.4.8) are (usually) orthogonal, as they are eigenstates of the same two-cluster Hamiltonian, whereas the expansion states in Eq. (9.4.6) are nonorthogonal, in general. We have, however, no clue as how to include correction terms for the neglected portion of $G_\alpha^{(+)}(E)$ in Eq. (9.4.8). A direct attempt to include any portion of this continuum part in Eq. (9.4.7) gives rise to the nonuniqueness problem that was circumvented through the BSA.

9.4a.3. BSA in Connected-Kernel Formalism for the Wave Function

The same approximation when made in the wave-function versions of C-K integral equations developed in Ch. 5 leads to a substantially different set of equations than those obtained from the Schrödinger or the LS equations.

We demonstrate this by considering the wave-function version of the three-particle Faddeev equations:

$$(E - H_\alpha)|\psi_\alpha^{\mathrm{F}}\rangle = V_\alpha \sum_\beta \bar{\delta}_{\alpha\beta} |\psi_\beta^{\mathrm{F}}\rangle, \tag{9.4.9}$$

where the full wave-function $|\psi\rangle$ is given by

$$|\psi\rangle = \sum_{\alpha=1}^{3} |\psi_\alpha^{\mathrm{F}}\rangle, \tag{9.4.10}$$

and satisfies Eq. (9.4.2). Equations (9.4.9–9.4.10) should be compared with Eqs. (9.4.1) and (9.4.3). If we now use the same BSA prescription that led to the Eqs. (9.4.5), we obtain

$$(E - P_\alpha H_\alpha P_\alpha) P_\alpha |\hat{\psi}_\alpha^{\mathrm{F}}\rangle = \sum_\beta P_\alpha V_\alpha \bar{\delta}_{\alpha\beta} P_\beta |\hat{\psi}_\beta^{\mathrm{F}}\rangle, \tag{9.4.11}$$

where we have made the approximation

$$|\psi_\alpha^{\mathrm{F}}\rangle \simeq P_\alpha |\hat{\psi}_\alpha^{\mathrm{F}}\rangle. \tag{9.4.12}$$

Though Eqs. (9.4.5) and (9.4.11) have the same structure, the potential matrices, appearing on their right-hand sides, are different. In the Eqs. (9.4.5), the potential matrix has nonzero terms in all channels β, whereas in the Eqs. (9.4.11) it has nonzero terms only in channels $\beta \neq \alpha$. Consequently, the potential matrix in the Eqs. (9.4.11) is diminished in some sense; we shall see later that numerically this leads to an unacceptably weak potential matrix. The direct term ($\alpha = \beta$) that is absent in the Eqs. (9.4.11), in accord with the Faddeev structure (9.4.9), seems to be indispensable for a reasonable description of the physical process in the BSA. In the Eqs. (9.4.11), the direct transition only occurs in second order through an iteration of the potential matrix. We have illustrated these points with the Faddeev equations, but they are generally valid for all types of C-K equations.

We note that the Eqs. (9.4.11) can also be derived by making the BSA

$$G_\alpha^{(+)}(E) \simeq \sum \frac{|\phi_\alpha\rangle\langle\phi_\alpha|}{E + B_\alpha - h_{0\alpha}}, \tag{9.4.13}$$

for the Green function (9.4.8) in the integral form of the C-K Faddeev equations:

$$|\psi_\alpha^{\mathrm{F}}\rangle^\gamma = \delta_{\alpha\gamma}|\phi_\alpha\rangle + G_\alpha^{(+)}(E) V_\alpha \sum_\beta \bar{\delta}_{\alpha\beta} |\psi_\beta^{\mathrm{F}}\rangle^\gamma, \tag{9.4.14}$$

which is Eq. (5.2.37) for $N = 3$ with the potential matrix (5.3.10). The set (9.4.14), with Eq. (9.4.13), has connected kernels and possesses a unique solution, in contrast to Eq. (9.4.7). It can be solved, and the contribution due to the neglected part of the Green function can be determined.

9.4b. Numerical Test of the BSA

The CRC and RGM methods in nuclear physics, and the CC method in atomic and molecular physics are very useful for describing two-cluster atomic, nuclear, and molecular collisions.[1–3] The usual derivation of these methods employs an expansion of the wave function in a nonorthogonal and overcomplete basis. This leads to serious conceptual and numerical problems and provides no means of calculating correction terms.

On the other hand, the BSA, as applied to certain types of C-K equations, has been tested by various workers in solvable three-body models. The results of these studies indicate that the BSA, when implemented by the truncation of the channel Green functions occurring in these equations, leads to a poor description of the actual state of affairs. This conclusion was reached by Eyre and Osborn[4] and by Adhikari[5] using the Faddeev equations for three identical particles. Lewanski and Tobocman[6] also reached the same conclusion with the Kouri–Levin version of the channel-coupling array (CCA) equations. Glöckle and Ueta[7] tested the BSA with a C-K version of the three-body LS equations. They found that the continuum spectrum of the channel Green function $G_\alpha^{(+)}(E)$ was essential for a reasonable description of the three-body bound state. Greben and Levin[8] studied stripping reactions in the three-body Mitra model[8] and found that the BSA to the Green function has limited success.

When the BSA is used in the *homogeneous* many-body LS equation,

$$|\psi_\alpha\rangle = G_\beta^{(+)} V^\beta |\psi_\alpha\rangle, \tag{9.4.15}$$

it leads to a complete suppression of the incoming wave. This follows because $|\psi_\alpha\rangle$ contains the incoming wave $|\phi_\alpha\rangle$, and this must be generated via the continuous spectrum of $G_\beta^{(+)}$. In the BSA, this part of $G_\beta^{(+)}$ is eliminated and, therefore, so is the incoming wave $|\phi_\alpha\rangle$. The application of the BSA using both the homogeneous and the inhomogeneous LS equations, leads to the suppression or the elimination of flux for certain rearrangement processes; in general, it expected to be a poor approximation.[9]

At this point it is interesting to refer to the extensive comparative numerical, three-body-model study by Kuruoglu and Levin[10] of the BSA applied to the Schrödinger equation (CRC method), and to four different versions of C-K equations. They also calculated successive improvements of these approximation schemes by discretizing the continuum of the channel Green functions, and found that the CRC ansatz responds better than any of the C-K collision theories. Adhikari, Kuruoglu, and Levin[10] gave plausible arguments for this behavior. They argue that the non-Hermitian nature of the model Hamiltonian in the case of Eq. (9.4.11) and, consequently, the absence of the direct $\alpha \rightarrow \alpha$ channel transition in lowest-order, leads to physically unacceptable results. The direct $\alpha \rightarrow \alpha$ channel transition is allowed only as a higher-order process via coupling to other channels. In some of the C-K cases, e.g., the CCA formalism, some of the rearrangement transitions, in addition to the direct transition, also occur as higher-order processes. So, after the BSA is made in a C-K formalism, the flux in the asymptotic region may be suppressed, leading to a poor description of the actual physical process. It is now generally accepted that when the BSA is applied to the Schrödinger equation it leads to results superior to those obtained when it is applied to C-K equations.

A way to circumvent the flux-suppression problem is suggested by the investigations of Greben and Levin[8] and Tobocman[11]. Greben and Levin show that the BSA to C-K equations yields good results if one uses a model containing absorptive interactions. The effect of the absorptive parts of the interactions in the Greben–Levin study of stripping reaction in the Mitra model is to mock up the continuum contributions of the channel Green functions in the C-K equations that the BSA eliminates.

In this context, it is also appropriate to mention the interesting study by Tobocman[11] of the three-identical-boson system interacting via pair two-body potentials. He confirms that the BSA when applied to the C-K, CCA equations leads to poor results. He also considered the BSA in distorted-wave C-K, CCA equations and found that this approximation is good compared to the exact numerical calculation. From this, one again concludes that the distortion potential simulates the continuum contributions of the channel Green functions. Finally, Adhikari[5] observed in his numerical analysis that when the two-body binding is artificially increased by increasing the strength of the interaction, the BSA becomes a good approximation, in the context of the Faddeev equations, for studying low-energy elastic scattering in two-fragment channels. From this, one concludes that, for energies much below the breakup threshold, the BSA applied to C-K equations leads to a good description of the underlying physics.

From the preceding discussion, one can define a "safe" domain of applicability of the BSA to C-K integral equations. First, distortion potentials, if properly chosen, may simulate the effect of the continuum. (Unfortunately, these distortion potentials are chosen in an *ad hoc* way.) Second, the BSA works better at very low energies, far away from the breakup threshold.

The second condition is easily realized in atomic systems, where there are a very large number of excited states in the two-cluster channels and the breakup (ionization) threshold is usually well above the lowest scattering threshold. This is usually not the case for nuclear systems, where there is a very small number of excited states of each of the two clusters and the lowest scattering and the breakup thresholds are very close. However, it is possible to introduce distortion potentials in both cases. It is for these two reasons that the BSA has been applied successfully to C-K equations of the general CCA type for problems in atomic and molecular physics. We next present an account of these calculations.

9.4c. *BSA to Channel-Coupling Array Equations*

The CCA equations for transition operators were formulated by many workers in the early seventies to deal with reactions involving different channels. The foundations[12–15] of this approach were established by Baer, Kouri, Levin, and Tobocman; new versions of this approach have been suggested recently.[16–18] Consistent with our philosophy (Section 1.6), we only present a coupled set of equations for the transition operators, which automatically incorporate the necessary boundary conditions. Similar equations can also be written for the wave functions, which are usually solved in coordinate space after transforming them to coupled differential equations.

Typically, one uses the post and prior forms of the transition operators

$$T_{\alpha\lambda}^{(+)} = V^{\alpha} + V^{\alpha} G^{(+)} V^{\lambda}, \tag{9.4.16}$$

and

$$T_{\alpha\lambda}^{(-)} = V^{\lambda} + V^{\alpha} G^{(+)} V^{\lambda}, \tag{9.4.17}$$

respectively. Following Baer[19], with the partition-degenerate equation

$$G^{(+)} = G_{\beta}^{(+)} + G_{\beta}^{(+)} V^{\beta} G^{(+)}, \tag{9.4.18}$$

and a channel-coupling array W satisfying

$$\sum_{\beta} W_{\bar{\alpha}\beta} = 1, \qquad \text{for any } \bar{\alpha}, \tag{9.4.19}$$

we obtain

$$G^{(+)} = \sum_{\beta} W_{\bar{\alpha}\beta} G_{\beta}^{(+)} + \sum_{\beta} W_{\bar{\alpha}\beta} G_{\beta}^{(+)} V^{\beta} G^{(+)}. \tag{9.4.20}$$

Substituting Eq. (9.4.20) in Eq. (9.4.16), we find,[19] after some straightforward manipulation,

$$T_{\alpha\lambda}^{(+)} = V^{\alpha}\left\{1 + \sum_{\beta} W_{\bar{\alpha}\beta} G_{\beta}^{(+)}(V^{\lambda} - V^{\beta})\right\} + \sum_{\beta} V^{\alpha} W_{\bar{\alpha}\beta} G_{\beta}^{(+)} T_{\beta\lambda}^{(+)}. \quad (9.4.21)$$

With the use of the Lippmann identity, we can reduce this equation to half-on-shell form when operating on the channel state $|\phi_{\lambda}\rangle$:

$$T_{\alpha\lambda}^{(+)} = V^{\alpha} W_{\bar{\alpha}\lambda} + \sum_{\beta} V^{\alpha} W_{\bar{\alpha}\beta} G_{\beta}^{(+)} T_{\beta\lambda}^{(+)}. \quad (9.4.22)$$

Similarly, except that the Lippmann identity is not required (*cf.* Eq. (5.2.27d)), we find that Eqs. (9.4.17) and (9.4.20) lead to

$$T_{\alpha\lambda}^{(-)} = V^{\lambda} + \sum_{\beta} V^{\alpha} W_{\bar{\alpha}\beta} G_{\beta}^{(+)} T_{\beta\lambda}^{(-)}. \quad (9.4.23)$$

Equations (9.4.22–9.4.23) are the channel-coupling array equations. The Eqs. (9.4.22) have the structure of a set of LS equations, whereas the Eqs. (9.4.23) do not have this form.

Now neither the index $\bar{\alpha}$ nor the matrix elements of W have been specified yet. Usually, $\bar{\alpha}$ is chosen to be equal to α; other choices are possible. The LS structure of the Eqs. (9.4.22) is explicit if this equation is written in matrix form

$$\underline{\tilde{T}}^{(+)} = \mathscr{V} + \mathscr{V}\underline{G}^{(+)}\underline{\tilde{T}}^{(+)}, \quad (9.4.24)$$

where half-on-shell

$$(\underline{\tilde{T}}^{(+)})_{\alpha\beta} = T_{\alpha\beta}^{(+)},$$

and

$$(\mathscr{V})^{\alpha\beta} = V^{\alpha} W_{\bar{\alpha}\beta} \equiv \mathscr{V}^{\alpha\beta}, \quad (9.4.25)$$

$$(\underline{G}^{(+)})_{\alpha\beta} = \delta_{\alpha\beta}\, G_{\alpha}^{(+)}.$$

An important advantage of the LS form (9.4.24) is that a reactance operator $\underline{\hat{K}}$ can be introduced by[13,20]

$$\underline{\hat{K}} = \mathscr{V} + \mathscr{V}\underline{G}^{\mathrm{P}}\underline{\hat{K}}, \quad (9.4.26)$$

where P denotes the principal-value prescription. The $\underline{T}$-matrix is determined by means of the Heitler equation

$$\underline{\tilde{T}}^{(+)} = \underline{\hat{K}} + \underline{\hat{K}}\underline{G}^{\delta}\underline{\tilde{T}}^{(+)}, \quad (9.4.27a)$$

with

$$(\underline{G}^{\delta})_{\alpha\beta} = -i\pi\, \delta_{\alpha\beta}\, \delta(E - H_{\alpha}). \quad (9.4.27b)$$

This means that only real algorithms are needed to solve the integral equation (9.4.24), and because of this only Eq. (9.4.26) is used in recent numerical applications of the BSA. Note that the BSA implies that $\underline{G}^{\delta}$ in Eq. (9.4.27a) is appropriately truncated as well.

Another interesting set of coupled integral equations has been introduced recently by Zhang *et al.*[16] This set is known as the Fock coupling scheme; it is equivalent to the method used in the multiconfiguration Hartree–Fock method for electron–atom scattering and atom–molecule reactions. Summing up both sides of Eq. (9.4.22) over α we obtain[19]

$$\sum_{\alpha} T_{\alpha\lambda}^{(+)} = \sum_{\alpha} V^{\alpha} W_{\bar{\alpha}\lambda} + \sum_{\beta} \left(\sum_{\alpha} V^{\alpha} W_{\bar{\alpha}\beta} \right) G_{\beta}^{(+)} T_{\beta\lambda}^{(+)}, \qquad (9.4.28)$$

where $\bar{\alpha}$ and W are still unspecified. Generally, $\bar{\alpha}$ may be α or some function of α. The matrix W is chosen such that

$$\sum_{\alpha} V^{\alpha} W_{\bar{\alpha}\beta} = V^{\beta},$$

so Eq. (9.4.28) becomes

$$T_{\alpha\lambda}^{(+)} = V^{\lambda} + \sum_{\beta} V^{\beta} G_{\beta}^{(+)} T_{\beta\lambda}^{(+)} - \sum_{\beta \neq \alpha} T_{\beta\lambda}^{(+)}. \qquad (9.4.29)$$

Equation (9.4.29) is the basic equation employed in the Fock coupling scheme.

In the general N-channel case, the choice given by Tobocman[14] for the matrix W appearing in Eqs. (9.4.22–9.4.23) is

$$W_{\alpha\beta} = \begin{cases} 1, & \beta = \alpha + 1, \alpha = 1, \ldots, N-1, \\ 1, & \beta = 1, \alpha = N, \\ 0, & \text{otherwise}, \end{cases} \qquad (9.4.30)$$

which leads to a connected kernel in these equations after $2^{N-1} - 1$ iterations. In the three- and two-channel cases, $\underline{W}$ as given by Eq. (9.4.30) is

$$\underline{W} = \begin{pmatrix} 0 & 1 & 0 \\ 0 & 0 & 1 \\ 1 & 0 & 0 \end{pmatrix} \quad \text{or} \quad \underline{W} = \begin{pmatrix} 0 & 0 & 1 \\ 1 & 0 & 0 \\ 0 & 1 & 0 \end{pmatrix}, \qquad (9.4.31)$$

and

$$\underline{W} = \begin{pmatrix} 0 & 1 \\ 1 & 0 \end{pmatrix}, \qquad (9.4.32)$$

respectively.

The first applications of the preceding versions of the BSA were to the two-channel problem in electron–atom scattering by Kouri and coworkers.[12,15]

It was found that, in the two-channel problem, when the BSA was applied to Eqs. (9.4.22–9.4.23) with choice (9.4.32) for the W-matrix, it led to good numerical results. The reason for this success is not understood. It was later found that the same approximation when applied to the three-channel case with choice (9.4.31) for the W-matrix gives poor numerical results;[16,19] the reason for this is conjectured to be the asymmetry of this W-matrix, which leads to weak coupling between the channels after the BSA has been introduced.[16–19] Also it has been noted that employing distorted-wave Green functions produces vastly improved results once the BSA has been introduced.[11,16,19]

Neuhauser and Baer[18] have introduced a W-matrix that leads to a symmetric generalized potential matrix $\mathscr{V}$ in Eqs. (9.4.24–9.4.27). They require the W-matrix elements in these equations to obey the conditions

$$\begin{aligned} \mathscr{V}^{\alpha\beta} &= \mathscr{V}^{\beta\alpha}, \\ V^{\alpha}W_{\alpha\beta} &= V^{\beta}W_{\beta\alpha}, \end{aligned} \tag{9.4.33}$$

where they take $\bar{\alpha} = \alpha$ in Eq. (9.4.25). Equations (9.4.25) and (9.4.33) do not determine the W-matrix uniquely. In the two- (three-)channel case, we have three (six) equations to constrain the four (nine) elements of the W-matrix.

In the three-channel case, the diagonal elements of the W-matrix are chosen to be $W_{\alpha\alpha} = 1 - \Gamma$. We then find from Eqs. (9.4.19) and (9.4.33) that

$$\mathscr{V}^{\alpha\beta} = \tfrac{1}{2}\Gamma(V^{\alpha} + V^{\beta} - V^{\gamma}), \tag{9.4.34}$$

$\alpha \neq \beta \neq \gamma$. Similarly, in the two-channel case one obtains

$$\begin{aligned} \mathscr{V}^{\alpha\alpha} &= V^{\alpha} - \Gamma V^{\alpha}V^{\beta}, \\ \mathscr{V}^{\beta\beta} &= V^{\beta} - \Gamma V^{\alpha}V^{\beta}, \\ \mathscr{V}^{\alpha\beta} &= \mathscr{V}^{\beta\alpha} = \Gamma V^{\alpha}V^{\beta}. \end{aligned} \tag{9.4.35}$$

In Eqs. (9.4.34–9.4.35), Γ is undetermined.

Baer and coworkers have applied the BSA to Eq. (9.4.24) with the potential matrix given by Eqs. (9.4.34–9.4.35) for three- and two-channel problems, respectively.[17–19] They studied atom–diatom collisions in one dimension with two channels, and in three dimensions with three channels. It was found that only with an appropriate Γ and the use of distorted waves does the BSA lead to accurate results. The results, however, are Γ-dependent. Even though some values of Γ give good results, it is not known how to choose either Γ or the distortion potential in a given problem.

In another study, Baer and Shima[17] considered W-matrix elements that are functions of the internal orientation of the three bodies in a three-dimensional, three-cluster channel. This gives rise to angular-dependent W-matrix elements. It has been shown that the resulting equation is connected after the BSA has been introduced. This choice, however, does not lead to a completely symmetric potential operator $\mathscr{V}$ as in Eqs. (9.4.34–9.4.35) but, nevertheless, leads to a more symmetric $\mathscr{V}$ than that given by Eqs. (9.4.31–9.4.32).

9.4d. *Applications of the BSA*

The first applications of the BSA to the CCA scheme was carried out for two-channel systems of the kind

$$A + BC \rightarrow AB + C,$$

where B was assumed to be infinitely heavy, which led to some numerical simplifications. These calculations used Eq. (9.4.23), and determined the transition probabilities for the collinear wave-guide model in good agreement with other methods. These equations were also successfully applied to low-energy e–H scattering. Later on, use was made of Eq. (9.4.22); because of its LS form and the associated calculational advantages mentioned before, only Eq. (9.4.22) has been used since. Early applications of Eq. (9.4.22) were limited to the collinear wave-guide model and e–H scattering.[15]

More recently, Eq. (9.4.22) together with the BSA has been applied to a number of atom–diatom systems with smooth interatomic potentials, and all with three atoms having finite masses. For the (H + FH), ($H + H_2$), ($D + H_2$), and ($F + H_2$) systems, the results are in good agreement with those obtained by other methods.[21]

Some tests have also been made with the completely symmetric choice for $\mathscr{V}$. However, there are still problems with it—it is not clear how one should choose the parameter Γ.[19]

The most successful and realistic application of the BSA using distorted Green functions is for the three-dimensional-model calculation of transition probabilities for the processes[19]

$$H + H_2(n_i, j_i|J = 0) \rightarrow H_2(n_f, j_f|J = 0) + H,$$

where n_i, j_i are the initial vibrational and rotational states and n_f, j_f are the final ones, and J is the total angular momentum. Angle-dependent W-matrix elements were used. Though there are some undetermined parameters in the approach, all choices of them led to accurate results. However, the rate of convergence was different with the different choices.

Good results were obtained after making the BSA using the more symmetric choices of the *W*-matrix. But, since the three-channel, three-body dynamical equation to which the BSA was applied did not have a connected kernel, correction terms could not be introduced in a controlled way. The usual CCA choices (9.4.30–9.4.31), however, lead to C-K equations.

Independently of the efforts of Baer *et al.*, Kouri, Truhlar, and coworkers[16,22] also realized the deficiency of the CCA (9.4.30) and the BSA. They performed successful calculations with the BSA using a wave-function version of the Fock coupling scheme (9.4.29) with distorted-wave Green functions. The resulting equations were solved with the separable-expansion and degenerate-kernel schemes of Ch. 3. They applied their approach to three-dimensional atom–diatom collision processes, and their calculations for the $H + H_2$ and $D + H_2$ systems produced accurate results.

This discussion would be incomplete without comment on the benchmark calculations of atom–diatom reactions by Kouri, Truhlar, Miller, and coworkers.[23] This work explained some puzzling experimental data and provided new physical insight. To quote some of the authors: "We are all jumping up and down..." because for the first time it is demonstrated for chemical reactions that "...the world really is quantum mechanical on the molecular level and not classical..."—a remarkable achievement of dynamical collision theory.

9.5. Generalized Reaction Theories

A reaction theory can be loosely defined as a few-body model for the collision of objects involving many constituent particles. Except for OP models, the two-body multichannel equations that result from the BSA reviewed in the preceding section define the only reaction theories in substantial calculational use at the present time.[1] Various three-cluster models have been proposed, but these are used much less frequently.[2] Both types of models have evolved using conventional methods.

The preceding situation has presented two problems for more comprehensive approaches to collision theory such as those reviewed in Ch. 5. The first is to explain how conventional reaction theories result from C-K theories, while the second is to use the latter methods to improve upon the former. These challenges have been only partially met. Generally speaking, the complexity of multiparticle collision theory has up to now frustrated the attainment of more sophisticated, yet practical, approximation methods when the number of constituent particles is larger than four. In this section, which represents sort of a culmination of the strategy articulated throughout the preceding chapters, we review some of what has been accomplished in the way of formulating generalized reaction theories. We do this by giving brief descriptions of three different approaches to reaction theory. In all three instances, the full analyses that are needed to flesh out these theories are extraordinarily sophisticated, even though some of the final results may be simple.

There are quite a few proposals for obtaining reaction theories from N-particle collision theory that we did not have space to elaborate upon.[3] None of these proposals has progressed any further towards calculational fruition than the efforts we have chosen to review. In assessing both the work we have reviewed and the work we have not discussed, it seems plainly obvious that there are extraordinary opportunities for further research in the substantiation and practical implementation of general reaction theories. Our intention is to provide an introduction that suffices to stimulate further pursuit of this fascinating field.

9.5a. *Chandler–Gibson Theory*

The N-particle scattering formalism proposed by Chandler and Gibson (CG)[4] is quite different from those we discussed in Ch. 5. For example, although the transition-operator integral equations obtained are partition-labeled, they do not have connected kernels. These (CG) equations are distinguished in other respects as well, not least of which is the unprecedented amount of functional analysis that attended their inception. Nevertheless, they seem to provide such a natural generalization of conventional reaction theories that our discussion would be incomplete without a review of their properties.

The formalism that is used to obtain the CG equations is often called the two-Hilbert-space method.[5] This is because besides the usual N-particle Hilbert space $\mathscr{H}(N)$, it is often found expedient to introduce another space $\mathscr{H}_{\text{asy}}(N)$ that is a direct sum of all possible cluster subspaces, $\mathscr{H}_a(N)$, of $\mathscr{H}(N)$. The cluster subspaces are spanned by the cluster states $|\phi_a\rangle$ that are eigenstates of the cluster Hamiltonians H_a, and correspond to each of the groups of particles in each cluster being in a separate bound state. The projectors on $\mathscr{H}_a(N)$ are denoted as P_a. Explicit communication between $\mathscr{H}_{\text{asy}}(N)$ and $\mathscr{H}(N)$ is provided by an injector operator J and its adjoint, which is conventionally denoted as J^*; we do not use this notation in what follows, but it is needed to make contact with the original literature.

There are a number of ways that one can motivate the introduction of CG theory. Most of them can perhaps be better appreciated after we obtain the CG equations by an elementary argument.[6,7]

First, we recall the expression for the prior off-shell extension of the transition operator:

$$T_{ba}^{(-)} = V^a + V^b I G V^a, \tag{9.5.1}$$

where we have ignored all dependences on the parametric energy z, and we have inserted the identity operator I. Equation (9.5.1) is defined on $\mathscr{H}(N)$. Second, we state as our goal the development of integral equations for the projected transition operators

$$T_{ba} \equiv P_b T_{ba}^{(-)} P_a. \tag{9.5.2}$$

To this end we define

$$X^{-1} \equiv \left(\sum_a^A P_a\right)^{-1}, \tag{9.5.3}$$

where the restriction A on the sum over all partitions is consistent with the existence of the inverse. Then we use the following expression for the identity:

$$I = X^{-1} \sum_c^A G_c P_c G_c^{-1}, \tag{9.5.4}$$

in the projected form of Eq. (9.5.1) to obtain

$$T_{ba} = P_b V^a P_a + P_b V^b X^{-1} \sum_c^A G_c P_c T_{ca}. \tag{9.5.5}$$

In obtaining Eq. (9.5.5), we have used $P_c G_c^{-1} G V^a P_a = P_c T_{ca}$ and we have supposed that the directionality of the inverse of G_c is unimportant. In the original CG theory, $A =$ all partitions; it has been proven that X^{-1} exists and is bounded in this case.[4]

Generally, the Eqs. (9.5.5) have disconnected kernels just like the LS equations. An interesting feature of the Eqs. (9.5.5) is that they form a generalization of the BSA equations because only the bound-cluster parts $G_c P_c$ of the cluster Green functions appear. This resemblance to the BSA equations becomes much stronger if A is restricted to the set of all two-cluster partitions in which case the Eqs. (9.5.5) are C-K equations because $P_\alpha V^\alpha$ is a connected operator. The crucial question of under what conditions such a restricted X^{-1} exists will be taken up in Section 9.5c.

A C-K version of the original CG equations has been found,[6] but this result seems to play no part in the solution strategy proposed by CG. A distinctive feature, and possibly the potential power, of the CG approach is to circumvent the C-K problem via the construction of an appropriate sequence of *approximate* transition operators that do satisfy C-K equations.[8] This unusual dynamical strategy has not yet yielded practical calculational procedures that are competitive with conventional methods. Nevertheless, there has been an interesting attempt to incorporate permutation symmetries and obtain an OP formalism.[9] Perhaps most importantly, CG theory has been used to investigate the validity and limitations of the conventional version (*cf.* Section 9.4) of the CRC approximation.[10] A new set of coupled dynamical equations is derived in Ref. 10 that include the CRC equations as a special case. This is certainly competitive with the work reviewed in Section 9.5c that employs C-K starting points. What is required next is a calculable series of corrections to the CRC and comprehensive numerical tests.

The motivation of CG theory was the perception that the number of the difficulties of N-body collision theory are magnified by working in the full Hilbert space. (The restricted domains and ranges of the transition operators that appear in the Eqs. (9.5.5) illustrate, in part, how this problem is dealt with, but at the price of the appearance of the very complicated operator X^{-1}.) It has now evolved into a complete collision theory replete with a well-defined and consistent set of approximations that constitute its dynamical strategy, all overlaid with a support structure of rigorous mathematics. At this point, however, some definitive calculations, even for $N = 3$, are urgently needed to assess the effectiveness of this scheme.

9.5b. C.H.U.C.K. Theory

Few-cluster models for reactions derive their physical validity in circumstances where a few ($n_{\bar{0}}$) clusters of particles corresponding to a dominant partition,[11] $\bar{0}$, are so tightly bound that the possibility of their breakup can be ignored in the explicit dynamics. Breakup is accounted for implicitly and approximately within effective interactions among the clusters. The practical appeal of such a model is that it can be applied to describe the collisions of systems of arbitrarily large numbers of the constituent particles that make up the clusters.

About a decade ago, an ambitious attempt was made by Polyzou and Redish to formulate a consistent theory of few-cluster models.[12] This theory is characterized by *c*ombinatoric *H*amiltonians that enter as input to a formalism that generates a *u*nitary dynamics by means of *c*onnected-*k*ernel collision integral equations (C.H.U.C.K.). After a flury of activity,[12–17] the further development and application of this theory has nearly abated. However, many of the ideas, such as that of a reaction mechanism, that were introduced in the course of formulating C.H.U.C.K. theory have had a significant impact upon current research in the applications of multiparticle collision theory. Therefore, it is essential that we provide a review of at least its rudiments.

Our usual cluster states $|\phi_a\rangle$ are eigenstates of the Hamiltonian H_a and have n_a bound clusters; the partitions a for which such states exist are often called "stable." The cluster states are specified by a set q_a, called a channel, of the internal quantum numbers of each of the clusters of the stable partition; they are also specified by the momenta $\mathbf{P}_i$, $i = 1, \ldots, n_a$, of the c.m. of each of the clusters, but we suppress this dependence and simply write $|\phi_a(q_a)\rangle$. The scattering eigenstates of H_a, $|\psi_a^{(+)}(q_b)\rangle$, evolve from the cluster states $|\phi_b(q_b)\rangle$, where b is a stable partition such that $a \supset b$, $a \neq b$. The cluster and scattering states of H_a will be denoted collectively as $|\hat{\phi}_a(q_b)\rangle$ with $a \supseteq b$. Then, the operator

$$P_a(q_b) \equiv \left\{ \prod_{i=1}^{n_b} \int d^3 P_i \right\} |\hat{\phi}_a(q_b)\rangle \langle \hat{\phi}_a(q_b)| \tag{9.5.6}$$

defines a projection on $\mathscr{H}_a(q_b)$, the Hilbert space spanned by the $|\hat{\phi}_a(q_b)\rangle$ for fixed a, b, and q_b. The resolution of the identity on the N-particle Hilbert space $\mathscr{H}(N)$ is then

$$I = \sum_b \sum_{q_b}^{A_0} \Delta_{ab} P_a(q_b), \tag{9.5.7}$$

where the sum over q_b is restricted to the set A_0 of physical scattering channels defined by the set of all stable partitions b with $n_b \geq 2$.

A crucial concept is that of a *reaction mechanism* (RM), which is formally defined as a subset, $A \subset A_0$. This notion of an RM becomes a physically useful one if A is identified with a set of dominant reaction channels, and this identification permits a simplification of the N-particle collision integral equations. Associated with a given RM, A, is a set of partitions $r(A)$, consisting of all partitions $a \supseteq b$, $a \neq \underline{1}$, such that $q_b \in A$.

Now, in the absence of N-particle forces, which we assume henceforth, the full Hamiltonian H can be decomposed in terms of the cluster Hamiltonians as

$$H = \sum_a \bar{\delta}_{a\underline{1}} C_a H_a. \tag{9.5.8}$$

This decomposition is obtained by multiplying both sides of the sum rule (A.5.7) by $[H]_b$, the b-connected part of H, and then summing over all b, where, if there are no N-particle forces,

$$H = \sum_b \bar{\delta}_{b\underline{1}} [H]_b. \tag{9.5.9}$$

Given Eq. (9.5.8), we see that

$$H(A) = \sum_a \bar{\delta}_{a\underline{1}}\, C_a H_a P_a(A), \tag{9.5.10}$$

where

$$\begin{aligned} P_a(A) &= \sum_b \sum_{q_b}^{A} \Delta_{ab}\, P_a(q_b), \qquad && a \in r(A), \\ &= 0, && a \notin r(A), \end{aligned} \tag{9.5.11}$$

defines a model total Hamiltonian for each RM. Comparing Eq. (9.5.8) with Eq. (9.5.10), we are led to define a projected cluster Hamiltonian for each A:

$$H_a(A) \equiv P_a(A) H_a = H_a P_a(A), \tag{9.5.12}$$

so that we have

$$H(A) = \sum_a \bar{\delta}_{a\underline{1}}\, C_a H_a(A), \tag{9.5.13}$$

where $H_a(A) \equiv 0$ for $a \notin r(A)$. One cannot argue by analogy that

$$H_a(A) = \sum_b \Delta_{ab}\, [H(A)]_b, \tag{9.5.14}$$

where $[H(A)]_b$ is the b-connected part of $H(A)$, even though its validity seems plausible. However, Eq. (9.5.14) is true, as can be shown by a straightforward but lengthy calculation.[12] Given Eq. (9.5.14), we can now decompose $H(A)$ into an RM cluster-Hamiltonian and an RM interaction given by

$$V^a[A] \equiv \sum_b \bar{\Delta}_{ab}\, [H(A)]_b \tag{9.5.15}$$

for each $a \in r(A)$ completely analogous to the expression $H = H_a + V^a$, namely,

$$H(A) = H_a(A) + V^a(A). \tag{9.5.16}$$

Now given this carefully crafted analogy to the usual Hamiltonian structure of collision theory, the C.H.U.C.K. dynamical strategy is obvious. One introduces RM Green functions

$$G(A) \equiv [z - H(A)]^{-1}, \tag{9.5.17}$$

$$G_a(A) \equiv [z - H_a(A)]^{-1}, \tag{9.5.18}$$

for some complex parametric energy with $z = E + i0$ in the physical limit. It is pertinent to remark that these Green functions are defined in the full

N-particle Hilbert space; indeed, we have

$$G_a(A) = P_a(A)G_a + \frac{Q_a(A)}{z}, \tag{9.5.19}$$

where $Q_a(A) = I - P_a(A)$. Of course, the two Green functions (9.5.17) and (9.5.18) are connected by the resolvent identity

$$G(A) = G_a(A) + G_a(A)V^a(A)G(A), \tag{9.5.20}$$

which suggests that we introduce an RM transition operator by

$$T_{ba}^{(+)}(A) = V^b(A) + V^b(A)G(A)V^a(A). \tag{9.5.21}$$

Then, if it is possible to define an RM channel-coupling scheme

$$V^b(A) = \sum_a \{\mathscr{V}(A)\}^{ba}, \tag{9.5.22}$$

and obtains the coupled integral equations

$$T_{ba}^{(+)}(A) = \sum_c \{\mathscr{V}(A)\}^{bc} G_c(A)[\overleftarrow{G}_b(A)^{-1} + T_{cb}^{(+)}(A)]. \tag{9.5.23}$$

The principal result of C.H.U.C.K. theory is is that RM potential matrices (9.5.22) exist such that the Eqs. (9.5.23) are C-K equations and their solutions satisfy a unitarity condition consistent with the RM. In the original treatment,[12] the appropriate potential matrix is the RM version of $\mathscr{V}_\mathrm{C}$ given by Eq. (5.4.18):

$$\mathscr{V}_\mathrm{C}(A) \equiv \bar{V}_\mathrm{M}(A)\underline{C}, \tag{9.5.24}$$

where

$$\{\bar{V}_\mathrm{M}(A)\}_{ab} \equiv \sum \bar{\delta}_{d\underline{1}}\,\bar{\Delta}_{ad}\,[H(A)]_d\,\Delta_{bd}. \tag{9.5.25}$$

The resulting equations can be put into a form that makes the multiple-scattering structure of the RM model manifest. An RM version of the KLT potential matrix (5.4.53) yields a different set of $T^{(+)}$-matrix equations, but which also have connected kernels and satisfy an RM unitary condition.[15]

C.H.U.C.K. theory is formally analogous to a "normal" partition-labeled set of C-K integral equations. Because $\mathscr{V}(A)\underline{P}(A) = \mathscr{V}(A)$, where $[\underline{P}(A)]_{ab} = \delta_{ab}P_a(A)$, these equations can be reduced to a set only for the reaction-set operators $P_b(A)T_{ba}^{(+)}(A)P_a(A)$ with no coupling outside of the reaction set. The dynamical strategy is designed to result in a model with the structure of a correct collision theory; in particular, unitarity is maintained.

There have been attempts to formulate RM models with $n_{\bar{0}} = 3$;[13,17] this may be interpreted as a formal recovery of *ad hoc* three-cluster reaction theories.[2] C.H.U.C.K. theory merits a critical assessment as to its practicality. Independently of this, it is likely that some of its basic ideas are of permanent value.

9.5c. *Recovery of the Coupled-Reaction Channels and Resonating-Group Methods from Connected-Kernel Theories*

A productive episode in the history of C-K collision theories was the "recovery" of the CRC and RGM methods. In this subsection, we review the highlights of this saga and what seems to have been learned from it.

We mean by the CRC, the seemingly *ad hoc* approximation $|\psi\rangle \simeq |\psi_{\mathrm{CRC}}\rangle$ to the Schrödinger wave function given by (*cf.* Eq. (9.4.4))

$$|\psi_{\mathrm{CRC}}\rangle = \sum_{\alpha} P_{\alpha}|\hat{\psi}_{\alpha}\rangle. \tag{9.5.26}$$

The components $P_{\alpha}|\hat{\psi}_{\alpha}\rangle$ satisfy the CRC equations (9.4.5). At the present time there are at least four questions[18] that can be asked concerning the CRC:

(i) Is the CRC actually effective, or does it merely serve as a framework for phenomenology?
(ii) If a CRC calculation is successful, what does it mean in dynamical terms?
(iii) Is the CRC an approximation one really wishes to emulate?
(iv) Can the CRC be "recovered" by means of consistent approximations to a C-K theory?

We will be concerned only with question (iv). The other questions remain unanswered at present and deserve further investigation because they bear heavily upon the scientific credibility of the CRC technique.

The "recovery" of the CRC from a C-K theory is meaningful if the CRC is itself free of internal inconsistencies. If it is consistent, the problem is then to indicate a definite approximation scenario within the context of a given C-K theory that yields the CRC. Both aspects of question (iv) were addressed in a series of articles by Birse *et al.*,[18–20] and we closely follow that work in some of the following discussion.

Let us consider first the matter of the internal consistency of the CRC as an approximation to the Schrödinger equation. The state $|\psi_{\mathrm{CRC}}\rangle$ is defined on the model space $\mathscr{H}_{\pi}$ corresponding to the union of the ranges of the projections P_{α} on the two-cluster subspaces. The projection P_{π} on $\mathscr{H}_{\pi}$ therefore satisfies

$P_\pi P_\alpha = P_\alpha P_\pi = P_\alpha$ for each α. On the other hand, $(I - P_\alpha)P_\pi = (P_\pi - P_\alpha)$, so there is some overlap of the space spanned by the scattering states of the cluster Hamiltonian H_α with the model space. This contradicts the naive perception of the CRC as purely a bound-state or cluster approximation. This is one facet of the nonorthogonality $P_\alpha P_\beta \neq 0, \alpha \neq \beta$, of the different channel spaces.

Given P_π, any model-space wave function $|\psi^{\mathrm{M}}\rangle = P_\pi|\psi^{\mathrm{M}}\rangle$ can always be expanded in terms of some components $P_\alpha|\psi_\alpha^{\mathrm{M}}\rangle$:

$$|\psi^{\mathrm{M}}\rangle = \sum_\alpha P_\alpha|\psi_\alpha^{\mathrm{M}}\rangle. \tag{9.5.27}$$

This expansion is not necessarily unique because the P_α may not be linearly independent, a condition referred to as overcompleteness. The CRC equations follow from the assertion that $|\psi^{\mathrm{M}}\rangle$ is determined by the model Hamiltonian $P_\pi HP_\pi$:

$$(E - P_\pi HP_\pi)P_\pi|\psi^{\mathrm{M}}\rangle = 0. \tag{9.5.28}$$

A set of components satisfying Eq. (9.5.27) always exists:

$$|\hat{\psi}_\alpha\rangle \equiv P_\alpha X_\pi^{-1}|\psi^{\mathrm{M}}\rangle, \tag{9.5.29}$$

where X_π^{-1} is a generalized (Moore-Penrose (MP)) inverse[10,16] that satisfies

$$X_\pi^{-1} = \left(\sum_\alpha P_\alpha\right)X_\pi^{-1} = P_\pi, \tag{9.5.30}$$

which is similar to the object (*cf*. Eq. (9.5.3)) that appears in CG theory. The MP inverse exists for any bounded operator that has a domain equal to the entire Hilbert space, and is defined even if the operator has a nontrivial null space. (The MP inverse of an operator A is defined as an ordinary inverse on the range of A when it is restricted to the orthogonal complement of its null space, and as zero on the remainder of the range space.) From Eq. (9.5.30), we see that

$$P_\alpha X_\pi^{-1} = P_\alpha - \sum_\beta \bar{\delta}_{\alpha\beta}\, P_\alpha P_\beta(P_\beta X_\pi^{-1}) \tag{9.5.31}$$

which, since $\bar{\delta}_{\alpha\beta}\, P_\alpha P_\beta$ is a compact operator, forms a set of C-K integral equations for the operators $P_\alpha X_\pi^{-1}$.

Now when the P_α are all independent, so are the components that are uniquely defined by definition (9.5.29). We find in this case, using Eq. (9.5.27), that

$$P_\alpha X_\pi^{-1} P_\beta = \delta_{\alpha\beta}\, P_\alpha. \tag{9.5.32}$$

We see that X_π^{-1}, in effect, creates an orthogonal projection, and its deviation from the identity is a measure of the nonorthogonality among the projectors. When we have overcompleteness, Eq. (9.5.32) must be supplemented by a matrix, with elements $Q_{\alpha\beta}$, that is a projector on the space of components $|\chi_\alpha\rangle$ for which $\sum_\alpha P_\alpha|\chi_\alpha\rangle = 0$. A major open question is whether overcompleteness appears for physically realistic situations. There are arguments that suggest it does not,[21] and no entirely convincing counterexamples seem to have been given to the contrary.[22] We will ignore henceforth, in the interests of simplicity, the possibility of overcompleteness. Then, with the unique components (9.5.29) it is possible to restate the CRC equations so that all nonorthogonality effects are contained in X_π^{-1}, and we can regard the CRC as an approximation that is internally consistent.[18]

It remains to see how the CRC can be obtained from a C-K theory. The essence of what is involved in doing this is completely contained in the case when $N = 3$, and we confine ourselves to this example. From the numerical analyses reported in Section 9.4, one would suspect that some combination of the BSA and a distorting potential within a C-K formalism will yield the CRC. This is precisely what happens.[18–20] A convenient starting point for demonstrating this result is the set of Faddeev equations for the auxiliary wave functions $|\psi_\alpha\rangle$, with

$$|\psi\rangle = \sum_{\alpha=1}^{3} |\psi_\alpha\rangle, \tag{9.5.33}$$

in terms of distorting potentials:[23,24]

$$(E - H_\alpha - V_1^\alpha)|\psi_\alpha\rangle = \sum_{\beta=1}^{3} (V_\alpha \bar{\delta}_{\alpha\beta} - V_1^{\alpha\beta})|\psi_\beta\rangle, \tag{9.5.34}$$

where

$$V_1^\alpha = \sum_{\beta=1}^{3} V_1^{\beta\alpha}. \tag{9.5.35}$$

It is supposed that the $V_1^{\alpha\beta}$ are compact, i.e., connected, operators and that $V_1^{\alpha\alpha} = 0$, in order to avoid ambiguity in the choice of the distorting potentials V_1^α in each channel α.

The BSA-form of Eq. (9.5.34) is obtained by replacing $|\psi_\alpha\rangle$ by $P_\alpha|\psi_\alpha\rangle$ for all α:

$$(E - H_\alpha)P_\alpha|\psi_\alpha\rangle = \sum_{\beta=1}^{3} [\bar{\delta}_{\alpha\beta} V_\alpha - V_1^{\alpha\beta} + \delta_{\alpha\beta} V_1^\beta] P_\beta|\psi_\beta\rangle. \tag{9.5.36}$$

We obtain two sets of equations by projecting the Eqs. (9.5.36) alternatively with P_α and P_π to obtain

$$(E - H_\alpha)P_\alpha|\psi_\alpha\rangle = \sum_{\beta=1}^{3} P_\alpha F_{\alpha\beta} P_\beta |\psi_\beta\rangle, \tag{9.5.37a}$$

$$(E - H_\alpha)P_\alpha|\psi_\alpha\rangle = \sum_{\beta=1}^{3} P_\pi F_{\alpha\beta} P_\beta |\psi_\beta\rangle, \tag{9.5.37b}$$

respectively. The operator

$$F_{\alpha\beta} \equiv \bar{\delta}_{\alpha\beta} P_\pi V_\alpha P_\beta - V_1^{\alpha\beta} + \delta_{\alpha\beta} V_1^\alpha \tag{9.5.38}$$

embodies the freedom we have in the choice of the distorted-potential matrix, and satisfies

$$\sum_{\alpha=1}^{3} F_{\alpha\beta} = P_\pi V^\beta P_\beta. \tag{9.5.39a}$$

Suppose that we demand that Eqs. (9.5.37a, b) are equivalent. This is possible only if

$$P_\alpha F_{\alpha\beta} = F_{\alpha\beta}. \tag{9.5.39b}$$

It is clear from Eq. (9.5.30) that

$$\tilde{F}_{\alpha\beta} \equiv P_\alpha P_\pi X_\pi^{-1} V^\beta P_\beta \tag{9.5.40}$$

is a solution of Eqs. (9.5.39a, b). Thus, the general solution has the form[18]

$$F_{\alpha\beta} = \tilde{F}_{\alpha\beta} + A_{\alpha\beta}, \tag{9.5.41}$$

where

$$\sum_{\alpha=1}^{3} A_{\alpha\beta} = 0, \tag{9.5.42a}$$

$$P_\alpha A_{\alpha\beta} = A_{\alpha\beta}. \tag{9.5.42b}$$

For arbitrary vectors $|\chi_\beta\rangle$, the Eqs. (9.5.42b) imply that

$$\sum_{\alpha=1}^{3} P_\alpha(A_{\alpha\beta}|\chi_\beta\rangle) = 0, \tag{9.5.43}$$

which is precisely the condition for overcompleteness, so that in its absence we have $A_{\alpha\beta} \equiv 0$.

With the unique solution (9.5.40) for $F_{\alpha\beta}$, both V_1^α and $V_1^{\alpha\beta}$ follow from (9.5.38):[18]

$$V_1^{\alpha\beta} = \bar{\delta}_{\alpha\beta}[P_\pi V_\alpha P_\beta - P_\alpha X_\pi^{-1} V^\beta P_\beta] \tag{9.5.44a}$$

$$V_1^\alpha = P_\alpha X_\pi^{-1} V^\alpha P_\alpha. \tag{9.5.44b}$$

More to the point, we see that Eqs. (9.5.37a, b) both become

$$(E - H_\alpha)P_\alpha|\psi_\alpha\rangle = \sum_\beta P_\alpha X_\pi^{-1} V^\beta P_\beta|\psi_\beta\rangle, \tag{9.5.45}$$

which are exactly the orthogonalized CRC equations for the unique components (9.5.29) that follow from Eq. (9.5.28) by operating upon it from the left by $P_\alpha X_\pi^{-1}$. This establishes that, with uniquely defined distorting potentials, the C-K Faddeev equations reduce to the CRC equations in the BSA. We see that the demand that Eqs. (9.5.37a, b) be equivalent is purely of heuristic value.

The generalization of the preceding argument to the N-particle case is straightforward if one uses the form of the Eqs. (5.2.37) corresponding to the Eqs. (9.5.34) that one obtains using the combinatorial potential matrix $\mathscr{V}_C$ for the pair interactions, and after introducing compact distorting potentials.[18] Permutation symmetries are readily incorporated to establish the connection to the RGM.[20]

It is worth commenting upon the preceding results. Superficially, it appears that the "simple" CRC approximation is recovered only in a very complicated manner from C-K theory. However, we recall that it is not always easy to recover a specific approximation from a complete theory. For example, some of the highly successful methods discussed in Ch. 3 for solving two-body problems do not reproduce something as simple as the Born series in a direct way.

The introduction of even a uniquely defined distorting potential would appear to conceal rather than reveal the dynamical approximation that the CRC represents. An open problem is whether it is possible to formulate an equivalent approximation to the undistorted C-K equations. Thus, this "recovery" still leaves open to question the physical circumstances under which the CRC is a good approximation and how this is translated into a cogent approximation scenario with a given set of C-K equations.

It would be interesting to see how the CRC is "recovered" in alternative C-K formalisms. Is the choice of distorting potential unique in some sense? Is the BSA prescription altered? How formalism-dependent are the Birse *et al.*[18–20] results? Answers to these questions and many others that suggest themselves would provide a lot of insight into the dynamics of C-K theories as well as the physics of the CRC.

Given the comparatively small amount of collective intuition that has been developed concerning C-K theories, the possibility remains open that there is some very simple, plausible low-order approximation to a suitable C-K formalism that is superior to the CRC. A variety of different C-K approaches to coupled reaction theory already exist, but they have been investigated only superfically in regard to realizable approximations. Birse[19] has extended some of the results obtained in Section 9.2 to the case of arbitrary sets of two-cluster channels to find interrelationships among the Feshbach formalism, the CRC, and effective-interaction formalisms, especially those employing the AGS off-shell extension for the transition operators.[25] However, just as in the identical-particle case, the existence of these interrelationships does not suggest the optimum vehicle for successive approximation.

The further investigation of the dynamical underpinnings of the CRC and RGM, particularly from the vantage points of several alternative C-K theories, promises to yield important insights into the application of dynamical collision theory.

References and Notes (9)

Section 9.1

1. Itzykson (1980).
2. Kadanoff (1962), Negele (1988).
3. Isihara (1971).
4. Fetter (1971).
5. Barrett (1983).
6. Mahaux (1969); see also Johnson (1987).
7. See Kowalski (1979a) for an elaboration of this point of view.
8. Hodgson (1971), p. 59.
9. For a review of the Coulomb problem see Chandler (1981b, 1987).
10. Faddeev methods do have many useful applications in atomic problems nevertheless. See Micha (1985).
11. Temkin (1985b).
12. Baer (1989).

Section 9.2

1. The literature concerning the optical potential is immense. A good source for the discussion of the wide variety of meanings of the OP concept in nuclear scattering is von Geramb (1979). Discussions of the state of the art a decade earlier along with extensive documentation are contained in Fetter (1965) and Lipperheide (1966). For a review of the use of the OP in atomic physics, see Thirumalai (1987).
2. The optical model is commonly understood to be realized when the elastic scattering of a projectile and a target can be described in terms of a local, complex potential

whose shape and range can be parameterized with a few, possibly energy-dependent, parameters. An OP of this sort may or may not exist for a particular situation.

3. The original work on the Dirac approach is contained in Arnold (1981), Clark (1983a, b), McNeil (1983a, b) and Shepard (1983); for a recent review of the field, see Wallace (1987) and Ray (1988).

4. The discussion in Celenza (1984) is particularly useful in its delineation of some of the theoretical uncertainties of the Dirac approach.

5. The OP in pion–nucleus interactions is reviewed in Eisenberg (1980). For more recent work, see Karaoglu (1986).

6. This two-body problem has been investigated both in configuration and momentum space. The first setting follows the ancient tradition of the optical model and allows the easy introduction of Coulomb effects, but is a preferred strategy only when the EI is local. Momentum-space calculations have been performed in detail only relatively recently—Landau (1973, 1975, 1986), Páez (1984a, b), Picklesimer (1984)—but are superior in their capacity to utilize arbitrary representations of the two-body subsystem collisions.

7. See Goldflam (1980a, b) and Picklesimer (1980).

8. An energy-independent OP was introduced in Kuo (1980) and Lee (1981), and was subsequently investigated vigorously, e.g., Buck (1981), Bencze (1982c), until the work by Hussein (1984), where serious shortcomings in the use of such a potential either for direct microscopic construction or phenomenological parameterization were pointed out. No example of an energy-independent OP in a C-K, N-particle theory has yet been found.

9. von Geramb (1979).

10. Feshbach (1958).

11. In this regard, the remarks of Hodgson on p. 459 in von Geramb (1979) are especially pertinent.

12. The historical origin of these criteria is obscure. These conditions are consistent with the feeling that in choosing a phenomenological OP there is no imperative to do otherwise. For discussions of the stipulations of the Hermiticity requirement for microscopic OP's see, e.g., Feshbach (1958), Fetter (1965), Lipperheide (1966) and Kowalski (1984, 1986). For some other questions concerning the definition of the OP, see Bencze (1982b, 1985), MacDonald (1983) and Johnson (1984).

13. Kerman (1959).

14. The purpose of the alternative derivation is to avoid working with LS equations that have nonunique solutions, although the same answers are obtained with such equations.

15. There is a considerable literature concerning Feshbach resonances especially, in atomic physics. See Temkin (1985a, b), Berk (1986), and references cited therein.

16. Feshbach (1971).

17. Ernst (1977), Siciliano (1977), Tandy (1980).

18. Picklesimer (1981a).

19. Essentially the complete original literature on this facet of OP theory consists of

Kowalski (1979a, 1981a, 1986), Goldflam (1980a, b) and Adhikari (1984a). For reviews, see Levin (1981), Kowalski (1984) and Cattapan (1985a).
20. Kowalski (1981a).
21. Kowalski (1982b).
22. Adhikari (1984a).
23. Picklesimer (1980, 1981a, b), Kowalski (1981b, c, e; 1982a).
24. The boson case, or instances of mixed symmetries, can be handled in a similar fashion.
25. Feshbach (1962). In point of fact, Feshbach considers nucleon–nucleus elastic scattering rather than the general two-fragment case.
26. The explicit operator representations obtained for $P(\Lambda)$ in the case of electron scattering from an N-electron atom, Hahn (1968, 1970), Temkin (1985a, b), are not idempotent on $\mathscr{H}$ for $N > 1$, but are so on $\mathscr{H}(\Lambda)$, which is the meaning of Eq. (9.2.33).
27. Temkin (1985a, b).
28. See, e.g., Temkin (1985a). This identification is entirely harmless except for the confusion it creates with the conventional definition of the OP, whether permutation symmetries are included or not.
29. Saito (1969) and, e.g., Leeb (1980).
30. Kowalski (1982a).
31. As pointed out by Birse (1983b), this restriction is necessary in order that Eqs. (9.2.39a, b) define the EI operators uniquely.
32. Kowalski (1981c).
33. For examples of just how complicated things can get in the application of the Feshbach formalism, see Fiebig (1978, 1981).
34. A detailed discussion of multiple-scattering algorithms for the scattering of a nucleon from a nucleus for both the AGS and the post off-shell extensions is given in Picklesimer (1981b). We should remark that what we have called the post extension is called "prior" in that reference and in much of the related literature, *cf.* Refs. 19, 23. Evidently, one can assign "priority" to the occurrence of the full wave function rather than the interaction. This option, and a few others, give rise to a variety of "post-prior" ambiguities.
35. Goldflam (1980e, 1982).
36. Bell (1959, 1962).
37. For an example of the e^+–H problem, see Varracchio (1983).
38. Eisenberg (1980). For a recent example, see Johnson (1987).
39. Fetter (1965).
40. Blaizot (1986).
41. Negele (1988).
42. Eisenberg (1989).
43. Lipperheide (1966).
44. Redish (1970).
45. Celenza (1979, 1983, 1984, 1986) and references cited therein. For an introduction to the Bethe–Salpeter approach, see Lurie (1968).

46. Clark (1983a).
47. Wallace (1987).
48. Bleszynski (1987).
49. Hippchen (1988).
50. See Serot (1986) and Celenza (1986) for general reviews. Recent work on the application of the bound-state picture to the determination of nucleus–nucleus OP's is reported in Ohtsuka (1988).
51. Hynes (1984, 1985).
52. See Wallace (1987) for a review of the status of these controversies circa 1986. For later developments, see Cooper (1986, 1988), Picklesimer (1987, 1988) in regards to the work by Thies (1985, 1986).
53. Thies (1985, 1986), Cooper (1986, 1988).
54. Keister (1986).
55. Huang (1975), Kowalski (1979b, 1980a, b).

Section 9.3

1. Foldy (1945).
2. Watson (1953).
3. Most of the subject is either covered in or can be traced from the reviews: Goldberger (1964), Fetter (1965), Rodberg (1967), Jochain (1975), Austern (1977), Eisenberg (1980), Wallace (1981) and Tandy (1986). Nonstandard approaches include Siciliano (1977), Kowalski (1979a, 1982b), Goldflam (1980a) and Picklesimer (1981a, b; 1983a, b).
4. Watson (1954, 1956).
5. Riesenfeld (1956).
6. Takeda (1955).
7. Kerman (1959).
8. Glauber (1959).
9. The works by Austern (1977), Eisenberg (1980), Wallace (1981) are especially concerned with these questions.
10. One of the assumptions entering into Glauber theory is that the target nucleons are fixed. This is known as the fixed-scatterer approximation (FSA), which has had an interesting history in MS theory. There are several versions of the FSA including one (the "frozen" approximation) where the c.m. of the target is allowed to move, but in the c.m. frame the nucleons are fixed; the final amplitude is then obtained by averaging over all the target nucleon positions as weighted by the nuclear wave functions. See Foldy (1969) and Kujawski (1972, 1973). Attempts to justify the FSA have relied on closure with respect to the target states (Foldy (1969)); the validity of this closure argument requires that the two-body potentials be local (Kowalski (1971b), Benayoun (1976b)). Generally, one encounters kinematical ambiguities that arise from the lack of Galilean invariance of the FSA.

11. A critical comparison of the Watson and KMT formalisms is given in Kowalski (1982b). A careful discussion of what are believed to be the realms of applicability of the Watson and KMT approximation scenarios is given in Tandy (1986).
12. Picklesimer (1983a).
13. Liu (1979).
14. Chew (1951, 1952).
15. Some notable exceptions are the calculations of Páez (1984a, b) and Landau (1986) for $A = 3$ target nuclei.
16. The Watson and KMT approaches are compared for pion–deuteron scattering ($A = 2$) in Afnan (1981) using separable pion–nucleon interactions. Considerable differences were found between the two. The failure to represent the diffractive components of the pion–nucleon transition matrix may be a prime reason for these differences. See also Garcilazo (1982).
17. Tandy (1977).
18. A brief summary of these arguments along with pertinent references up to about 1978 can be found in Kowalski (1979a). For later work, see Gurvitz (1986).
19. Picklesimer (1984).
20. Takeda (1955). The analysis in this paper is apparently unique in one respect. Namely, despite the considerable amount of redundancy in the literature on MS formalisms, the particular details of the argument of Takeda and Watson have never, to our knowledge, been retraced except essentially verbatim in Goldberger (1964). A crucial step in Takeda (1955) is a particle-label ordering convention that appears to be a less direct way to ignore target exchange effects. If this convention is ignored, strange multiple-scattering groupings appear (Kowalski (1963c)). Other approaches, e.g., Gurvitz (1980, 1981), but still using conventional operator formalisms, arrive at results different than the TW prescription for low-order inclusion of the Pauli principle. There are some open questions here that deserve further study.
21. Some of the corrections to the TW prescription appear to be small even at relatively low incident-nucleon energies. See Ray (1989). In this last work, no direct estimates are made of the replacement of $G_\alpha R_\alpha$ by a two-particle, unsymmetrized Green function, which is a crucial aspect of the TW prescription, unless some of the target-exchange effects that are evaluated can be considered to simulate part of this replacement.
22. Picklesimer (1981a, b).
23. The modifications that are entailed when one assumes relativistic kinematics are nicely detailed in Wallace (1981, 1987).
24. Fulton (1959).
25. Kowalski (1971b).
26. Remler (1974).
27. Arellano (1989).
28. Kowalski (1963a), Kottler (1965).
29. Landau (1975, 1986), Paez (1984a, b).
30. Wallace (1981).

31. Weiss (1982).
32. A similar strategy has been used to arrive at an "optimal" prescription for the energy-shift to be used in conjunction with the IA. See Gurvitz (1979, 1986), Wallace (1981).
33. Picklesimer (1984).
34. Ernst (1983), Giebink (1988).
35. Micha (1985).
36. Goldflam (1980b, d).
37. Adhikari (1984a).
38. Kowalski (1979a, 1982b). The detailed calculations underlying the results quoted in Section 9.3c are given in these references.

Section 9.4

1. Tobocman (1961), Ohmura (1969, 1970), Udagawa (1973), Imanishi (1974), Austern (1970), Satchler (1980, 1983), Cotanch (1976).
2. Wheeler (1937), Wildermuth (1958a, b; 1977), Saito (1969), Tang (1978, 1987), Hofmann (1984), Schmid (1984).
3. Burke (1963, 1971), Oberoi (1973), Nesbet (1975), Truhlar (1974).
4. Eyre (1979).
5. Adhikari (1981c).
6. Lewanski (1978).
7. Glöckle (1970a, b; 1973).
8. Greben (1979), Mitra (1965a).
9. Raphael (1976), Adhikari (1980b).
10. Kuruoglu (1982, 1985), Adhikari (1985).
11. Tobocman (1981a).
12. Baer (1971–1973).
13. Kouri (1974a, b; 1975a, b).
14. Tobocman (1974a, b; 1975a, b).
15. Kouri (1974c, d).
16. Zhang (1988a).
17. Baer (1987a, b).
18. Neuhauser (1988a, b).
19. Baer (1989). The analysis underlying Eqs. (9.4.18–9.4.23) is developed in detail in Sections 5.2 and 5.4.
20. It should be noted that this is not the only possible form of the three-particle $\hat{K}$-matrix.
21. Shima (1983a–c, 1986), Abu Salbi (1984).
22. Schwenke (1987, 1988).
23. Blais (1990), Zhang (1989), Kliner (1990). Quote from *Science*, 26 Jan., 1990, p. 413.

Section 9.5

1. See the citations to Section 9.4 for representative works.
2. Works on three-body models that contain extensive citations to the literature are Greben (1982), Bencze (1984), Austern (1987), and Kozack (1987).
3. The following represents a partial list of citations to work concerning the derivation of reaction theories from N-particle C-K equations, and further works by the same authors as well as others can be traced from these references: Narodetskii (1971), Redish (1974b), Schwager (1976), Cattapan (1979b, 1982, 1983b, 1984, 1985a), Goldflam (1979, 1980a), Hahn (1980, 1982), Bencze (1985, 1988).
4. Chandler (1977, 1978a).
5. For readable introductions to the two-Hilbert space method, see Bencze (1982b) and Cattapan (1985a).
6. Polyzou (1982).
7. Tobocman (1983b).
8. Chandler (1983, 1984, 1989).
9. Bencze (1982b).
10. Bencze (1982d).
11. Dixson (1980).
12. Polyzou (1979a, b).
13. Goldflam (1980a).
14. Polyzou (1980).
15. Goldflam (1980c).
16. Polyzou (1981).
17. Bencze (1982e).
18. Birse (1983a).
19. Birse (1983b).
20. Adhikari (1984c).
21. Cotanch (1976).
22. See Schmid (1989) for a recent discussion of nonorthogonality and overcompleteness problems in the context of contemporary practical realizations of the CRC and RGM methods.
23. Connected N-particle forces were introduced into N-particle collision theory by means of channel-coupling schemes involving only connected operators in Kowalski (1978a). It was established that the structural characteristics of partition-labeled C-K theories remain unaltered. Various sets of coupled-reaction channel equations including distorting potentials are established there.
24. For the introduction of compact distorting potentials into Faddeev wave equations for $N = 3$, see Hahn (1968, 1972), Bencze (1980), and Redish (1980).
25. Goldflam (1980e, 1982).

Bibliography

Aaron, R., R. D. Amado, and Y. Y. Yam (1964) *Phys. Rev.* **136**, B650.

———, ———, and ——— (1965) *Phys. Rev.* **140**, B1291.

——— and ——— (1966) *Phys. Rev.* **150**, 857.

———, ———, and J. E. Young (1968) *Phys. Rev.* **174**, 2022.

——— and ——— (1973) *Phys. Rev. Lett.* **31**, 1157.

———, R. H. Thompson, R. D. Amado, R. A. Arndt, D. C. Teplitz, and V. L. Teplitz (1975) *Phys. Rev. D* **12**, 1984.

——— (1977) in "Modern Three-Hadron Physics," ed. A. W. Thomas, Springer-Verlag, Berlin, p. 139.

Abdallah, Jr., J., and R. L. Smith (1974) *Adv. Chem. Phys.* **25**, 211.

Abdel-Raouf, M.A. (1982) *Phys. Rep.* **84**, 163.

——— (1984a) *Phys. Rep.* **108**, 1.

——— (1984b) *Phys. Rev. A* **30**, 2794.

Abramowitz, M., and I. Stegun (1965) "Handbook of Mathematical Functions," National Bureau of Standards, pp. 821–825.

Abu Salbi, N., D. J. Kouri, Y. Shima, and M. Baer (1984) *J. Chem. Phys.* **81**, 1813.

Adhikari, S. K., and R. D. Amado (1971) *Phys. Rev. Lett.* **27**, 485.

——— and ——— (1972) *Phys. Rev. C* **6**, 1484.

——— (1973) *Phys. Rev. D* **8**, 1195.

——— (1974a) *Phys. Rev. C* **10**, 1623.

——— and R. D. Amado (1974b) *Phys. Rev. D* **9**, 1467.

——— and I. H. Sloan (1975a) *Nucl. Phys.* **A241**, 429.
——— and ——— (1975b) *Nucl. Phys.* **A251**, 297.
——— and ——— (1975c) *Phys. Rev. C* **11**, 1133.
——— and ——— (1975d) *Phys. Rev. C* **12**, 1152.
——— (1978a) *Phys. Rev. D* **18**, 4250.
——— (1978b) *Phys. Rev. D* **18**, 4264.
——— (1979a) *Phys. Rev. C* **19**, 1729.
——— and W. Glöckle (1979b) *Phys. Rev. C* **19**, 616.
——— (1979c) *Phys. Rev. C* **19**, 2121.
——— and ——— (1980a) *Phys. Rev. C* **21**, 54.
——— and ——— (1980b) *Phys. Rev. C* **22**, 309.
——— and A. C. Fonseca (1981a) *Phys. Rev. D* **24**, 416.
——— and L. Tomio (1981b) *Phys. Rev. C* **24**, 1186.
——— (1981c) *Phys. Rev. C* **24**, 379.
———, A. C. Fonseca, and L. Tomio (1982a) *Phys. Rev. C* **26**, 77.
——— and L. Tomio (1982b) *Phys. Rev. C* **26**, 83.
——— and J. R. A. Torreão (1983a) *Phys. Lett.* **132B**, 257.
——— (1983b) *Phys. Lett.* **95A**, 21.
——— (1983c) *Phys. Lett.* **126B**, 489.
———, R. Kozack, and F. S. Levin (1984a) *Phys. Rev. C* **29**, 1628.
——— (1984b) *Phys. Rev. C* **30**, 31.
———, M. C. Birse, R. Kozack and F. S. Levin (1984c) *Phys. Rev. C* **30**, 780.
———, Z. C. Kuruoglu, and F. S. Levin (1985) *Ann. Phys.* (N.Y.) **163**, 149.
——— and L. Tomio (1986) *Phys. Rev. C* **33**, 467.
——— and L. Tomio (1987a) *Phys. Rev. C* **35**, 1991.
——— and L. Tomio (1987b) *Phys. Rev. C* **36**, 1275.
——— and L. Tomio (1988a) *Phys. Rev. C* **37**, 41.
——— and T. K. Das (1988b) *Phys. Rev. C* **37**, 1376.
———, A. Delfino, T. Frederico, I. D. Goldman, and L. Tomio (1988c) *Phys. Rev. A* **37**, 3666.
——— and K. L. Kowalski (1989) *Phys. Rev. A* **40**, 2270.
——— and T. Fredrico (1990a) *Phys. Rev. C.* **42**, 128.
——— (1990b) *Phys. Rev. A* **42**, 6.
Afnan, I. R., and J. M. Read (1973a) *Phys. Rev. C* **8**, 1294.
——— and ——— (1973b) *Aust. J. Phys.* **26**, 725.
——— and A. W. Thomas (1974) *Phys. Rev. C* **10**, 109.
——— and N. D. Birrell (1977a) *Phys. Rev. C* **16**, 823.
——— and A. W. Thomas (1977b) in "Modern Three-Hadron Physics," ed. A. W. Thomas, Springer-Verlag, Berlin, p. 1.
——— and B. Blankleider (1980) *Phys. Rev. C* **22**, 1638.
——— and A. T. Stelbovics (1981) *Phys. Rev. C* **23**, 845.
——— (1984) *Nucl. Phys.* **A416**, 257c.
——— and B. Blankleider (1985a) *Phys. Rev. C* **32**, 2006.
——— and B. C. Pearce (1985b) *Phys. Rev. C* **31**, 986.

——— (1988) *Phys. Rev. C* **38**, 1792.

Ahmadzadeh, A., and J. A. Tjon (1965) *Phys. Rev.* **139**, B1085.

——— and V. Chung (1967) *Phys. Rev.* **161**, 1602.

Aitchison, I. J. R. (1965) *Phys. Rev.* **137**, B1070.

——— and C. Kacser (1968) *Phys. Rev.* **173**, 1700.

——— and R. J. A. Golding (1975) *Phys. Lett.* **59B**, 288.

Alaylioglu, A., D. S. Lubinsky, and D. Eyre (1984) *J. Comput. and Appl. Maths.* **11**, 353.

———, D. Eyre, M. Brannigan, and J. P. Svenne (1986) *J. Comput. Phys.* **62**, 383.

Alessandrini, V. A., and R. L. Omnes (1965) *Phys. Rev.* **139**, B167.

——— (1966) *J. Math. Phys.* **7**, 215.

———, C. A. Garcia, and H. Fanchiotti (1968) *Phys. Rev.* **170**, 935.

Alhassid, Y., and S. E. Koonin (1981) *Phys. Rev. C* **23**, 1590.

———, F. Gürsey, and F. Iachello (1986) *Ann. Phys.* (N.Y.) **167**, 181.

Alt, E. O., P. Grassberger, and W. Ṣandhas (1967) *Nucl. Phys.* **B2**, 167.

———, ———, and ——— (1970) *Phys. Rev. C* **1**, 85.

———, ———, and ——— (1972) Joint Institute for Nuclear Research Report No. E4-6688, Dubna, unpublished.

Amado, R. D. (1963) *Phys. Rev.* **132**, 485.

——— (1967) *Phys. Rev.* **158**, 1414.

——— (1969) *Ann. Rev. Nucl. Sci.* **19**, 61.

——— and M. H. Rubin (1970a) *Phys. Rev. Lett.* **25**, 194.

——— (1970b) in "Elementary Particle Physics and Scattering Theory," eds. M. Chretien and S. S. Schweber, Gordon and Breach, New York, Vol. 2, p. 1.

——— and J. V. Noble (1971a) *Phys. Lett.* **35B**, 25.

———, D. F. Freeman, and M. H. Rubin (1971b) *Phys. Rev. D* **4**, 1032.

——— and J. V. Noble (1972) *Phys. Rev. D* **5**, 1992.

——— and F. C. Greenwood (1973) *Phys. Rev. D* **7**, 2517.

——— (1974) *Phys. Rev. Lett.* **33**, 323.

——— (1975) *Phys. Rev. C* **11**, 719.

——— (1977) in "Modern Three-Hadron Physics," ed. A. W. Thomas, Springer-Verlag, Berlin, p. 85.

———, M. P. Locher, and M. Simonius (1978) *Phys. Rev. C* **17**, 403.

——— (1979) *Phys. Rev. C* **19**, 1473.

——— (1981) *Comments Nucl. Part. Phys.* **10**, 131.

Amos, K., L. Berge, F. A. Brieva, A. Katsogiannis, L. Petris, and L. Rikus (1988) *Phys. Rev. C* **37**, 934.

Amrein, W. O., J. M. Jauch, and K. B. Sinha (1977) "Scattering Theory in Quantum Mechanics," W. A. Benjamin, Inc., Reading, Massachusetts.

Andrews, G. E. (1976) in "Encyclopedia of Mathematics and Its Applications," Addision-Wesley, Reading, Massachusetts, Vol. 2, "The Theory of Partitions."

Antunes, A. C., V. L. Baltar, and E. M. Ferreira (1976) *Nucl. Phys.* **A265**, 365.

Araki, M., and T. Ueda (1982) *Nucl. Phys.* **A379**, 449.

——— and I. R. Afnan (1988) *Phys. Rev. C* **38**, 213.

Arellano, H. F., F. A. Brieva, and W. G. Love (1989) *Phys. Rev. Lett.* **63**, 605.

Arkhipov, A. A., and V. I. Savrin (1985) *Sov. J. Part. Phys.* **16**, 485. [*Fiz. Elem. Chastits. At. Yadra.* **16**, 1091 (1985)].
Arnold, L. G., B. C. Clark, R. L. Mercer, and P. Schwandt (1981) *Phys. Rev. C* **23**, 1949.
Arriaga, A., and F. D. Santos (1984) *Phys. Rev. C* **29**, 1945.
Arvieux, J. (1974) *Nucl. Phys.* **A221**, 253.
Ascoli, G., and H. Wyld (1975) *Phys. Rev. D* **12**, 43.
Austern, N. (1970) "Direct Nuclear Reaction Theory," John Wiley, New York.
———, F. Tabakin, and M. Silver (1977) *Am. J. Phys.* **45**, 361.
———, Y. Iseri, M. Kamimura, M. Kawai, G. Rawitscher, and M. Yahiro (1987) *Phys. Rep.* **154**, 125.
Avishai, Y., W. Ebenhoh, A. S. Rinat-Reiner (1969) *Ann. Phys.* (N.Y.) **55**, 341.
——— (1971) *Nucl. Phys.* **A161**, 621.
——— and T. Mizutani (1979) *Nucl. Phys.* **A326**, 352.
——— and ——— (1981) *Nucl. Phys.* **A352**, 399.
Baer, M., and D. J. Kouri (1971) *Phys. Rev. A* **4**, 1924.
——— and ——— (1972) *J. Chem. Phys.* **56**, 1758; *ibid.*, 4840.
——— and ——— (1973) *J. Math. Phys.* **14**, 1637.
——— (1987a) *J. Phys. Chem.* **91**, 5846.
——— and Y. Shima (1987b) *Phys. Rev. A* **35**, 5252.
——— (1989) *Phys. Rep.* **178**, 99.
Bagchi, B., T. O. Krause, and B. Mulligan (1977) *Phys. Rev. C* **15**, 1623.
——— and B. Mulligan (1979) *Phys. Rev. C* **20**, 1973.
Baker, Jr., G. A., and J. L. Gammel eds., (1970) "The Pade Approximant in Theoretical Physics," Academic Press, New York.
Baldo, M., L. S. Ferreira, and L. Streit (1985) *Phys. Rev. C* **32**, 685.
———, ———, ———, and T. Vertse (1986) *Phys. Rev. C* **33**, 1587.
———, ———, and ——— (1987a) *Phys. Rev. C* **36**, 1783.
———, ———, and ——— (1987b) *Nucl. Phys.* **A467**, 44.
Ball, J. S., and W. R. Frazer (1965) *Phys. Rev. Lett.* **14**, 746.
Ballot, J. L., and F. Becker (1967) *Phys. Rev.* **164**, 1285.
Baranger, M., B. Giraud, S. K. Mukhopadhyay, and P. U. Sauer (1969) *Nucl. Phys.* **A138**, 1.
Barbour, I. M., and A. C. Phillips (1967) *Phys. Rev. Lett.* **19**, 1388.
——— and ——— (1970) *Phys. Rev. C* **1**, 165.
——— and J. A. Hendry (1972) *Phys. Lett.* **38B**, 151.
Barlette, V. E., and S. K. Adhikari (1991) *Int. J. Mod. Phys.* **A6**.
Barrett, E. F., B. A. Robson, and W. Tobocman (1983) *Rev. Mod. Phys.* **55**, 155.
Bartnik, E. A., H. Haberzettl, and W. Sandhas (1986) *Phys. Rev. C* **34**, 1520.
———, ———, Th. Januschke, U. Kerwath, and W. Sandhas (1987) *Phys. Rev. C* **36**, 1678.
Barton, G., and A. C. Phillips (1969) *Nucl. Phys.* **A132**, 97.
Basdevant, J. L., and R. E. Kreps (1966) *Phys. Rev.* **141**, 1398; *ibid.*, 1404; *ibid.*, 1409.
Bateman, H. (1922) *Proc. R. Soc. London Ser.* **A100**, 441.

Baur, G., and D. Trautmann (1976) *Phys. Rep.* **25**, 293.
———, F. Rosel, D. Trautmann, and R. Shyam (1984) *Phys. Rep.* **111**, 333.
Bell, J. S., and E. J. Squires (1959) *Phys. Rev. Lett.* **3**, 96.
——— (1962) in "Lectures on the Many-Body Problem," ed. E. R. Caianiello, Academic Press, Inc., New York, p. 91.
Benayoun, J. J., J. Chauvin, C. Gignoux, and A. Laverne (1976a) *Phys. Rev. Lett.* **36**, 1438.
———, C. Gignoux and J. Gillespie (1976b) *Nucl. Phys.* **A274**, 525.
Bencze, Gy. (1973a) *Nucl. Phys.* **A210**, 568.
——— and P. Doleschall (1973b) *Phys. Lett.* **44B**, 235.
——— and E. F. Redish (1975) *Nucl. Phys.* **A238**, 240.
——— (1976) *Phys. Lett.* **61B**, 139.
——— and P. C. Tandy (1977a) *Phys. Rev. C* **16**, 564.
———, G. Cattapan, and V. Vanzani (1977b) *Lett. Nuovo Cim.* **20**, 248.
——— (1977c) *Phys. Lett.* **72B**, 155.
——— (1978a) in "Few-Body Nuclear Physics," IAEA, Vienna, p. 113.
——— and E. F. Redish (1978b) *J. Math. Phys.* **19**, 1909.
——— and ——— (1980) *Phys. Lett.* **91B**, 1.
——— and C. Chandler (1982a) *Phys. Lett.* **90A**, 162.
——— and ——— (1982b) *Phys. Rev. C* **25**, 136.
——— and ——— (1982c) *Phys. Lett.* **112B**, 295.
———, ———, and A. G. Gibson (1982d) *Nucl. Phys.* **A390**, 461.
———, W. N. Polyzou, and E. F. Redish (1982e) *Nucl Phys.* **A390**, 253.
——— and P. Doleschall (1984) in "Few-Body Problems in Physics," ed. B. Zeitnitz, Elsevier, Amsterdam, Vol. II, p. 461.
——— and C. Chandler (1985) *Phys. Lett.* **154B**, 347.
——— and ——— (1988) *Phys. Lett.* **210B**, 23.
Benoist-Gueutal, P. (1975) *Phys. Lett.* **56B**, 413.
———, M. L'Huillier, E. F. Redish, and P. C. Tandy (1978) *Phys. Rev. C* **17**, 1924.
——— and ——— (1982) *J. Math. Phys.* **23**, 1823.
——— (1986) *Phys. Rev. C* **33**, 412.
Berk, A., A. K. Bhatia, B. R. Junker, and A. Temkin (1986) *Phys. Rev. A* **34**, 4591.
Berthold, G. H., and H. Zankel (1986) *Phys. Rev. C* **34**, 1203.
———, A. Stadler, and H. Zankel (1988a) *Phys. Rev. Lett.* **61**, 1077.
———, ———, and ——— (1988b) *Phys. Rev. C* **38**, 444.
Bethe, H. A., and J. Goldstone (1957) *Proc. Roy. Soc.* **A238**, 551.
———, B. H. Brandow, and A. G. Petschek (1963) *Phys. Rev.* **129**, 221.
Betz, M., and T. S. H. Lee (1981) *Phys. Rev. C* **23**, 375.
Bhakar, B. S., and A. N. Mitra (1965) *Phys. Rev. Lett.* **14**, 143.
Bhalerao, R. S., L. C. Liu, and C. M. Shakin (1980) *Phys. Rev. C* **21**, 1903.
Bhasin, V. S., and V. K. Gupta (1982) *Phys. Rev. C* **25**, 1675.
Bianchi, L., and L. Favella (1964) *Nuovo Cimento* **34**, 1825.
Birkhoff, G. (1967) "Lattice Theory," Amer. Math. Soc., Providence, Rhode Island.

Birrell, N. D., and I. R. Afnan (1978) *Phys. Rev. C* **17**, 326.

Birse, M. C., and E. F. Redish (1983a) *Nucl. Phys.* **A406**, 149.

——— (1983b) *Phys. Rev. C* **28**, 2204.

Bishop, R. F. (1973) *Phys. Rev. C* **7**, 479.

Bjorken, J. D., and S. D. Drell (1964) "Relativistic Quantum Mechanics," McGraw-Hill, New York.

——— and ——— (1965) "Relativistic Quantum Fields," McGraw-Hill, New York.

Blais, W. C., M. Zhao, D. G. Truhlar, D. W. Schwenke, and D. J. Kouri (1990) *Chem. Phys. Lett.* **166**, 11.

Blaizot, J.-P., and G. Ripka (1986) "Quantum Theory of Finite Systems," The M.I.T. Press, Cambridge, Massachusetts.

Blankenbecler, R. (1959a) *Nucl. Phys.* **14**, 97.

———, M. Goldberger, and F. Halpern (1959b) *Nucl. Phys.* **12**, 629.

——— and R. Sugar (1964) *Phys. Rev.* **136**, B472.

——— and ——— (1966) *Phys. Rev.* **142**, 1051.

Blankleider, B., and I. R. Afnan (1981) *Phys. Rev. C* **24**, 1572.

——— (1987) *Nucl. Phys.* **A463**, 77c.

Blasczak, D., and M. G. Fuda (1973) *Phys. Rev. C* **8**, 1665.

Bleszynski, B., M. Bleszynski, and T. Jaroszewicz (1987) *Phys. Rev. Lett.* **59**, 423.

Blum, K., and P. G. Burke (1977) *Phys. Rev. A* **16**, 163.

Bollé, D., and K. L. Kowalski (1970) *Nuovo Cim.* **67A**, 523.

——— (1981) *Nucl. Phys.* **A353**, 377c.

Bolsterli, M. (1969) *Phys. Rev.* **182**, 1095.

——— (1974) *Phys. Rev. Lett.* **32**, 436.

——— and A. S. Rinat (1977) *Phys. Rev. C* **15**, 40.

Bornard, M., G. R. Plattner, R. D. Viollier, and K. Alder (1978) *Nucl. Phys.* **A294**, 492.

Bower, R. H. J. (1972) *Ann. Phys.* (N.Y.) **73**, 372.

Brady, T. J., and I. H. Sloan (1972) *Phys. Lett.* **40B**, 55.

——— and ——— (1974) *Phys. Rev. C* **9**, 4.

Brandenburg, R. A., Y. E. Kim, and A. Tubis (1975) *Phys. Rev. C* **12**, 1368.

———, G. S. Chulick, R. Machleidt, A. Picklesimer, and R. M. Thaler (1988a) *Phys. Rev. C* **37**, 1245.

———, ———, Y. E. Kim, D. J. Klepacki, R. Machleidt, A. Picklesimer, and R. M. Thaler (1988b) *Phys. Rev. C* **37**, 781.

Brannigan, M., and D. Eyre (1983) *J. Math. Phys.* **24**, 177.

Brayshaw, D. D. (1968a) *Phys. Rev.* **167**, 1505.

——— (1968b) *Phys. Rev.* **176**, 1855.

——— (1971) *Phys. Rev. C* **3**, 35.

——— (1974) *Phys. Rev. Lett.* **32**, 382.

——— (1976) *Phys. Rev. C* **13**, 1024.

——— (1977) in "Modern Three-Hadron Physics," ed. A. W. Thomas, Springer-Verlag, Berlin, p. 105.

Broido, M., and J. G. Taylor (1969a) *J. Math. Phys.* **10**, 184.

——— (1969b) *Rep. Prog. Phys.* **32**, 493.

Brown, G. E., and A. D. Jackson (1976) "The Nucleon–Nucleon Interaction," North-Holland Pub. Co., Amsterdam.
Brown, L., D. I. Fivel, B. W. Lee, and R. F. Sawyer (1963) *Ann. Phys.* (N.Y.) **23**, 187.
Bruinsma, J., W. Ebenhoh, J. H. Stuivenberg, and R. Wageningen (1974) *Nucl. Phys.* **A228**, 52.
Buck, B., and R. Lipperheide (1981) *Nucl. Phys.* **A368**, 141.
Bund, G. W., and M. C. Tijero (1980) *Nuovo Cim.* **57A**, 234.
——— (1985) *Phys. Rev. C* **31**, 2022.
Burke, P. G., H. M. Schey, and K. Smith (1963) *Phys. Rev.* **129**, 1258.
———, A. Hibbert, and W. D. Robb (1971) *J. Phys. B* **4**, 153.
Burov, V. V., and V. K. Lukyanov (1987) *Nucl. Phys.* **A463**, 263c.
Byron, Jr., F. W., F. J. de Heer, and C. J. Joachain (1975) *Phys. Rev. Lett.* **35**, 1147.
Cahill, R. T., and I. H. Sloan (1971) *Nucl. Phys.* **A165**, 161.
——— (1972) *Nucl. Phys.* **A194**, 599.
——— (1974) *Phys. Rev. C* **9**, 473.
Callaway, J. (1978) *Phys. Rep.* **45**, 89.
——— (1980) *Phys. Lett.* **77A**, 137.
Campbell, W. B., P. Finkler, C. E. Jones, and M. N. Misheloff (1975) *Phys. Rev. D* **12**, 2353.
Canton, L., G. Cattapan, and G. Pisent (1987) *Nuovo Cim. A* **97**, 319.
———, ———, and ——— (1988) *Nucl. Phys.* **A487**, 333.
Carew, J., and L. Rosenberg (1973) *Phys. Rev. D* **7**, 1113.
Casel, A., H. Haberzettl, and W. Sandhas (1982) *Phys. Rev. C* **25**, 1738.
Cattapan, G., and V. Vanzani (1975) *Lett. Nuovo Cim.* **14**, 465.
——— and ——— (1979a) *Phys. Rev. C* **19**, 1168.
——— and ——— (1979b) *Nuovo Cim.* **50A**, 97.
——— and ——— (1979c) *Lett. Nuovo Cim.* **24**, 391.
——— and ——— (1982) *Nuovo Cim.* **68A**, 368.
——— and ——— (1983a) *Nuovo Cim.* **77A**, 263.
——— and ——— (1983b) *Nuovo Cim.* **78A**, 485.
——— and ——— (1984) *Nuovo Cim.* **81A**, 447.
——— and ——— (1985a) "Recent Developments in N-Body Scattering Theory," talk at Xth European Symposium on Dynamics of Few-Body Systems.
——— and ——— (1985b) *Nuovo Cim.* **89A**, 29.
——— and ——— (1986) in "Few-Body Approaches to Nuclear Reactions in Tandem and Cyclotron Regions," eds. S. Oryu and T. Sawada, p. 200.
Celenza, L. S. (1979) *Phys. Rev. C* **19**, 447.
——— and C. M. Shakin (1983) *Phys. Rev. C* **28**, 1256.
——— and ——— (1984) *Phys. Rev. C* **29**, 1784.
——— and ——— (1986) "Relativistic Nuclear Theories of Structure and Scattering," World Scientific, Singapore.
Chandler, C., and A. G. Gibson (1977) *J. Math. Phys.* **18**, 2336.
——— and ——— (1978a) *J. Math. Phys.* **19**, 1610.
——— (1978b) *Nucl. Phys.* **A301**, 1.

——— and I. H. Sloan (1981a) *Nucl. Phys.* **A361**, 521.
——— (1981b) *Nucl. Phys.* **A353**, 129c.
——— and A. G. Gibson (1983) *J. Func. Analy.* **52**, 80.
——— and ——— (1984) *J. Math. Phys.* **25**, 1841.
——— (1987) *Nucl. Phys.* **A463**, 181c.
——— and ——— (1989) *J. Math. Phys.* **30**, 1533.
Charnomordic, B., C. Fayard, and G. H. Lamot (1977) *Phys. Rev. C* **15**, 864.
Chauvin, J., D. Garreta, and M. Fruneau (1975a) *Nucl. Phys.* **A247**, 335.
——— and J. Arvieux (1975b) *Nucl. Phys.* **A247**, 347.
Chen, C. R., G. L. Payne, J. L. Friar, and B. F. Gibson (1985) *Phys. Rev. C* **31**, 2266.
———, ———, ———, and ——— (1986a) *Phys. Rev. C* **33**, 401.
———, ———, ———, and ——— (1986b) *Phys. Rev. C* **33**, 1740.
———, ———, ———, and ——— (1989) *Phys. Rev. C* **39**, 1261.
Chew, G. (1951) *Phys. Rev.* **80**, 196.
——— and G. C. Wick (1952) *Phys. Rev.* **85**, 636.
——— and S. Mandelstam (1960) *Phys. Rev.* **119**, 467.
Christie, I., and D. Eyre (1986) *Few Body Syst.* **1**, 111.
Clark, B. C., S. Hama, P. L. Mercer, L. Ray, and B. D. Serot (1983a) *Phys. Rev. Lett.* **50**, 1644.
———, ———, ———, ———, G. Hoffman, and B. D. Serot (1983b) *Phys. Rev. C* **28**, 1421.
Coelho, H. T., T. K. Das, and M. R. Robilotta (1983) *Phys. Rev. C* **28**, 1812.
Coester, F. (1971) *Phys. Rev. C* **3**, 525.
——— and W. Polyzou (1982) *Phys. Rev. D* **26**, 1348.
Cohen, H., and A. Pagnamenta (1969) *Phys. Rev.* **181**, 2098.
Collins, P. D. B. (1977) "An Introduction to Regge Theory and High Energy Physics," Cambridge University Press, New York.
Considine, D. M. (ed.) (1976) "Van Nostrand's Scientific Encyclopedia," 5th Ed., Van Nostrand Reinhold Co., New York, pp. 612, 1967.
Coon, S. A., M. D. Scadron, P. C. McNamee, B. R. Barrett, D. W. E. Blatt, and B. H. J. McKellar (1979) *Nucl. Phys.* **A317**, 242.
——— and W. Glöckle (1981) *Phys. Rev. C* **23**, 1790.
Cooper, E. D., and B. K. Jennings (1986) *Nucl. Phys.* **A458**, 711.
——— and ——— (1988) *Phys. Rev. C* **37**, 1332.
Cornelius, T., and W. Glöckle (1986) *J. Chem. Phys.* **85**, 3906.
Cotanch, S. R., and C. M. Vincent (1976) *Phys. Rev. C* **14**, 1739.
Dashen, R., S. Ma, and H. J. Bernstein (1969) *Phys. Rev.* **187**, 345.
——— and ——— (1970) *J. Math. Phys.* **11**, 1136.
——— and ——— (1971) *J. Math. Phys.* **12**, 689.
de Boor, C. (1978) "A Practical Guide to Splines," Springer-Verlag, New York.
de Goey, L. P. H., J. P. J. Driessen, B. J. Verhaar, and J. T. M. Walraven (1984) *Phys. Rev. Lett.* **53**, 1919.
———, T. H. M. v. d. Berg, N. Mulders, H. T. C. Stoof, B. J. Verhaar, and W. Glöckle (1986) *Phys. Rev. B* **34**, 6183.

Delfino, A., T. Frederico, and L. Tomio (1988) *Phys. Rev. C* **38**, 11.
Delves, L. M. (1972) *Adv. Nucl. Phys.* **5**, 1.
Dereziński, J. (1987) *J. Math. Phys.* **28**, 1080.
de Tourreil, R., and D. W. L. Sprung (1973) *Nucl. Phys.* **A201**, 193.
———, B. Rouben, and D. W. L. Sprung (1975) *Nucl. Phys.* **A242**, 445.
Dilg, W., L. Koester, and W. Nistler (1971) *Phys. Lett.* **36B**, 208.
Dixson, R. M., and E. F. Redish (1980) *J. Math. Phys.* **21**, 372.
Dodd, L. R. (1977) in "Modern Three-Hadron Physics," ed. A. W. Thomas, Springer-Verlag, Berlin, p. 49.
Doleschall, P. (1973) *Nucl. Phys.* **A201**, 264.
———, W. Grüebler, V. Konig, P. A. Schmelzbach, F. Sperisen, and B. Jenny (1982) *Nucl. Phys.* **A380**, 72.
Donellan, T. (1968) "Lattice Theory," Pergamon, New York.
Doolen, G., M. Jain, and T. Burnett (1974) *Phys. Rev. C* **9**, 23.
Duck, I. (1968) in "Advances in Nuclear Physics," eds. M. Baranger and E. Vogt, Vol. I, p. 343.
Dzhibuti, R. I., and Sh. M. Tsiklauri (1984) *Yad. Fiz.* **39**, 1115 [*Sov. J. Nucl. Phys.* **39**, 704 (1984)].
Eden, R. J., P. V. Landshoff, D. I. Olive, and J. C. Polkinghorne (1966) "The Analytic S-Matrix," Cambridge University Press, Cambridge.
Edmonds, A. R. (1957) "Angular Momentum in Quantum Mechanics," Princeton University Press, Princeton.
Efimov, V. (1970a) *Phys. Lett.* **33B**, 563.
——— (1970b) *Yad. Fiz.* **12**, 1080 [*Sov. J. Nucl. Phys.* **12**, 589 (1971)].
——— (1973) *Nucl. Phys.* **A210**, 157.
——— (1976) in "Few-Body Dynamics," eds. A. N. Mitra I. Slaus, V. S. Bhasin, and V. K. Gupta, North-Holland, Amsterdam, p. 126.
——— (1981) *Nucl. Phys.* **A362**, 45.
——— and E. G. Tkachenko (1985) *Phys. Lett.* **157B**, 109.
——— and ——— (1988) *Few Body Syst.* **4**, 71.
Eisenberg, J. M., and D. S. Koltun (1980) "Theory of Meson Interactions with Nuclei," John Wiley, New York.
——— and ——— (1989) "Quantum Many-Particle System," John Wiley, New York.
Ekstein, H. (1956) *Phys. Rev.* **101**, 880.
Epstein, S. T. (1957) *Phys. Rev.* **106**, 598.
Ericson, T. E. O., and M. P. Locher (1970) *Nucl. Phys.* **A148**, 1.
——— and M. Rosa-Clot (1983) *Nucl. Phys.* **A405**, 497.
——— (1984) *Nucl. Phys.* **A416**, 281c.
——— and M. Rosa-Clot (1985) *Ann. Rev. Nucl. Part. Sci.* **35**, 271.
Erkelenz, K., and K. Holinde (1972) *Nucl. Phys.* **A194**, 161.
Ernst, D. J., C. M. Shakin, and R. M. Thaler (1973a) *Phys. Rev. C* **8**, 46.
———, ———, ———, and D. Weiss (1973b) *Phys. Rev. C* **8**, 2056.
———, J. T. Londergan, G. A. Miller, and R. M. Thaler (1977) *Phys. Rev. C* **16**, 537.
———, G. A. Miller, and D. L. Weiss (1983) *Phys. Rev. C* **27**, 2733.

Eyre, D., and T. Osborn (1979) *Phys. Rev. C* **20**, 869.

Faddeev, L. D. (1960) *Zh. Exp. Teor. Fiz.* **39**, 1459 [*Sov. Phys. JETP* **12**, 1014 (1961)].

——— (1961) *Dokl. Akad. Nauk. USSR* **138**, 565 [*Sov. Phys.-Doklady* **6**, 384 (1961)].

——— (1962) *Dokl. Akad. Nauk. USSR* **145**, 301 [*Sov. Phys.-Doklady* **7**, 600 (1963)].

——— (1965) "Mathematical Aspects of the Three-Body Problem in Quantum Scattering Theory," Davey, New York.

——— (1970) in "Three-Body Problem in Nuclear and Particle Physics," ed. J. S. C. McKee and P. M. Rolph, North-Holland, Amsterdam, p. 154.

——— and O. A. Yakubovskii (1981) *Yad. Fiz.* **33**, 634 [*Sov. J. Nucl. Phys.* **33**, 331 (1981)].

Fano, U., and A. R. P. Rau (1986) "Atomic Collisions and Spectra," Academic Press, Inc., Orlando, Florida.

Fayard, C., G. H. Lamot, and T. Mizutani (1980) *Phys. Rev. Lett.* **45**, 524.

Fearing, H. W., G. R. Goldstein, and M. J. Moravcik (1984) *Phys. Rev. D* **29**, 2612.

——— (1987) *Nucl. Phys.* **A463**, 95c.

Federbush, P. G. (1966) *Phys. Rev.* **148**, 1551.

——— (1967) *J. Math. Phys.* **8**, 2415.

Ferreira, L. S. (1986) Lecture Notes in Physics **273**, 100 Springer-Verlag, Berlin.

Feshbach, H. (1958) *Ann. Phys.* (N.Y.) **5**, 357.

——— (1962) *Ann. Phys.* (N.Y.) **19**, 287.

———, A. Gal, and J. Hufner (1971) *Ann. Phys.* (N.Y.) **66**, 20.

Fetter, A. L., and K. M. Watson (1965) in "Advances in Theoretical Physics," ed. K. A. Brueckner, Academic Press, New York, p. 113.

——— and J. D. Walecka (1971) "Quantum Theory of Many-Particle Systems," McGraw-Hill, New York.

Fiebig, H. R., M. Stingl, and W. Timm (1978) *J. Phys. G. Nucl. Phys.* **4**, 1291.

——— and W. Timm (1981) *Ann. Phys.* (N.Y.) **134**, 141.

Fiedeldey, H. (1987) *Nucl. Phys.* **A463**, 335c.

Finkel, R. W., and L. Rosenberg (1968) *Phys. Rev.* **168**, 1841.

Fishbane, P. M., and J. Noble (1967) *Phys. Rev.* **160**, 880.

Foldy, L. L. (1945) *Phys. Rev.* **67**, 107.

——— and W. Tobocman (1957) *Phys. Rev.* **105**, 1099.

——— and J. D. Walecka (1969) *Ann. Phys.* (N.Y.) **54**, 447.

Fonseca, A. C., and P. E. Shanley (1976a) *Phys. Rev. D* **13**, 2255.

——— and ——— (1976b) *Phys. Rev. C* **14**, 1343.

———, E. F. Redish, and P. E. Shanley (1979) *Nucl. Phys.* **A320**, 273.

———, H. Haberzettl, and E. Cravo (1983) *Phys. Rev. C* **27**, 939.

——— (1984) *Nucl. Phys.* **A416**, 421c.

——— and T. K. Lim (1985) *Phys. Rev. Lett.* **55**, 1285.

——— (1986a) Lecture Notes in Physics **273**, 161, Springer-Verlag, Berlin.

——— (1986b) in "Few Body Methods: Principles and Applications," eds. T. K. Lim, C.-G. Bao, D.-P. Hou, and S. Huber World Scientific, p. 111.

——— (1986c) *Few Body Syst.* **1**, 69.

——— (1989a) *Phys. Rev. C* **40**, 1390.
——— (1989b) *Phys. Rev. Lett.* **63**, 2036.
Frank, T. W., H. Haberzettl, Th. Januschke, U. Kerwath, and W. Sandhas (1988) *Phys. Rev. C* **38**, 1112.
Frautschi, S. C. (1963) "Regge Poles and S-Matrix Theory," W. A. Benjamin Inc., New York.
Frederico, T., and I. D. Goldman (1987) *Phys. Rev. C* **36**, 1661.
———, S. K. Adhikari, and M. S. Hussein (1988a) *Phys. Rev. C* **37**, 364.
———, I. D. Goldman, and S. K. Adhikari (1988b) *Phys. Rev. C* **37**, 949.
———, I. D. Goldman, and A. Delfino (1988c) *Phys. Rev. C* **37**, 497.
Freedman, D. Z., C. Lovelace, and J. M. Namyslowski (1966) *Nuovo Cim.* **43A**, 258.
Friar, J. L. (1979) *Phys. Rev. C* **20,** 325.
———, B. F. Gibson and, G. L. Payne (1983) *Phys. Rev. C* **28**, 983.
———, ———, ———, and C. R. Chen (1984a) *Phys. Rev. C* **30**, 1121.
———, ———, and ——— (1984b) *Ann. Rev. Nucl. Part. Sci.* **34**, 403.
———, ———, ———, and C. R. Chen (1986) *Phys. Rev. C* **34**, 1463.
——— (1987a) *Nucl. Phys.* **A463**, 315c.
———, B. F. Gibson, and G. L. Payne (1987b) *Phys. Rev. C* **35**, 1502.
———, ———, D. R. Lehman, and G. L. Payne (1988a) *Phys. Rev. C* **37**, 2859.
———, ———, G. L. Payne, E. L. Tomusiak, and M. Kimura (1988b) *Phys. Rev. C* **37**, 2852.
Frye, G., and R. L. Warnock (1963) *Phys. Rev.* **130**, 478.
Fuda, M. G. (1967) *Phys. Rev.* **178**, 1682.
——— (1968) *Nucl. Phys.* **A116**, 83.
——— (1969) *Phys. Rev.* **186**, 1078.
——— (1971) *Phys. Rev. C* **3**, 485.
——— (1976) *Phys. Rev. C* **14**, 1336.
———, B. A. Girard, and J. S. Whiting (1980) *Phys. Rev C* **21**, 18.
Fulton, T., and P. Schwed (1959) *Phys. Rev.* **115**, 973.
Garcilazo, H. (1980) *Phys. Rev. Lett.* **45**, 780.
——— and G. Mercado (1982) *Phys. Rev. C* **25**, 2596.
——— (1984) *Phys. Rev. Lett.* **53**, 652.
——— (1987) *Phys. Rev. C* **35**, 1820.
——— (1988a) *Phys. Rev. Lett.* **61**, 1457.
——— (1988b) *Phys. Rev. C* **37**, 2022.
Gell-Mann, M., and F. E. Low (1951) *Phys. Rev.* **84**, 350.
——— and M. L. Goldberger (1953) *Phys. Rev.* **91**, 398.
Gerjuoy, E., and J. Schwinger (1942) *Phys. Rev.* **61**, 138.
——— and D. S. Saxon (1954) *Phys. Rev.* **94**, 478.
——— (1958) *Phys. Rev* **109**, 1806.
——— and N. A. Krall (1960) *Phys. Rev.* **119**, 705.
——— and ——— (1962) *Phys. Rev.* **127**, 2105.
——— (1971) *Philos. Trans. Roy. Soc.* (London) **270**, 197.

——— and C. M. Lee (1978) *J. Phys. B* **11**, 1137.
———, A. R. P. Rau, and L. Spruch (1983) *Rev. Mod. Phys.* **55**, 725.
——— and S. K. Adhikari (1985) *Phys. Rev. A* **31**, 2005.
——— and ——— (1986a) *Phys. Rev. C* **34**, 1.
——— (1986b) *Phys. Rev. C* **34**, 1143.
——— and S. K. Adhikari (1987a) *Phys. Rev. C* **35**, 415.
——— (1987b) *Phys. Rev. C* **36**, 1.
Gerry, C. C., and V. A. Singh (1980) *Phys. Rev. D* **21**, 2979.
Gibson, B. F., and D. R. Lehman (1975) *Phys. Rev. C* **11**, 29.
——— and ——— (1976) *Phys. Rev. C* **13**, 477.
——— and ——— (1977) *Phys. Rev. C* **15**, 545.
——— (1981) *Nucl. Phys.* **A353**, 85c.
———, B. C. Pearce, and G. L. Payne (1989) *Phys. Rev. C* **40**, 2877.
Gibson, W. G. (1972) *Phys. Rev. A* **6**, 2469.
——— (1983) *Mol. Phys.* **49**, 103.
——— (1986) "Few Body Methods: Principles and Applications," eds. T. K. Lim C.-G. Bao, D.-P. Hou, and S. Huber World Scientific, Singapore, p. 637.
——— (1987) *Phys. Rev. A* **36**, 564.
Giebink, D. R., and D. J. Ernst (1988) *Comput. Phys. Comm.* **48**, 407.
Gillespie, J. (1964) "Final-State Interactions," Holden-Day, Inc., San Francisco.
Girard, B. A., and M. G. Fuda (1979a) *Phys. Rev. C* **19**, 583.
——— and ——— (1979b) *Phys. Rev. C* **19**, 579.
Giraud, B. G., and M. A. Nagarajan (1980) *J. Physique* **41**, 477.
Glauber, R. J. (1959) in "Lectures in Theoretical Physics," eds. W. E. Brittin and L. G. Dunham, Interscience, New York, Vol. 1, p. 315.
Glendenning, N. K. (1983) "Direct Nuclear Reactions," Academic Press, New York.
Glöckle, W. (1970a) *Nucl. Phys.* **A141**, 620.
——— (1970b) *Nucl. Phys.* **A158**, 257.
——— and K. Ueta (1973) *Z. Phys.* **258**, 64.
——— (1974) *Z. Phys.* **271**, 31.
——— (1982a) *Nucl. Phys.* **A381**, 343.
———, G. Hasberg, and A. R. Neghabian (1982b) *Z. Phys.* **A305**, 217.
——— (1983a) "The Quantum Mechanical Few-Body Problem," Springer, New York.
——— and R. Brandenberg (1983b) *Phys. Rev. C* **27**, 83.
——— (1988) *Phys. Rev. C* **37**, 6.
Goldberger, M. L. (1960) in "Dispersion Relations and Elementary Particles," Les Hoches 1960, ed. C. DeWitt and R. Omnes, John Wiley, New York, p. 15.
——— and K. M. Watson (1964) "Collision Theory," John Wiley, Inc., New York.
Goldflam, R., and W. Tobocman (1978) *Phys. Rev. C* **17**, 1914.
——— and ——— (1979) *Phys. Rev. C* **20**, 904.
——— and K. L. Kowalski (1980a) *Phys. Rev. C* **21**, 483.
——— and ——— (1980b) *Phys. Rev. Lett.* **44**, 1044.
——— and ———, and W. Tobocman (1980c) *J. Math. Phys.* **21**, 1888.
——— and ——— (1980d) *Phys. Rev. C* **22**, 949.

——— and ——— (1980e) *Phys. Rev. C* **22**, 2341.
———, W. Tobocman, and R. M. Thaler (1981) *Nucl. Phys.* **A359**, 122.
——— (1982) *Phys. Rev. C* **26**, 34.
Grassberger, P., and W. Sandhas (1967) *Nucl. Phys.* **B2**, 181.
——— and ——— (1969) *Z. Physik* **220**, 29.
Graves-Morris, P. R. (1966) *Phys. Rev. Lett.* **16**, 201.
——— (1967) *Ann. Phys.* (N.Y.) **41**, 477.
Greben, J. M., and L. P. Kok (1976a) *Phys. Rev. C* **13**, 489.
——— and ——— (1976b) *Phys. Rev. C* **13**, 1352.
——— and Yu. A. Simonov (1978) *Phys. Rev. C* **18**, 642.
——— and F. S. Levin (1979) *Phys. Rev. C* **20**, 437.
——— (1982) *Phys. Rev. C* **25**, 446.
Green, M. B., J. H. Schwarz, and E. Witten (1987) "Superstring Theory," Vols. I and II, Cambridge University Press, Cambridge.
Gross, F. (1969) *Phys. Rev.* **186**, 1448.
Guennegues, J. Y. (1966) *Nuovo Cim.* **42A**, 549.
Gurvitz, S. A., J.-P Dedonder, and R. D. Amado (1979) *Phys. Rev. C* **19**, 142.
——— (1980) *Phys. Rev. C* **22**, 964.
——— (1981) *Phys. Rev. C* **24**, 29.
——— (1986) *Phys. Rev. C* **33**, 422.
Haberzettl, H., and W. Sandhas (1981a) *Phys. Rev. C* **24**, 359.
——— (1981b) *Phys. Rev. Lett.* **47**, 1367.
Haftel, M. I., and F. Tabakin (1970a) *Nucl. Phys.* **A158**, 1.
——— (1970b) *Phys. Rev. Lett.* **25**, 120.
——— and E. L. Petersen (1974) *Phys. Rev. Lett.* **33**, 1229.
——— and H. Zankel (1981) *Phys. Rev. C* **24**, 1322.
——— and T. K. Lim (1982) *J. Chem. Phys.* **77**, 4515.
——— and ——— (1986) *J. Chem. Phys.* **84**, 4407.
Hahn, Y. (1968) *Phys. Rev.* **169**, 794.
——— (1969) *Nucl. Phys.* **A132**, 353.
——— (1970) *Ann. Phys.* (N.Y.) **58**, 137.
——— and K. M. Watson (1972) *Phys. Rev.* **A5**, 1718.
———, D. J. Kouri, and F. S. Levin (1974) *Phys. Rev. C* **10**, 1615.
——— (1976) *Phys. Rev. C* **14**, 1709.
——— (1977) *Phys. Rev. C* **16**, 2449.
——— (1980) *Phys. Lett.* **97B**, 1.
——— (1982) *Nucl. Phys.* **A389**, 1.
Haidenbauer, J., and W. Plessas (1983) *Phys. Rev. C* **27**, 63.
——— and ——— (1984) *Phys. Rev. C* **30**, 1822.
———, Y. Koike, and W. Plessas (1986a) *Phys. Rev. C* **33**, 439.
———, ———, and ——— (1986b) in "Proceedings of the International Workshop on Few-Body Approaches to Nuclear Reactions in Tandem and Cyclotron Energy Regions," Tokyo 1986, eds. S. Oryu and T. Sawada, World Scientific, 1987, p. 23.

Hajduk, C., and P. U. Sauer (1981) *Nucl. Phys.* **A369**, 321.
Hamilton, J., and W. S. Woolcock (1963) *Rev. Mod. Phys.* **35**, 737.
Harms, E. (1970) *Phys. Rev. C* **1**, 1667.
Hartt, K., and P. V. A. Yidana (1987) *Phys. Rev. C* **36**, 475.
Haymaker, R. W. (1968) *Phys. Rev.* **165**, 1790.
Hendry, J. A., and A. C. Phillips (1973) *Nucl. Phys.* **A211**, 533.
Hepp, K. (1969) *Helv. Phys. Acta.* **42**, 425.
——— (1972) "Quantum Scattering Theory of Multiparticle Systems: A Guide to the Literature," in "Mathematics of Contemporary Physics," ed. R. F. Streater, Academic Press, London, p. 183.
Herndon, D. J., P. Soding, and R. J. Cashmore (1975a) *Phys. Rev. D* **11**, 3165.
———, R. Longacre, L. R. Miller, A. H. Rosenfeld, G. Smadja, P. Söding, R. J. Cashmore, and D. W. G. S. Leith (1975b) *Phys. Rev. D* **11**, 3183.
Hess, H. F., D. A. Bell, G. P. Kochanski, D. Kleppner, T. J. Greytak (1984) *Phys. Rev. Lett.* **52**, 1520.
Hetherington, J. H., and L. H. Schick (1965) *Phys. Rev.* **137**, B935.
Hippchen, Th., and K. Holinde (1988) *Phys. Rev. C* **37**, 239.
Hodgson, P. E. (1971) "Nuclear Reactions and Nuclear Structure," Oxford University Press, London.
Hofmann, H. M. (1984) *Nucl. Phys.* **A416**, 363c.
Hoheisel, G. (1968) "Integral Equations," Ungar, New York.
Holz, J., and W. Glockle (1988) *Phys. Rev. C* **37**, 1386.
Hořaček, J., and T. Sasakawa (1983) *Phys. Rev. A* **28**, 2151.
——— and ——— (1984) *Phys. Rev. A* **30**, 2274.
——— and ——— (1985) *Phys. Rev. C* **32**, 70.
Huang, K., and H. A. Weldon (1975) *Phys. Rev. D* **11**, 257.
Huber, H. S., T. K. Lim, and D. H. Feng (1978) *Phys. Rev. C* **18**, 1534.
——— (1985) *Phys. Rev. A* **31**, 3981.
Hulthén, L. (1944) *Kgl. Fysiograf. Sällskap. Förh.* **14**, 257.
——— (1948) *Ark. Mat. Astron. Fys.* **35A**, No. 25.
Hunziker, W. (1965) *J. Math. Phys.* **6**, 6.
——— (1968) in "Lectures in Theoretical Physics," eds. A. O. Barut and W. E. Brittin, Gordan and Breach, New York, Vol. X-A, p. 1.
——— (1977) *Acta. Phys. Aus. Suppl.* **17**, 43.
Hussein, M. S., and E. J. Moniz (1984) *Phys. Rev. C* **29**, 2054.
——— and K. W. McVoy (1985) *Nucl. Phys.* **A445**, 124.
Hynes, M. V., A. Picklesimer, P. C. Tandy, and R. M. Thaler (1984) *Phys. Rev. Lett.* **52**, 978.
———, ———, ———, and ——— (1985) *Phys. Rev. C* **31**, 1438.
Imanishi, B., M. Ichimura, and M. Kawai (1974) *Phys. Lett.* **52B**, 267.
Ishikawa, S., and T. Sasakawa (1986a) *Phys. Rev. Lett.* **56**, 317.
——— and ——— (1986b) *Few Body Syst.* **1**, 143.
——— and ——— (1987a) *Phys. Rev. C* **36**, 2037.

——— (1987b) *Nucl. Phys.* **A463**, 145c.

Isihara, A. (1971) "Statistical Physics," Academic Press, New York.

Itzykson, C., and J-B Zuber (1980) "Quantum Field Theory," McGraw-Hill Book Co., New York.

Jackson, D. F. (1970) "Nuclear Reactions," Methuen, London.

Jackson, J. D. (1965) in "High Energy Physics," Les Houches 1965, eds. C. DeWitt and M. Jacob, Gordon and Breach, New York, p. 367.

Jauch, W., A. Konig, and P. Kroll (1984) *Phys. Lett.* **143B**, 509.

Joachain, C. K. (1975) "Quantum Collision Theory," North-Holland, Pub. Co., Amsterdam.

Johnson, C. H., and N. M. Larson, C. Mahaux, and R. R. Winters (1984) *Phys. Rev. C* **29**, 1563.

Johnson, M. B., and M. K. Singham (1987) *Ann. Phys.* (N.Y.) **180**, 254.

Jost, R., and A. Pais (1951) *Phys. Rev.* **82**, 840.

——— (1965) "The General Theory of Quantized Fields," American Math. Soc., Providence.

Jourdan, J., M. Baumgartner, S. Burzynski, P. Egelhof, A. Klein, M. A. Pickar, G. R. Plattner, W. D. Ramsay, H. W. Roser, and I. Sick (1985) *Phys. Lett.* **162B**, 269.

——— (1986) *Nucl. Phys.* **A453**, 220.

Kadanoff, L. A., and G. Baym (1962) "Quantum Statistical Mechanics," Benjamin/Cummings, London.

Kadychevsky, V. G. (1968) *Nucl. Phys.* **B6**, 125.

Kagan, Yu., I. A. Vartan'yants, and G. V. Shlyapnikov (1981) *Zh. Eksp. Teor. Fiz.* **81**, 113 [*Sov. Phys. JETP* **54**, 590 (1981)].

Karaoglu, B., and E. Moniz (1986) *Phys. Rev. C* **33**, 974.

Karlsson, B. R., and E. M. Zeiger (1974a) *Phys. Rev. D* **9**, 1761.

——— and ——— (1974b) *Phys. Rev. D* **10**, 1291.

——— and ——— (1975) *Phys. Rev. D* **11**, 939.

——— and ——— (1977a) *Phys. Rev. D* **16**, 2553.

——— (1977b) *Phys. Lett.* **69B**, 13.

——— (1978) *Phys. Rev. D* **17**, 642.

——— (1982) *J. Math. Phys.* **23**, 421.

Karner, G. (1987) *Few Body Syst.* **3**, 7.

Kazaks, P. A., and K. R. Greider (1970) *Phys. Rev. C* **1**, 856.

Keister, B. D., and R. B. Wiringa, (1986) *Phys. Lett.* **173B**, 5.

Kerman, A. K., H. McManus, and R. M. Thaler (1959) *Ann. Phys.* (N.Y.) **8**, 551.

——— and K. W. McVoy (1979) *Ann. Phys.* (N.Y.) **122**, 197.

Kharchenko, V. F., N. M. Petrov, and S. A. Storozhenko (1968) *Nucl. Phys.* **A106**, 464.

——— and V. P. Levashev (1976) *Phys. Lett.* **60B**, 317.

——— and ——— (1980) *Nucl. Phys.* **A343**, 249.

Kholyavin, I. I., and O. A. Yakubovskii (1977) *Yad. Fiz.* **26**, 181 [*Sov. J. Nucl. Phys.* **26**, 94 (1977)].

Kim, Y. E., and A. Tubis (1970) *Phys. Rev. C* **1**, 414.

——— and ——— (1974) *Ann. Rev. Nucl. Sci.* **24**, 69.
King, S. E., N. R. Roberson, H. R. Weller, D. R. Tilley (1983) *Phys. Rev. Lett.* **51**, 877.
Kitching, P., D. A. Hutcheon, K. Michaelian, R. Abegg, G. H. Coombes, W. K. Dawson, H. Fielding, G. Gaillard, P. Green, L. G. Greeniaus, M. Hugi, C. A. Miller, G. C. Neilson, W. C. Olsen, J. Soukup, N. R. Stevenson, J. Wesick, H. W. Fearing, and R. L. Workman (1986) *Phys. Rev. Lett.* **57**, 2363.
Klein, A. (1953) *Phys. Rev.* **90**, 1101.
——— (1954) *Phys. Rev.* **94**, 1052.
——— and C. Zemach (1957) *Phys. Rev.* **108**, 126.
——— and ——— (1958a) *Nuovo Cim.* **10**, 1078.
——— (1958b) *Prog. Theor. Phys.* **20**, 257.
———, and T.-S. H. Lee (1974) *Phys. Rev. D* **10**, 4308.
Kliner, D. A. V., K. D. Rinnen, and R. N. Zare (1990) *Chem. Phys. Lett.* **166**, 107.
Kloet, W. M., and J. A. Tjon (1973a) *Nucl. Phys.* **A210**, 380.
——— and ——— (1973b) *Ann. Phys.* (N.Y.) **79**, 407.
———, R. R. Silbar, R. Aaron, and R. D. Amado (1977) *Phys. Rev. Lett.* **39**, 1643.
——— and ——— (1980) *Nucl. Phys.* **A338**, 281; *ibid.*, 317.
——— and ——— (1981) *Nucl. Phys.* **A364**, 346.
Kohn, W. (1948) *Phys. Rev.* **74**, 1763.
Koike, Y. (1978a) *Nucl. Phys.* **A301**, 411.
——— (1978b) *Prog. Theor. Phys.* **59**, 87.
——— (1980) *Nucl. Phys.* **A337**, 23.
———, W. Plessas, and H. Zankel (1985) *Phys. Rev. C* **32**, 1796.
——— and Y. Taneguchi (1986) *Few Body Syst.* **1**, 13.
———, J. Haidenbauer, and W. Plessas (1987a) *Phys. Rev. C* **35**, 396.
——— and ——— (1987b) *Nucl. Phys.* **A463**, 365c.
Kok, L. P., and A. S. Rinat (1970) *Nucl. Phys.* **A156**, 593.
Komarov, V. V., and A. M. Popova (1969) *Phys. Lett.* **28B**, 476.
——— and ——— (1970) *Yad. Fiz.* **10**, 1080 [*Sov. J. Nuc. Phys.* **10**, 621 (1970)].
———, V. L. Shablov, A. M. Popova, and T. A. Osborn (1980) *Phys. Rev. C* **22**, 976.
Kopaleishvili, T. I., A. I. Machavariani, and G. A. Emelyanenko (1978) *Nucl. Phys.* **A302**, 423.
Kottler, H., and K. L. Kowalski (1965) *Phys. Rev.* **138**, B619.
Kouri, D. J., and F. S. Levin (1974a) *Phys. Lett.* **50B**, 421.
——— and ——— (1974b) *Phys. Rev. A* **10**, 1616.
———, M. Craigie, and D. Secrest (1974c) *J. Chem. Phys.* **60**, 1851.
———, F. S. Levin, M. Craigie, and D. Secrest (1974d) *J. Chem. Phys.* **61**, 17.
——— and ——— (1975a) *Nucl. Phys.* **A253**, 395.
——— and ——— (1975b) *Phys. Rev. C* **11**, 352.
———, F. S. Levin, and W. Sandhas (1976) *Phys. Rev. C* **13**, 1825.
Kowalski, K. L., and D. Feldman (1961) *J. Math. Phys.* **2**, 499.
——— and ——— (1963a) *Phys. Rev.* **130**, 276.
——— and ——— (1963b) *J. Math. Phys.* **4**, 507.

———, (1963c) *Nuovo Cim.* **30**, 266.
——— (1965a) *Nuovo Cim.* **39**, L1002.
——— (1965b) *Phys. Rev. Lett.* **15**, 798, 908 (E).
——— (1966a) *Phys. Rev.* **144**, 1239.
——— (1966b) *Nuovo Cim.* **45**, 769.
——— (1967) *Phys. Rev.* **163**, 1030.
——— and J. Krauss (1968) *Nuovo Cim.* **57B**, 30.
——— (1969) *Phys. Rev.* **188**, 2235; *ibid. D* **2**, 812 (E) (1970).
———, J. E. Monahan, C. M. Shakin, and R. M. Thaler (1971a) *Phys. Rev. C* **3**, 1146.
——— and S. C. Pieper (1971b) *Phys. Rev. C* **4**, 74.
——— and ——— (1972a) *Phys. Rev. C* **5**, 324.
——— (1972b) *Phys. Rev. D* **5**, 395.
——— (1972c) *Nucl. Phys.* **A190**, 645.
——— (1972d) *Phys. Rev. D* **6**, 3705.
——— (1973a) *Phys. Rev. D* **7**, 1806.
——— (1973b) *Phys. Rev. D* **7**, 2957.
——— (1973c) *Phys Rev. C* **8**, 1973.
——— (1974a) *Phys. Rev. C* **10**, 366.
——— (1974b) *Phys. Rev. D* **10**, 1271.
——— (1975) *Phys. Rev. C* **11**, 2094.
——— (1976a) *Phys. Rev. D* **13**, 2352.
——— (1976b) *Nucl. Phys.* **A264**, 173.
——— (1977a) *Phys. Rev. C* **15**, 42.
——— (1977b) *Phys. Rev. C* **16**, 7.
——— (1977c) *Phys. Rev. C* **16**, 1735.
——— (1977d) *Phys. Rev. C* **16**, 2073.
———, E. R. Siciliano, and R. M. Thaler (1978a) *Phys. Rev. C* **18**, 2017.
——— (1978b) *Lett. al Nuovo Cim.* **22**, 531.
——— (1978c) in "Few-Body Systems and Nuclear Forces," Lecture Notes in Physics, eds. H. F. Zingl, M. I. Haftel and H. Zankel, Springer-Verlag, Berlin, **87**, 393.
——— (1979a) *Ann. Phys.* (N.Y.) **120**, 328.
———, E. R. Siciliano, and R. M. Thaler (1979b) *Phys. Rev. C* **19**, 1843.
——— (1979c) *Phys. Rev. D* **20**, 2526.
——— (1980a) *Phys. Rev. C* **21**, 2122.
——— (1980b) "Field-Theoretic Methods in Pion-Nucleus Interactions," Tri-State Nuclear Theory Conf., Ohio State U., CWRU-80/27.
——— (1981a) *Phys. Rev. C* **23**, 597.
——— and A. Picklesimer (1981b) *Phys. Rev. Lett.* **46**, 228.
——— and ——— (1981c) *Nucl. Phys.* **A369**, 336.
———, W. Polyzou, and E. F. Redish (1981d) *J. Math. Phys.* **22**, 1965.
——— (1981e) *Phys. Rev. C* **24**, 1915.
——— and A. Picklesimer (1982a) *Ann. Phys.* (N.Y.) **139**, 215.
——— (1982b) *Phys. Rev. C* **25**, 700.

——— and A. Picklesimer (1982c) *Phys. Rev. C* **26**, 1835.
——— and ——— (1983a) *J. Math. Phys.* **24**, 284.
——— (1983b) *Phys. Rev. C* **27**, 489.
——— (1984) in "Few-Body Problems in Physics," Vol. I, ed. H. Zeitnitz, Elsevier, Amsterdam, p. 465; *Nucl. Phys.* **A416**, 465c.
——— (1986) *Nuovo Cim.* **92A**, 289.
Kozack, R., and F. S. Levin (1987) *Phys. Rev. C* **36**, 883.
Krall, N. A., and E. Gerjuoy (1960) *Phys. Rev.* **120**, 143.
Kujawski, E. (1972) *Ann. Phys.* (N.Y.) **74**, 567.
——— (1973) *Ann. Phys.* (N.Y.) **81**, 591.
Kuo, T. T. S., F. Osterfeld, and S. Y. Lee (1980) *Phys. Rev. Lett.* **45**, 786.
Kuruoglu, Z. C., and F. S. Levin (1982) *Phys. Rev. Lett.* **48**, 899.
——— and ——— (1985) *Ann. Phys.* (N.Y.) **163**, 120.
Lacombe, M., B. Loiseau, J. M. Richard, R. Vinh Mau, J. Côte, P. Pirès, and R. de Tourreil (1980) *Phys. Rev. C* **21**, 861.
Landau, R., S. C. Phatak, and F. Tabakin (1973) *Ann. Phys.* (N.Y.) **78**, 299.
——— (1975) *Ann. Phys.* (N.Y.) **92**, 205.
——— and M. Sagen (1986) *Phys. Rev. C* **33**, 447.
Larsen, S. Y., and P. L. Mascheroni (1970) *Phys. Rev. A* **2**, 1018.
Larson, N. M., and J. H. Hetherington (1974) *Phys. Rev. C* **9**, 699.
Laughlin, R., and B. L. Scott (1968) *Phys. Rev.* **171**, 1196.
Laverne, A., and C. Gignoux (1973) *Nucl. Phys.* **A203**, 597.
Lee, S. Y., F. Osterfeld, K. Tam, and T. T. S. Kuo (1981) *Phys. Rev. C* **24**, 329.
Lee, T. D., and C. N. Yang (1959) *Phys. Rev.* **116**, 25.
Lee, T. S. H. (1983) *Phys. Rev. Lett.* **50**, 157.
——— (1984) *Phys. Rev. C* **29**, 195.
——— (1987) *Nucl. Phys.* **A463**, 61c.
Leeb, H., and E. W. Schmid (1980) *Z. Phys.* **A296**, 51.
Lehman, D. R. (1982) *Phys. Rev. C* **25**, 3146.
——— (1987) *Nucl. Phys.* **A463**, 117c.
Lehmann, H., K. Symanzik, and W. Zimmerman (1955) *Nuovo Cim.* **97A**, 77.
Levin, F. S. (1980) *Ann. Phys.* (N.Y.) **130**, 139.
——— (1981) *Nucl. Phys.* **A353**, 143c.
——— and W. Sandhas (1984) *Phys. Rev. C* **29**, 1617.
——— (1990) in "Trends in Theoretical Physics," Vol. I, ed. P. J. Ellis and Y. C. Tang, p. 1.
Levine, M., J. A. Tjon, and J. Wright (1966) *Phys. Rev. Lett.* **16**, 962.
———, ———, and ——— (1967) *Phys. Rev.* **154**, 1435.
Levinger, J. S. (1974) in "Springer Tracts in Modern Physics," Springer-Verlag, Berlin, **71**, 88.
Levy, M. M. (1952) *Phys. Rev.* **88**, 72.
Lewanski, A. J., and W. Tobocman (1978) *Phys. Rev. C* **17**, 423.
L'Huillier, M., E. F. Redish, and P. C. Tandy (1976) University of Maryland Tech. Rept. No. 76-068.

———, ———, and ——— (1978) *J. Math. Phys.* **19**, 1276.
Lieber, M., L. Rosenberg, and L. Spruch (1972) *Phys. Rev. D* **5**, 1330; *ibid.*, 1347.
Lim, T. K., S. K. Duffy, and W. C. Damert (1977) *Phys. Rev. Lett.* **38**, 341.
——— and P. A. Maurone (1980a) *Phys. Rev. B* **22**, 1467.
——— and B. Shimer (1980b) *Z. Phys.* **A297**, 185.
———, ———, and M. I. Haftel (1981) *Chem. Phys. Lett.* **81**, 87.
——— (1984) *Nucl. Phys.* **A416**, 491c.
Lipperheide, R., (1966) *Nucl. Phys.* **89**, 97.
Lippmann, B. A., and J. Schwinger (1950) *Phys. Rev.* **79**, 469.
——— (1956) *Phys. Rev.* **102**, 264.
Liu, L. C., and C. M. Shakin (1979) *Phys. Rev. C* **20**, 2339.
Locher, M. P. (1970) *Nucl. Phys.* **B23**, 116.
——— (1972) *Nucl. Phys.* **B36**, 634.
——— (1975) *Nucl. Phys.* **A251**, 493.
——— and T. Mizutani (1978) *Phys. Rep.* **46**, 43.
——— (1984) *Nucl. Phys.* **A416**, 243c.
Lock, J. A. (1975) *Phys. Rev. D* **12**, 3319.
——— (1976a) *Phys. Rev. D* **13**, 1009.
——— (1976b) *Nucl. Phys.* **A271**, 458.
Logunov, A. A., and A. N. Tavkhelidze (1963) *Nuovo Cim.* **29**, 380.
Loiseau, B., L. Mathelistch, W. Plessas, and K. Schwarz (1985) *Phys. Rev. C* **32**, 2165.
Londergan, J. T., C. E. Price, and E. J. Stephenson (1987) *Phys. Rev. C* **35**, 902.
Lovelace, C. (1964a) *Phys. Rev.* **135**, B1225.
——— (1964b) in "Strong Interactions and High Energy Physics," ed. R. G. Moorhouse, Oliver and Boyd, London, p. 437.
Lovitch, L., and V. Vanzani (1979) *Lett. Nuovo Cim.* **26**, 65.
——— (1982) *Nuovo Cim.* **68A**, 81.
Lucchese, R. R., and V. McKoy (1980) *Phys. Rev. A* **21**, 112.
———, K. Takatsuka, and V. McKoy (1986) *Phys. Rep.* **131**, 147.
Lucchese, R. R. (1989) *Phys. Rev. A* **40**, 6879.
Lurie, D. (1968) "Particles and Fields," Interscience, New York.
MacDonald, W. M., and M. C. Birse (1983) *Nucl. Phys.* **A403**, 99.
Machleidt, R., K. Holinde, and Ch. Elster (1987) *Phys. Rept.* **149**, 1.
——— (1988) in "Advances in Nuclear Physics," eds. J. M. Negele and E. Vogt, Vol. 19, p. 189.
Macke, W. (1953) *Z. Naturforsch* **89**, 599; *ibid.*, 615.
Mahaux, C., and H. A. Weidenmuller (1969) "Shell-Model Approach to Nuclear Reactions," Wiley-Interscience, New York.
Malfliet, R. A., and J. A. Tjon (1968) *Nucl. Phys.* **A127**, 161.
——— and ——— (1970) *Ann. Phys.* (N.Y.) **61**, 425.
Mandelstam, S. (1965) *Phys. Rev.* **140**, B375.
Martin, A. (1962) *Nuovo Cim.* **23**, 641.
Mascheroni, P. L. (1970) *Phys. Rev. Lett* **25**, 726.

McCurdy, C. W., T. N. Rescigno, and B. I. Schneider (1987) *Phys. Rev. A* **36**, 2061.

McGuire, J. B. (1964) *J. Math. Phys.* **5**, 622.

McKellar, B. H. J., and W. Glöckle (1984) *Nucl. Phys.* **A416**, 435c.

McNeil, J. A., J. R. Shepard, and S. J. Wallace (1983a) *Phys. Rev. Lett.* **50**, 1439.

———, L. Ray, and S. J. Wallace (1983b) *Phys. Rev. C* **27**, 2123.

Mdlalose, T. E., H. Fiedeldey, and W. Sandhas (1986) *Phys. Rev. C* **33**, 784.

———, ———, and ——— (1988) *Nucl. Phys.* **A480**, 215.

Mennessier, G., J.-Y. Pasquier, and R. Pasquier (1972) *Phys. Rev. D* **6**, 1351.

Merkuriev, S. P. (1973) *Teor. Mat. Fiz.* **17**, 221 [*Theor. Math. Phys.* (USSR) **17**, 115 (1973)].

———, C. Gignoux, and A. Laverne (1976) *Ann. Phys.* (N.Y.) **99**, 30.

——— and S. L. Yakovkev (1984) *Theor. Math. Phys.* **56**, 673.

Messiah, A. (1965) "Quantum Mechanics," Vol. I, II, North-Holland, Amsterdam.

Micha, D. A. (1985) in "Theory of Chemical Reaction Dynamics," ed. M. Baer, CRC Press, Akron, Ch. 3, p. 181.

Mikhlin, S. G. (1964) "Integral Equations," 2nd Ed., Springer-Verlag, New York.

Miller, W. H., and B. M. D. D. Jansen op de Haar (1987) *J. Chem. Phys.* **86**, 6213.

Mitra, A. N. (1962) *Nucl. Phys.* **32**, 529.

——— (1965a) *Phys. Rev.* **139**, B1472.

———, J. Gillespie, R. Sugar, and N. Panchapakesan (1965b) *Phys. Rev.* **140**, B1336.

——— and B. S. Bhasin (1966) *Phys. Rev. Lett.* **16**, 523.

——— (1969) *Adv. Nucl. Phys.* **3**, 1.

Mizutani, T., and D. Koltun (1977) *Ann. Phys.* (N.Y.) **109**, 1.

———, C. Fayard, G. H. Lamot, and R. S. Nahabetian (1981) *Phys. Lett.* **107B**, 177.

Mongan, T. R. (1969) *Phys. Rev.* **178**, 1597.

Mukherjee, S. (1978) *Phys. Lett.* **80B**, 73.

——— (1981a) *Phys. Lett.* **81A**, 207.

——— (1981b) *Phys. Lett.* **83A**, 1.

Mulligan, B., and S. B. Qadri (1981) *Phys. Rev. C* **24**, 874.

Muskhelishvili, N. I. (1953) "Singular Integral Equations," P. Nordhoff, Groningen.

Muslim, Y., E. Kim, and T. Ueda (1983) *Nucl. Phys.* **A393**, 399.

Mutze, U. (1984) *Phys. Rev. D* **29**, 2255.

Nagarajan, M. A., and B. G. Giraud (1983) *Phys. Rev. C* **27**, 232.

Nakaichi-Maeda, S., and T. K. Lim (1982) *Phys. Rev. A* **28**, 692.

Nakanishi, N. (1969) *Suppl. Prog. Theor. Phys.* **43**, 1.

Nambu, Y. (1950) *Prog. Theor. Phys.* **5**, 614.

Narodetskii, I. M., and O. A. Yakubovskii (1971) *Yad. Fiz.* **14**, 315 [*Sov. J. Nucl. Phys.* **14**, 178 (1972)].

———, ——— (1980) "The Integral Equations for *N*-Body Scattering," I.T.E.P. Report No. 1, Moscow.

——— (1981) *Rivista del Nuovo Cim.* **4**, N.7, pp. 1–125.

Negele, J., and H. Orland (1988) "Quantum Many-Particle Systems," Addison-Wesley Pub. Co., Redwood City, California.

Nelson, B. (1983) *Phys. Rev. D* **27**, 841.
Nesbet, R. K. (1975) *Adv. Quantum Chem.* **9**, 215.
——— (1979) *Adv. At. Mol. Phys.* **13**, 315.
Neuhauser, D., and M. Baer (1988a) *J. Chem. Phys.* **88**, 2856.
——— and ——— (1988b) *Phys. Rev. C* **38**, 1571.
Newton, R. G. (1967) *Phys. Rev.* **153**, 1502.
——— (1982) "Scattering Theory of Waves and Particles," 2nd Ed., Springer-Verlag, New York.
Niskanen, J. A. (1978) *Nucl. Phys.* **A298**, 417.
Noyes, H. P., and D. Y. Wong (1959) *Phys. Rev. Lett.* **3**, 191.
——— (1965) *Phys. Rev. Lett.* **15**, 538.
——— (1967) in Proc. of the International Conf. on Polarized Targets and Ion Sources, Saclay, La Documentation Francaise, Paris. p. 309.
——— (1969) in Progress in Nuclear Physics, eds. D. M. Brink and J. H. Mulvey, Pergamon Press, London, Vol. 10, 355.
——— (1982) *Phys. Rev. C* **26**, 1858.
Nuttall, J. (1967) *J. Math. Phys.* **8**, 4.
Oberoi, R. S., and R. K. Nesbet (1973) *Phys. Rev. A* **8**, 2969.
Ohmura, T., B. Imanishi, M. Ichimura, and M. Kawai (1969) *Prog. Theor. Phys.* **41**, 391.
———, ———, ———, and ——— (1970) *Prog. Theor. Phys.* **43**, 347.
Ohtsuka, N., M. Shabshiry, R. Linden, H. Muther, and, A. Faessler (1988) *Nucl. Phys.* **A490**, 715.
Omnes, R. (1964) *Phys. Rev.* **134**, B1358.
——— (1968) *Phys. Rev.* **165**, 1265.
Oryu, S. (1974) *Prog. Theor. Phys.* **52**, 550.
——— (1983) *Phys. Rev. C* **27**, 2500.
——— (1986) *Lecture Notes in Phys.* **273**, 123, Springer-Verlag.
Osborn, T. A. (1969) *Nucl. Phys.* **A138**, 305.
——— and K. L. Kowalski (1971) *Ann. Phys* (N.Y.) **68**, 361.
——— (1973) *J. Math. Phys.* **14**, 373.
——— and D. Eyre (1979) *Nucl. Phys.* **A327**, 125.
Ouchinnikov, Yu. N., and I. M. Sigal (1979) *Ann. Phys.* (N.Y.) **123**, 274.
Padé, H. (1892) *Ann. Ec. Norm.* **9**, Suppl. 1.
Páez, M. J., and R. H. Landau (1984a) *Phys. Rev. C* **29**, 2267.
——— and ——— (1984b) *Phys. Lett.* **142B**, 235.
Pagels, H. (1965) *Phys. Rev.* **140**, B1599.
Pagnamenta, A. (1968) in "Lectures in Theoretical Physics," eds. W. E. Britten and K. T. Mahanthappa, Gordon and Breach, New York, Vol. XI-D, p. 479.
Pagorezelski, W. (1966) "Integral Equations and their Applications," Pergamon Press, Oxford.
Payne, G. L., J. L. Friar, B. F. Gibson, and I. R. Afnan (1980) *Phys. Rev. C* **22**, 823.
Pearce, B. C. (1987) *Phys. Rev. C* **36**, 471.
Peccei, R. D., and R. D. Viollier (1977) *Ann. Phys.* (N.Y.) **103**, 29.

Petrov, N. M. (1988) *Yad. Fiz.* **48**, 50 [*Sov. J. Nucl. Phys.* **48**, 31 (1988)].
Phillips, A. C. (1965) *Phys. Rev.* **142**, 984.
——— (1968) *Nucl. Phys.* **A107**, 209.
——— (1977) *Rep. Prog. Phys.* **40**, 905.
Picker, H. S., E. F. Redish, and G. J. Stephensen (1971) *Phys. Rev. C* **4**, 287.
——— and G. J. Stephenson (1975) *Nucl. Phys.* **A241**, 443.
Picklesimer, A., and K. L. Kowalski (1980) *Phys. Lett.* **95B**, 1.
——— and R. M. Thaler (1981a) *Phys. Rev. C* **23**, 42.
——— (1981b) *Phys. Rev. C* **24**, 1400.
———, P. Tandy, and R. M. Thaler (1982a) *Phys. Rev. C* **25**, 1233.
———, ———, and ——— (1982b) *Phys. Rev. C* **26**, 315.
———, ———, and ——— (1983a) *Ann. Phys.* (N.Y.) **145**, 207.
——— (1983b) *Phys. Rev. C* **27**, 1927.
———, P. C. Tandy, R. M. Thaler, and D. H. Wolfe (1984) *Phys. Rev. C* **30**, 1861.
——— and ——— (1987) *Phys. Rev. C* **35**, 1174.
——— and ——— (1988) *Phys. Rev. C* **37**, 1335.
Pieper, S. C., L. Schlessinger, and J. Wright (1970a) *Phys. Rev. D* **1**, 1674.
———, J. Wright, and L. Schlessinger (1970b) *Phys. Rev. D* **2**, 1561.
——— (1971) *Phys. Rev. Lett.* **27**, 1738. [*ibid.* **28**, 1154(E)(1972)]
——— and K.L. Kowalski (1972a) *Phys. Rev. C* **5**, 306.
——— (1972b) *Nucl. Phys.* **A193**, 529.
——— (1974) *Phys. Rev. C* **9**, 883.
Plattner, G. R., R. D. Viollier, D. Trautmann, and K. Alder (1973) *Nucl. Phys.* **A206**, 513.
———, ———, and K. Alder (1975) *Phys. Rev. Lett.* **34**, 830.
———, M. Bornand, and R. D. Viollier (1977) *Phys. Rev. Lett.* **39**, 127.
Plemelj, J. (1908) *Monatshette fur Math. u. Phys.* **19**, 205.
Plessas, W. (1986a) Lecture Notes in Physics **273**, 137.
——— (1986b) in "Few Body Methods: Principles and Applications," eds. by T. K. Lim, C.-G. Bao, D.-P. Hou, and S. Huber, World Scientific, Singapore, p. 43.
——— and J. Haidenbauer (1987) *Few Body Syst. Suppl.* **2**, 185.
Polyzou, W. N. (1979a) University of Maryland, Ph.D. thesis.
——— and E. F. Redish (1979b) *Ann. Phys.* (N.Y.) **119**, 1.
——— (1980) *J. Math. Phys.* **21**, 506.
——— (1981) *J. Math. Phys.* **22**, 798.
———, A. G. Gibson, and C. Chandler (1982) *Phys. Rev. C* **26**, 1878.
Press, W. H., B. P. Flannery, S. A. Teukolsky, and W. T. Vettering (1986) "Numerical Recipes," Cambridge University Press, Cambridge.
Pretorius, L., and D. Eyre (1987) *J. Comput. and Appl. Maths.* **18**, 235.
Prugovečki, E. (1971) "Quantum Mechanics in Hilbert Space," Academic, New York.
Raphael, R., P. C. Tandy, and W. Tobocman (1976) *Phys. Rev. C* **14**, 1355.
Rarita, W., and J. Schwinger (1941) *Phys. Rev.* **59**, 436.
Rawitscher, G. H. (1982) *Phys. Rev. C* **25**, 2196.

——— and G. Delic (1984) *Phys. Rev. C* **29**, 747.
——— (1987) *Nucl. Phys.* **A475**, 519.
——— (1989) *Phys. Rev. C* **39**, 440.
Ray, L., G. W. Hoffman, M. L. Bartlett, and N. Ottenstein (1988) *Phys. Rev. C* **37**, 224.
——— (1989) *Phys. Rev. C* **39**, 1170.
Redish, E. F., and F. Villars (1970) *Ann. Phys.* (N.Y.) **56**, 355.
———, G. J. Stephenson, Jr., and H. S. Picker (1972) *Phys. Rev. C* **5**, 707.
——— (1974a) *Nucl. Phys.* **A225**, 16.
——— (1974b) *Nucl. Phys.* **A235**, 82.
——— (1980) *Phys. Lett.* **90B**, 188.
——— and K. Stricker-Bauer (1987a) *Phys. Rev. C* **36**, 513.
——— (1987b) *Nucl. Phys.* **A463**, 417c.
Redmond, P. J., and J. L. Uretsky (1960) *Ann. Phys.* (N.Y.) **9**, 106.
Reed, M., and B. Simon (1979) "Methods of Modern Mathematical Physics III: Scattering Theory," Academic Press, Inc., Orlando, Florida.
Regge, T. (1959) *Nuovo Cim.* **14**, 951.
Reid, Jr., R. V. (1968) *Ann. Phys.* (N.Y.) **50**, 411.
Reiner, M. J. (1976) *Ann. Phys.* (N.Y.) **100**, 131; *ibid.*, 149.
Reinhardt, H. (1982) *Nucl. Phys.* **A390**, 70.
Remler, E. A., and R. A. Miller (1974) *Ann. Phys.* (N.Y.) **82**, 189.
Riahi, F. (1969) *Helv. Phys. Acta.* **42**, 299.
Riesenfeld, W. B., and K. M. Watson (1956) *Phys. Rev.* **102**, 1157.
Rinat, A. S., and M. Stingl (1971) *Ann. Phys.* (N.Y.) **65**, 141.
———, L. P. Kok, and M. Stingl (1972) *Nucl. Phys.* **A190**, 328.
——— and A. W. Thomas (1977a) *Nucl. Phys.* **A282**, 365.
——— (1977b) *Nucl. Phys.* **A287**, 399.
——— and Y. Starkand (1983) *Nucl. Phys.* **A397**, 381.
Robilotta, M. R., and M. P. Isidro Filho (1984) *Nucl. Phys.* **A414**, 394.
———, M. P. Isidro Filho, H. T. Coelho, and T. K. Das (1985) *Phys. Rev. C* **31**, 646.
Rodberg, L. S., and R. M. Thaler (1967) "Introduction to the Quantum Theory of Scattering," Academic Press, New York.
Roriz, L. A. L., and A. Delfino (1988) *Phys. Rev. C* **38**, 607.
Rosenberg, L. (1965) *Phys. Rev.* **140**, B217.
——— (1968) *Phys. Rev.* **168**, 1756.
Rota, G. C. (1964) *Z. Wahrsch. Vern. Gebiete* **2**, 340.
Rubin, M., R. Sugar, and G. Tiktopoulos (1966) *Phys. Rev.* **146**, 1130.
———, ———, and ——— (1967a) *Phys. Rev.* **159**, 1348.
———, ———, and ——— (1967b) *Phys. Rev.* **162**, 1555.
Rupp, G., and J. A. Tjon (1988) *Phys. Rev. C* **37**, 1729.
Saito, S. (1969) *Prog. Theor. Phys.* **41**, 705.
Salpeter, E. E., and H. A. Bethe, (1951) *Phys. Rev.* **84**, 1232.
Sandhas, W. (1966) *Comm. Math. Phys.* **3**, 358.
——— (1974) *Acta. Phys. Aus. Suppl.* **13**, 679.

——— (1975) *Czech. J. Phys.* **B25**, 251.
——— (1976) in "Few-Body Dynamics," eds. A. N. Mitra, I. Slaus, V. S. Bhasin, and V. K. Gupta, North-Holland, Amsterdam, p. 540.
——— (1978) in "Few-Body Nuclear Physics," IAEA-SMR-43, p. 3.
——— (1986) in "Few-Body Methods: Principles and Applications," eds. T-K Lim, C.-G. Bao, D.-P. Hou, and S. Huber World Scientific, Singapore, p. 3.
Sandhya Devi, K. R., and S. F. Koonin (1981) *Phys. Rev. Lett.* **47**, 27.
Sasakawa, T. (1963) *Prog. Theor. Phys. Suppl.* **27**, 1.
——— (1973) *Nucl. Phys.* **A203**, 496.
——— and S. Ishikawa (1986) *Few Body Syst.* **1**, 3.
——— (1987) *Nucl. Phys.* **A463**, 327c.
Satchler, G. R. (1980) "Introduction to Nuclear Reactions," Halsted, New York.
——— (1983) "Direct Nuclear Reactions," Oxford University Press, New York.
Sauer, P. U. (1971) *Nucl. Phys.* **A170**, 497.
——— (1987) *Nucl. Phys.* **A463**, 273c.
Sawada, Y., and T. Sasakawa (1988) *Prog. Theor. Phys.* **79**, 1396.
Schmid, E. W., and H. Ziegelmann (1974) "The Quantum Mechanical Three-Body Problem," Pergamon Press, F. Vieweg, Braunscheig.
——— (1984) *Nucl. Phys.* **A416**, 347c.
——— and H. Fiedeldey (1989) *Phys. Rev. C* **39**, 2170.
Schrenk, G. L., and A. N. Mitra (1967) *Phys. Rev. Lett.* **19**, 530.
Schwager, J. (1976) *Ann. Phys.* (N.Y.) **98**, 14.
Schwarz, K., J. Haidenbauer, and J. Fröhlich (1986) *Phys. Rev. C* **33**, 456.
Schweber, S. S., and E. C. G. Sudarshan (1962a) *Ann. Phys.* (N.Y.) **19**, 351.
——— (1962b) *Ann. Phys.* (N.Y.) **20**, 61.
Schwenke, D. W., D. G. Truhlar, and D. J. Kouri (1987) *J. Chem. Phys.* **86**, 2772.
———, K. Haug, M. Zhao, D. G. Truhlar, Y. Sun, J. Z. H. Zhang, and D. J. Kouri (1988) *J. Chem. Phys.* **92**, 3202.
Schwinger, J. (1947a) Lecture Notes, Harvard University (unpublished).
——— (1947b) *Phys. Rev.* **72**, 742.
——— (1951) *Proc. Natl. Acad. Sci.* (U.S.A.) **37**, 452; *ibid.*, 455.
Serot, B. D., and J. D. Walecka (1986) in "Advances in Nuclear Physics," eds. J. Negele and E. Vogt, Plenum, Vol. 16, p. 1.
Shanley, P. E. (1969) *Phys. Rev.* **187**, 1328.
Shapoval, D. V., I. V. Simenog, and A. I. Sitnichenko (1987) *Phys. Lett.* **199B**, 322.
Sharma, D. K. (1973) Case Western Reserve University, Ph.D. thesis.
Shepard, J. R., J. A. McNeil, and S. J. Wallace (1983) *Phys. Rev. Lett.* **50**, 1443.
Shima, Y., M. Baer, and D. J. Kouri (1983a) *J. Chem. Phys.* **78**, 6666.
———, ———, and ——— (1983b) *Chem. Phys. Lett.* **94**, 321.
——— and ——— (1983c) *J. Phys. B* **16**, 2169.
———, ———, and D. J. Kouri (1986) *Phys. Rev. A* **34**, 846.
Shirato, S., K. Saitoh, N. Koori, and R. T. Cahill (1973) *Nucl. Phys.* **A215**, 277.
Siciliano, E. R., and R. M. Thaler (1977) *Phys. Rev. C* **16**, 1322.

Sigal, I. M., and A. Soffer (1987) *Ann. Math.* **126**, 35.
Simon, B. (1970) *Helv. Phys. Acta.* **43**, 607.
——— (1977) *Commun. Math. Phys.* **58**, 205.
Sitenko, O. G., and V. F. Kharchenko (1963) *Nucl. Phys.* **49**, 15.
———, ———, and N. M. Petrov (1966) *Phys. Lett.* **21**, 54.
Skornyakov, G. V., and K. A. Ter-Martirosian (1956) *Zh. Exp. Teor. Fiz.* **31**, 755 [*Sov. Phys. JETP* **4**, 648 (1957)].
Sloan, I. H. (1968a) *Phys. Rev.* **165**, 1587.
——— (1968b) *J. Comp. Phys.* **3**, 332.
——— (1969) *Phys. Rev.* **185**, 1361.
——— and T. J. Brady (1972a) *Phys. Rev. C* **6**, 701.
——— (1972b) *Phys. Rev. C* **6**, 1945.
——— (1972c) *Nucl. Phys.* **A182**, 549.
——— and J. C. Aarons (1972d) *Nucl. Phys.* **A198**, 321.
——— and J. D. Gray (1973) *Phys. Lett.* **44B**, 354.
——— and S. K. Adhikari (1974) *Nucl. Phys.* **A235**, 352.
——— (1981) *Phys. Rev. C* **23**, 1289.
Smithies, F. (1958) "Integral Equations," Cambridge University Press, Cambridge.
Sobel, M. I. (1967) *Phys. Rev.* **156**, 1553.
Sofianos, S., N. J. McGurk, and H. Fiedeldey (1979) *Nucl. Phys.* **A318**, 295.
———, H. Fiedeldey, and W. Sandhas (1985) *Phys. Rev. C* **32**, 400.
Sohre, F., and H. Ziegelmann (1971) *Phys. Lett.* **34B**, 579.
Sperisen, F., W. Gruebler, V. Konig, P. A. Schmelzbach, B. Jenny, K. Elsener, C. Schweizer, J. Ulbricht, and P. Doleschall (1981) *Phys. Lett.* **102B**, 9.
———, ———, ———, ———, K. Elsener, B. Jenny, C. Schweizer, J. Ulbricht, and P. Doleschall (1984) *Nucl. Phys.* **A422**, 81.
Srivastava, M., and D. W. L. Sprung (1970) *Nucl. Phys.* **A149**, 113.
——— and ——— (1975) in "Advances in Nuclear Physics," eds. M. Baranger and E. Vogt, Vol. 8, p. 121.
Staszewska, G., and D. G. Truhlar (1987) *J. Chem. Phys.* **86**, 2793.
Stelbovics, A. T., and L. R. Dodd (1972) *Phys. Lett.* **39B**, 450.
——— and ——— (1974) *Phys. Lett.* **48B**, 13.
——— and M. Stingl (1978) *Nucl. Phys.* **A299**, 391.
Stichel, P. (1964) *Nucl. Phys.* **52**, 189.
Stingl, M. and A. S. Rinat (1974) *Phys. Rev. C* **10**, 1253.
——— and A. T. Stelbovics (1978) *J. Phys. G* **4**, 1371, 1389.
Stoof, H. T. C., L. P. H. de Goey, W. M. H. M. Rovers, P. S. M. Kop Jansen, and B. J. Verhaar (1988) *Phys. Rev.* **A38**, 1248.
Streater, R. F., and A. S. Wightman (1964) "PCT, Spin, and Statistics, and All That," Benjamin, New York.
Sugar, R., and R. Blankenbecler (1964) *Phys. Rev.* **136**, B472.
Sweeny, C. J., and P. L. De Vries (1989) *Comp. in Phys.*, Jan./Feb., p. 49.
Szasz, G. (1963) "Introduction to Lattice Theory," Academic Press, New York.

Takahashi, T., and N. Mishima (1965) *Prog. Theor. Phys.* **34**, 498.

——— and ——— (1966) *Prog. Theor. Phys.* **36**, 440.

Takatsuka, K., and V. McKoy (1981a) *Phys. Rev. A* **23**, 2352.

——— and ——— (1981b) *Phys. Rev. A* **23**, 2359.

———, R. R. Lucchese, and V. McKoy (1981c) *Phys. Rev. A* **24**, 1812.

Takeda, G., and K. M. Watson (1955) *Phys. Rev.* **97**, 1336.

Takemiya, T. (1987) *Nucl. Phys.* **A463**, 151c.

Tandy, P. C., R. T. Cahill, and I. E. McCarthy (1972) *Phys. Lett.* **41B**, 241.

———, E. F. Redish, and D. Bolle (1977) *Phys. Rev. C* **16**, 1924.

——— and R. M. Thaler (1980) *Phys. Rev. C* **22**, 232.

——— (1986) in "Relativistic Dynamics and Quark-Nuclear Physics," eds. M. B. Johnson and A. Picklesimer, John Wiley, New York, p. 3.

Tang, Y. C., M. LeMere, and D. R. Thompson (1978) *Phys. Rep.* **47**, 167.

——— (1987) *Nucl. Phys.* **A463**, 377c.

Taylor, J. G. (1963) *Nuovo Cim. Suppl.* **1**, 857.

Taylor, J. R. (1966) *Phys. Rev.* **142**, 1236.

——— (1967) *J. Math. Phys.* **7**, 181.

——— (1987) "Scattering Theory: The Quantum Theory of Nonrelativistic Collisions," R. E. Krieger Pub. Co., Malabar, Florida.

Temkin, A., and A. K. Bhatia (1985a) *Phys. Rev. A* **31**, 1259.

———, ——— (1985b) in "Autoionization: Recent Developments and Applications," ed. A. Temkin, Plenum, New York, p. 1.

Thacker, H. B. (1975) *Phys. Rev. D* **11**, 838.

——— (1976) *Phys. Rev. D* **14**, 3058.

Thies, M. (1985) *Phys. Lett.* **162B**, 255.

——— (1986) *Phys. Lett.* **166B**, 23.

Thirumalai, D., G. Staszewska, and D. G. Truhlar, (1987) *Comm. on At. and Mole. Phys.* **20**, 217.

Thomas, A. W. (1976) *Nucl. Phys.* **A258**, 417.

———, ed. (1977) "Modern Three-Hadron Physics," Springer-Verlag, Berlin.

Thomas, L. H. (1935) *Phys. Rev.* **47**, 903.

Thompson, R. H. (1970) *Phys. Rev. D* **1**, 110.

———, A. Gersten, and A. E. S. Green (1971) *Phys. Rev. D* **3**, 2069.

Tip, A. (1977) *J. Phys.* **B10**, L295.

Tjon, J. A., and H. M. Nieland (1970) in "The Padé Approximant in Theoretical Physics," eds. G. A. Baker, Jr. and J. L. Gammel, Academic Press, New York, p. 289.

——— (1975) *Phys. Lett.* **56B**, 217.

——— (1976) *Phys. Lett.* **63B**, 391.

——— (1978) *Phys. Rev. Lett.* **40**, 1239.

Tkachenko, E. G. (1981) *Phys. Lett.* **98B**, 328.

Tobocman, W. (1961) "Theory of Direct Nuclear Reactions," Oxford University Press, London.

——— (1974a) *Phys. Rev. C* **9**, 2466.

——— (1974b) *Phys. Rev. C* **10**, 60.
——— (1975a) *Phys. Rev. C* **11**, 43.
——— (1975b) *Phys. Rev. C* **12**, 741.
——— (1975c) *Phys. Rev. C* **12**, 1146.
——— (1981a) *Nucl. Phys.* **A369**, 77.
——— (1981b) Case Western Reserve University Rept., October, 1981, unpublished.
——— (1983a) *Phys. Rev. C* **27**, 88.
——— (1983b) *Phys. Rev. C* **27**, 1405.
Tomio, L., and S. K. Adhikari (1980a) *Phys. Rev. C* **22**, 28.
——— and ——— (1980b) *Phys. Rev. C* **22**, 2359.
——— and ——— (1981) *Phys. Rev. C* **24**, 43.
———, A. Delfino, and S. K. Adhikari (1987) *Phys. Rev. C* **35**, 441.
———, T. Frederico, and A. Delfino (1990) *Phys. Rev. C* **41**, 876.
Truhlar, D. G., J. Abdallah, Jr., and R. L. Smith (1974) *Adv. Chem. Phys.* **25**, 211.
Udagawa, T., H. H. Wolter, and W. R. Coker (1973) *Phys. Rev. Lett.* **31**, 1507.
Ueda, T., and A. E. S. Green (1968) *Phys. Rev.* **174**, 1304.
——— (1987) *Nucl. Phys.* **A463**, 69c.
van Leeuwen, J. M. J., and A. S. Reiner (1961) *Physica* **27**, 99.
van Winter, C. (1964) *Mat. Fis. Ser. Dan. Vid. Selski.* **2**, No. 8.
Vanzani, V. (1971) *Nuovo Cim.* **2A**, 525.
——— (1976) *Lett. Nuovo Cim.* **16**, 1.
——— (1978a) in "Few-Body Nuclear Physics," IAEA-SMR-45, p. 57.
——— (1978b) *Lett. Nuovo Cim.* **23**, 586.
Varracchio, E. F. (1983) *Ann. Phys.* (N.Y.) **145**, 131.
Vetterli, M. C., J. A. Kuehner, A. J. Trudel, C. L. Woods, R. Dymarz, A. A. Pilt, and H. R. Weller (1985) *Phys. Rev. Lett.* **54**, 1129.
Viollier, R. D., G. R. Plattner, D. Trautmann, and K. Alder (1973) *Nucl. Phys.* **A206**, 498; *ibid.*, 513.
von Geramb, H. V., ed. (1979) "Microscopic Optical Potentials," Lecture Notes in Physics, Vol. 89, Springer, Berlin.
Wallace, S. J. (1981) in "Advances in Nuclear Physics," eds. J. W. Negele and E. Vogt, Plenum, New York, Vol. 12, p. 135.
——— (1987) *Ann. Rev. Nucl. Part. Sci.* **37**, 267.
Watson, D. K., T. N. Rescigno, and V. McKoy (1979) *J. Phys.* **B14**, 1875.
Watson, K. M. (1953) *Phys. Rev.* **89**, 575.
——— (1954) *Phys. Rev.* **95**, 228.
——— (1956) *Phys. Rev.* **103**, 489.
——— and J. Nuttall (1967) "Topics in Several Particle Dynamics," Holden-Day, San Francisco.
Weinberg, S. (1963) *Phys. Rev.* **131**, 440.
——— (1964) *Phys. Rev.* **133**, B232.
Weiss, D. L., and D. J. Ernst (1982) *Phys. Rev. C* **26**, 605.
Weller, H. R., and D. R. Lehman (1988) *Ann. Rev. Nucl. Part. Sci.* **38**. 563.

Wellner, M. (1967) *Phys. Rev.* **158**, 1620.
Weyers, J. (1966) *Phys. Rev.* **145**, 1236; *ibid.* **151**, 1159.
Wheeler, J. A. (1937) *Phys. Rev.* **52**, 1083; *ibid.*, 1107.
Whiting, J. S., and M. G. Fuda (1976) *Phys. Rev. C* **14**, 18.
Wichmann, E., and J. H. Crichton (1963) *Phys. Rev.* **132**, 2788.
Wildermuth, K., and T. Kanellopoulos (1958a) *Nucl. Phys.* **7**, 150.
——— and ——— (1958b) *Nucl. Phys.* **9**, 449.
——— and Y. C. Tang (1977) "A Unified Theory of the Nucleus," Vieweg, Braunschweg.
Wiringa, R. B., R. A. Smith, and T. A. Ainsworth (1984) *Phys. Rev. C* **29**, 1207.
Witala, H., W. Glöckle, and Th. Cornelius (1987) *Few Body Syst. Suppl.* **2**, 555.
———, Th. Cornelius, and W. Glöckle (1988) *Few Body Syst.* **3**, 123.
———, W. Glöckle, and Th. Cornelius (1989a) *Phys. Rev. C* **39**, 384.
———, ———, and ——— (1989b) *Nucl. Phys.* **A496,** 446.
Woloshyn, R., and A. D. Jackson (1973) *Nucl. Phys.* **B64**, 269.
Workman, R. L., and H. W. Fearing (1986) *Phys. Rev. C* **34**, 780.
Wu, J., A. Stahlhofen, L. C. Biedenharn, and F. Iachello (1987) *J. Phys.* **A20**, 4637.
Yaes, R. J. (1971) *Phys. Rev. D* **3**, 3086.
Yakubovskii, O. A. (1967) *Yad. Fiz.* **5**, 1312 [*Sov. J. Nucl. Phys.* **5**, 937 (1967)].
Zankel, H., W. Plessas, and J. Heidenbauer (1983a) *Phys. Rev. C* **28**, 538.
——— and L. Mathelitsch (1983b) *Phys. Lett.* **132B**, 27.
——— and W. Plessas (1984) *Z. Phys.* **A317**, 45.
Zhang, J. Z. H., D. J. Kouri, K. Haug, D. W. Schwenke, Y. Shima, and D. G. Truhlar (1988a) *J. Chem. Phys.* **88**, 2492.
Zhang, J. Z. H., S.-I. Chu, and W. H. Miller (1988b) *J. Chem. Phys.* **88**, 6233.
——— and W. H. Miller (1989) *J. Chem. Phys.* **91**, 1528.
Zmora, I., and A. Gersten (1977) *Phys. Rev. D* **16**, 3581.
Zubarev, A. L. (1976) *Yad. Fiz.* **23**, 77 [*Sov. J. Nucl. Phys.* **23**, 40 (1976)].

Index

F

G

H

P

T